Nuclear Proliferation and Terrorism in the Post-9/11 World

David Hafemeister

Nuclear Proliferation and Terrorism in the Post-9/11 World

David Hafemeister
California Polytechnic State University
San Louis Obispo, CA
USA

ISBN 978-3-319-25365-7 ISBN 978-3-319-25367-1 (eBook)
DOI 10.1007/978-3-319-25367-1

Library of Congress Control Number: 2016934195

Printed on acid-free paper

This Springer imprint is published by Springer Nature
The registered company is Springer International Publishing AG Switzerland

I dedicate this book to my eclectic and wonderful six grand-children: Matthieu, Adeline, Alexandre, Lydia, Grace and Elijah. Your diverse strengths are part of the solutions to issues raised in this text.

Preface

The 9/11/2001 and 11/13/2015 attacks

Four days after the attack on Paris, the New York Times headlines showed signs of panic:

- Call to Arms in France Amid Raids and a Manhunt
- Hollande Asks Parliament to Act Quickly to Help in Eradicating ISIS
- Distrust, Even Fear, as Secular France Cools on Muslims
- GOP Governors Vow to Close Doors to Syrian Refugees
- Gloves Off, Director of the CIA Faults Surveillance Curbs
- After Paris Attacks, Vilifying Refugees
- Paris Changes Everything
- Parties Split on Response but United Behind France
- Obama says Strategy Against Militants Will Work
- US Airstrikes Destroy Scores of ISIS Oil Trucks in Syria

It is *deja vu (all over again)*, we are reminded that modern society is vulnerable to modest-sized attacks, that then evoke big responses. Death of 129 at the Baticlan music hall and elsewhere caught the attention of the democracies and dictatorships. Should we install boots on the ground, as we did in Afghanistan and Iraq? Is that what Osama bin Laden wanted us to do in the first place? Are we being provoked into future Armageddon confrontations? Or, are we moving too slowly? Will the 1,800 metric tons of global weapons useable plutonium and uranium be diverted to nuclear weapons? There should be no doubt that ISIS would use crude nuclear weapons if they can gather the materials. But then we followed that false siren call into Iraq in 2003.

The death of 129 is very serious business, but the following issues are greater in scope:

- Nuclear arms races with China, Russia, and the US;
- Proliferation of nuclear weapons to India, Iran, Israel, North Korea, Pakistan, and other states;
- Control of weapons-usable fissile materials by the International Atomic Energy Agency (IAEA);
- Nuclear power and its various cycles;
- Improvised nuclear weapons;
- Radioactive dirty bombs;
- Drone attacks;
- Biological and chemical weapons;
- Improvised chemical explosions;
- Cyber attacks on the Internet; and
- Privacy issues.

Two approaches can be used to describe our uncertain world. We will apply both historical and technical approaches. Because we participated in crafting legislation in the Senate, negotiating arrangements in the State Department and studying at the National Academy, we cannot help but describe its rich history. The first chapter of our text examines the historical context, as well as many sections of chapters, and especially Chaps. 7 and 8 and Appendices A and B.

We want to go beyond particular historical examples because history is continually being reinvented with new arrangements in new ways. The Sunni–Shiite rivalry is a major cause of conflict in Syria. Bashar al–Assad, Syria's dictator, is strongly supported by Shiite Iran and its Hezbollah terrorists, as well as expansionist Russia. Syria is under attack by the barbarous, overly religious ISIS and Al–Qaeda, as well as the "democratic" Free Syrian Army. US participation is further complicated because of Turkey's disdain for Kurdish forces. Plus, the US and Russia disagree because of Russian adventurism in Crimea and eastern Ukraine. At the same time, Iran and the P5+1 (the five UN veto nations plus Germany) have agreed to reduce both centrifuges and financial sanctions. And don't forget that Israel, Saudi Arabia, and Egypt live in this neighborhood. Our text is not an analysis of this complicated witch's brew of global politics, because of this large degree of uncertainties. Our task is to understand the weapons of mass destruction first, then use this data to proceed to political understanding, if that is possible.

New technical paths can be different from past actions. If we wait long enough, all empires will collapse. The laws of science will continue during these changes and give terrorists and proliferators new ways to attack. I have not abandoned my heritage of physics. It is helpful to have training in this area to describe nuclear events. But it is also advantageous to remember to live the two-culture conflict between science and humanism. I've striven to do that in my life, which began with a year's hitchhiking 25,000 miles before age twenty-one. One culture isn't enough.

Nuclear Proliferation and Terrorism in the Post-9/11 World

This text is intended for upper-division undergraduates (juniors and seniors) of all majors, including the liberal arts. It also is a useful reference book, but that is not its primary purpose. The chapters contain ample homework questions and seminar topics. It is a lone *text of this type*. The text describes technical devices but without mathematics. The text is reduced to that which would be needed by a foreign service officer in the State Department with a bachelor's degree in history. Foreign service officers are expected to know many things from culture to history to politics to economics to geography to military affairs, with basic science. Finally, we are human beings trying to learn, no matter what your major or your role in life.

This text evolved over 43 years at Cal Poly University, San Luis Obispo, California in a course that has been variously named *the Nuclear Arms Race* or *Nuclear Weapons Proliferation and Terrorism in the Post-9/11 World*. Students initially are concerned about having to learn some science in a general education science class. They calmed down when they learned that the students were going to present 25 % of the course material in the seminar portion of the class. My students knew little of these matters, but they were able to look quite sophisticated with the help of Google and Wikipedia. And they liked to hear each other as it became less threatening. We do have a society that is shy about science and more lecturing won't solve it. A 4-minute talk a week is not a great burden. If you can't verbalize the issue, you won't understand the issue.

Implementation of This Text

I departed from campus for a dozen years of involvement with arms control matters, in the Senate Foreign Relations Committee and elsewhere for 5 years, in the Arms–Control and Disarmament Agency for 1 year, in the State Department for 3 years, and in the National Academy of Sciences as study director for 2 years, as well as 4 years in the national laboratories at Los Alamos, Argonne, Berkeley and Oak Ridge, and for 2 years at Stanford and Princeton. This allowed me to make contact with the START Treaty, the Nuclear Nonproliferation Treaty, the Nuclear Nonproliferation Act of 1978, the ABM Treaty, the Comprehensive Nuclear-Test-Ban Treaty, the Conventional Forces in Europe Treaty, the collapse of the Soviet Union while in Moscow, ending the US plutonium economy, responses to the 9/11 attack, and chairing the External Review Panel of Los Alamos's Nonproliferation Division. Hopefully these diverse experiences will make this text a description of actual work, not just an academic theory.

San Luis Obispo, CA, USA David Hafemeister

Acknowledgment

I greatly benefited from frequent contact on these matters with colleagues in the American Physical Society's *Forum on Physics and Society*, the *APS Panel on Public Affairs* and co-workers in the Senate, the State Department, the Arms Control and Disarmament Agency and the National Academy of Sciences. I deeply thank those who encouraged me on this manuscript and the precursor articles in the *American Journal of Physics*, *Scientific American*, *Science and Global Security,* and *Physics Today*. I thank Springer editors Tom Spicer and Cindy Zitter for their encouragement with this project. I thank Julie Frankel for the thoughtful and artful cover that captures the spirit of this book. I am indebted to Roger Longden for his help and good humor on the 50-campus tour in 2003. Lastly, without my beloved wife and companion, Gina, this book would not have been completed. What adventures we have had over these 55 years. I am proud to have all of you as my friends.

Contents

Introduction

Process and Substance: It is in the character of growth that we should learn from both pleasant and unpleasant experiences.

Nelson Mandela, President of South Africa (1994–1999)

Arms Control: Both sides in the arms race are confronted by the dilemma of steadily increasing military power and steadily decreasing national security. It is our considered professional judgment that this dilemma has no technical solution. If the great powers continue to look for solutions in the area of science and technology only, the result will be to worsen the situation.

Jerome Wiesner and Herb York, *Scientific American* 211(4), 27 (1964)

Proliferation: Pakistan will fight, flight for a thousand years. If India builds the (atomic) bomb… (Pakistan) will eat grass or (leaves), even go hungry but we (Pakistan) will get one of our own (atom bomb). We (Pakistan) have no choice.

Pakistani Prime Minister and Foreign Minister, Zulfikar Ali Bhutto, 1965

Terrorism: I am in charge of the 19 brothers…. I was responsible for entrusting the 19 brothers…. with the raids.

Osama bin Laden, 23 May 2006

My T-shirt shows a picture of Earth's Western Hemisphere, with the caption, "Still in Beta?" There is much we don't know about our untested experiment on Earth. We do not have the luxury of doing accurate computer simulations to determine the future. What appears to be an excellent move can turn out to be horrible in the future. We can all suggest an event that should not have been done in the past, but continues to live with us in negative ways every day.

The lack of permanence becomes evident when reading our text, *Nuclear Proliferation and Terrorism in the Post–9/11 World*. Cal Poly, a technocratic university, requires its students to take one upper-division course on *science and society issues* to graduate. One science and society course is only a beginning but it is a journey that must begin somewhere. The impacts from technology are huge. It pays for society to use technology assessments to understand what the future can bring.

Three Main Thrusts

The quotations above point to the three main themes of Nuclear Proliferation and Terrorism as follows:

- ***Major–power arms race***: US, Russia, UK, France, China, and others (Chaps. 1–7). This issue continues.
- ***Proliferation of nuclear weapons*** to additional states: India, Iran, Israel, North Korea, Pakistan, and others (Chaps. 8–11). Who will be next?
- ***Terrorism after 9/11 attack***: Fossil fuel explosives, dirty nuclear bombs, improvised nuclear devices, drones, cyber attacks, biological and chemical weapons and more (Chaps. 12–15). If ISIS had highly-enriched uranium or plutonium, they would use it to make weapons of mass destruction.

The elegant book cover by Julie Frankel points to the progression of nuclear weapons from big superpower states, to smaller proliferant states, to the possibility for terrorist regimes. The chessboard implies that logic is supreme. But the quote by Jerome Wiesner and Herb York tells us the opposite, that "this dilemma has no technical solution. If the great powers continue to look for solutions in the area of science and technology only, the result will be to worsen the situation."

Strategic Arms Control (Chaps. 1–7)

The strategic arms race is not just an issue from the past. If nations are too isolated, they will consider reaching for more nuclear weapons to feel more secure, even when these weapons may make them less secure. At one point there were 70,000 nuclear weapons between the US and the Soviet Union. This number cannot be justified on the basis of science and logic. With a lack of agreement between Obama and Putin during the Ukraine invasion, it is difficult to develop new agreements.

Politically it is important to be strong and strident. It is difficult for you to run for president if it appears that you would compromise too easily. If we display more military prowess, Iran would probably ignore it. If we showed strength through financial sanctions, Iran did not ignore us. But alternatively, mindless rigidity maintains the status quo, as the decades pass. If we display unbending strength, that does not mean that Iran will do all that we desire.

Chapter 1 describes the historical underpinnings of nuclear weapons. The main issues are summarized, along with a lengthy chronology for those wishing more details.

Chapter 2 describes fission and fusion weapons, and some that are more exotic. Impacts from nuclear explosions are discussed, along with the Stockpile Stewardship program, meant to ensure confidence in the reliability of weapons under a global testing ban.

Chapter 3 describes nuclear reactors; their proliferation, safety issues, and nuclear waste. The biological effects of low-dose radiation are described.

Chapter 4 describes missile systems and the triad of land-, sea-, and air-based nuclear weapons. The potency of nuclear weapons is described in terms of their parameters.

Chapter 5 describes ballistic missile-defense systems, including President Reagan's *Star Wars* program and the present approach to defend against smaller attacks or accidental launches.

Chapter 6 describes the complex fabric of arms control treaties, including the ways to monitor compliance with the terms of treaties.

Chapter 7 describes the wind down of the Cold War with the Soviet Union. The annexation of Crimea by Russia in 2014 and other events show problems between East and West.

Proliferation of Nuclear Weapons (Chaps. 8–11)

In 2015, seven decades after Hiroshima and Nagasaki, there are nine generally recognized nuclear weapon states, the P5 of the Nuclear Nonproliferation Treaty (China, France, Russia, UK, and US) and the four *defacto* nuclear states (India, Israel, North Korea, and Pakistan). Over 27 nations started nuclear weapon programs, although many of these have stopped. Will we be able to constrain states striving to become nuclear weapon states in an era of easier uranium enrichment? It could be worse than nine nuclear states, if you consider that over 27 nations began this journey.

Chapter 8 describes the incentives and disincentives of various nonproliferation policies. The first six Articles of the NPT are described, revealing a more complex situation than bilateral arms control. Sanctions can only be effective if the vast majority of nations participate.

Chapter 9 describes proliferation technologies, particularly the enrichment technologies, monitoring technologies, and the missile technology control regime.

Chapter 10 describes the individual proliferation paths of 27 nations. It is a race between technology transfer, the adoption of the NPT/IAEA regime, or a new regime.

Chapter 11 describes the intersection between the NPT Treaty and the Comprehensive Nuclear-Test-Ban Treaty. The P5 called for an NPT without a time limit. The NPT regime would be strengthened if CTBT entered into force without a time limit.

Terrorism After 9/11 (Chaps. 12–15)

The 9/11 destruction of Twin Towers in New York and the attack on the Pentagon were horrible events. But, the invasion of Iraq in 2003 was a flawed response. The new world of the Department of Homeland Security and the more invasive examination of email and phone messages, as allowed by the Foreign Intelligence Surveillance Act and the Patriot Act, is now part of our lives.

Chapter 12 describes vulnerabilities and responses to terrorism. Some of these responses are useful, but others are not, as terrorists learn to change their modus operandi, by using advanced encryption and by avoiding cell phones and email.

Chapter 13 describes radioactive terrorism from radiological dispersal devices (dirty bombs), from improvised nuclear devices, from improvised explosive devices, and from drones. Significant terrorist nuclear terror events have not happened over the past 15 years, but the attack on Paris in 2015 shows they can happen with just bullets and smaller explosions.

Chapter 14 describes cyber terrorism, which includes the Stuxnet virus that attacked Iran's centrifuges, denial-of-service attacks, phishing to gain access to Internet materials, and attacks to digital infrastructures. Measures are suggested to strengthen the Internet, but these are only as strong as their weakest links.

Chapter 15 describes biological and chemical weapons. The current treaty regimes have made the use of biological and chemical weapons more difficult. Most primary nations (not Egypt and Israel) are members to the Biological and Chemical Weapon Conventions. But the relative ease of using BW/CW, even if ineffective, makes them a concern. New BW research can complicate this.

A Course on Nuclear Weapons, Proliferation and Terrorism

I taught a course based on this text at Cal Poly University over a period of 43 years, from 1972 to 2015. This text is the output of these courses. One-quarter of the course was a seminar, with students making weekly presentations, reporting back on their Internet research. It is important for the students to verbalize to help students succeed with this material.

My background began with a basic-research physics postdoc at Los Alamos National Laboratory. With the many nuclear conflicts between the US and the Soviet Union, it seemed like the nuclear enterprise was a freight train, heading down the tracks, with no one to slow the momentum. Because of this, I wrote a series of ten papers for the *American Journal of Physics* on the nuclear, energy, and environment issues. These papers led to a variety of positions in the US Senate (Senator John Glenn, Senate Committees on Foreign Relations, Energy and Governmental Affairs), State Department (Office of Undersecretary on Arms Control and Technology, Bureau of Politico-Military Affairs, Bureau on Oceans, Environment and Science), Arms Control and Disarmament Agency (Office of

Strategic Negotiations), National Academy of Sciences (Study director for *Beyond START*, technical staff member for the Comprehensive Nuclear-Test-Ban Treaty), as well as positions at Stanford, Princeton, MIT, University of Maryland, Carnegie Mellon University, Cal Poly University, Arms Control Association, and Federation of American Scientists. The teaching of Physical Science 307 started me on the road to negotiations in Moscow in the last weeks of the Soviet Union.

Quo Vadis Planet Earth?

The US is the world's remaining military superpower, but with limited financial assets. There will be shifts in how the US deals with the world community. What issues will be decided on our watch?

- Will China equal and then eclipse the US militarily?
- Will drones give a lasting edge to some US attacks?
- Will cyberwar and struggles over big data lead to stable rules of conduct on information technology?
- Will rights to privacy be diminished?
- Will the war on terror be a never-ending war?
- Will the number of nuclear weapon states expand or contract?
- Will the US act unilaterally or multilaterally?
- Will reduced dependence on Middle East oil reduce US military in that region?
- What international structures can reduce terrorism by weapons of mass destruction and by conventional terrorism?
- Can a two-tiered world of nuclear weapon states and non-nuclear weapon states remain viable in the asymmetric NPT?
- Will Iran and the P5+1 constrain Iran's nuclear program? Will the Sunni nations be satisfied with the result?
- Can religious extremism of the Islamic State be constrained?
- Will Russia's foray into Ukraine prevent diplomatic and nuclear cooperation between Russia and the US/EU?

Technology at Nuremberg and Now

Albert Speer was Hitler's favorite architect and, more importantly, he was the lead minister for military procurement for Nazi Germany. In his defense at Nuremberg, Speer commented on the unique role of technology in the Third Reich. These remarks conclude Speer's 1970 book, *Inside the Third Reich,* written during his 20-year imprisonment in Spandau:

> Hitler's dictatorship was the first dictatorship of an industrial state in this age of modern technology, a dictatorship which employed to perfection the instruments of technology to dominate its own people…. By means of such instruments of technology as the radio and public address systems, 80 million persons could be made subject to the will of one individual. Telephone, teletype, and radio made it possible to transmit the commands of the highest levels directly to the lowest organs where because of their high authority they were executed uncritically. Thus many offices and squads received their evil commands in this direct manner. The instruments of technology made it possible to maintain a close watch over all citizens and to keep criminal operations shrouded in a high degree of secrecy. To the outsider this state apparatus may look like the seemingly wild tangle of cables in a telephone exchange; but like such an exchange it could be directed by a single will. Dictatorships of the past needed assistants of high quality in the lower ranks of the leadership also—men who could think and act independently. The authoritarian system in the age of technology can do without such men. The means of communication alone enable it to mechanize the work of the lower leadership. Thus the type of uncritical receiver of orders is created.

The power of technology over society was amplified a year later by Henry Stimson, former Secretary of War (1911–1913, 1940–1945) and former Secretary of State (1929–1933) in *Foreign Affairs* (1947).

> We must never forget, that under modern conditions of life, science, and technology, all war has become greatly brutalized, and that no one who joins in it, even in self-defense, can escape becoming also in a measure brutalized. Modern war cannot be limited in its destructive method and the inevitable debasement of all participants…. A fair scrutiny of the last two World Wars makes clear the steady intensification in the inhumanity of the weapons and methods employed by both, the aggressors and the victors. In order to defeat Japanese aggression, we were forced, as Admiral Nimitz has stated, to employ a technique of unrestricted submarine warfare, not unlike that which 25 years ago was the proximate cause of our entry into World War I. In the use of strategic air power the Allies took the lives of hundreds of thousands of civilians in Germany and Japan…. We as well as our enemies have contributed to the proof that the central moral problem is war and not its methods, and that a continuance of war will in all probability end with the destruction of our civilization.

These remarks on technology from seven decades ago are relevant today. The advance of technology makes democracy more difficult. Jack Gibbons, former director of Office of Technology Assessment (1979–1993) and Presidential Science Advisor (1993–1999), was concerned that the public was losing faith in science. His 2003 comments reflect this, as he called for strengthening the role of science to help resolve social ills:

> People thought of science as a cornucopia of goodies. Now they have to choose between good and bad….I hope rationality will triumph….but you can't count on it.

I believe that science and technology, acting together, is a main driving force of history. I hope that *Nuclear Proliferation and Terrorism in the Post-9/11 World* can be useful to present facts on the world that we live in.

Chapter 1
History of the Nuclear Age

Those who do not study history will relive it.
Talk softly please. I have been engaged in experiments, which suggest that the atom can be artificially disintegrated. If it is true, it is of far greater importance than a war.
[Ernest Rutherford, 1919].

1.1 Survey of Events

Why this history? This initial chapter summarizes the historical events of the nuclear age from Rutherford's 1909 discovery that the nucleus is very small to our era in 2016. Nuclear weapons have had a major impact on global thinking, beyond that of Hiroshima and Nagasaki. The next dozen pages are similar to drinking water from a fire hose. I apologize for the density of information, but it is necessary to put the rest of the book into perspective. This text is not a history book, one can easily spend the entire term on following the events, from Hiroshima to 9/11. Our goal describes the proliferation and terrorism technologies that threaten us, and the attempts to control these technologies. Today's students didn't grow up with this information, it is my task to summarize it, while not sinking you in details. Why study this history? I believe it is necessary for our survival, it is as simple as that.

For me, it began with reading the August 1945 headline in the Winona Republican Herald, "*Japan Attacked with Atomic Bomb.*" For those of you, born after 1990, this is a complicated story of the Manhattan Project, the SALT/START treaties, proliferation by more than 20 states, the 9/11 attack, the Afghanistan and Iraq wars, the war on terrorism, the Islamic State of Iraq and Syria (ISIS), and more. If we are to explore our planet's future, we must first learn the basics from both history and science.

Beginnings of nuclear physics. The nuclear age began in 1909, when Ernest Rutherford (New Zealand, England) discovered that nuclei are much smaller than atoms. How did he do that? The heavy nuclei at the bottom of the periodic chart are changed by radioactive decay in which alpha particles (ionized helium nuclei, $^4He^{++}$) are emitted and the remaining nucleus is left with two fewer protons

D. Hafemeister, *Nuclear Proliferation and Terrorism in the Post-9/11 World*,
DOI 10.1007/978-3-319-25367-1_1

(+1 charge each) and two fewer neutrons (charge of 0). The new nucleus has four fewer mass units. For example, uranium-235 (^{235}U) alpha decays to thorium-231 (^{231}Th).

From plum pudding to infinitesimal nuclei. Rutherford showed in 1909 that the nucleus was very small in the following way: He placed a source of alpha particles near a thin foil of gold metal. He noted that some alpha particles scattered backwards towards the direction of the alpha source. This can only be explained by the fact that the nucleus is very, very small. The alpha particles that came very close to the small nucleus experienced a very large repulsive force, sufficient to make some alpha particles turn around by as much as 180°. Rutherford showed that the radius of gold nucleus is one millionth of a millionth of a centimeter, which is one ten-thousandth the size of the electronic cloud around the gold nucleus. Before the discovery of the nucleus, it was thought that atoms consisted of negatively point-charged electrons embedded in a constant density of positive-charge. This incorrect model was called the *plum-pudding model* of the atom with electrons playing the role of the plums and mythical positive charge spread uniformly, playing the role of the pudding.

Alchemy of element production. Rutherford made another major discovery in June 1919 when he created the first manmade, stable nucleus by combining two stable nuclei. Rutherford fulfilled the dream of the alchemists, who tried to make gold from other elements. In fact alchemists cheated by first putting gold inside the firewood to fool the king. Rutherford created oxygen-17 (^{17}O) nuclei and a proton by placing an alpha source, such as radium, in nitrogen gas. The alpha particles stuck to the nitrogen (^{14}N). This was revolutionary since one could change one element into another element. This took place just at the end of First World War I. Rutherford's 1919 quote at the beginning of the chapter was a response to questions about his research.

Neutron predicted. Within a year, Rutherford made news again with a startling speculation. Looking at the periodic chart, Rutherford noted that chlorine had a mass of 35.5 protons, which is a non-integer, while the elements at the top of the chart had integer-proton masses. The 35.5 mass could be explained by speculating that chlorine was made up of three parts of chlorine-35 with mass 35 (^{35}Cl) combined with one part chlorine with mass 37 (^{37}Cl), which is correct. Within a decade, Rutherford accomplished the physics *hat trick* by showing that (1) the nucleus was very small, (2) by creating stable isotopes and (3) by speculating on the existence of the neutron. These results began the atomic era of nuclear bombs and nuclear reactors.

Nuclear physics. Nuclear explosions rely on fission, the splitting of either uranium or plutonium nuclei into two fission fragments, plus several neutrons. To call a nuclear weapon an *atomic bomb* is a misnomer since atoms are electron clouds surrounding a small-sized nucleus, the electrons mainly experience the electrostatic force from the positive nuclear charge. Nuclear weapons have little to do with the electronic, atomic cloud, the electrons depend on the tiny, attractive nuclei at the center of the atomic solar system. The science of the atom began with Niels Bohr's (Denmark) theory that explained the wavelengths of light emitted from the simplest of all atoms, the hydrogen atom, made up of a proton and an electron. A decade later Ernest Schrodinger and Werner Heisenberg (Germany) used higher mathematics to obtain quantum mechanics to explain atoms, molecules, solids and nuclei.

Discovery of isotopes. In 1931 Harold Urey (University of Chicago) detected heavy hydrogen (deuterium, d, ^{2}H), consisting of a nucleus with a proton (p) and a neutron (n). Urey accomplished this by observing the faint optical lines of deuterium, a minor constituent of natural hydrogen. The proof of isotopes strengthened Rutherford's prediction for the neutron, since the extra mass of ^{2}H was due to the added neutron. This evidence did not prove discovery the neutron.

Discovery of the neutron. One year later James Chadwick (England) discovered the neutron in 1932. Chadwick did this by placing an alpha source near beryllium (^{9}B). When an alpha particle of mass four strikes a beryllium nuclei of mass nine, it produces carbon-12 and a neutron. Neutrons do not leave tracks in photographic emulsions since they don't have an electric charge to interact. However, Chadwick observed proton tracks that began a short distance from the disintegrated beryllium nucleus. When neutrons collide with hydrogen nuclei (protons), they transfer momentum to the struck particle, similar to a cue ball (neutron) hitting another ball of equal mass. If the collision is head on, the cue ball stops dead and the struck ball takes up the motion. He observed the proton and not the neutron, but he discovered the neutron.

Political landscape of the 1930s. These physics events were quickly followed by traumatic political events. On 28 January 1933, Adolph Hitler became chancellor of Germany. This was followed on 4 March 1933 when Franklin Delano Roosevelt become president of the United States. These two leaders did not know that they would start secret research that would lead to a successful US atomic bomb and to a failed German atomic bomb. Within four months after taking office, Hitler started firing Jewish physics professors at the University of Gottingen and elsewhere. It soon became forbidden to teach the theory of relativity in Germany because relativity was discovered by Albert Einstein, a Jew. In place of *Jewish Physics*, Hitler's government demanded that *Aryan Physics* be taught by Nazi notables, Jonathan Stark and Philipp Lenard. Anti-Semitism transferred significant physics brainpower from Germany to the United States. This ironic blunder hurt Hitler's nuclear weapons program, while helping the United States.

Concept of a nuclear weapon. Leo Szilard (Hungary) played a complex role in the development of nuclear weapons. Szilard was the first to realize and patent the idea of a run-away chain reaction, producing a nuclear explosion. Szilard attempted to keep scientists from publishing facts that would lead others to the bomb. Szilard authored two letters to Roosevelt, with Albert Einstein's signature, that launched the US Manhattan Project. At the close of the war Szilard authored a third Einstein letter that suggested nuclear restraint against Japan, as a way to begin international control. Later Szilard lobbied for nuclear constraints with the new, civilian-controlled Atomic Energy Commission. He was instrumental in creating peace groups, such as the *Pugwash meetings* with the Soviets and the *Council for a Livable World.* His career spanned both sides of the nuclear aisle, to convince governments to build nuclear weapons and, after they were built, to constrain them. This synopsis is only a tip of the iceberg of his complicated life.

First thoughts on nuclear weapons. Szilard got the idea for an atomic bomb while crossing a street in London in October 1933! He realized it would be possible to start a chain reaction for a nuclear bomb, if the absorption of one neutron released more than one neutron, giving exponential growth in a chain reaction. He did not know which elements could give neutron multiplication. At that time, ^{235}U fission had not been discovered, so he incorrectly speculated that beryllium might accomplish a chain reaction. He realized that nuclear weapons would be very dangerous if Hitler obtained them, recognizing the ideas must remain secret. Strangely enough, Szilard took his idea to the British Navy and obtained a classified patent in 1935. This may seem strange since you wouldn't expect the military not to act upon them.

Artificial radioactivity. Artificial radioactivity was first produced from stable elements in Paris and Rome. Irene Curie and her husband Pierre Joliot (France) created artificial radioactivity using alpha particles from natural radioactivity in 1934. Enrico Fermi (Italy) used neutrons from radium-beryllium sources. Fermi produced considerable artificial radioactivity when bombarding natural uranium with neutrons. He speculated that the level of radioactivity would be reduced if uranium and the neutron source were placed underwater in a pond outside the laboratory. He believed that water would reduce the flow of neutrons to the uranium target. Fermi was surprised when he obtained much *more* radioactivity with the water-bath experiment, rather than *less* radioactivity. What Fermi did not realize was that slow neutrons (thermal neutrons) interact 1000 times more strongly than fast neutrons from fission. When fast neutrons hit protons in water, they remove momentum (velocity times mass) from the neutrons, giving them reduced, thermal energy (slow velocity). Fermi, not knowing of fission, incorrectly claimed he produced transuranic elements, with nuclear charge greater than uranium's 92 protons. He unknowingly discovered the fission of uranium-235. Fermi had also produced very small amounts of transuranic elements, such as plutonium, but his equipment couldn't detect this. Nonetheless, this discovery led to Fermi's 1938 Nobel Prize in physics. Fermi had done very significant other research in physics, more than enough for a Nobel Prize. Fermi traveled to Stockholm to receive the prize, which allowed Fermi and his Jewish wife to escape fascist Italy. They immigrated to the U.S., first to Columbia University, where he observed neutron multiplication with Szilard. He then moved to the University of Chicago, where he produced the world's first nuclear chain-reaction with the uranium-graphite pile at Stagg Field (2 December 1942).

Barium implies fission. In 1938–39, physicists in Europe and America began experiments that led to the discovery of fission. The German group led by Otto Hahn repeated Fermi's *transuranic* work, but couldn't make sense of their results. They told a Jewish-exile member of their group, Lisa Meitner, residing in Sweden, that they detected the presence of barium after bombarding uranium with neutrons. Meitner and her nephew, Otto Frisch, realized on 22 December 1938 that the element barium was created when the uranium nucleus split into two new nuclei. After calculations they realized the presence of barium (and most of the periodic chart) were a natural result of splitting of uranium nuclei. The large amount of released energy from fission results from the large electrostatic repulsive energy between the two, positively-charged, new nuclei (the fission fragments) at very close distance where they were created. The 20 kilotons of released energy from a fission weapons is a million times larger than the energy released in a chemical reaction of the same mass. This reduction in mass by a factor of a million allows bombers and missiles to deliver massive destructive energy.

Two neutrons per fission. Experiments from January to May of 1939 showed that more than two neutrons are released when a uranium nucleus fissions (splits) with a neutron, fulfilling Szilard's requirement for an atomic bomb. Frederic Joliot (France) also observed this fact and wanted to publish quickly. Szilard tried to discourage Joliot from publishing. When he failed to stop Joliot, Szilard and Fermi published their similar results. A lesson for history is that powerful science knowledge can be controlled, but only for a limited time. If nuclear weapons are to be controlled, it will take more than secrecy to do this. It was clear to the physicists who read the 1939 *Physical Review* and *Nature* that sufficient neutrons are released per fission event, fulfilling a necessary condition for a nuclear weapon. Basic information for reactors and uranium enrichment was in the open literature when Hitler invaded Poland on 1 September 1939.

Germany begins the race. Despite expelling an excellent group of Jewish physicists, Germany still had a potent group of physicists. The German government was not sitting idle as it convened a conference in Berlin on 29 April 1939 to consider building an atomic bomb. Because they lacked significant facts, the Germans knew they had to learn more through basic research, and they knew they must create new technologies to attain their goal. This political impetus encouraged German Professor Paul Hartek to try to produce neutron multiplication with a primitive reactor (185 kg uranium, CO_2 dry-ice moderator) in Hamburg in June 1940. Hartek failed in this effort, but we note that his attempt took place 2.5 years *before* Fermi's success at Stagg Field. A second attempt at criticality failed in January 1941 at the Berlin "virus-house" reactor, using 300 kg of uranium with a graphite moderator. This reactor could have gone critical, but too many neutrons were absorbed in impure graphite, which contained boron, a strong neutron absorber. These failed results convinced the Germans to base their reactor program on heavy water moderators. Since heavy water is hard to produce, Germany confiscated Norwegian heavy water supplies to begin their reactor program. Finally, Werner Heisenberg, the director of the German atom bomb project, obtained 13 % neutron multiplication in May 1942. The allies foreclosed this effort by destroying the Norwegian Vermork heavy-water plant in February 1943. The Germans were aware that plutonium could be used for nuclear weapons and that it can be made in reactors, but they made no progress on this approach.

German enrichment of uranium. Hartek started an alternative approach in August 1941 by building ultracentrifuges to enrich the ^{235}U content. The early centrifuges exploded at high rotational velocities. Hartek ultimately obtained 7 % enriched uranium by March 1943, which was a long way from the 90 % enrichment needed for nuclear weapons. Thus, German mistakes on reactors and uranium enrichment took them out of the race that they began. The US became certain of the German lack of progress when Samuel Goudsmit (Netherlands) followed US forces into Germany to examine captured German files on the ALSOS mission. By November 1944 Goudsmit determined that the German program was not a threat. Historians have conjectured on the reasons for the German failure: Was it stupidity or arrogance, lack of funding, or a desire by the German scientists to undermine Hitler?

Manhattan project. The world's second nuclear weapons program began after Roosevelt received the second Einstein-Szilard letter on 6 December 1941, the day before Pearl Harbor. The British Maud Committee had determined by July 1941 that 10 kg of highly-enriched uranium could produce a nuclear explosion. The US National Academy of Sciences endorsed a many-tracked program. It took but a year for Fermi to produce the world's first nuclear chain reaction on 2 December 1942. Three months later, J. Robert Oppenheimer established the weapon design and construction laboratory at Los Alamos, New Mexico. Large reactors (natural uranium, graphite moderated, water cooled) were built at Hanford, Washington to produce plutonium. The chemical reprocessing plants produced separated plutonium at Hanford, Washington by January 1945. Six months later, sufficient plutonium was available for the first nuclear explosion at Alamogordo, NM and the third explosion at Nagasaki, Japan. Three enrichment technologies were used to obtain weapons grade, highly-enriched uranium (HEU): gravitational diffusion (to 2 %), gaseous diffusion (to 23 %) and electromagnetic isotope separation (to 84 %), while gaseous centrifuges were abandoned. Some 50 kg of HEU was obtained for the second nuclear explosion at Hiroshima, Japan.

Pu-240 complications. In March 1943, Glenn Seaborg realized that plutonium-240 (^{240}Pu), a spontaneous neutron-emitter, complicates the use of plutonium in nuclear weapons. The explosion of uranium with gun-type weapons is not greatly affected by spontaneous neutron emission as they are assembled quickly enough when one uranium part is projected into the other uranium part. This is not true for a plutonium warhead since plutonium spontaneous neutrons appear in large numbers early on, before it is fully assembled. This causes the chain reaction to begin too early, explosively disassembling the bomb at too early a time, dramatically reducing the yield. The isotope ^{240}Pu is a large neutron-emitting contaminate in ^{239}Pu fuel. The problem was solved by Seth Neddermeyer and George Kistiakowsky by imploding a spherical weapon with segmented explosive lenses in a 32-sided polyhedron, an icosododecahedron soccer ball. The soccer ball is made up with 20 hexagons and 12 pentagons, which is more spherical than the icosododecahedron made up with 20 triangles and 12 pentagons (Fig. 1.1).

Proliferation and international control. By August 1944, Niels Bohr knew that US, England and Germany had nuclear weapons programs and other nations would join the nuclear arms race in the future. Bohr wrote a memorandum to President Roosevelt and Prime Minister Winston Churchill, describing proliferation problems, along with a possible solution, the international control of nuclear weapons. Churchill didn't appreciate the briefing and asked not to see Bohr again. Roosevelt was more courteous but took no action. These great leaders were in the middle of a traumatic war, knowing that a proliferated world was in the distant future, which is now. Ultimately twenty-one Nobel Laureates participated in the development of nuclear weapons, mostly during the Second World War.

Fig. 1.1 Two different icosododecahedrons of 32-sides (and a tetrahedron) using Jackson Pollack's painting style (D. Hafemeister)

Baruch plan. In 1945 a group of scientists, led by James Franck (Germany, US) at the University of Chicago wrote the *Franck Report*, calling for restraint against Japan and for international control of nuclear weapons. Secretary of War Stimson and President Truman ignored this warning as the atomic bomb was dropped on Japan on August 6th (Hiroshima) and August 9th (Nagasaki), ending World War II on August 15th. However, these concerns and ideas penetrated the government. On 14 June 1946, Bernard Baruch presented the Acheson-Lilienthal plan to internationalize the atom to the UN General Assembly. He began with the famous quotation, "We are here to make a choice between the quick and the dead." Agreement could not be reached with the Soviets, who objected to the lack of a veto, to the wide-ranging inspections and to the fact that it maintained the US nuclear supremacy until after the inspections were completed. US nuclear testing continued at Bakini and Eniwetok. These atmospheric tests were followed by massive public demonstrations, opposing further nuclear testing. In November 1948 the UN General assembly adopted the US plan for international control by a vote of 42 to 6, which the Soviets opposed.

Hydrogen bombs. In August 1949 the Soviets detonated their first atomic bomb, Joe-I, in the Ustyurt Desert. This event encouraged the US to explore more powerful nuclear weapons. The General Advisory Committee of the Atomic Energy Commission, chaired by J. Robert Oppenheimer, advised against the hydrogen bomb and in favor of the construction of more powerful fission weapons with yields up to 500 kton from hydrogen boosting. The fusion process is similar to fusion reactions on the sun, but it is done in one step, by combing deuterium (^{2}H, d, heavy hydrogen) with tritium (^{3}H, t, heavy-heavy hydrogen) to give ^{4}He plus a neutron. The *George* test of a boosted weapon produced 225 kton on 9 May 1951.

Klaus Fuchs. It was discovered in January 1950 that Klaus Fuchs (Germany, England) gave nuclear secrets to the Soviets. This encouraged President Truman to announce on 31 January 1950 that the US would not stop with boosted fission weapons, but commence building hydrogen bombs. The US opened a second nuclear weapons laboratory in Livermore, California to expedite this research. The invasion of South Korea by North Korea on 25 June 1950 further inflamed the Cold War.

The US exploded its first fusion device *Mike* on 1 November 1952, using liquid deuterium in a huge thermos-bottle at Eniwetok. The first fusion device was not a deliverable weapon, as it was too large to be delivered. The huge yield of 10.4 Mtons was a thousand times larger than the Hiroshima bomb. The Soviets followed suit eight months later. The 1 March 1954 Castle Brovo test gave a fatal radiation dose to a Japanese fisherman on the Lucky Dragon boat and polluted the Marshall Islands. Soon both sides learned to miniaturize warheads, using lithium-6 deuteride ($^{6}Li^{2}H$). This solid gives *both deuterium and tritium* after being exposed to the neutrons from the primary fission weapon ($^{6}Li + {}^{1}n \Rightarrow {}^{4}He + {}^{3}H$).

Atoms for peace. On 8 December 1953, President Eisenhower delivered his famous *Atoms for Peace* speech before the UN General Assembly. This speech paved the way for peaceful use of nuclear energy to make electricity, declassifying some secrets and accelerating technology transfer. It also began the process of establishing international controls administered by the International Atomic Energy Agency under the Nuclear Non-Proliferation Treaty. This tradeoff enhanced near-term proliferation, but it did encourage several nations to abandon their nuclear weapons programs. Hopefully the NPT-regime enhances longer-term controls as the technological barriers to be bomb are reduced as technologies improve. For the long haul, proliferation policy hopes to reduce proliferation, but it will be difficult to roll it back to five nations, or less.

Nuclear arms race. Other technical advances accelerated the nuclear arms race. Initially nuclear bombs were placed on long-range heavy bombers. In January 1954 the first the nuclear-powered submarine, the USS Nautilus, was launched. By 1957 the US, the USSR, and Great Britain had exploded deliverable hydrogen bombs. The first Soviet satellite, Sputnik I, was placed into orbit by the Soviets in October 1957. Missile technology advanced in parallel with satellite technology, as land-based intercontinental ballistic missiles (ICBM) were deployed to deliver warheads in thirty minutes to targets 10,000 km away. In time, missiles carried 10 or more warheads, using multiple-independently-targetable re-entry vehicles (MIRV) on ICBMs on land and on submarine-launched ballistic missiles (SLBM). This was further complicated with the following topics:

- tactical and theater nuclear weapons,
- cruise missiles,
- stealth bombers,
- anti-ballistic missile systems,
- anti-satellite warfare systems,
- maneuverable re-entry vehicles (MRV) on Pershing II,
- leadership decapitating earth penetrating warheads, and
- electromagnetic pulse weapons to shut down command and electrical power.

By the end of the Cold War, the US had spent $5.5 trillion on nuclear weapons, with 30,000 warheads at the peak of the Cold War, while the Soviets had 42,000 warheads at its 1986-peak level. Eisenhower's Farewell Speech of January 1961 warned of the military-industrial-scientific establishment that made the world a much-more dangerous place.

Nuclear-testing treaties. The process of arms control began in the area of nuclear testing. The Eisenhower-Khrushchev nuclear testing moratorium of 1958–61 ushered in the first negotiations. Russia broke the moratorium in September 1961, along with the creation of the Berlin Wall. US and USSR leaders, John Kennedy and Nikita Khrushchev, almost came to agreement to ban all nuclear tests with a Comprehensive Test–Ban Treaty (CTBT) in 1963. If they had succeeded with a CTBT, they could have constrained the development of miniaturized weapons needed for MIRV as well as other new types of weapons. They did partially fulfilled the goals of a CTBT with modest progress in 1963 by agreeing to constrain nuclear tests to underground locations in the Limited Test Ban Treaty. Three decades later, in 1996, the five nuclear weapon states (China, France, Russia, UK, US) and 166 other states signed the CTBT. The US failed to ratify CTBT in 1999. Following the Vienna Convention on Treaties, the US is committed to obey the terms of the treaty by refraining from nuclear tests.

Strategic arms control. President Lyndon Johnson and Soviet Premier Alexei Kosygin agreed in 1968 in Glassboro, New Jersey to begin the process to constrain both offensive and defensive weapons. In 1972 President Richard Nixon and Premier Brezhnev agreed on numerical limits on offensive strategic weapons with the SALT (Strategic Arms Limitation Talks) executive agreement, as well numerical limits on the number of defensive weapons with the Anti Ballistic Missile Treaty (ABM treaty). At the Glassboro meeting, the US and USSR agreed to place the Nuclear Non-Proliferation Treaty (NPT) for signature in the United Nations. The NPT is the most important arms control treaty since it affects the nuclear status of its 188 states-parties by establishing basic rules of the road for nuclear weapons in an era of nuclear power that uses enriched uranium as a fuel and produces plutonium. The NPT facilitated nuclear technology transfer, but it also established global norms monitored by the IAEA. The 1995 extension of the NPT by 183 non-nuclear weapon states politically joined the missions of the CTBT and NPT treaties.

Anti-ballistic-missile weapons. Defensive weapons to attack incoming ICBMs have long complicated the diplomacy of offensive weapons. Both the US and USSR deployed defensive systems to defend their cities and weapons in the 1960s and 1970s. The Russian Galosh system continues to operate today, but the US system in North Dakota was shut down in 1976 after only four months of its activation, because of doubts over its effectiveness in a confrontation and its continuing costs. The American ABM weapons used 5 Mton nuclear weapons on Spartan, to attack the mid-course phase, and kton-class on Sprint, to attack the re-entry phase, similar to the Russian systems.

Strategic defense initiative. In March 1983 President Ronald Reagan called for defensive weapons "to make nuclear weapons impotent and obsolete." The

Strategic Defense Initiative (SDI) program was soon dubbed *Star Wars*, after the George Lucas fantasy movie trilogy of 1977–2005. The original SDI weapons were mostly directed energy weapons systems (DEWS) that use a laser or particle beam to attack missiles in their boost or mid-course phases. The sine qua non (necessary condition) for SDI was to have space-based weapons attack the boost phase of the rising missiles from the USSR. In 1987, the American Physical Society carried out the pre-eminent study on SDI, showing that the SDI concepts were technically immature. Shortly after the APS study, the Reagan administration shifted from beam weapons to kinetic-kill vehicles (KKV) to destroy incoming missiles by physically colliding, or by using a nearby explosion. At its best, SDI was not capable of stopping a vigorous Soviet first-strike. To stop an attack of 10 incoming missiles is also difficult.

Wind-down of the Cold War. The December 1979 Soviet invasion of Afghanistan stopped progress on arms control in the Carter Administration. The election of Ronald Reagan as US President raised tensions with the Soviets during his first term but harsh rhetoric was followed with improved diplomacy in Reagan's second term. Both leaders favored a commitment to zero nuclear weapons at Reykjavik in October 1986, but disagreements over defensive weapons blocked this thrust. It is not clear that either leader could have sold this goal to their governments without haggling. The combination of the rise of Mikhail Gorbachev as Soviet president and the commitment of President Reagan to progress in arms control helped close out the Cold War. This progress was made possible by the fall of the Berlin Wall in 1989 and the Intermediate Nuclear Forces Treaty (INF) in 1988 (ratification), the first arms control treaty since the 1972 ABM and SALT Treaties. Suddenly Gorbachev accepted on-site inspections (OSI) of all types, so much so that the US began to regret its past offers to accept OSI. I agree with many in the literature who believe that the threat of the US SDI Program did not change the end date of the Cold War by more than 6 months and probably less. The agreement on the Conventional Forces in Europe Treaty (CFE) paved the way to end the Cold War. The CFE treaty cut the conventional forces of the Warsaw Pact by 60 %, without changing the situation for NATO forces. Soviet numerical superiority on conventional forces in Europe had long put fear into NATO, but much of the Soviet equipment was older equipment that should have been scrapped. With conventional force issues removed, control of nuclear forces became possible. The START treaty was ratified in 1992, cutting strategic nuclear weapons in half to about 6000 each, including 50 % cuts in the feared Soviet/Russian SS-18. START created an intensive inspection system to make sure both sides adhere to their commitments.

Nuclear weapon cuts. START I was followed with the 1993 signing of the START II Treaty, lowering the limit to 3500 strategic warheads and banning land-based MIRVed missiles, such as the US Peacekeeper and the Russian SS-18. In 1997 this was followed by START III pre-negotiations between Presidents Bill Clinton and Boris Yeltsin to reduce to 2500 strategic warheads. The START II and III agreements did not enter into force because of disagreements over defensive weapons, covered by the ABT Treaty. The 2002 Strategic Offensive Reductions Treaty (SORT) and the 2010 New START Treaty lowered the limit on deployed warheads to 2200 and 1550, respectively. President Obama indicated he would like a lower limit of 1000 deployed warheads, but disagreements over US ballistic missile defense programs, Ukraine and the "Snowden" affair dim this goal.

Effective verification. Monitoring is the gathering of data related to an arms control treaty. Verification is the quasi-judicial interpretation of the monitored data to determine if the data is in compliance to the terms of the treaty. Monitoring sensors detect nuclear particles, electromagnetic signals of all types, seismic signals, infrared and visible signals and photographs, infrasound and hydroacoustic signals, SAR radar, laser signals, and other electronic data. During ratification of the INF Treaty, Ambassador Paul Nitze defined *effective verification*:

> if the other side moves beyond the limits of the treaty in any militarily significant way, we would be able to detect such violation in time to respond effectively and thereby deny the other side the benefit of the violation.

Any cheating must be detected in a timely manner before it can threaten national security. Since nations are already threatened with a plethora of *legal* nuclear weapons, the "effective" criterion concerns the degree of marginal threat due to cheating beyond the legal level of threat agreed to by the treaty. Another way to look is the following: As long as the U.S. maintains a lethal second-strike attack, there was no reason for the Soviets to attack first. The 1992 Senate Foreign Relations Committee Report on START-I ratification showed there was a survivable US force of 3600 weapons after a very major Soviet attack. This was a worst-case analysis, as it chose parameters favorable to the Soviets. Many of the Soviet targets would be empty after their attack and no longer be worth attacking. The extensive hearings on START and the Triad showed that the submarines were invulnerable to a massive attack by the Soviets.

Patriot missiles in first Gulf War. Defensive weapons were given new political life with the apparent success of the Patriot missile in the 1991 Gulf War in Iraq. This success was greatly exaggerated by the Army. Very few Scud missiles were destroyed by the Patriot, which was designed to attack airplanes. These events gave a political impetus to shift the focus of defensive weapons from attacking strategic missiles to attacking theater and tactical missiles with shorter ranges. The election of President George W. Bush in 2000 increased the emphasis on defensive systems, propelling the US to withdraw from the ABM Treaty in 2001 and considerably raising ballistic missile defense (BMD) budgets. Although the Soviet Union pioneered them, Russia always opposed deployment of US defensive systems since they realized the great complexity of BMD weapons prevented Russia from succeeding in this defensive race. In general, offensive weapons are far cheaper than the defensive weapons needed to knock them down, as dictated by Paul Nitze's ABM acceptance criteria. After the Cold War, Russia was not able to build significant offensive or defensive nuclear weapons because of fiscal problems. For this reason the Soviets/Russians conditioned their approval of all strategic-arms treaties on continuing constraints on deployed defensive systems. During this period the US and Russia worked together on technologies to detect and count nuclear weapons, under the Biden Amendment, as a way of moving to agreed levels below 500–1000 warheads. This work continued at a lower level for some time, but it was totally cancelled with the Russian intervention in the Ukrainian War in 2014–15.

SORT and New START treaties. President George W. Bush and President Vladimir Putin agreed in 2002 to a limit on operational, strategic warheads of 2200. The Strategic Offense Reduction Treaty (SORT) was based on compromises that accept Bush's withdraw from the ABM Treaty, while Russia retained 138 SS-18 s (1380 warheads), which would have had to been destroyed under START II. This agreement backtracked from the ABM Treaty of Richard Nixon, the START II Treaty of George H. Bush (ban of land-based MIRV) and the START III agreement of Bill Clinton (further verification). The New START of 2010 was necessary to continue the monitoring provisions of START with a lowered limit of 1550.

Former-Soviet weaponry. At the end of the Cold War, the Nunn-Lugar legislation helped control and maintain the accounting of 1200 tonnes of HEU and 130 tonnes of plutonium in the former Soviet Union. Without this financial and technical assistance, it is doubtful that these materials would have stayed beyond the reach of terrorists or other nations, as the Soviet's million-person nuclear payroll could not be met. With the demise of the KGB, there was no longer fear in the hearts of potential smugglers. This was exacerbated since the Soviets had not developed sufficient physical security and nuclear accounting. We will describe some of the monitoring technologies used to help account for these materials.

Proliferation to other nations. During his presidency, President John Kennedy warned that 15–25 nations might obtain the bomb within a decade, which was accurate as to the number of nations that made *some movement towards the bomb*. But it took three decades to reach the number of 25 attempters, with nine actual nuclear weapon states in 2015. Let us look at this prediction. At first Germany, the US, and UK had nuclear weapons programs. The Soviet Union and Japan, during World War II, had very small research efforts to explore the concepts needed for nuclear weapons. As soon as World War II was over, the USSR took up nuclear weapons with vigor, exploding its first weapon in 1949. UK followed in 1952. France joined the nuclear club in 1960, followed by China in 1964. Interestingly enough, the first five weapon states are the Big-5 winners of World War II and the permanent members to the UN Security Council. This was followed by India in 1974, which was the first state after the Big 5 states to explode a nuclear device. Pakistan followed with its first explosion in 1998. Israel is widely reported to have had nuclear weapons since the 1960s. Four nations gave up their nuclear weapons in the early 1990s; South Africa and the former Soviet states of Belarus, Kazakhstan and Ukraine. North Korea carried out four small tests in 2006, 2009, 2011 and 2016, which were detected by the International Monitoring System of the CTBT Organization. Iran was moving towards nuclear weapons, after years of negotiations and sanctions, on a program that dates back to the Shah of Iran in the 1970s. Lines have been drawn in the sand by both sides. The biggest fear from Iranian nuclear weapons is that the Islamic religious split would encourage Sunni nations to catch up to Shiite Iran. Policy making on proliferation issues is far more complicated than it is on strategic arms because it covers 190 nations and is involved with the commercial sector.

Response to proliferation. The Indian nuclear test of May 1974 ushered in the era of further nuclear proliferation. Also in 1974, France contracted to export reprocessing plants to Japan, Taiwan, South Korea and Pakistan; and Germany contracted to export uranium-enrichment technology to Brazil. France and Germany rightfully complained that the U.S. had a nuclear-power market advantage since the US sold long-term uranium fuel contracts with their reactor sales. Control was lacking on plutonium from commercial reprocessing. In addition, the ability to produce highly-enriched uranium was proliferating with centrifuge designs stolen by A.Q. Kahn (Pakistan), sold to North Korea, Iran and Libya. The Carter administration and the Congress (Senators John Glenn, Abe Ribicoff and Charles Percy) tightened nuclear export laws in 1978 and created sanctions against non-nuclear nations that explode a nuclear device or imported sensitive fuel facilities. Sanction legislation was tightened by Senators Glenn, Pell and Helms in 1994. Sanctions can only be truly effective when the full family of nations abides by them in concert. If solidarity is weakened, control by sanctions is weakened.

Plutonium economy. To be consistent with US policy against commercial plutonium separation abroad, the Carter administration mothballed the Allied General Nuclear Services reprocessing plant at Barnwell, South Carolina and it halted plans for the Clinch River Liquid Metal Fast Breeder Reactor (LMFBR) at Oak Ridge, Tennessee. These events were attacked at the time in the US and abroad, but with the passage of time the anti-breeder position has become the acceptable norm, saving billions of dollars. Not even the French have made progress on their breeder reactors in the past 40 years, as the Superphoenix proved costly and unreliable. Additional export rules from the Nuclear Suppliers Group are now tighter, but there are signs of weakened standards.

9/11 terrorism, drones and cyberwar. Global terrorism increased with the 11 September 2001 Al-Qaeda terrorist attack on the World Trade Center and the Pentagon. The U.S. responded with wars on Afghanistan and bystander Iraq. The 9/11 attack showed vulnerabilities in modern societies to attacks by fewer individuals with relatively smaller assets. Al-Qaeda showed an interest in nuclear weapons, but not much action beyond that. There will always be other avenues of attack that can't be anticipated. The presence of weapons of mass destruction in Iraq was the stated reason to invade Iraq, but in fact this premise was not true. The situation was further complicated by North Korea's 2003 departure from the NPT treaty after its 1994 agreement to abandon its nuclear weapons program. Further complications arose with the *Arab Spring* rebellions, which are a work in progress. The movement towards democratization appears halted.

Questions about the future. Now that the US is the world's only military superpower, with limited financial assets, will there be shifts in how the US deals with the world community. What future issues will be decided using the materials of this text?

- Will China eclipse the US militarily?
- Will drones give a lasting edge to US attacks in the Middle East?

- Will cyberwar and struggles over big data lead to stable rules of conduct between nations for the information technologies?
- Will our future rights to privacy be forever diminished?
- Will a war on terror be a never-ending war?
- Will the US be more or less unilateral in its actions?
- Will reduced dependence on oil from the Middle-East reduce the need for a strong Middle-East policy?
- What international structures will be used to prevent terrorism from weapons of mass destruction and from conventional, non-nuclear terrorism?
- Can a two-tiered world of Nuclear Weapons States and Non-Nuclear Weapon States remain viable in the asymmetric NPT?
- Will Iran and the P5+1 constrain Iran's nuclear program? Will the Sunni nations be satisfied with the result?
- Can religious extremism of the Islamic State of Syria and Iraq be constrained in a today's world?
- Will Russia's foray into Ukraine curtail future nuclear cooperation between Russia and the US/EU?

Global-zero nuclear weapons. Former Secretaries of State Henry Kissinger and George Schultz, former Secretary of Defense Bill Perry and former Chair of the Senate Armed Services Committee San Nunn are committed to work towards the goal of removing all nuclear weapons from the Earth. This goal was considered at the 1986 Reykjavik summit by President Reagan and Soviet leader Mikhail Gorbachev. Both leaders pushed hard to do this, but they were undone by Presidnet Reagan's commitment to SDI and Gorbachev's unwillingness to live with it. The Op-Ed pieces in the Wall Street Journal by the *Gang of Four* on 4 January 2007, 15 January 2008, 7 March 2011 and 5 March 2013 laid out steps that would move the Earth towards zero nuclear weapons. They readily admit they do not have all the answers in terms of verification and nuclear stability, but they also realize that the recommended steps are very useful even if we do not get to zero:

> But the risks from continuing to go down the mountain or standing pat are too real to ignore. We must chart a course to higher ground where the mountaintop becomes more visible.

Why would these retired leaders, risking political flack from some of their peers, recommend the goal of zero nuclear weapons? My guess is they see the problems of the world as so severe that they are willing to take political smears to make some progress, with zero as a very distant goal. Will the *gang-of-four* with a goal of zero nuclear weapons strengthen the drive towards arms control or will it polarize events by being considered too idealistic?

1.2 Conflict Literature

The bards of literature portray the glamour and sufferings of war. William Shakespeare portrayed Julius Caesar as a good leader, dying for a good cause, but yet he overthrew the power of Rome's Senate. Tolstoy reminds us in *War and Peace* of the pain that Napoleon caused during his 1812 attack on Russia. Emile Zola tells a similar story in *The Debacle* about the misguided French soldiers in the Franco-Prussian War of 1870–71. The text before you is not great literature, but it is all I have to offer. It mostly contains hard facts from science, history, law, sociology and international relations. To do great literature takes special skills that I do not possess.

As a small step in the literary direction, I recommend the play *Copenhagen* by Michael Frayn. This well-written play shows the inner conflicts of a scientist who was working on nuclear weapons (Werner Heisenberg, Germany) and a scientist who would soon join the US nuclear weapons effort (Niels Bohr, Denmark). As the story opens, Werner Heisenberg visits his physics mentor, Niels Bohr, in Nazi-occupied Copenhagen in October 1941. Now their status is reversed, Heisenberg is head of the German atom bomb project and Bohr is the subject of an occupied state, soon to escape and become Nicholas Baker in Los Alamos. The play tries to gain insight into Heisenberg's motives for the visit. Heisenberg takes considerable personal risk to talk to Bohr on secret matters, since Bohr opposes the Nazi occupation and their methods.

Questions: Frayn suggests three possible motives for Heisenberg's trip to Copenhagen with Bohr.

- Was it to recruit Bohr into the German program?
- Was it to send a message to the West saying that the bomb was too difficult to complete during the war?
- Was Heisenberg trying to obtain information from Bohr on whether the Allies were pursuing a bomb, obtained from Bohr's contacts with US and UK physicists?

None of us know the complete answer to this question. Heisenberg was constrained to speak cautiously as he didn't want to risk being caught as a traitor to Germany. Heisenberg did not say that Germans were too moral to build the bomb. Bohr was not an eloquent speaker, and his emotions might have clouded the facts. At any rate, Bohr became quite angry, writing an angry, unsent letter. But later after the war he seemed to have softened somewhat towards Heisenberg. Heisenberg answered the question of his intentions many times, but there is always some confusion about his answers. Did the Germans take the heroic wrong path to the bomb to defy Hitler, or was he arrogant and brilliant and not wise?

Heisenberg and other German nuclear scientists were captured in the spring of 1945 and then interned in an English farmhouse. Not surprisingly, their conversations were secretly recorded and taped. These discussions were later published as *The Farmhouse Tapes*. When Heisenberg learned from the radio that the atom bomb was dropped on Hiroshima, he seemed surprised. He was not able to answer a question on the correct size of the critical mass of a uranium bomb. But the next morning after he took time to consider the issue more deeply, he corrected himself in his discussions with his peer group. Does this mean that he was an arrogant theorist who did give enough attention to computational details, or does it mean that he did not cross the theoretical bridge to the bomb, denying Hitler of his goal?

In the seminar portion of the class on *Nuclear Proliferation and Terrorism in the Post 9/11 World*, we have spent a little time watching portions of the video of *Copenhagen*. A few students portrayed the characters; Werner Heisenberg, Niels Bohr and Margrethe Bohr. In this way they gained some feeling for these issues that were new to them. If you are a budding writer and looking for a topic, I have a suggestion for you. Many of the events and topics in this book are extremely important, but yet they are not well known. Before Frayn's Copenhagen, very few knew about the October visit of Heisenberg to the Bohr household. There is much to consider about this visit, it was a good topic for Frayn to explore. Perhaps some of the pages in this book can be a starting place for literary efforts. But, first do the proper homework on the topic. Then you will be able to better portray the events and the players that formed these events. Ultimately you will have to deal with the intentions of the characters. Let the truth be your guide. Good luck.

1.3 Nuclear Arms Chronology

See Figs. 1.2, 1.3, 1.4, 1.5 and 1.6.

Fig. 1.2 *Left* Albert Einstein and Leo Szilard re-enact 1939 letter to FDR. *Center* Los Alamos main gate. *Right* Manhattan Project Director J. Robert Oppenheimer (These figures are from the Atomic Heritage Foundation (AHF))

Fig. 1.3 *Left* Enrico Fermi built first US reactor in 1942. *Center* Alsos Mission dismantles German Haigerloch reactor in 1945. *Right* Hans Bethe, theory group director (AHF)

Fig. 1.4 *Left* Fat Man plutonium bomb. *Center* Jumbo to contain Pu at Trinity (not used). *Right* Enola Gay crew that dropped the Hiroshima bomb (AHF)

Fig. 1.5 *Left* Hiroshima after August 6, 1945. *Right* Operation Crossroads 1946 (AHF)

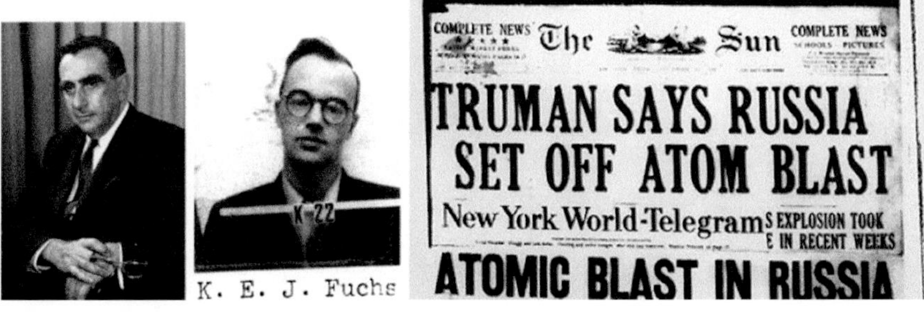

Fig. 1.6 *Left* Edward Teller, father of H bomb. *Center* Klaus Fuchs, Soviet spy. *Right* First Soviet nuclear test, August 29, 1949 (AHF)

Chronology of the Nuclear Age

1763

- July 13: British give blankets with small-pox to Indians that attacked Fort Pitt.

1909

- Ernest Rutherford, Hans Geiger and Ernest Marsden discover the very small size of nuclei by scattering alphas particles from a gold foil.

1915

- April 22: Germans use chlorine gas weapons at Ypes.

1919

- June: Soon after World War I, Ernest Rutherford creates nuclei (^{4}He + ^{14}N ⇒ ^{1}H + ^{17}O). "Talk softly please. I have been engaged in experiments, which suggest that the atom can be artificially disintegrated. If it is true, it is of far greater importance than a war."

1920

- Rutherford speculates on existence of the neutron at the Royal Society of London.

1931

- November: Harold Urey discovers deuterium in hydrogen's optical spectra.

1932

- February 27: James Chadwick observes knock-out protons to discover the neutron.
  (^{4}He + ^{9}Be ⇒ ^{12}C + ^{1}n)

1933

- January 28: Adolph Hitler becomes Chancellor of Germany.
- March 4: Franklin Roosevelt becomes President of the United States.
- April: Eleven scientists who had won the Nobel Prize or would win it later were fired from German Universities because they were Jewish. Johannes Stark and Philip Lenard's *Aryan physics* is not accepted by physicists.
- October: Leo Szilard recollects, "It occurred to me in October 1933, that a chain reaction might be set up if an element could be found that would emit two neutrons when it swallowed one neutron." This idea becomes a classified British patent in 1935 before uranium fission is discovered.

1934

- Artificial radioactivity discovered by Irene Curie and Frederic Joliet (alpha particles) and by Enrico Fermi (neutrons).
- June 30: Adolph Hitler leads assassination of storm-trooper rival, Ernst Rohm and others, army taken over.

1938

- December: Fermi receives Nobel prize for physics for discovery of transuranic elements (actually fission of uranium) and departs for the "new world" of the United States.
- December 22: Otto Hahn and Fritz Strassman identify barium isotopes and a few days later, Lise Meitner and Otto Frisch conclude that uranium nuclei fission by neutron capture.

1939

- January to May: Many experiments on uranium fission.
- April 29: Conference in Berlin to consider a German bomb and nuclear reactor.
- August 2: Szilard and Edward Teller obtain a letter from Einstein on the possibility of uranium weapons, which Roosevelt receives from Alexander Sachs on October 11, 1939.
- September 1: Hitler invades Poland. US Ambassador to Germany William Dodd (1933–37) writes Roosevelt (Sept.18), "If the [democracies] had co-operated, they would have succeeded. Now it is too late."

1940

- June 3: Paul Hartek fails to produce neutron multiplication in Hamburg reactor (185 kg natural UO_2, 15 tonnes CO_2 ice).

1941

- January: Experiments at the Berlin "virus house," using a 300-kg uranium reactor with impure graphite, cause incorrect rejection of graphite as a moderator in favor of heavy water from Vermork, Norway.
- July: British "Maud" Committee reports that a weapon could be made with 10 kg ^{235}U; US National Academy Sciences endorses bomb program.
- August: Hamburg group begins construction of ultracentrifuges to obtain ^{235}U. Centrifuges explode in April 1942, but they attained 7 % enrichment by March, 1943.
- October: Niels Bohr and Werner Heisenberg indirectly discuss nuclear weapons, described in Michael Frayn's play *Copenhagen*.
- December 6: Roosevelt directs substantial financial and technical resources to construct uranium bombs.
- December 7: Japan attacks Pearl Harbor.

1942

- May: Germans observe first neutron multiplication (13 %) in Leipzig using 570 kg U and 140 kg heavy ice.
- October 2: First of 3700 German-V2 flights.
- December 2: First nuclear chain reaction at Chicago's Stagg Field by Fermi.

1943

- February 28: Vermork heavy water factory destroyed by Allies after a failed attack on November 19, 1942.
- March 15: J. Robert Oppenheimer moves to Los Alamos, New Mexico.
- March: Glen Seaborg suggests Pu weapons might be jeopardized by ^{240}Pu as a spontaneous neutron emitter.
- Resumption of Soviet nuclear experiments.

1944

- Enigma Code decrypted at Bletchly Park, UK.
- August: Bohr presents memorandum on international control of nuclear weapons to British Prime Minister Winston Churchill and Roosevelt.
- November: First batch of spent fuel obtained from Hanford reactors.
- November: Samuel Goudsmit's Alsos mission obtains documents in Strasbourg, which imply that German weapons program made little progress.

1945

- January: First plutonium reprocessing at Hanford, Washington.
- January 20: First ^{235}U separated at the K25 gaseous diffusion plant, Oak Ridge, Tennessee.
- April 25: UN Charter signed by 50 nations in San Francisco.
- May 4: End of war in Europe.
- June 11: The Franck Report on the demonstration of the bomb and its international control sent to US Secretary of War.
- July 16: US explodes first atomic bomb, 19-kton *Trinity*, at Alamogordo, NM with electronics shielded to avoid EMP pulse, 300-m crater.
- August 6, 9: Atomic bombs destroy Hiroshima (Thin Man uranium, 13 kton, 3 m by 0.7 m diameter) and Nagasaki (Fat Man plutonium, 22 kton, 3.3 m by 1.5 m). Each weapon kills 70,000 prompt deaths and 140,000 within a year.
- August 15: End of war in the Pacific.

1946

- June 14: Bernard Baruch presents the Acheson-Lilienthal plan to the UN to internationalize the atom. "We are here to make a choice between the quick and the dead."
- June 30: First underwater detonation on ships by US at Bikini Atoll.

- July: Demonstrations in Times Square, New York City, against nuclear testing.
- December 31: Atomic Energy Commission takes over nuclear weapons program from US Army.

1948

- April, May: US atomic tests, Eniwetok Atoll in the Marshall Islands.
- November 4, 1948: UN General Assembly adopts US plan for international atomic control; USSR opposed.

1949

- April 4: NATO established.
- August 29: First Soviet atomic test, 22 kton of a US design at Semipalatinsk.
- October 30: AEC General Advisory Committee gives approval to build more powerful U bombs rather than the H bomb.

1950

- January 27: Klaus Fuchs confesses he transmitted atomic secrets to the Soviets.
- January 31: US President Harry Truman announces the decision to proceed with H bombs.
- March: Worldwide peace offensive to "ban the bomb" (the Stockholm Appeal) signed by more than 500 million people.
- June 25: North Korean army crosses 38th parallel, starting a war that lasted until July 1953.

1951

- January 21: First of 928 US tests at Nevada Test Site.
- May 9: *George,* DT-boosted atomic bomb, 225 kton.

1952

- January: UN Security Council establishes Disarmament Committee.
- June: Lawrence Livermore Laboratory established.
- October 3: First of 12 British nuclear tests at Monte Bello Islands, W. Australia.
- November 1: US explodes first fusion device *Mike*, 10.4 Mton at Eniwetok (liquid deuterium, not deliverable).

1953

- April 18: *Badger* test in front of 2800 troops at Nevada Test Site.
- May: Utah officials ask AEC to stop Nevada tests because of fallout on St. George, Utah.
- August 12: First Soviet fusion device exploded on a tower in Siberia (probably not deliverable).
- Fall: India and Australia propose a total ban on nuclear weapons to the UN.
- December 8: "Atoms for Peace" speech by US President Dwight Eisenhower at UN.

1954

- January 21: First nuclear-powered submarine, USS Nautilus, launched.
- March 1: Castle-Bravo, 15-Mton H bomb test, affects Marshall Islanders and Lucky Dragon.
- April: British Parliament petitions Churchill, Eisenhower, and Malenkov to control nuclear weapons.
- April 12 to May 6: AEC hearings deny Oppenheimer his access to nuclear secrets.
- August 30: Atomic Energy Act of 1954 emphasizes peaceful uses of atomic energy.

1955

- January–December: Reports of increasing global nuclear fallout.
- May 6: West Germany joins NATO.
- May 14: Warsaw Pact Organization established.
- August 8–20: First International Conference on Peaceful Uses of Atomic Energy in Geneva.
- November 22: First deliverable, high yield, Soviet H bomb, 1.6 Mton, Semipalatinsk.

1956

- US National Academy of Sciences panel finds genetic effects from fallout of nuclear tests to be slight compared with effects from natural radiation background. Adlai Stevenson makes fallout an issue in US presidential campaign.

1957

- May 15: First British H bomb explosion of 0.3 Mton at Christmas Island in the Pacific Ocean.
- July 6–11: First Pugwash Conference advocates test ban; Soviet scientists attend.
- August: AEC inaugurates Plowshare Program for peaceful nuclear explosions.
- September 19: Rainier, the first underground test at 1.7 kilotons at Nevada Test Site.
- September 29: Soviets evacuate 10,000 from Mayak reprocessing plant explosion. US keeps the secret.
- October 1: International Atomic Energy Agency inaugurated in Vienna.
- October 4: First satellite Sputnik I put into orbit by USSR.
- December 2: Shippingport electric power reactor reaches full power of 60 MW.

1958

- January 1: European Atomic Energy Community (Euratom) established.
- January 31: Explorer I, first US satellite, put into orbit.
- November1958–September 1961: US, UK, USSR observe informal moratorium on nuclear tests.

1959

- September 2: Vela Uniform Seismic Project established by US Department of Defense.
- November 24: US/USSR sign a memorandum of cooperation for utilization of nuclear energy.

1960

- February 13: First nuclear test by France, 70 kton, in the Sahara Desert.
- May 1: U-2 spy plane shot down over USSR.
- July 26: USSR suggests 3 on-site inspections/year as part of a test ban.
- November 15: First Polaris missile launched from a US submarine.

1961

- January 17: Eisenhower's farewell address discourses on *the military-industrial complex*.
- February 1: US launches Minuteman-I missile.
- March 31: Pravda suggests the use of H-bombs to obtain fresh water from Soviet glaciers.
- April 12: Yuri Gagarin becomes the first cosmonaut in orbit.
- May 29: US and UK agree to draft Comprehensive Test Ban (CTB) Treaty with 12 on-site inspections/year.
- August: Installation of first US seismic stations in a network of 60 countries.
- September 1: USSR resumes nuclear tests.
- September 15: US resumes nuclear tests.
- October 30: Soviet *Tsar* bomb, 58-Mton partial test of 100–150 Mton, Novaya Zemlya.

1962

- February 20: John Glenn becomes first US astronaut in orbit.
- July 6: *Sedan* excavation experiment as a *Plowshare peaceful nuclear explosion*; 10-Mton of earth displaced.
- July 9: Electromagnetic pulse from *Fishbowl* test, 1.4 Mton, 400 km above Johnston Atoll, turns off 300 streetlights on Oahu, Hawaii (1200 km away).
- September 3–7: Tenth Pugwash Conference proposes "black box" monitoring for CTBT verification.
- October: US reduces requirement to 8–10 inspections/year for a CTBT.
- October 15 to November 20: Cuban missile crisis, 162 warheads, and blockade.
- December 19: Communist Party Secretary Nikita Khrushchev accepts 2–3 inspections/year and 3 unmanned seismographic stations (black boxes) in USSR.

1963

- January: Secret test ban talks between US and USSR.
- February 19: US accepts 7 inspections/year, provided any mysterious event may be challenged.
- April: Khrushchev withdraws offer of 3 inspections/year.
- June 20: US and USSR sign "hot line" agreement.
- August 5: US/USSR/UK sign Limited Test Ban Treaty, banning nuclear explosions in atmosphere, space and underwater.

1964

- August 4: Gulf of Tonkin resolution sends US troops to Vietnam under President Lyndon Johnson.
- October 16: China explodes first nuclear bomb.
- October 22: Salmon, a US cavity-explosion, muffles seismic signal in a salt dome, near Hattiesburg, Mississippi.

1965

- June: Large Seismic Array (LASA, Montana) starts detecting nuclear explosions.

1966

- January 17: B-52 bomber crashes near Palomares, Spain with 4 unarmed H-bombs, 2 by sea and 2 by land.
- September 24: First French H-bomb, in Tuamato Archipelago, Polynesia.

1967

- January 27: Outer Space and Celestial Bodies Treaty bans nuclear weapons in orbit and space.
- February 14: Treaty for Prohibition of Nuclear Weapons in Latin America signed in Mexico City (Tlatelolco); all nations must ratify for the treaty to enter into force.
- June 17: First Chinese H bomb, 3.3 Mton, Lop Nor.
- June 23–25: President Johnson and Soviet Premier Alexi Kosygin begin SALT and ABM talks in Glassboro, NJ,
- December 10: Gasbuggy Plowshare explosion in New Mexico to stimulate natural gas production.

1968

- July 1: Nuclear Non-Proliferation Treaty (NPT) opened for signature.
- August: Soviet GALOSH ABM system deployed around Moscow.
- August 24: French H bomb, 2.6 Mton, Fangataufa Atoll.

1969

- July 20: US Apollo 11 lands on moon.
- November 3: UN Committee on Disarmament proposes global exchange of seismic data to monitor a CTB treaty.
- November–December: Preliminary SALT talks in Helsinki.

1970

- January: Administration of President Richard Nixon cuts Plowshare budget.
- March 5: NPT enters into force with 46 nations in 1970 and 190 in 2013.
- November 30: Atlantic-Pacific Inter-oceanic Canal Commission rejects use of nuclear explosives.
- December 18: Largest US release of radioactivity from *Baneberry* test.

1971

- March 30: US deploys Poseidon SLBM.
- June 15: Three Russian nuclear explosions set off to increase oil productivity.

1972

- May 26: SALT-I and ABM treaties signed by Nixon and Soviet First Secretary Lenoid Brezhnev in Moscow.
- June: UN Conference votes 48 to 2 to halt nuclear testing; France/China oppose, US/UK abstain, USSR absent.
- October: US detects Soviet SS-18 flight test.
- November: SALT II negotiations begin.

1973

- May 30: Protocol establishes the ABM/SALT Standing Consultative Commission.
- June 22: World Court issues injunction to France to refrain from testing on Mururoa in Pacific Ocean.
- October 17, 1973 to March 17, 1974: OPEC embargoes oil to US/Netherlands.

1974

- May 18: India sets off a 12-kton, Pu-device under Rajasthan Desert, expands nuclear club beyond *WWII P-5*.
- November 24: US/USSR agree to limit strategic launchers (to 2400) and MIRV launchers (to 1320).

1975

- January 19: AEC reorganized into NRC (regulatory) and ERDA (development, became DOE on August 4, 1977 and weapons became NNSA in 1999).

- October 1: Grand Forks, North Dakota Safeguard ABM site is activated. It was closed 4 months later because of its ineffectiveness against Soviet MIRVs at a cost of $21 billion from 1968–78 (1998$).

1976

- Public key encryption invented.
- January 29: France cancels its reprocessing exports for South Korea and Pakistan (1978).
- October 28: US President Gerald Ford postpones reprocessing of spent fuel to obtain Pu.

1977

- April 7: US President Jimmy Carter postpones indefinitely commercial reprocessing, halts Clinch River Breeder Reactor and calls for an International Fuel Cycle Evaluation (INFCE) to make the nuclear cycle more resistant to proliferation.
- October 19, 1977, to February 26, 1980: 40 nations prepare INFCE report.

1978

- January 11: Nuclear Suppliers guidelines forwarded to IAEA.
- March 10: Nuclear Nonproliferation Act (NNPA) tightens nuclear export criteria.
- April 4: Anwar Saddat of Egypt and Menachem Begin of Israel sign Camp David Accord in Washington.
- July 13: Euratom ends US embargo of nuclear fuel to Europe (NNPA required renegotiation) with a brief meeting.

1979

- April, 6: US cuts economic and military aid to Pakistan because it is making weapons-grade uranium.
- June 8: First Trident with SLBM launched.
- June 18: Brezhnev and Carter sign SALT II Treaty in Vienna.
- September 22: Vela satellite bhangmeter picks up a signal over the South Atlantic, perhaps a 3 kton explosion.
- December 12: NATO agrees to negotiate INF Treaty, while building Pershing-II and ground launched cruise missiles.
- December 26: USSR invasion of Afghanistan removes SALT II from US Senate consideration.

1980

- July 15: Office of Science and Technology Policy concludes the light signals recorded by a Vela satellite over the South Atlantic on September 22, 1979 were probably not from a nuclear explosion. Others disagree.
- September 1980–August 1988: Iran–Iraq war, with chemical weapons.

- December: Tomahawk submarine-launched cruise missile flies 600 miles to sea and returns to land with 5-m accuracy.

1981

- April 12–14: Space shuttle Columbia orbits Earth 36 times.
- June 7: Israel destroys Iraq's Osirak reactor, an act that was condemned but later appreciated after Gulf War.
- November 13: Senate ratifies Protocol I of Tlatelolco Treaty; US cannot deploy weapons in Puerto Rico, Virgin Islands, or Guantanamo Naval Base.
- November 30: US tables Zero-Zero INF proposal in Geneva.

1982

- June 14, Argentina surrenders to UK in the Falkland War.
- June 29: Strategic Arms Reduction Talks (START) begin in Geneva.
- July 16: Ambassadors Paul Nitze and Yuli Kvitsinsky walk in Geneva woods, trying to cut INF weapons by 50 %.
- 1982: USSR places 300 SS-20 missiles on its western border.
- November: US President Ronald Reagan adopts dense-pack basing for MX missiles.

1983

- March 23: Reagan calls for SDI "to make nuclear weapons impotent and obsolete."
- April 6: Scowcroft Commission recommends Midgetman (road-mobile, single-warhead) and 100 MX.
- Summer: Large phased-array radar construction discovered at Krasnoyarsk.
- November 23: Soviets leave Intermediate Nuclear Force (INF) negotiations as Pershing-II and GLCM are deployed.
- December 8: Soviets suspend START negotiations.

1984

- January: Strategic Defense Initiative Organization (SDIO) created.
- January 23: Reagan's first "Soviet Noncompliance" report released to Congress amid controversy.
- January 30: Soviets respond with an *aide memoire* to US State Department on US noncompliance.
- June 10: Homing Overlay Experiment (HOE) intercepts midcourse missile in limited test of kinetic kill vehicles.

1985

- January 8: Sec. State George Schultz and Ambassador Andrei Gromyko agree to negotiate on START/INF/ABM/SDI.
- February 20: Nitze-criteria for SDI deployment in terms of reliability, survivability and marginal cost.

- October 6: Reagan encourages broad interpretation of ABM Treaty to test other physical principles in space.
- October 14: Schultz states that a broad interpretation the ABM Treaty "is fully justified" to allow SDI testing.

1986

- July: Natural Resources Defense Council sets up seismographs in the Soviet Union.
- September: First B-1 bomber squadron becomes operational.
- October 7: Mordechai Vanunu publishes photos of the Israeli Dimone nuclear weapons facility in the London Times. He is drugged and taken to Israel for solitary confinement until April 22, 2004.
- October 11–12: Reagan and Soviet leader Mikhael Gorbachev agree to cut strategic weapons by 50 % at Reykjavik.
- November 28: The 131st B-52 bomber with air launched cruise missiles is deployed, exceeding the 1320 warhead limit on MIRVed launchers, negating SALT II. US is first country to publicly renounce an arms control agreement.
- December: First 10 Peacekeeper missiles become operational in Minuteman-III silos in Wyoming.

1987

- February: Soviets end unilateral nuclear testing moratorium after 18 months.
- April 7: Missile Control Technology Regime signed to slow missile proliferation.
- April 23: APS Directed Energy Weapons Study (DEWS) Group "estimates that even in the best of circumstances, a decade or more of intensive research would be required [for an informed decision on DEWS.]"
- August: First of six South African gun-type weapons is qualified.
- September 15: Schultz and Soviet Foreign Minister Eduard Shevardnadze sign Nuclear Risk Reduction Accord.
- December 8: Reagan and Gorbachev sign Intermediate Nuclear Force Treaty.

1988

- May 27: US Senate ratifies INF Treaty (93-5), the first ratification of an arms control treaty since 1972.
- June 28: Office of Technology Assessment testifies it disagrees with the charge of *likely violation of TTBT*.
- August–September: Joint Verification Experiment on nuclear tests carried out by US/USSR.

1989

- June 4: Chinese shoot 100's in Tiananmen Square. By 2012, many rebel students now national leaders.
- November 9: Fall of Berlin Wall.

1990

- October 24: Last Soviet nuclear test, as Gorbachev calls for a moratorium.
- November 19: NATO and Warsaw Pact sign Conventional Armed Forces in Europe (CFE) Treaty, limiting tanks/planes to same levels, a 60 % reduction for WTO and no reduction for NATO. Enters into force in 1992, ending the Cold War.

1991

- January 17 to February 28, 1991: *Operation Desert Storm*, 34 nations defeat Iraq.
- April 3: UN Resolution 687 requires destruction of Iraq's nuclear/bio/chem weapons.
- July 10: South Africa signs NPT and later destroys its 6 nuclear weapons.
- July 31: US and USSR sign START I, each cutting from 1200 to 7000 actual warheads.
- September 27/October 5: US President George H.W. Bush and Soviet leader Gorbachev reciprocal-unilateral measures place missiles to be cut by START I off alert, and remove surface-ship and ground-based tactical weapons.
- November 26: Last British test at NTS.
- November 27: Sam Nunn–Richard Lugar Act passes, giving Russia first $400 million to destroy weapons.
- December 1: Ukraine votes to become independent of USSR, followed by 13 other Soviet Republics and Russia.

1992

- March: US ends military plutonium reprocessing, HEU production stopped in 1964.
- March 9: China joins the NPT regime.
- March 24: Open Skies Treaty signed, allowing monitoring by over-flights of NATO and Warsaw Pact aircraft.
- May 23: START's Lisbon Protocol signed by US/Russia/Ukraine/Belarus/ Kazakhstan, all agreeing to join NPT.
- August 3: France joins NPT regime.
- September 23: Last US nuclear test, *Divider*. Bush-I signs testing ban legislation.
- October 1: US Senate ratifies START I with the Biden condition calling for "the adoption of cooperative measures to enhance confidence in reciprocal declarations [for warheads and fissile materials]."

1993

- January 3: US/Russia sign START II, cutting deployed strategic warheads from 7000 to 3500.
- February 18: US/Russia agree on sale of 500 tonnes of HEU, blended to LEU reactor fuel.

- March 3: France joins the NPT.
- April 1: IAEA declares North Korea in violation of the NPT.
- September: US/Russia agree to start Materials Protection Control and Accounting program with lab-to-lab assistance.
- September 27: President Bill Clinton proposes Fissile Material Cut-off Treaty banning production of weapons Pu/HEU.
- December: UN adopts resolution calling for a Fissile Material Cut-off Treaty.

1994

- April 30: Nuclear Proliferation Prevention Act (Glenn–Pell–Helms Act) provides for sanctions on nuclear testing.
- May 30: US/Russia detarget strategic missiles, but they can be quickly reprogrammed.
- June 23: US Vice President Gore and Russian Prime Minister Viktor Chernomyrdin agree to close last 3 Russian plutonium-production reactors by 2000, with US aid to establish another power source.
- September 20: Clinton and Russian President Boris Yeltsin agree to exchange data on warheads and Pu/HEU.
- October 1: Russia stops plutonium production, except at 3 reactors where waste heat is used to heat cities.
- October 23: US/N-Korea sign "Agreed Framework" to eliminate N. Korean nuclear program and get S. Korean reactors.

1995

- March 1: Clinton announces permanent removal of 100 tonnes of Pu/HEU from US stockpile.
- May 10: Clinton and Yeltsin agree to declarations/inspections on excess Pu/HEU from weapons.
- May 12: NPT extended indefinitely after the 5 nuclear weapon states promise to join a CTBT.
- August 11: Clinton announces US will support a zero-yield CTBT.
- December 19: Wassenaar Arrangement on export controls replacing Cold War controls.

1996

- January 26: Senate ratifies START II by 87-4, cutting to 3500 total warheads each and bans land-based MIRV.
- January 27: France conducts last test and supports a zero-yield CTBT.
- March 25: US joins South Pacific Nuclear-Free Zone.
- April 11: Pelindaba Treaty for Nuclear-Free Africa signed by 43 African states.
- July 29: China's last nuclear test, 2 months before signing CTBT.
- September 10: UN opens CTBT for signature, banning nuclear tests everywhere, for all time, at any yield. Only India, Bhutan and Libya are opposed.

- September 17: US/Russia/IAEA agree to trilateral inspections on excess SNM.
- September 24: All five nuclear weapon states sign the CTBT at the UN.
- 1996: Last nuclear warheads from former Soviet Republics returned to Russia.

1997

- April 29: Chemical Weapons Convention enters into force.
- March 21: Clinton and Yeltsin in Helsinki call for START III reductions to 2000–2500 deployed, strategic warheads and "measures relating to the transparency" for inventories and destruction of strategic warheads.
- May 15: IAEA adopts Additional Protocol to give IAEA more powers.
- May 27: Yeltsin signs NATO-Russian Founding Act.
- September 7: Russian General Alexander Lebed claimed "up to 100 suitcase bombs are not under the control of the Armed Forces of Russia."

1998

- May 11, 13, 28 and 30: India and Pakistan conduct multiple underground nuclear tests.
- August 31: North Korea launches a Dong-1 missile over Japan.
- September 22: US and Russia sign Nuclear Cities Agreement to help 10 Russian closed weapon cities.

1999

- October 13: US Senate rejects CTBT (51–48) in spite of 150 signatory nations.

2000

- April: Duma ratifies CTBT (April 21) and START II (April 14), contingent on constraint of ABM demarcation.
- May, 22: NPT review conference agrees to the 13 steps for progress on nuclear disarmament.
- September 1: Gore/Chernomyrdin agree to dispose of 34 tonnes Pu each, as MOX fuel or buried in a repository.

2001

- May 1: President George W. Bush states he will withdraw US from ABM Treaty to test/deploy ABM options.
- May 31: INF continuous Perimeter and Portal Monitoring ends as INF 13-year period ends.
- September 11: Osama bin Laden's al Qaeda uses US commercial planes to kill 3000 in destroying the World Trade Center and damaging the Pentagon. The US joins Afghan resistance forces to topple Taliban government, producing a new leader, Hamid Karzai. The 9-11 event creates a *major paradigm shift* to focus on terrorism over other matters.
- September–October: Letter-borne anthrax contamination in US postal system and buildings.

- December: G.W. Bush administration funds CTBT research, but not that which aids CTBT entry-into-force.
- December 4: US and Russia complete START I reductions to 6000-warheads.
- December 7: US ends discussion on verification of Biological Weapons Convention.
- December 13: G.W. Bush formally starts process to withdraw from ABM Treaty 6 months hence.
- December 14: Navy Area Theater Ballistic Missile Defense system, covering small areas, is cancelled.

2002

- January 4: DoD Ballistic Missile Defense Office elevated to Missile Defense Agency.
- January 8: Nuclear Posture Review sets path to 1700–2200 deployed, operational US warheads by 2012, while retaining up to 5000 warheads on "active and inactive reserve" that can be deployed in days to years. The "new triad" combines offense, defense, conventional arms and rapid information. Targeted countries are Russia, China, Iran, Iraq, Libya, N. Korea and Syria in possible conflicts between Israel/Iraq, N. Korea/S. Korea, China/Taiwan. The NPR suggests that new designs are needed to create a 5-kiloton earth-penetrating bomb to attack underground bunkers. Congress shortens preparation time for test-sites readiness for this weapon.
- January 23: Bush administration continues US/Russia Pu-MOX disposition but cancels geological immobilization.
- January 29: Bush labels North Korea, Iran and Iraq as an "axis of evil arming to threaten the peace of the world."
- March 14: India integrates the Agni-2 missile (2000 km, 1000 kg) into its armed forces.
- May 24: Presidents George W. Bush and Vladimir Putin sign the Strategic Offensive Reductions Treaty, which reduces *operational* warheads to 1700–2200. SORT allows response forces of some 2400 additional warheads, adds no new verification measures, and expires on December 31, 2012, the same date the limit of 1700–2200 is required. Russia retains 138 SS-18 s and many SS-19 s which would have been removed under START II..
- October 16: US announces North Korea admits having a clandestine centrifuge program to make HEU.
- November 4: Cuba accedes to the NPT, completing Western Hemisphere. UN conference lobbies nuclear weapon states to agree to a negative security assurance, so non-weapon states would not fear nuclear weapons.
- November 27: UN inspections begin in Iraq, pursuant to UN resolution 1444 that required Iraq to disarm its weapons of mass destruction, or be in "material breach." In 1997, Iraq declared 8 presidential palaces off limits, leading to an end of inspection.
- December 8: Iraq declares its lack of weapons of mass destruction as required by UN resolution.

2003

- January 10: North Korea withdraws from NPT after IAEA inspectors are removed (Dec. 27) and Governing Board "deplores in the strongest terms" (Jan. 6) the reprocessing restart, claimed two weapons and beginning centrifuges.
- March 2: Department of Homeland Security established with a budget of $30 billion.
- March 6: Senate ratifies Strategic Offensive Reduction Treaty. Iraq-war anger delays Russian Duma to May 14.
- March 19: US/UK invade Iraq for a 3-week war, without explicit support from the UN, which wanted more time for more inspections. Unrest in Iraq continues after the war, without evidence for weapons of mass destruction.
- March 31: US mildly sanctions Pakistan's Kahn Research Institute for exporting centrifuge equipment to N. Korea in trade for missile technology.
- April 22: Los Alamos builds first US Pu pit in 15 years, a beginning for a plant to make 125 pits/year by 2018.
- April 25: North Korea tells US in Beijing that it has Nuclear weapons.
- June 6: IAEA Board of Governors charges Iran with a violation over clandestine centrifuges discovered on Feb. 20.
- August 28: North Korea informs China, Japan, Russia, South Korea and the US that it may test in the near future.
- December 14: Saddam Hussain captured.
- December 19: Libya announces it will dismantlement its nuclear weapons programs.

2004

- February 4: Pakistan President Pervez Musharraf pardoned Abdul Qadeer Khan for supplying centrifuge technology to Iran, Libya and North Korea.
- February 20: IAEA report states that Libya had ordered 10,000 centrifuge tubes, but was now cooperating.
- July 9: The Senate Intelligence Committee Report rebukes the October 2002 National Intelligence Estimate that "Baghdad has chemical and biological weapons by stating that this conclusion "overstated both what was known and what intelligence analysis judged abut Iraq's chemical and biological weapons holdings." The CIA did not invite DOE to test alleged centrifuge tubes. The administration's charge of Iraq's purchase of uranium was incorrect.
- July 22: The *9/11 Commission Report* states "we have seen no evidence that these or other earlier contacts ever developed into a collaborative operational relationship [Al Qaeda/Iraq]."
- September 30: Report on Iraq Weapons of Mass Destruction released by Charles Duelfer (Chief Iraq WMD Inspector). It agreed with the results of David Kay ("we were all wrong"). The report concluded Iraq's nuclear program ended in 1991, that Iraq unilaterally destroyed its chemical weapons in 1991 and that Iraq abandoned its ambitious biological weapon plans, but Saddam retained his desire for WMD.

- October 19: Brazil agrees to IAEA inspections of centrifuges, hoped as a precedent for Iran.
- December 7: US intelligence organized under centralized National Intelligence Director.

2005

- January 31: Elections in Iraq with high Shiite/Kurd turn-out but negligible Sunni participation.
- February: Iran nuclear program evident as FR/FRG/UK try diplomacy and US exerts pressure.
- February 10: North Korea claims it has nuclear weapons and refuses to join 6-party talks.
- March 6: China questions US data on DPRK nuclear weapons, suggests 2-party, US/NK talks.
- July 29: State department absorbs its arms control bureau into the non-proliferation bureau.
- September 19: North Korea commits to abandon its nuclear weapons program, returning to the NPT and safeguards.
- September 24: IAEA finds Iran in noncompliance to the NPT.

2006

- March 2: President George W. Bush and Prime Minister Manmohen of India agree to cooperate on peaceful nuclear power, this requires amending the 1978 NNPA nonproliferation law and obtaining unanimous consent from the Nuclear Suppliers Group. India would place 14 of its 22 reactors under India-specific IAEA safeguards.
- April: Iran declares it has 164 centrifuges. US joins France, Russia and UK to dissuade Iran from enriching uranium on its soil. Brazil continues its work on its enrichment plant since 2004.
- October 9: DPRK test of 0.6 kton detected by 22 IMS seismic stations (with 30 certified). Radio-Xe detected by IMS and portable sensors.
- December 10: President Ehud Olmert lets slip that Iran is "aspiring to have nuclear weapons as America, France, Israel, Russia."

2007

- January 11: China uses a kinetic-kill, anti-satellite weapon to destroy a Chinese weather satellite. Arms race in space or a nudge to control ASAT weapons?
- February 13: Six party talks yields "initial actions" plan to implement September 2005 pledge by DPRK to abandon nuclear weapons and programs. This later fails.
- March 2: NNSA chooses the LLNL to design Reliable Replacement Warheads. Congress denies funding in 2008.
- July 14: Russia suspends CFE observance, followed by US in 2011.
- July: First W-88 produced in 18 years at Los Alamos, to build 20–50 per year.

2009

- April 5: President Obama states in Prague his intention to move towards abolition of nuclear weapons, with NEW START and CTBT ratifications. He stated his intention to remove troops from Iraq, but send more troops to Afghanistan, until perhaps 2013.
- May 25: DPRK 3-kton test detected by 61 IMS seismic stations, but not the radio-Xe signal.
- September 25: US, UK and France announce that Iran is building a covert enrichment facility inside a mountain near Qom. Previously Iran indicated it might restrain its enrichment program.
- November 29: Indian Prime Minister Singh states India is willing to join NPT as NWS.

2010

- April 5: Nuclear Posture Review removes 1st US response of nuclear attacks on NNWS bio-chem attacks, but retains this response for the 9 states, which have nuclear weapons. No new ICBMs proposed.
- April 8: New START signed in Prague with limits of 1550 operational strategic weapons and 700 deployed launchers. Tactical weapons not covered, as usual, but they consist of about 250 for the US and 2000 for RF, down from the Cold War numbers of 8000 and 23,000. Prompt Global Strike conventional weapon ICBMs to count as nuclear systems, not to be deployed until 2015–20.
- May: NPT Review Conference obtains consensus, with modest progress in a time of difficulty.
- June: Stuxnet cyberwar attack on Iran, destroying 1000 of 5000 centrifuges.
- June 9: UN Security Council sanctions cut military imports to Iran and restrict proliferating companies on a 12–3 vote. Brazil and Turkey offer a partial solution to allow reactor fuel, but not constrain higher enrichments.

2011

- February 5: New START enters into force, passing Senate Dec. 22, 2010 on 71–26 vote, 1550 operational warheads.
- November 22: US suspends implementation of CFE because Russian transfers troops to Georgia and Moldova.

2012

- NNSA and JASON declare Pu pit lifetimes is over 100 years, using ^{238}Pu fast decay.
- April 15, 2012: Saudi Arabian Armaco oil company under cyber attack.
- September 25: 100th Joint Actinide Shock Physics Experimental Research (JASPER) to obtain Pu equation of state from Pu projectiles.
- November 29: UN votes 138/9/41 to recognize Palestine as a nation.

– November: Oak Ridge's Titan computer does 17.6 petaflops, same as Earth's population each doing 3 M/s.

2013

– January 23: Iran notified IAEA it is installing IR-2M centrifuges. Iran has 160 kg of 20 %-HEU, needing 240 kg for a warhead's input steam.
– February: China's People's Liberation Army cyber-attacks over 60 US high tech companies.
– February 12: DPRK third nuclear test, 7 kton, under Kim Jong Un.
– May 13: CTBT ratified by 159 nations (not US, China) of 183 signatories (not India, N. Korea, Pakistan), with 36 of 44 nuclear capable nations ratified. Monitoring capability believed to be at 0.1-kton level for tamped explosions, and 1–2 kton for cavity shots. Certified IMS stations (275), testing (20), under construction (20), planned (22).
– March 11: N. Korea nullifies 1953 Korean armistice. Restarts Yongbyon reactor on April 2.
– March 13: 2 million small pox vaccines prepared, only US and RF have small-pox samples.
– April: 12 more ground-based interceptors sent to Alaska in response to N. Korea.
– April 2: UN votes 154/3/23 to regulate global conventional arms trade of $70 billion/year, U.S. uncertain.
– April 9: US ship-board lasers shoot down slow drones.
– April 15: Tsarnaev brothers attack Boston Marathon. US follows UK lead with 30 million cameras to observe.
– May 23: Command of US drones moved from CIA to DoD.
– May 23: Iran has 14,244 of 50,000 planned centrifuges, 700 are IR-2, more potent. Obama's red line of 240 kg of 20 % medium enriched uranium may be crossed in the fall of 2013.
– May 31: Nuclear Regulatory Commission gives license for laser enrichment in North Carolina to GE-Hitachi.
– June 3: Supreme Court allows arrested suspects of serious crimes to be DNA tested for unknown other crimes.
– June 6: Edward Snowden leaks are published in the Washington Post and Manchester Guardian. NSA *PRISM* program tracks global email, web searches and other internet traffic. On June 9, *Boundless Informant* program tracks global phone calls. Snowden fights extradition from Hong Kong to Russia, but fails (July 2) to obtain passage to Equador or other states.
– June 6: Obama acknowledges phone-call and email lists of US citizens are kept, but he states they are not accessed for US domestic citizens until evidence is presented to the FISA court.
– June 19: In Berlin, Obama states he is willing to reduce New START limit of 1550 by 1/3 in a reciprocal, unilateral agreement (not a treaty), monitored under

New START. Putin wants ABM concessions to do this. The level of 1000 was last seen in 1954. Nunn Lugar renewed after 20 years with emphasis on securing nuclear materials, and not on missile and CW destruction.

- July 3: Prime Minister Mohamed Morsi of Egypt, representing the Muslim Brotherhood, is overthrown by Egyptian military with the support of secular parties. The Arab Spring oscillates away from democracy, without major regrets by the democracies.
- July: National Security Administration admits that domestic phone call, email and postal information is collected, but only read after approval by the Foreign Intelligence Surveillance Court (FISC). There are calls to expand oversight of FISC, which has been accused of leniency.
- August 9: UN panel of experts from 25 nations suggest ways to enhance cyber security with cooperative measures and with enhancements to international law.
- September 6: The National Security Agency foils encryption by hacking into computers via permitted backdoors. On September 10, Yahoo, Facebook, Google and Microsoft file suit against NSA.
- September 21: Old reports were released that describe a near catastrophe from the crash of a B52 bomber near Goldsboro, NC on 23 January 1961, 3 days after Kennedy's inauguration. Only a single switch prevented the 4-Mton bomb from exploding.
- September 26: Iraq ratifies the CTBT, after two US-Iraq wars and much chaos.
- September 27: Obama and Iranian President Hasssan Rouhani talk on the phone to attempt to start meaningful nuclear negotiations. This was the first leader-to-leader contact since 1979 when President Carter and the Shah of Iran talked, just before the 444-day hostage crisis began.
- October 2: UN dismantlement inspections begin in response to the death of 1400 from Sarin chemical weapons used by the Syrian government near Damascus on March 19 and August 21. Agreement between Russia, U.S., Syria and the UN Security Council (September 27) to end the Syrian CW program within a year came after threats of force by Obama and the lack of Congressional support to carry out such an attack. President Bashir al-Assad is helpful.
- November 24: Iran and *P5-plus-1 group* agree to a *Joint Plan of Action* to constrain Iran's nuclear production with instructive verification, in trade for a 10 % reduction in economic sanctions. Inspections began on 20 January 2014, for six months, renewable for another six months. This could be followed with a more comprehensive agreement if the *Joint Plan of Action* is successful.
- December 18: Presidential review panel urges restraint on National Security Agency wiretaps.

2014

- January 17: Congress fully funds the B-61 bomb, with plans for force structures with interchangeable warheads.
- February 22: Ukrainian Parliament declares President Viktor Yanukovych unable to govern after riots, following his refusal to signing a European Union trade agreement. On March 1, Russian President Vladimer Putin deploys

Russian troops into Crimea, which votes to join the Russian Federation on March 16, with accession on March 18.

- March 14: US mixed oxide fuel plant in South Carolina is mothballed.
- June 14: President Obama sends 5 nuclear bombers to EU to give assurance on the Russia/Ukraine issue.
- July: US State Department announces Russia violates INF Treaty by testing ground-based cruise missiles, that are not yet deployed. Russia claims US BMD deployments in Romania and Poland would be CFE violations.
- September 5: Ceasefire agreement in Ukraine's Donetsk and Luihansk regions. Putin makes German Chancellor Angela Merkel wait four hours in Milan after visit to Belgrade amid Western economic sanctions and falling oil prices.
- September 5: IAEA stated that Iran is complying with the interim agreement, but Iran failed to provide information about its past nuclear weapons activities. Iran had provided some information on exploding bridge detonators. Iran/P5+1 discussions continue.
- September 8: Syria exports its last CW stocks, but Syria has not destroyed its 12 remaining production facilities.
- September 11: US declares Russia in violation to INF treaty on its construction of a ground-based cruise missile with a range between 500 and 5,500 km.
- November 14: Russia terminates US-RF programs that prevent loss of nuclear materials.
- December 24: *Arms Trade Treaty* for transparent conventional weapon sales enters into force with 3 UN negative votes (Iran, Syria, DPRK), but US/RF/PRC and others have not ratified.

2015

- January 7: Charlie Hebdo magazine attacked in Paris by Al Qaeda.
- January 19: Google Glass camcorder no longer for sale to the general public.
- January 19: Cyber attack on Sony attributed to DPRK because of Kim Jong-un movie.
- January 26: Three-pound package delivered to White House by drone quadcopter.
- February 18: Federal Aviation Administration issues draft regulations for drones.
- March 3: Israeli Prime Minister Benjamin Netanyahu addresses Congress to undercut Iran-P5+1 agreement.
- March 10: Russia withdraws from the INF treaty.
- May 11: China deploys 3-warhead MIRV on 20 DF-5 ICBMs, capable of reaching US. All P5 nations now have MIRV.
- May 13: Saudi Arabian Prince Turki bin Faisal states, "Whatever the Iranians have, we will have, too."
- May 18: ISIS takes Ramadi, capital of Sunni Anbar Province in Iraq.

- May 26: NPT Review Conference fails to agree, as US, UK and Canada block a call for negotiations on Nuclear Free Zone in the Middle East.
- July 10: 20 million US security clearance forms for government jobs hacked, probably by China, includes all who applied over the past 15 years.
- July 14: Joint Plan of Action between Iran and P5+1 signed, constrain Iran's enrichment, extending break-out time from a few months to a year, while reducing economic sanctions.

Homework

1.1 **Small nucleus**. Why do back-scattered alpha particles imply the nucleus is very small?
1.2 **Neutron prediction**. Why does the mass of chlorine imply existence of neutrons?
1.3 **Atomic age misnamed?** Why is the name *atomic age* a misnomer?
1.4 **First isotope**. How does hydrogen atomic spectra imply existence of heavy hydrogen?
1.5 **Jewish and Aryan physics**. What is the difference?
1.6 **What Szilard didn't know**. Was he guessing when he created the idea of nuclear weapons?
1.7 **Fermi's error**. What did Fermi claim for his Nobel? What did he observe?
1.8 **Barium means fission**. Why did the observation of barium imply fission?
1.9 **Joliet's two neutrons**. Why did two neutrons per fission imply the atomic age was here?
1.10 **Szilard's letters**. How did Szilard's letters both promote both war and peace?
1.11 **German errors**. What errors did the German atom project make?
1.12 **Pu-240**. Why is Pu-240 an impediment to plutonium weapons?
1.13 **Baruch Plan**. What was the Baruch plan? Why did the Soviets block it?
1.14 **Eisenhower's Atoms for Peace**. What did Eisenhower offer? Good and bad results?
1.15 **Political exaggeration**. Give two examples on peaceful arms by each political party.
1.16 **First arms control**. What was the first nuclear arms treaty. Why was it not arms control?
1.17 **Defensive deterrent**. In what way can defensive weapons be destabilizing?
1.18 **MIRV deterrent**. In what way can multiple-warheads ICBMs be destabilizing?
1.19 **Nunn-Lugar and Cooperative Threat Reduction**. Why created? How successful?
1.20 **Bohr and Heisenberg**. Compare and contrast these men. Why did Heisenberg visit?

Seminar Topics: Early People

Bernard Baruch, Hans Bethe, Niels Bohr, Marie Curie, Albert Einstein, Enrico Fermi, Richard Feynman, Klaus Fuchs, Samuel Goudsmit, Richard Garwin, Leslie Groves, Paul Hartek, Werner Heisnberg, Frederic Joliot, George Kistiakowsky, Lisa Meitner, Seth Nedemeier, J. Robert Oppenheimer, Eugene Rabinowitch, the Rosenbergs, Ernest Rutherford, Andrei Sakharov, Glen Seaborg, Leo Szilard, Edward Teller, Harry Truman, Stanislaus Ulam, Evgeny Velikhov.

Bibliography

Beschloss, M., & Talbott, S. (1993). *At the highest levels*. Boston, MA: Little Brown.

Burns, R., & Siracusa, J. (2013). *A Global history of the nuclear arms race*. New York, NY: Praeger.

Close, F. (2015). *Half-life: The divided life of Bruno Pontecorvo*. New York, NY: Basic.

Cochran, T., Norris, R., & Bukharin, O. (1995). *Making the Russian bomb*. Boulder, CO: Westview.

Corden, P., Hafemeister, D., & Zimmerman, P. (Ed.). (2014). Nuclear weapons issues in the 21st century. In *AIP Conference Proceedings 1396, Melville, NY.*

Evangelista, M. (1999). *Unarmed forces*. Ithaca, NY: Cornell University Press.

Farmelo, G. (2013). *Churchill's bomb*. New York, NY: Basic.

Frayn, M. (1998). *Copenhagen*. New York, NY: Random House.

Graham, T. (2002). *Disarmament sketches*. Seattle, WA: University of Washington Press.

Hafemeister, D. (Ed.). (1991). *Physics and nuclear arms today*. New York, NY: American Institute Physics Press.

Hewlett, R., & Anderson, O. (1966). *The new world: 1936–46*. Washington, DC: US Atomic Energy Commission.

Hewlett, R., & Duncan, F. (1969). *Atomic Shield: 1947–52*. College Park, PA: Penn State University Press.

Hewlett, R., & Holl, J. (1989). *Atoms for peace and war: 1953–61*. Berkeley, CA: University of California Press.

Holloway, D. (1994). *Stalin and the bomb*. New Haven, CT: Yale University Press.

Lanouette, W. (1992). *Genius in the shadows*. New York, NY: Scribners.

Lourie, R. (2002). *Sakharov*. New England, Hanover, NH: Brandeis University Press.

Muller, R. (2008). *Physics for future presidents*. New York, NY: W.W. Norton.

Panetta, L. (2014). *Worthy fights*. New York, NY: Penguin.

Rhodes, R. (1988). *The making of the atom bomb*. New York, NY: Simon and Schuster.

Rhodes, R. (1995). *Dark sun: The making of the hydrogen bomb*. New York, NY: Simon and Schuster.

Rhodes, R. (2007). *Arsenals of folly*. NY: Knoopf.

Rhodes, R. (2010). *The twilight of the bombs*. NY: Knopf.

Talbott, S. (1979). *Endgame: The inside story of SALT-II*. New York, NY: Harper.

Talbott, S. (1984). *Deadly gambits*. New York, NY: Knopf.

Talbott, S. (1988). *The Master of the game*. New York, NY: Knopf.

Talbott, S. (2002). *The Russia hand*. New York, NY: Random House.

Chapter 2
Nuclear Weapons

Our world faces a crisis as yet unperceived by those possessing power to make great decisions for good or evil. The unleashed power of the atom has changed everything, save our modes of thinking, and we thus drift toward unparalleled catastrophe.

[Albert Einstein 1946]

2.1 The Nuclear Age

The fission age began in 1932 when James Chadwick discovered neutrons by observing knock-out proton tracks in a cloud chamber.

Observe neutrons with no electric charge and no tracks? An alpha particle (helium-4 nucleus from radioactive decay) combines with a beryllium-9 nucleus to make carbon-12 and a neutron. The super-script number on the isotope is the mass number, the sum of the neutrons and protons in the nucleus.

$$^{4}\mathrm{He} + {}^{9}\mathrm{Be} \Rightarrow {}^{13}\mathrm{C} \Rightarrow {}^{12}\mathrm{C} + {}^{1}\mathrm{n}.$$

The neutron without an electric charge doesn't leave a track, but it hits and propels a proton with electric charge, which does leave a track.

Fluke of nature with a rare isotope: Nuclear weapons are, essentially, a fluke of nature. Fission of uranium in nuclear weapons and nucleaer reactors is caused by a rare isotope, ^{235}U, of a moderately rare element, uranium. There are no easy replacements for ^{235}U, until mankind made plutonium from uranium. Uranium is, essentially, the only economic path to start-up reactors and nuclear weapons. One could use expensive particle accelerators or other isotopes produced in reactors, but that came later.

D. Hafemeister, *Nuclear Proliferation and Terrorism in the Post-9/11 World*,
DOI 10.1007/978-3-319-25367-1_2

Leo Szilard, the first to consider nuclear bombs, had this recollection in September 1933:

> As I was waiting for the light to change and as the light changed to green and I crossed the street, it suddenly occurred to me that if we could find an element which is split by neutrons and which would emit two neutrons when it absorbed one neutron, such an element, if assembled in sufficiently large mass, could sustain a nuclear chain reaction. I didn't see at the moment just how one would go about finding such an element.

Chain Reaction and Neutron Multiplication

Szilard thought that neutron multiplication might take place with this reaction:

$$^{1}\mathrm{n} + {}^{9}\mathrm{Be} \Rightarrow 2\,{}^{4}\mathrm{He} + 2\,{}^{1}\mathrm{n}.$$

If this worked, the initial neutron would make 2n, and these 2n would make 4n, to make 8n, finally to some 10^{24} neutrons. This idea is correct for uranium fission, since an average of 2.5 neutrons are released for each fission event.

Szilard's beryllium nuclear weapon can't be built, since beryllium-fission neutrons lack sufficient energy to fission ^{9}Be nucleui. Still, Szilard realized the military importance of chain reactions to the British military, even though uranium fission was not discovered for another 6 years. After Joliot submitted an article on the number of neutrons given off in a fission event to *Nature* on March 8, 1939, Szilard and Fermi published in the 1939 *Physical Review* that uranium fission produced "about two neutrons." Thus, physicists who read the *Physical Review* and *Nature* deduced that nuclear weapons could become a reality.

Scientific details of nuclear weapons are classified "top secret" and "restricted data," but the basic science of nuclear weapons was declassified in Robert Serber's *The Los Alamos Primer*, which presents basic equations and concepts, which was considered worrisome when it became publicized. In the 1970s the *Primer* was considered unclassified but don't talk about it. In the 1990s the *Primer* was considered basic science that is widely known, but without secret technical details. Lastly, *The Effects of Nuclear Weapons* by S. Glasstone and P. Dolan is considered a classic on describing the many–parameter results of nuclear explosions (Fig. 2.1).

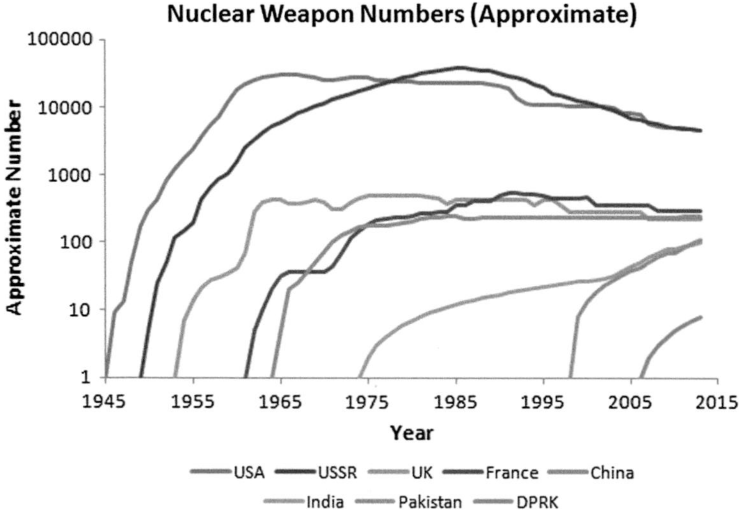

Fig. 2.1 Numbers of nuclear weapons from 1945 to 2012 for eight nations. We define a weapon as either deployed or in reserve, giving the total stockpile. The 8 curves are plotted in the following order, left to right, keyed to the date of the first nuclear test for the state: US (2015 totals; 4650 stockpile, 2150 operational), Russia (5000 stockpile, 1800 operational), UK (225), France (300), China (250), India (90–110), Pakistan (100–120) and N. Korea (5–10). The logarithm of the number of nuclear warheads is used in order to display the large differences in numbers between states. The US curve peaked in 1965 and the Soviet/Russian curve peaked in 1985. The US and Soviet/Russian curves show the drop in numbers over time (P. Corden and D. Hafemeister, *Physics Today*, April 2014. Pierce Corden and Derek Updegraff (AAAS), adapted for graphics from the warhead numbers from Hans Kristensen (FAS) 2013)

2.2 Nuclear Proliferation

At least 25 nations have attempted to develop nuclear weapons, beginning with Germany and the five nuclear weapon states (NWSs), the "big five" of World War II (the United States, United Kingdom, Russia, France and China). Nuclear tests by India in 1974 and 1998, Pakistan in 1998 and North Korea in 2006, 2009, 2013 and 2016 increased the list to eight. South Africa built six uranium weapons, but dismantled them in 1992. Israel is generally reported and believed to have nuclear weapons. During the Gulf War of 1991, the UN and the IAEA (International Atomic Energy Agency) discovered Iraq's large nuclear program. North Korea produced enough plutonium for a half-dozen weapons, encouraging South Korea, Japan, and the United States to give North Korea two commercial reactors in exchange for ending its program and allowing inspections. This 1994 agreement collapsed in 2002 with the announcement that North Korea restarted its weapons program in the late 1990s and ejected IAEA inspectors. Iran moves towards nuclear prowess with its enrichment program, but is now in "negotiations" with the *P5 + 1*. In the past, Argentina, Brazil, Germany, Japan, Libya, South

Korea, Sweden, Switzerland, Syria, Taiwan took steps to obtain nuclear weapons, but they stopped. The news is not all dark, as four weapons-owning states (South Africa, Ukraine, Belarus, and Kazakhstan) have given them up. This totals 23, but we know that doesn't end the list of nuclear-coveting states.

The 1970 Non-Proliferation of Nuclear Weapons Treaty (NPT) anchors a global regime that bans nuclear weapon technologies in non-nuclear weapons states (NNWS). The NPT regime relies on declarations by states-parties on their nuclear materials, which are monitored by IAEA inspections to determine their validity. In return for this loss of sovereignty, NNWSs expect NWSs to greatly reduce their reliance on nuclear weapons and to negotiate toward their ultimate elimination. The total elimination of nuclear weapons is hopeful, but unlikely, but substantial reduction is wise. As part of the bargain, NNWSs believe NWSs must stop testing nuclear weapons to conform to their obligation to stop the nuclear arms race. The NPT states agreed in 1995 to extend the NPT indefinitely, but only after the five NWSs stated that they would stop testing and join a Comprehensive Test Ban Treaty (CTBT). In addition, NWSs are expected to assist NNWSs with their peaceful nuclear power programs. Beyond these incentives, NNWSs prefer to live next to non-nuclear neighbors, which has led to the creation of nuclear-weapon-free zones in Latin America (Tlateloco), Africa (Pelindaba), Pacific (Rarotonga), SE Asia (Bangkok), Central Asia, Antarctic, Sea-Bed, and Outer Space.

The NPT regime was severely undercut when the US Senate rejected the CTBT by 51 to 48 in 1999 and the George W. Bush administration stated it would not seek CTBT ratification. The G.W. Bush Administration called for the development a 5-kton earth-penetrating weapon to attack underground bunkers. CTBT ratification has floundered since then, President Obama couldn't get enough Senate support for it.

2.3 Fission Energy

A 1-Mton weapon destroys houses at a distance of 5–10 km, gives third-degree burns at 10 km and releases lethal radioactive plumes beyond 100 km. The size of 1-Mton of conventional explosive could be likened to a train made up of 100 rail cars, each carrying 100 tons of coal. The length of this 1-Mton train is 200 km. However, the mass of a 1-Mton nuclear weapon is less than one-millionth of this trainload, making it far easier to deliver nuclear weapons over long distances with missiles and planes, as compared to conventional explosives of the same yield.

Fission of uranium and plutonium. Slow neutrons in reactors fission (split) ^{235}U nuclei, but do not have sufficient energy to fission ^{238}U. Fast neutrons from compact nuclear bombs that are not slowed (moderated) will fission ^{238}U. This is useful for nuclear weapons. Thermal (slow) neutrons produce these fission reactions:

$$n + {}^{235}U \Rightarrow FF_1 + FF_2 + 2.43n + 207\ MeV$$

$$n + {}^{239}Pu \Rightarrow FF_1 + FF_2 + 2.87n + 214\ MeV$$

where FF_1 and FF_2 are newly-created, binary-fission fragments. The total number of neutrons and protons is constant in these reactions. Fast neutrons produce an extra 0.1 neutron. The number of neutrons released in fission varies between zero and six, with energy between 1 and 10 MeV. Weapon neutrons remain very energetic since the explosion takes place in less than a microsecond.

After a ^{235}U or ^{239}Pu nucleus captures a neutron, the resultant ^{236}U or ^{240}Pu nucleus oscillates like a liquid drop, with repulsive forces from charged protons in the nucleus and attractive nuclear forces from the close-by neighbors. Oscillations split the nucleus into two fission fragments. Because target nuclei have an odd number of neutrons, the binding energy of the absorbed neutron includes pairing energy from combining spin-up and spin-down neutrons. On the other hand, ^{238}U with an even number of neutrons has a smaller neutron binding energy since the pairing energy is not available. Because less energy is available from neutron capture by ^{238}U nuclei, only fast neutrons over 1 MeV can fission ^{238}U. Thus ^{238}U cannot be used, by itself, to make a fission weapon. However, energetic 14-MeV neutrons from fusion can fission all isotopes of uranium. Thus ^{238}U and ^{235}U can be used in the secondary of a hydrogen bomb to gain fission energy along with fusion energy. Fission weapons can also be made with ^{233}U, but the energetic gamma rays from ^{233}U make it difficult to turn into a weapon. The U.S. produced 1500 kg of ^{233}U, and made a few weapons from it. Because it has no mission, it is being buried in Nevada in 2012. The IAEA is placing ^{237}Np and two americium isotopes under safeguards, since they also can, in principle, be made into weapons.

2.4 Critical Mass

The largest deliverable conventional bomb, made with conventional explosives, was the 6-ton (0.006 kton) BLU-82 bomb the United States used in Vietnam and Afghanistan. Nuclear weapon yields vary from the 0.01 kton backpack weapons, to destroy bridges and dams, to the huge Soviet 100–150 Mton weapon, that was tested at the 58-Mton level in 1962. The yield ratio between these two weapons is 10 million! To explain the concept of critical masses, we use scaling laws to consider the effects of simply changing the size of an object.

Scaling laws answer pragmatic questions, such as, "Why do cows eat grass and mice eat grains?" Consider animals as simple spherical shapes whose *heat loss* through skin is proportional to skin surface *area* (radius squared). The *amount of*

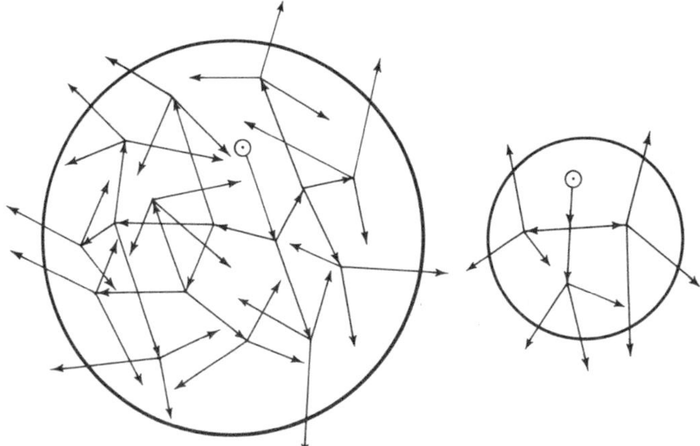

Fig. 2.2 Critical Mass. A larger mass of fissile material contains relatively more neutrons for future fission events, while a smaller sample loses more neutrons through its surface, halting the growth of future neutrons. The critical mass is the minimum amount of fissile material needed to sustain exponential growth of the neutron population, which depends on the isotope composition, its density and geometrical shape and situation (Atomic Heritage Foundation)

stored food energy is approximately proportional to the animal's *volume* (radius cubed) and the type of food it eats and stores. If an animal has thick skin fur, it can be smaller and still gain sufficient energy from eating grass. On the other hand, mice must eat high quality grains since their area/volume ratio (loss/storage) is large. Scaling shows that small animals (mice and humming birds) eat often and they must eat energetic grains to overcome their large ratio of area to stomach volume, and their thin fur. For the opposite reason, large animals eat less often and eat less energetic foods, such as grass. Scaling arguments also show that big animals must have relatively large diameter bones. Similarly scaling arguments show that you need three sticks to start a fire, by making a protective cavity for the fire (Fig. 2.2).

Scaling to obtain a critical mass. Simple geometry is a powerful tool to understand how much nuclear material is needed to "just" make a nuclear weapon. Neutrons are lost from a spherical bomb through its surface area, $4\pi r^2$. Thus the neutron loss rate is proportional to the square of the radius. At the same time neutrons are produced in a volume at rate proportional to the volume of a sphere, $4\pi r^3/3$. The production rate is proportional to the radius cubed. The rate of production divided by the rate of loss is r cubed over r squared, or it is proportional to the radius. More material gives you a better chance of having enough for a critical mass. Less material will make it harder to make a critical mass.

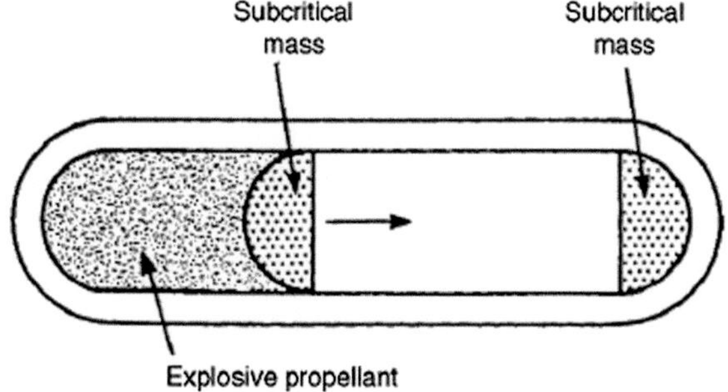

Fig. 2.3 Hiroshima weapon, ^{235}U gun-assembly nuclear device. *Little Boy* was long and thin because it used a 2 m long cannon that projected the smaller mass into the larger spherical mass with a velocity of 900 m/s to avoid pre-detonation. There was some certainty that the *Little Boy* would work as planned and there was little ^{235}U to spare, so it was not pre-tested (Glasstone and Dolan 1977)

Table 2.1 Critical mass

SNM	Bare sphere	U tamper
^{235}U (kg)	47.9	15
^{239}Pu (kg)	10.2	5

The ^{235}U and ^{239}Pu critical masses are at normal densities for the case of bare spheres and for the case of a natural uranium tamper surrounding the bare sphere. The thorium cycle makes ^{233}U, with a tampered critical mass of 2 kg, but it is harder to deal with because of its high radiation level. (*TOPS Task Force*, Nuclear Energy Research Advisory Committee, Department of Energy 2000)

Robert Serber in the *Los Alamos Primer* estimated a bare-sphere uranium critical mass with a 9-cm radius and a mass of 55 kg (too large by a factor of two). Reflectors made of beryllium (giving two neutrons) or uranium reduce the critical mass from the "bare-sphere" values. Implosion increases the density of ^{239}Pu or ^{235}U, reducing the critical mass (Figs. 2.2 and 2.3) (Table 2.1).

2.5 Neutron Generations and Bomb Yield

One kilogram of fissile material contains a total possible energy of 17 kton. The Hiroshima uranium gun-weapon released 13 kton of energy, consuming 0.8 kg of ^{235}U. *Little Boy* consumed only 1.3 % of its 60 kg of ^{235}U. The 22-kton Nagasaki implosion weapon obtained a much higher efficiency of 20 %, consuming 1.3 kg of 6 kg of ^{239}Pu. The higher efficiency of *fat man* was obtained because it was imploded, increasing the density of ^{239}Pu.

The growth of neutrons in a warhead is analogous to folding a piece of paper, as each folding doubles the thickness. Folding a sheet of paper 51 times gives a folded thickness of 2×10^{15} sheets of paper, which is 170 million km, the distance to the sun. A paper folded eighty times (corresponding to a 15-kton warhead) has a thickness of 10^{24} sheets of paper, or 1000 light years, 200 times the distance to the nearest star, Alpha Centuri.

2.6 Plutonium–Implosion Weapons

Advances in plutonium weapons have been dramatic. The 1945 *Fat Man* was a 22-kton bomb with a diameter of 1.5 m, while the Peacekeeper's (W-87) 300-kton warhead has a diameter of only 0.6 m, volume reduction of about ten, with a gain in yield of 15. Nuclear artillery shells are only 0.16 m in diameter. Plutonium is produced when ^{238}U absorbs a neutron to become ^{239}U, which beta-decays in minutes to neptunium (^{239}Np), which in turn beta-decays in days to plutonium (^{239}Pu), with a half-life of 24,000 years.

Prior to the atomic age, plutonium was produced naturally just below the surface of the earth. Strangely, the uranium at Oklo, Gabon, in Africa, contains only 0.4 % ^{235}U, rather than the usual 0.7 %. This apparent anomaly is explained because the ^{235}U content has been changed for two reasons:

1. ***Decay***: ^{235}U naturally decays during the 1.8 billion years, when ^{235}U was 4 % of all uranium. Today ^{235}U is 0.7 % of all uranium, when correcting for ^{235}U's half-life of 700 million years. The original 4 % level is similar to today's power-reactor fuel.
2. ***Natural Fission***: ^{235}U content was also depleted from that distant time by fission-consumption in a natural nuclear reactor. The natural reactor operated for several hundred thousand years. The rich uranium deposit was in a damp place with enough water to moderate neutrons, creating a natural reactor without human effort. The natural reactor was about 5–10 m thick and 600–900 m wide. It operated at an average power of 100 kW over 150,000 years. In practice the power cycled on and off as the water moderator evaporated with the seasons and power production and then was replenished. Nature managed to make plutonium without much effort long before mankind, but the plutonium direct-evidence has decayed.

2.6.1 *Pu Preferable to HEU for Weapons*

Plutonium is favored over *highly enriched uranium* (HEU with 90 % ^{235}U) for weapons since it emits more neutrons per fast fission ($\nu = 2.94$ versus 2.53), more neutrons per neutron capture (η = 2.35 versus 1.93), has a higher fast fission cross-section (1.7 barns versus 1.2 barns) and releases slightly more energy (214 MeV versus 207 MeV). For these reasons, plutonium makes smaller primaries, which are essential for multiple–independently–targetable reentry–vehicles (MIRV) on intercontinental ballistic–missiles (ICBM) and submarine–launched ballistic–missiles.

However, plutonium is more difficult to make into nuclear explosives because of the high rate of spontaneously-emitted neutrons that are emitted by ^{240}Pu. these precursor neutrons can begin a chain reaction before plutonium reaches its most compact form. We call this *preinitiation*, which is similar to preignition in automobiles, when the spark ignites the gasoline before maximum compression. Severe preignition prevents cars from operating; the technical fix is to delay the spark. In a similar fashion, slow-moving, plutonium gun-type weapons would preinitiate and lose considerable yield. This problem can be overcome by explosive implosion using multipoint detonation on a 32-sided soccer–ball sphere. Hollow pits of plutonium enhance efficiency and allow volume for deuterium–tritium gas to give a fusion boost. The hydrogen secondary is located near the primary and, in some cases, a *dial-a-yield* feature is used to tailor the yield to the mission.

2.6.2 Plutonium Spontaneous Neutron Emission

During a long reactor residency, a considerable fraction of ^{239}Pu captures a second neutron to become ^{240}Pu. The length of stay in a reactor determines whether the plutonium is *weapons-grade* Pu (6 % ^{240}Pu, made in months) or *reactorgrade* Pu (>20 % ^{240}Pu, made in years). The isotopic contents of the five most common types of plutonium are listed in Table 2.2. Plutonium metallurgy is complicated by the fact that it exists in six different phases, but it is stable in the delta phase with 2 % gallium. Note that plutonium from breeder-reactor blankets is excellent weapons-grade plutonium, while mixed-oxide (MOX) fuel, used in thermal reactors, is not. Reactor grade plutonium can be made into viable nuclear weapons by mature nuclear nations. Difficulties can arise because of the extra dose rate of spontaneous neutrons and the excess heat that can damage high explosives.

The rate of spontaneous neutrons from ^{240}Pu is 100,000 times higher than it is for ^{239}Pu. (The spontaneous neutron rate of ^{235}U is about 1 % of the ^{239}Pu rate.) The spontaneous neutrons from 5 kg of weapons-grade plutonium come primarily from ^{240}Pu, not ^{239}Pu.

Table 2.2 Isotopic composition of various grades of plutonium

Pu grade	(% Pu isotope)				
	^{238}Pu	^{239}Pu	^{240}Pu	^{241}Pu	^{242}Pu
Super-grade	–	98	2	–	–
Weapons-grade	0.01	93.8	5.8	0.4	0.02
Reactor-grade	1.3	60	24	9	5
MOX-grade	2	40	32	18	8
Breeder blanket	–	96	4	–	–

(Mark 1993)

Neutron multiplication and (α, n) reactions on light impurities (oxygen and nitrogen) marginally increase the neutron rate in metallic plutonium. The spontaneous neutron rate is 4 times larger in reactor-grade plutonium (24 % ^{240}Pu), which itself would be doubled if it were in plutonium oxide form from the (α, n)-reaction. Clearly it is more difficult to make a warhead with reactor-grade plutonium, but it can be done.

Assembly Time in Implosion. The assembly time of an implosion bomb is reduced by a factor of 100 compared to the gun-barrel design; this is due to two things:

- The critical implosion sizes are less than 10 % the length of the gun barrel.
- The implosion velocity is a factor of 10 higher than the projectile velocity of a *Little Boy* gun-type weapon. During the implosion time, about 0.5 neutrons are generated. This can be reduced by a factor of three by using super-grade 2 % ^{240}Pu.

2.6.3 Summary of ^{239}Pu versus ^{235}U for Weapons

Plutonium gives an extra one-half neutron on average from each fission event, reducing the number of fission generations needed, allowing for more complete fission of the bomb material and a smaller critical mass of 4 kg, compared to 20 kg for uranium This reduces the amount of fissile material needed, assisting miniaturization, which is useful for smaller primaries and for multi-warhead ICBMs and SLBMs. Pu is made in reactors and chemically reprocessed, which is easier to detect than uranium enrichment. Pu passively emits many more intense gamma rays than U, thus it is easier to detect. Pu needs implosion to make a viable bomb, this is more difficult to accomplish than a uranium gun-type weapon. Uranium-weapons (both implosion and gun-barrel) are easier to make than for Pu warheads, since pre-initiation is much less of an issue for U. Plutonium mixed-oxide fuel is not as a good as enriched uranium fuel. Pu MOX has a net negative economic value because its radioactivity, which complicates mixed oxide fuel production.

Uranium used to be more difficult to enrich in the past, as compared to the production of Pu in reactors. The advent of better centrifuges makes this no longer true. U production is harder to locate since U emits much less radiation and U does not need reactors for production. There is a shift in preference for wannabe nuclear states to use U for weapons as compared to Pu. U gun-barrel or implosion weapons are easier to make and HEU is less radioactive so easier to work with. There is more weapons grade uranium as compared to plutonium. The extra weight of U weapons is only relevant for missile delivery. HEU has a valuable market, when diluted for reactor fuel. After the break-up of the Soviet Union, prompt action was needed, more for HEU protection than Pu since HEU can be made more easily into a weapon, and it is harder to detect (Table 2.3).

Table 2.3 Weapon problems and solutions

• Neutron initiation: From ^{210}Po-Be radioactivity to DT initiator tubes
• Extra neutron multiplication: Add beryllium reflector (n to 2n)
• Reduce mass: Bare sphere to heavy tamper and implosion
• Reduce mass: Levitated hollow pit
• Reduce mass: Boost with DT to fission more efficiently
• Fission increase: ^{238}U tamper and casing (FFF for H bomb)
• H bomb: Radiation compression (not mechanical), U in secondary
• Deliverable H bomb: ^{6}Li^{2}H instead of Big Mike cryostat
• Safety: Two-point triggering, ENDS, fire-resistant pits, insensitive high explosives
• Theft: Permissive Action Links (PAL)

2.7 Boosted Primaries and H–Bombs

Basic Physics. Fusion of 4 hydrogen nuclei into helium sustains the Sun and its planet Earth. Shortly after physicists began on fission bombs, they realized that much more explosive energy could be available from fusion, which combines hydrogen isotopes into helium. The sun fuses four hydrogen isotopes into helium, releasing 27 MeV in a three-step, *proton-burning* process. (The carbon-nitrogen fusion process is more likely, but it gives the same result.) The mass loss going from four ^{1}H nuclei (4 × 1.008 atomic mass units = 4.032 amu) to one ^{4}He (4.003 amu) is 0.029 amu, This converts 0.7 % of the original mass to energy, an amount that is much less than 100 % conversion from antimatter conversion, but much more than that from chemical explosives (50 eV). The sun's gravity confines the energetic hydrogen/helium plasma to high pressures and temperatures. Since the sun has a life span of 10 billion years, it slowly uses a three-step process for *gravitational confinement* fusion. *In this 3 step process*, hydrogen atoms 1 and 2 combine, say a billion years ago. Another billion years goes by and #3 proton combines (1–2 to obtain 1–2–3). Another billion years goes by and #4 combines (1–2–3 to get 1–2–3–4), with great patience.

DT Burning. The little time available for nuclear weapons requires a *one-step process* to obtain ^{4}He from deuterium and tritium, or "d plus t gives He plus n,"

$$^{2}\mathrm{H} + {}^{3}\mathrm{H} \Rightarrow {}^{4}\mathrm{He} + {}^{1}\mathrm{n} + 17.6\mathrm{MeV}.$$

DT fusion develops 5 times more energy/mass than fission. DT is not limited in its storage size, it doesn't have a self-sustaining critical mass. The neutron carries 14.4 MeV, enough to fission all the isotopes of U and Pu, as well as ^{6}Li to produce tritium for further fusion,

$$^{1}\text{n} + {}^{6}\text{Li} \Rightarrow {}^{3}\text{H} + {}^{4}\text{He}.$$

Hydrogen bombs do not have large gravitational or magnetic fields to confine hot plasmas. They use *inertial confinement* since all the neutron generations take place in a microsecond before the weapon blows apart. For fusion to take place, d and t must have sufficient kinetic energy of motion to overcome coulomb repulsion between the two nuclei.

A boosted primary contains $d + t$ gas (heavy hydrogen and heavy-heavy hydrogen) to magnify its yield. The extra energy from *dt* "burning or fusing" is small, but the extra neutrons, released early in the cycle, allow the fission cycle to skip many generations, increasing the fraction of nuclei that fission. This increases the efficiency of burning *dt*. The energy released per warhead is small at one-third a kton, which is not significant. But a pulse of a mole of neutrons (6×10^{23}) rapidly advances the number of neutron generations, increasing fission yield and lowers the amount of fissile material. Without tritium, modern nuclear weapons would not function since *dt* reactions are needed to raise the yield to ignite the secondary stage.

Since tritium decays with a half-life of $T_{1/2}$ = 12.3 year. Tritium must be manufactured periodically to maintain nuclear arsenals. Tritium is produced by the absorption of neutrons by ^{6}Li in thermal reactors. The United States has not produced tritium since 1988. After considering proposals to make tritium in dedicated accelerators or reactors, the Department of Energy (DOE) opted to make tritium at an existing Tennessee nuclear power plant operated by the Tennessee Valley Authority (TVA). As the stockpile has dropped from 30,000 to 5000 warheads, the need for tritium was reduced greatly, balancing losses from decay. Under the *New START* treaty, we estimate tritium production will be needed in 2030, assuming a supply of 10 kg of tritium in 2005 for 2000 warheads.

Plutonium pit lifetime. The natural decay of plutonium propels alpha particles into the surroundings, which dislodges the plutonium atoms from their spots in the lattice. Will this jeopardize the quality of the plutonium pits? Experiments have been conducted with faster decaying ^{238}Pu, to speed up the decay cycle. The Department of Energy and the JASON group determined that pit lifetime is in excess of 150 years in 2012. The Los Alamos Pit Production Facility has a capability to make 6–10 pits per year. This can be increased to 40 per year with reuse capacity and investments. The New START Treaty might require a total inventory of 2000 warheads. This could be maintained with a capacity of about (2000 warheads)/(150 year lifetime) = 15 pits/year, much less than the original planned capacity of about 200/year.

2.7.1 Radiation Compression

From the beginning of the Manhattan Project, Edward Teller wanted to develop the hydrogen bomb so much so that he refused to work on pure fission bombs. The initial idea was to compress the secondary with mechanical shock waves. This would not work because the primary explosion destroys the secondary before sufficient mechanical compression of the secondary takes place. However, Stanislaus Ulam and Teller solved the problem when they deduced that x-rays from the primary would reflect (re-emit) from the casing on to the secondary. The absorbed x-rays heat the casing and other materials to such high temperatures that they produce black-body emission, x-rays as well as reflected x-rays. Since x-rays travel with the speed of light, they are absorbed by the secondary before the mechanical shock arrives. The high temperature rise at the surface of the secondary creates radiation pressure as the hot surface of the secondary evaporates. Since evaporating ions travel outward, perpendicular to the surface, this compresses the surface since the emission of x-rays carries away momentum, which equally is applied to the surface. When you push on the swimming pool wall to the North, you travel to the South! This compression of the hydrogen secondary heats the secondary. The bath of neutrons at high temperatures ignites both U for fission and ^{6}Li^{2}H to start fusion. In the late 1970s, Howard Moreland published rough drawings of the hydrogen bomb. The government's case to prevent publication was greatly weakened when it was discovered that it had already declassified these facts and they were publicly available at the Los Alamos laboratory library.

Radiation Compression. Mechanical compression was unsuccessful in its attempt to produce nuclear fusion, which was obtained with radiation compression. A primary generates some 10 kton in 100 ns, producing thermal power of 4×10^{20} W, considerably greater than the US electrical grid of 10^{12} W. This tremendous of amount of heat in a small volume is radiated away, proportional to the fourth power of the temperature, T^4. The temperature at the surface of the primary is about 15 million K, similar to the sun's core temperature, and a thousand times greater than sun's surface temperature. The energy of the x-rays is about 6 keV, similar to some medical x-rays. The x-rays are absorbed by the secondary to reradiate new x-rays, which vaporize outward and implode inward. The casing and other materials also absorb x-rays which re-radiate inwardly to further compress the secondary.

2.7.2 Lithium Deuteride for Large Hydrogen Bombs

The first hydrogen device, "Big Mike," was detonated on 31 October 1952. Mike was a large thermos bottle containing liquid deuterium, giving a yield of 10 Mton.

Mike could not be delivered with bombers or ICBMs because of the extreme size of its cryogenics. A deliverable H bomb was soon developed with deuterium in the LiD salt in the form of $^6Li^2H$. Since salt is a solid, there is no need for a cryostat. A neutron interacting with ^6Li gives an instant tritium to interact with deuterium ($^6Li + {}^1n \Rightarrow {}^3H + {}^4He$). Note that the tritium supplies from ^6Li are not dependent on tritium's 12.3 year half-life as it is used within microseconds of production. ^6Li is relatively cheap to separate since its mass is 15 % smaller than ^7Li. Thus, $^6Li^2H$ gives

$$d + t \Rightarrow He + n \quad \text{or} \quad {}^2H + {}^3H \Rightarrow {}^4He + {}^1n.$$

There is no problem with storing too much $^6Li^2H$. There will be no dangerous, unintended criticalities from $^6Li^2H$ stockpiles as there can be with ^{235}U and ^{239}Pu, which sets upper limits on sizes of unboosted fission weapons. The largest Soviet explosion of 58 Mton took place on 31 October 1961. This explosion was only one-third or one-half of the design size of 100–150 Mton. The 58 Mton explosion used an incomplete secondary with less nuclear fuel to reduce the yield.

Of what purpose are large US bombs of 20 Mton and those of that 100–150 Mton? There are four possible uses for large yield weapons, none of which is needed:

1. *Hard Target warheads* to attack ICBM silos and command posts with first-strike warheads. In Chap. 4 we address this issue, but for now we can state that the accuracy of the warhead and the hardness of the target dictates the choice of yield. As accuracy has gotten better, the yield of hard-target warheads decreased.
2. *Area destruction.* At some point it is inefficient to attack area (people) with large weapons. Since the blast pressure falls as the cube of the distance, the kill radius from large bombs is less effective per unit of yield. If the yield is reduced by a factor of 8, the kill radius is reduced by a factor 2, and the kill area is reduced by a factor of "only" 4.
3. *Electromagnetic pulses* can disrupt command and control and military communications and shut down the electrical grid.
4. Anti-ballistic-missile weapons (Chap. 5).

2.8 Neutron Bombs

The western nations were greatly concerned about the possibility of a Soviet invasion from the East. It is true that the Warsaw Pack had a substantial numerical advantage in conventional arms, but the threat was not as great as stated because the Soviets kept older, inferior equipment in place, and the risk of invasion to the attacker was very large. The tactical battlefield weapons were delivered with artillery pieces, or howitzers. At one time the US had some 7000 tactical weapons

Fig. 2.4 Tactical nuclear weapons. The first live nuclear artillery test of 15 kton on 25 May 1953 at the Nevada Test Site. The W33, 8 inch, howitzer projectile had a yield of either sub-kton or 5–10 kton, with a range of 18 km. The US and seven European nations (Belgium, Italy, Greece, Netherlands, Turkey United Kingdom and West Germany) controlled 1800 W33's. The smaller 6 inch W48 had a range of 14 km with 3000 deployed in Europe and S. Korea. The W82 enhanced radiation projectile was scheduled to replace the W48, but Congress did not agree (Atomic Heritage Foundation)

and the Soviets had far more, some 20,000. The neutron bomb was the follow-up weapon to these tactical weapons (Fig. 2.4).

Neutron bombs produces smaller blast yields, reducing collateral damage. At the same time, neutron bombs enhance neutron emissions to more effectively kill troops and tank drivers, producing prompt death in 5 min with fast neutron doses of 80 Sv (8000 rem). The introduction of the neutron bomb was a shift from pure fission tactical weapons to weapons that were approximately equally divided between fission and fusion at 1 kton. This was accomplished by replacing the uranium tamper with chromium, to let 50 % of the neutrons escape. In addition, extra *dt* boosting gas was used to produce fusion to produce more neutrons per energy released. Fusion neutrons are much more energetic at 14 MeV, as compared to fission neutrons at 1–2 MeV.

The political debate on deploying neutron bombs was sharply contested in the United States and Europe in the late 1970s. Those who wanted to deploy neutron bombs were concerned that tactical weapons would not be used in Europe because their yields would be too damaging, particularly in Germany. They wanted 1-kton weapons that would incapacitate tank crews at a distance of 850 m, compared to 375 m for a pure fission 1-kton weapon. Proponents believed deployment of the neutron bomb would increase Soviet perception that the US would actually use it, a result that would deter a Soviet invasion. On the other hand, it could be envisioned that the deployment of neutron bomb lowered the psychological and bureaucratic threshold for first-use of nuclear weapons. Such a deployment would increase the probability of its first-use by local commanders, thus starting a more general nuclear war. Lastly, there already was considerable deterrence to discourage an invasion because the United States had other nuclear weapons in Europe. The view of the opponents carried the day as Congress blocked its deployment. Those who lost the debate commented that a different name for the weapon, such as the *reduced-blast and—yield bomb*, would have helped their case.

Radiation dose and reduced blast. Neutron bombs produce only one-half as much radioactivity as pure fission bombs, but they give a much larger radiation dose to close-in troops. The radiation dose from a neutron bomb is considerably greater because it produces 6 times more neutrons *and* fusion neutrons are 7 times more energetic than fission neutrons. These effects are multiplicative, and considerable. Most fission energy appears as fission-fragment kinetic energy, which heats bomb debris to produce a blast wave. Pure fusion contributes less blast energy since escaping neutrons carry considerable energy away from the weapon. The neutron bomb is about 9 times (per unit area destroyed) more effective at killing tank drivers as compared to tactical nuclear weapons.

2.9 Exotic Weapons

Isomer EMP warhead. A nuclear isomer is a nuclear excited state of a nucleus with a long half-life. It appears stable in spite of it not being in the ground state, awaiting decay. If an isotope can be found that has a high-energy excited-state with a long half-life, it might be used as a weapon. Such a metastable state might be induced to discharge with an x-ray machine, releasing massive gamma rays in a short time. The isomer usually suggested is hafnium-178 m, ^{178m}Hf, with a 31 year half-life at 2.4 MeV. The isomer bomb obtained funding, but was canceled as nonsensical. See *Imaginary Weapons* by Sharon Weinberger (Nation Books 2006) for the history of the isomer bomb through the bureaucracy and the laboratory, where x-rays did not readily discharge ^{178m}Hf.

Trans-actinide warhead. The nuclear shell model shows that there are potential regions of stability for massive nuclei, considerably heavier than the actinides. These isotopes could be stable, similar to the lead region of the nuclear chart. This is *analogous* to the full shells of atomic rare gas elements, which are stable and reluctant to make chemical bonds. If proton/neutron closed-shell nuclei are made, it would be possible to have critical masses of *grams, rather than kilograms*. Such nuclei would emit many more than 3 neutrons per fissile event. Not to worry, such nuclei have not been discovered.

Robust nuclear earth–penetrating warhead. The bunker buster was devised by the Los Alamos National Laboratory in 1991. RNEP research was banned by Congress in 1994, but it reappeared in Congress in 2002, until it was terminated in 2005. RNEP warheads were to penetrate 10 m of rock, or 30 m in dirt, then explode, *increasing blast pressure by 30 times* since rock is hard to compress and air is not. There are disadvantages: (1) Considerable radioactive aerosols are produced when neutrons interact with rock and soil. There is much more radioactivity from a bunker buster than from an air-burst. A 10-kton warhead would have to penetrate 250 m to avoid spewing radioactivity, which is not possible. (2) A buried warhead can crush rock at a distance. A 10-kton crush–radius is 100 m, which is the accuracy of a good ICBM. But RNEP would be delivered as a bomb, guided with a laser to be more accurate, but not as much penetration as if it were from a missile. RNEP was intended to attack underground chem/bio stockpiles (or leadership), but it would have to be close to the CW/BW stockpiles for combustion. RNEP was finally terminated in 2006 (See reference by Rob Nelson).

Reliable replacement warhead. The Comprehensive Nuclear–Test–Ban Treaty bans nuclear warhead tests. Using more cautious designs, warheads can be built that are more robust, further from the reliability margins, and without the use of Be and BeO as neutron reflectors. In March 2007, NNSA chose Livermore's W-89 design since it already had been tested and did not need testing to be certified. It had a mass of 150 kg and a yield of 200 kton. The RRW was originally intended to be used in 2000 warheads, but this was scaled back. The stockpile stewardship program was designed to learn more than testing could ever provide, and RRW was terminated.

^{233}U warhead. About 1500 kg of ^{233}U was produced in reactors for nuclear weapons, at a cost of \$6–11 billion. Since ^{233}U is less useable for warheads, compared to ^{239}Pu and ^{235}U, it will be disposed of as waste. "DOE is to waive its own acceptance criteria to allow the direct, shallow-land disposal" in near-surface burial by 2014.

Suitcase bomb. In 1997, Russian General Alexander Lebed claimed that 50 RA-115 suitcase bombs "are not under the control of the armed forces of Russia." He was famous for allowing the 1991 military rebellion in Moscow to proceed. The US has made very small nuclear weapons. The smallest was 45 kg, 13 cm diameter and 62 cm long. The W-54 was backpack-carried for munitions demolition and on Davy Crockett short-range missiles.

X-Ray–Laser Pumped with a Nuclear Explosion (Chap. 5).

2.10 Nuclear Weapon Effects

Nuclear weapon energy appears as blast pressure waves, thermal radiation, and prompt/delayed radiation. The division of the total energy into these quantities depends on weapon yield, ratio of fusion to fission energies, and height of burst. Typically, 40–60 % of the yield appears as blast energy, 30–50 % appears as thermal radiation, 5 % as ionizing radiation and 5–10 % as residual radiation. This section discusses blast, thermal, and radiation effects, as well *nuclear winter* and electromagnetic pulses, while Chap. 3 discusses low-dose radiation effects. Nuclear weapons can destroy opponent's weapons, but it is far easier to devastate cities and people. The 13 and 22 kton weapons used on Hiroshima and Nagasaki killed

180,000 people, about 40 % of the inhabitants. Those that died from radiation, also died because they were within the lethal blast radius. Outside the lethal blast areas about 400 Japanese died from delayed, low-dose cancer. Citizens at 3-km distance had their eye-sight severely damaged. However, the 15-Mton Castle Bravo hydrogen bomb created a gigantic radioactive plume, killing a Japanese fisherman on the *Lucky Dragon*. The destructive effects of Mton-size weapons would be immense and it is possible with US and Russia forces operating under some launch-on-warning scenarios. We hope that the Launch–on–Warning scenarios have been abandoned. In the 1960s, Secretary of Defense Robert McNamara *defined mutual assured destruction* as the assured second strike that would kill 25 % of a nation's population and 50 % of its industry, as shown in Tables 2.4 and 2.5.

Some radiation rules-of-thumb effects are as follows:

- lethality from neutrons predominates up to a few kton;
- lethality from blast pressure waves predominates from 5–100 kton;
- lethality from thermal radiation predominates above 100 kton, but lethal radiation plumes can extend considerably beyond 100 miles.

Fusion makes much less radioactivity than fission since it does not produce fission fragments. Fusion neutrons are very harmful to people within 1 km of the blast, but this effect is much less significant than close-in blast effects and fission radioactivity. The yield of a secondary stage is about 50 % fusion and 50 % uranium–fission. Thus, a 1-Mton, 50–50 weapon has about 500 kton of fission, while a

Table 2.4 Assured destruction

	1999 population (M)	W88 warheads
China	1281	368
Iran	64	10
Iraq	21	4
North Korea	22	4
Russia	152	51
US	259	124

The number of hard-target warheads to kill 25 % of population and destroy 50 % of industry (Matt McKenzie, NRDC).

Table 2.5 LD-50 radii

1 Mton	Blast (5 psi)	Thermal (7 cal/cm^2)	Radiation (4.5 Sv, 450 rem)
Surface (km)	4.6	11	2.7
Air burst (km)	6.7	17	2.7

The distances from the explosion where 50 % of the affected population would die from blast, thermal radiation, and nuclear radiation from 1-Mton surface and air explosions. (Glasstone and Dolan 1977)

10-kton primary is 100 % fission if unboosted. Blast height is extremely important in determining the amount of radioactive fallout. If an explosion takes place at low altitudes, excess neutrons produce large amounts of radioactivity in the soil, which disperses in a plume. High altitude bursts make much less radioactivity since nitrogen and oxygen absorb neutrons, which decay quickly, but ^{14}C lingers. In addition, a high altitude burst directly disperses and dilutes the radioactivity. A particularly nasty target would be a nuclear reactor. One estimate for a 1-Mton bomb hitting a 1-GW_e reactor predicts an area of 34,000 km^2, which would give a lifetime dose of over 1 Sv (100 rem) to the affected population.

Overpressure of 5 psi (30 kPa) destroys wood and brick houses beyond repair. One might think that blast pressure would diminish as the inverse square of the distance, but it falls off faster, proportional the inverse cube on the distance (Fig. 2.5).

Fallout. The radioactive plume from a nuclear weapon depends on yield, height of blast, and wind conditions. A 1-Mton weapon can produce a plume that deposits radiation of 5 Sv (500 rem) over an area 30 miles wide and 1000 miles long. A prompt dose of 4.5 Sv (450 rem) is lethal to about 50 %, and essentially no one survives 1000 rem. If citizens stay inside buildings, the dose is reduced by a factor of three. Terrorist *dirty-bombs* are discussed in Chap. 13. Radioactive plumes from nuclear accidents are discussed in Chap. 3 (Figs. 2.6, 2.7 and 2.8).

Cratering. The radius of a crater in hard rock is about 150 m for 1 Mton and 75 m for 100 kton (OTA). The Russian 1908—asteroid of 30 to 50 m in diameter broke apart in the atmosphere, destroying 2000 km^2 of Siberian forest. Another example of danger from asteroids is the Barringer Crater of 1.2 km diameter and 0.2 km deep, created by a 50-m asteroid, 50,000 years ago near Winslow, Arizona. If astroids are discovered very early and very far away, one might use very large nuclear weapons to get a small deflection angle from (1) x-rays, (2) explosive push, (3) fragmentation. This is difficult to do.

Fig. 2.5 Thermal radiation of Hiroshima victim and valve shadow (AHF)

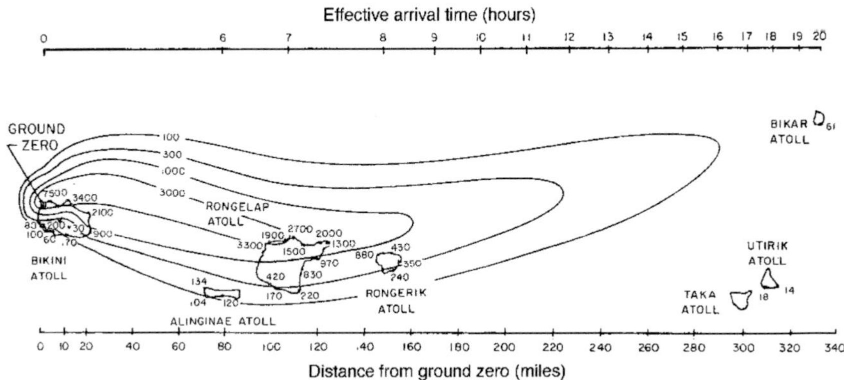

Fig. 2.6 Bikini fallout. Total accumulated dose contours in rads, 4 days after the BRAVO test explosion (Glasstone and Dolan, 1977)

Nuclear Winter. The volcanic eruption of 10 April 1815, on Tambora led to global cooling and June frost in 1816, the "year without a summer." Atmospheric physicists realized in 1982 that large nuclear attacks on cities would create massive amounts of micron-sized soot and raise them to the stratosphere, with effects similar to very large volcanic eruptions. It was projected that 10,000 0.5-Mton warheads could reduce light levels to a few percent of ambient levels and temperatures could *drop* by 30 °C for a month, warming to 0 °C for another 2 months. Hence, the name *nuclear winter*.

A key factor is *lofting of soot* to the upper troposphere. Weapons over 300 kton raise soot high enough to absorb sunlight and heat the upper atmosphere by 80 °C, which raises (lofts) the soot higher. A major effect of such weapon blasts would be destruction of much of the world's food supply by low temperatures. The US government carried out burning and chemical explosions to test some of these ideas, but it is difficult to test larger scenarios peacefully. The 1980s debate became that of a *matter of degrees* between a *nuclear winter* and a *nuclear autumn*.

After 3 decades, there is less concern about possible large-scale exchanges with the Russia, but dire predictions are still relevant for the case of accidental wars and regional wars involving 100 Hiroshima-sized weapons. Such a war between India and Pakistan, each with fifty 15-kton weapons could lead to about 44 million casualties and 6.6 trillion grams (Tg) of soot. A war using 4400 warheads, each with a yield of 100 kton, could cause 770 million casualties and 180 Tg of soot, enough to create ice-age conditions. Toon, Robock and Turco "estimate that most of the world's population, including that of the Southern Hemisphere, would be threatened by the indirect effects on global climate." Thus, nuclear winter scenarios are possible.

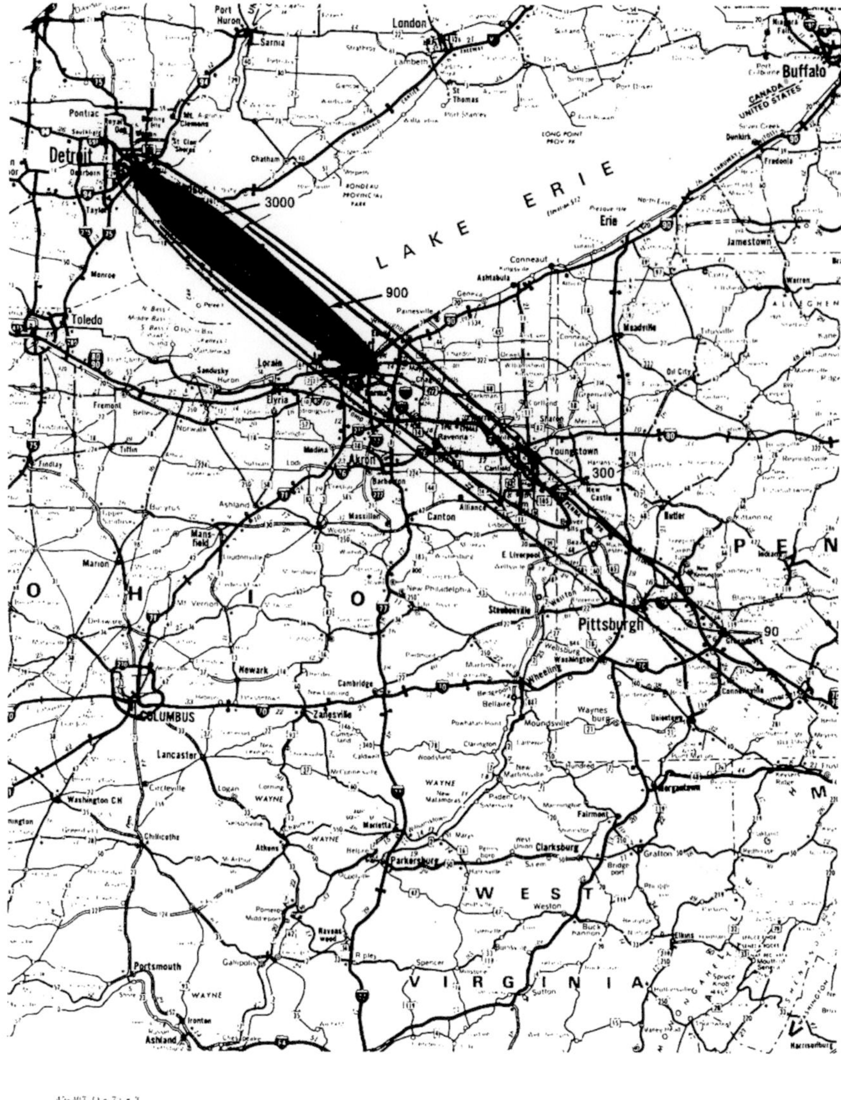

Fig. 2.7 Detroit fallout. Main fallout pattern after a 1-Mton surface explosion in Detroit, with a uniform, steady 15-mile/h wind from the Northwest. The 7-days accumulated dose contours (without shielding) are for 3000, 900, 300, and 90 rem. The constant wind would give lethal fallout in Cleveland and 100 rem in Pittsburgh (Office of Technology Assessment, 1979)

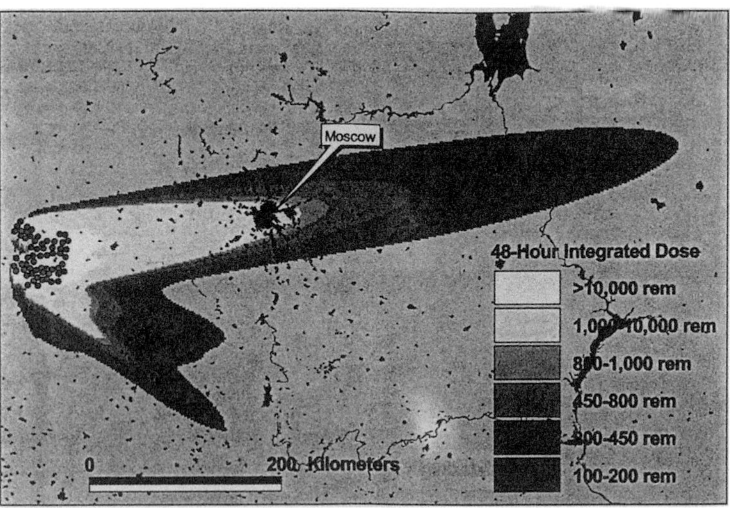

Fig. 2.8 Attack on SS-19's at Kozelsk. Under this scenario 13 million die from radiation received in the first two days (McKinzie et.al., NRDC, 2001)

2.11 Electromagnetic Pulse Attack on Power Grids

The electric power grid and the command, control, and communications (C^3) systems that control US strategic forces are vulnerable to large electromagnetic pulses (25,000 V/m), created by high-altitude (100–500 km) and low-altitude bursts of nuclear weapons. These effects were observed with our first nuclear tests, but they were formally tested on 9 July 1962 with the *Starfish Prime* explosion above Johnston Island. The 1.4-Mton weapon exploded at an altitude of 400 km on a Thor ballistic missile. The nighttime sky above Hawaii, 1300 km away, momentarily lit up with a white flash as if at noontime. This was followed by the sky turning green for about a second. The Hawaiian streetlights suddenly went out. Radio stations and telephone lines failed for a time. The US, Russia and China are probably not going to do such attacks these days, but EMP high altitude tasks might be used to attack sensors in orbit. Satellites are soft targets for EMP. It is difficult to shield sensitive detectors and soft silicon logic.

This vulnerability creates a possible instability in nuclear arms. If a country perceives its strategic forces could be negated by EMP, it might be tempted to adopt a "launch on warning" policy to use them rather than lose them to a preemptive first strike. The situation is not as unstable as I have characterized because these systems are hardened to partially withstand EMP and missiles based on submarines are not vulnerable to a first strike. However, perception of vulnerability in the land-based leg of the strategic triad creates pressure for a *launch on warning response*. Some of the C^3 facilities, such as the Air Force's Looking Glass Command Post in the sky, are more vulnerable than the strategic weapons.

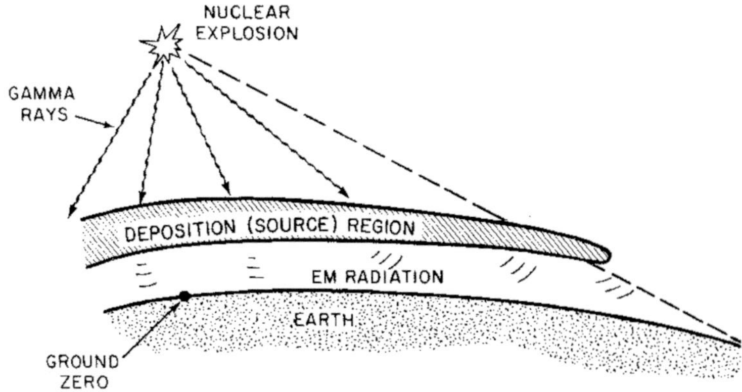

Fig. 2.9 Electromagnetic pulse. Schematic representation of an electromagnetic pulse (EMP) from a high-altitude burst. Fission fragments release prompt MeV gamma rays, which interact with the thin upper atmosphere creating Compton electrons. The electrons spiral with reasonable coherence in the Earth's magnetic field at about 1 MHz because they begin at essentially the same phase in the orbit. The October 1962 test at Johnston Island shut down the power grid in Hawaii and blocked radio and TV for several hours on the West Coast and throughout the Pacific region. See problems 1.17 and 1.18 (Glasstone and Dolan, 1977)

These issues have not ended with the end of the Cold War. EMP can be used by nations with few nuclear weapons against a nation with many nuclear weapons in an asymmetrical attack. "An EMP attack may degrade 70 % of the Nation's electrical service, in one instant." Extra-high-voltage transformers that are used for long-distance lines are vulnerable. These are not made domestically and might take 2 years to replace at considerable cost.

The adoption of variable power sources of wind and solar encourages upgrading of power grids. The smart grid, with digital sensors and control devices, could become targets for hackers and saboteurs. Incorporating shielding to protect against EMP might increase costs by 2–10 %. Installing resistors on the neutral–to–ground connections could reduce currents by 60 %. In recent years, non-nuclear EMP weapons have been down-scaled for delivery by planes and land vehicles. It is planned to use mini-EMP weapons to shut-down vehicles and boats (Fig. 2.9).

2.12 Stockpile Stewardship

The Comprehensive Nuclear–Test–Ban Treaty (CTBT) was open for signature in 1996. Since then there have been a half-dozen tests by India and Pakistan in 1998, which were easily detected. Four North Korean (DPRK) nuclear tests (2006, 2009, 2013 and 2016) were easily detected by the International Monitoring System (IMS) seismic stations. Nuclear radiation from the 2006 and 2013 tests was also detected by the IMS radionuclide network. Chap. 11 discusses CTBT monitoring

Table 2.6 DOE complex

	Employees	$B/year
Kansas City Plant	2700	0.6
Los Alamos National Laboratory	7300	1.6
Lawrence–Livermore National Lab.	5200	1.1
Nevada National Security Site	2400	0.3
Pantex Plant, Amarillo, TX	3100	0.6
Sandia National Lab., NM & CA	9800	1.7
Savannah River Site	5600	0.55
Y-12, Oak Ridge, TN	4700	1.2

Managing the DOE weapons complex involves over 40,000 employees with an annual budget of over $8 billion. The number of employees and budget are listed for the eight largest facilities.

and the connection between the Non-Proliferation Treaty and CTBT. In this section, we discuss the Stockpile Stewardship Program, which maintains nuclear weapon reliability and safety, without using nuclear testing for diagnostics.

Between 1958 and 1961, there was a moratorium on nuclear tests by the US and USSR, as Eisenhower/Kennedy and Khrushchev negotiated a test ban treaty. A few problems developed in untested US weapon types with major design innovations. This is not the situation today, as the US stockpile consists of 7 well-tested legacy warheads. When the US was conducting 20 tests per year, only one test per year was used to determine confidence in reliability, thus reliability was determined with very low accuracy. Statistics show that the weapon primary is much more vulnerable to change than the secondary (Table 2.6).

A major concern is that aging might reduce primary yields, preventing the triggering of the secondary. Aging effects could be caused by alpha decay of ^{239}Pu, causing thousands of lattice displacements for every Pu-decay.[1] Since 1997, NNSA and the JASONS studied Pu pits by combining short-lived ^{238}Pu with ^{239}Pu, 24,000-years lifetime, to speed up the aging process. By combining 8 % ^{238}Pu with 92 % ^{239}Pu the aging rate is increased by a factor of 16. An 11-years old Pu sample has an effective age of 170 years. The samples are examined for size changes with an accuracy of 0.1 microns out of 2 cm. The tests showed changes of only 0.25 % in volume in a 100-years period. The NNSA 2006 panel concluded that Pu pits have a minimum life of 85 years, "indicating that lattice damage and helium in-growth are not leading to catastrophic aging effects such as void swelling." In 2012, the DOE weapons labs reported more information after doing more experiments, concluding that the primaries (pits) "will function as designed up to 150 years after they were manufactured." Longer-lived pits imply that remanufacturing capabilities for

[1]A. Heller, Plutonium at 150 years: Going strong and aging gracefully, *Science and Technology Review* (LLNL), December 2012, 11–14.

Table 2.7 US enduring stockpile weapons

Weapon	System	Yield (kton)	Number Built	Laboratory
B61	Bombs	170	1,200	LANL
B61/11	Earth Penetrator	400	47	LANL
W76	SLBM Trident	100	3,200	LANL
W78	ICBM Minuteman-3	335	1,000	LANL
W80/1	ALCM (SLCM = 0)	150	1750	LANL
B83	Bomber	325	625	LLNL
W87	ICBM Minuteman-3	300	560	LLNL
W88	SLBM Trident	455	400	LANL

The US dismantled 11,751 warheads between 1990 and 1999. The US and Russia both had about 5000 operable warheads in 2015.
US (strategic deployed 1700, tactical 500, non-deployed 2800)
Russia (strategic deployed 1500, tactical 2000, non-deployed 2000)
(Kristensen and Norris 2014, 2015)

making pits can be reduced. Originally the plan was to make 125 to 450 pits per year (2004). Congress cancelled that plan in 2006. This was replaced with a Los Alamos facility, to make 50 to 80 pits per year, then lowered to 20 to 30 per year. Since Los Alamos can now make 10 to 20 per year, the new facility will not be needed. The Obama administration has suggested a five-year waiting period to determine actual needs. For a 1000-warhead stockpile, lasting 100 years each, a production rate of 15 per year would be sufficient.

Each of the enduring warheads in Table 2.7 is being refurbished under the individualized life-extension program (LEP). Their goal is not to make changes in the basic physics package design. Mostly these programs appear to be successful, with the exception of the B61 bomb, which is costing additional funds, but is progressing to be refurbished by 2022.

The Stockpile Stewardship Program is determining an improved equation of state for solid-state plutonium, used to calculate the properties of imploding primaries. The Joint Actinide Shock Physics Experiment Research (JASPER) Facility at the Nevada Test Site projects high–velocity plutonium samples into plutonium targets with a two-stage gas gun, 20-m long. JASPER achieves millions of atmospheres of pressure and temperatures of thousands of Kelvins, resulting from an initial velocity of 8 km/s.

Strategic nuclear bombers have had a number of major accidents, obtaining a much poorer safety record than that of ICBMs and SLBMs. This is no longer relevant since heavy bombers have been de-alerted, no longer carrying nuclear weapons, except for times of nuclear threat. In addition, the vulnerable US liquid-fueled, Titan ICBMs have been decommissioned, further reducing safety concerns.

In 2002 and 2012, the National Research Council released CTBT oversight reports by a panel of experts. They examined the US Stockpile Stewardship

Program (SSP), which is dedicated to making US warheads reliable and safe without nuclear testing.[2] The NRC committee concluded the following in 2012:

> Constraints placed on nuclear-explosion testing by the monitoring capabilities of the IMS, and the better capabilities of the US National Technical Means, will reduce the likelihood of successful clandestine nuclear-explosion testing, and inhibit the development of new types of strategic nuclear weapons. The development of weapons with lower capabilities, such as those that might pose a local or regional threat, or that might be used in local battlefield scenarios, is possible with or without the CTBT for countries of different levels of nuclear sophistication. However, such developments would not require the United States to return to testing in order to respond because it already has—or could produce—weapons of equal or greater capability based on its own nuclear explosion test history. Thus, while such threats are of great concern, the United States would be able to respond to them as effectively whether or not the CTBT were in force.
>
> A technical need for a return to nuclear-explosion testing would be most plausible if the United States were to determine that adversarial nuclear activities required the development of weapon types not previously tested. In such a situation, the United States could invoke the supreme national interest clause and withdraw from the CTBT.
>
> As long as the United States sustains its technical competency, and actively engages its nuclear scientists and other expert analysts in monitoring, assessing, and projecting possible adversarial activities, it will retain effective protection against technical surprises. This conclusion holds whether or not the United States accepts the formal constraints of the CTBT.
>
> Finding 1.1: The technical capabilities for maintaining the U.S. stockpile absent nuclear-explosion testing are better now than anticipated by the 2002 report.
>
> Finding 1.2: Future assessments of aging effects and other issues will require quantities and types of data that have not been provided by the surveillance program in recent years.
>
> Finding 1.3: The committee judges that Life-Extension Programs (LEPs) have been, and continue to be, satisfactorily carried out to extend the lifetime of existing warheads without the need for nuclear-explosion tests. In addition to the original LEP approach of refurbishment, sufficient technical progress has been made since the *2002 Report* that *re-use or replacement* of nuclear components can be considered as options for improving safety and security of the warheads.
>
> Finding 1.4: Provided that sufficient resources and a national commitment to stockpile stewardship are in place, the committee judges that the United States has the technical capabilities to maintain a safe, secure, and reliable stockpile of nuclear weapons into the foreseeable future without nuclear-explosion testing. [The three weapon laboratory directors indicated they agreed with this conclusion.] Sustaining these technical capabilities require at least the following:
>
> – *A **Strong Scientific and Engineering Base***. Maintaining both a strategic computing capability and modern non-nuclear-explosion testing facilities (for hydrodynamic testing, radiography, material equation-of-state measurements, high-explosives testing, and fusion testing) is essential for this purpose.
> – *A **Vigorous Surveillance Program***. An intensive surveillance program aimed at discovering warhead problems is crucial to the health of the stockpile.
> – ***Adequate Ratio of Margin to Uncertainty***. Performance margins that are sufficiently high, relative to uncertainties, are key ingredients of confidence in weapons performance.

[2]National Research Council, *The Comprehensive Nuclear Test Ban Treaty: Technical Issues for the United States*, (National Academy Press, Washington, D.C., 2012).

- *Modernized Production Facilities*.
- *A Competent and Capable Workforce*.

Hydronuclear testing refers to a test in which criticality is achieved but the nuclear yield is less than the energy released by the high explosive. In this report the committee distinguishes hydronuclear tests as a subset of nuclear-explosion tests, most of which have nuclear yield far greater than the energy released by the high explosive but all of which are banned under the CTBT.

Finding 4.2: Hydronuclear tests would be of limited value in maintaining the United States nuclear weapon program in comparison with the advanced tools of the stockpile stewardship program.

Finding 4.3: Based on Russia's extensive history of hydronuclear testing, such tests could be of some benefit to Russia in maintaining or modernizing its nuclear stockpile. However, it is unlikely that hydronuclear tests would enable Russia to develop new strategic capabilities outside of its nuclear-explosion test experience. Given China's apparent lack of experience with hydronuclear testing, it is not clear how China might utilize such testing in its strategic modernization.

Problems

2.1 **Neutron detection**. How did Chadwick detect this neutral-charged particle?
2.2 **Kennedy's 25 NWp states**. Name 25 that started nuclear weapons research.
2.3 **Fission energy**. Why is fission energy primarily electrostatic energy.
2.4 **235 versus 238**. Why does 235 work in atom bombs and 238 does not?
2.5 **Cow/Mice diets**. Why can cows live on grass, but mice cannot?
2.6 **Critical mass**. How is critical mass affected by neutrons/fission, geometry, isotope ratios, reflectors, compression, etc.?
2.7 **Tritium. How is tritium produced**?
2.8 **d + t = He + n**. What three ways is this reaction used in nuclear weapons?
2.9 **Radiation compression**. How does this squeeze and heat the secondary?
2.10 **Tritium supplies**. Because of tritium's 12.4–year half life, it needs replenishing. Why did our tritium resource become larger at the end of the Cold War?
2.11 **HEU versus Pu**. What are the relative difficulties to obtain Pu and HEU?
2.12 **Weaponization**. Is HEU or Pu easier to make into a weapon? Why?
2.13 **Ease of detection**. Is HEU or Pu easier to hide? Why?
2.14 **H-bomb**. Why is lithium-6 deuteride ($^{6}Li^{2}H$) useful for hydrogen weapons?
2.15 **H-bomb size**. Is there an upper limit on yield for a hydrogen secondary?
2.16 **Boosted primaries**. Why do they work, extra energy vs. extra neutrons?
2.17 **Neutron bomb**. Why were these lethal to enemy tank-drivers? Why dangerous?
2.18 **Anti-Matter bomb**. Why are anti-matter bombs unlikely? How efficient if exist?

2.19 **Nuclear winter**. What conditions might cause a temperature drop of 30 °C?
2.20 **Stockpile stewardship**. What are the legacy weapons? What is the Pu-pit lifetime? What tools is the SSP using to make weapons reliable?

Seminar: Nuclear Weapon Issues
artificial radioactivity, Ban the Bomb, Bikini Tests, Baruch Plan, Big Mike explosion, big vs. small nuclear weapons, boosted nuclear weapons, Castle Bravo Test, critical mass, Cuban Missile Crisis, Earth penetrating warhead, electronic initiator, electromagnetic pulse, Franck report, heavy water in Norwegian Vermok, Jewish physics, lithium deuteride, lost nuclear weapons, Manhattan Project, neutron bomb, peaceful nuclear explosions (PNE), permissive action link (PAL), polonium-210, Project Orion, Pugwash Conferences, safety features on nuclear weapons, tamper, thorium, uranium resources, uranium-233 stocks and use, Y-12 enrichment.

Bibliography

Ahlswede, J., & Kalinowski, M. (2012). Global plutonium production with civilian research reactors. *Science and Global Security, 20*(2), 69–96.

Albright, J., & Kunstel, M. (1997). *Bombshell: The secret story of America's unknown atomic spy conspiracy*. New York, NY: Times Books.

Alverez, R. (2013). Managing the ^{233}U stockpile of the United States. *Science and Global Security, 21*(1), 53–69.

Bergeron, K. (2002). *Tritium on Ice*. Cambridge, MA: MIT.

Bernstein, J. (2008). *Nuclear weapons: What you need to know*. Cambridge University Press.

Beschloss, M., & Talbott, S. (1993). *At the highest levels*. Boston, MA: Little Brown.

Bodansky, D. (2004). *Nuclear Energy* (2nd ed.). New York: AIP Press.

Brode, H. (1968). Review of nuclear weapon effects. *Annual Review of Nuclear Science, 18*, 153–202.

Chyba, C., & Milne, C. (2015). Simple calculation of critical mass of HEU/Pu. *American Journal of Physics, 82*, 977–980.

Cochran, T., Arkin, W., Hoenig, M., Norris, R., & Sands J. (1984–1989). *Nuclear Weapons Databooks*. Cambridge, MA: Ballanger.

Cochran, T., Norris, R., & Bukharin, O. (1995). *Making the Russian bomb*. Boulder, CO: Westview.

DeVolpi, A., et al. (2005). *Nuclear shadowboxing: Contemporary threats from cold war weapons*. Kalamazoo, MI: Fidlar Doubeday.

Dowling, J., & Harrel, E. (Eds.). (1986). *Civil defense: A choice of disasters*. New York, NY: AIP Press.

Duderstadt, J., & Moses, F. (1982). *Intertial confinement fusion*. New York, NY: Wiley.

Ford, K. (2015). *Building the H bomb*. Hackensack, NJ: World Scientific Publishing.

Glaser, A., & Mian, Z. (2008). Resource letter: Nuclear arms control. American Journal of Physics, *76*(1), 5–14.

Glasstone, S., & Dolan, P. (1977). *The effects of nuclear weapons*. Washington, DC: DoD/DOE.

Hafemeister, D. (1983). The arms race revisited: Science and society test VIII. *American Journal of Physics, 51*, 215–225.

Hafemeister, D. (Ed.). (1991). *Physics and nuclear arms today*. New York, NY: AIP Press.

Halloway, D. (1994). *Stalin and the bomb*. New Haven, CT: Yale University Press.

Hansen, C. (1999). *Swords of Armageddon: History of US development of nuclear weapons*. Sunnyvale, CA: Chuckela Publications. (2909 pages in 7 volumes). http://www.uscoldwar.com

Hanson, R. (2015a, January). Next generation manufacturing for the stockpile. *Science and Technology Review,* 4–11.

Hanson, R. (2015b, January). Building future modeling and uncertainty quantification for accelerated certification. *Science and Technology Review,* 12–18.

Harwell, M. (Ed.). (1984). *Nuclear winter*. New York, NY: Springer.

Hecker, S. (2000). Challenges in Plutonium science. *Los Alamos Science*, *26*(2).

Heller, A. (2012, December). Plutonium at 150 years. *Science and Technology Review*, 11–14.

Heller, A. (2014, June). Significant achievement on the path to ignition. *Science and Technology Review*, 3–10.

Hewlett, R., & Anderson, O. (1966). *The new world: 1936–46*. Washington, DC: US AEC.

Hewlett, R., & Duncan, F. (1969). *Atomic shield: 1947–52*. College Park, PA: Pennsylvania State University Press.

Hewlett, R., & Holl, J. (1989). *Atoms for peace and war: 1953–61*. Berkeley, CA: University of California Press.

Holloway, D. (1994). *Stalin and the bomb*. New Haven, CT: Yale University Press.

Kaplan, F. (1978). Enhanced-radiation weapons. *Scientific American, 238*(5), 44–51.

Kristensen, H., & Norris, R. S. (2014). Russian nuclear weapons. *Bulletin of the Atomic Scientists, 70*(2), 57–85.

Kristensen, H., & Norris, R. S. (2015). US nuclear weapons. *Bulletin of the Atomic Scientists, 71* (2), 107–119.

Lourie, R. (2002). *Sakharov*. Hanover, NH: Brandeis University Press, University. Press of New England.

MacCracken, J. (1988). The environmental effects of nuclear war, In D. Schroeer & D. Hafemeister (Eds.), *Nuclear Arms Technologies in the 1990s*: *AIP Conference Proceedings, New York* (Vol. 178, pp. 1–18).

Mark, J. (1993). Explosive properties of reactor-grade plutonium. *Science Global Security* 4(2), 111–128 (1993) and *SAGS*, *17*(2), 170–185 (2009).

McKinzie, M., Cochran, T., Norris, R., & Arkin, W. (2001). *The US nuclear war plan: A time to change*. Washington, DC: Natural Resources Defense Council.

National Research Council. (1989). *The nuclear weapons complex*. Washington, DC: National Academies Press.

Nelson, R. (2004). Nuclear bunker busters. *Science and Global Security, 12*(1), 69–89.

O'Brian, H. (2012, March). Extending the life of an aging weapon. *Science and Technology Review,* 4–11.

Office of Technology Assessment. (1979). *The effects of nuclear war*. Washington, DC: OTA.

Pearson, J. (2015, June). On the belated discovery of fission. *Physics Today*, 40–45.

Reed, C. (2011). *The Physics of the Manhattan Project*. New York, NY: Springer and *American Journal of Physics, 73*(9), 805–811 (2005).

Rhodes, R. (1995). *Dark sun: The making of the hydrogen bomb*. New York, NY: Simon and Schuster.

Rhodes, R. (1988). *The making of the atom bomb*. New York, NY: Simon and Schuster.

Sakharov, A. (1990). *Memories*. New York, NY: Knopf.

Schroeer, D., & Dowling, J. (1982). Physics and the nuclear arms race. *American Journal of Physics, 50*, 786–795.

Schroeer, D. & Hafemeister, D. (Eds.). (1988). *Nuclear Arms Tech. in the 1990s: AIP Conference, New York* (Vol. 178).

Serber, R. (1992). *The los alamos primer*. Berkeley, CA: University of California Press.

Talbott, S. (1979). *Endgame: The inside story of SALT II*. New York, NY: Harper.

Talbott, S. (1984). *Deadly gambits*. New York, NY: Knopf.

Talbott, S. (1988). *The master of the game*. New York, NY: Knopf.

Taylor, T. (1987). Third-generation nuclear weapons. *Scientific American, 256*(4), 30–38.

Toon, O., Robock, A., & Turco R. (2008, December). Environmental consequences of nuclear war. *Physics Today*, 37–42.

Turco, R., Toon, O., Ackerman, T., Pollack, J., & Sagan, C. (1983). Nuclear winter: Global consequences of multiple nuclear explosions. *Science, 222*, 1283–1292.

Turco, R., Toon, O., Ackerman, T., Pollack, J., & Sagan, C. (1990). Climate and smoke: An appraisal of nuclear winter. *Science, 247*, 166–176.

von Hippel, F., & Sagdeev, R. (Eds.). (1990). *Reversing the arms race*. New York, NY: Gordon-Breach.

Chapter 3
Nuclear Reactors and Radiation

The ingenuity of our scientists will provide special safe conditions under which such a bank of fissionable material can be made essentially immune to surprise seizure.
[President Dwight Eisenhower, UN Atoms for Peace Speech, 8 December 1953]

3.1 Nuclear Reactors

First nuclear reactor in Oklo. Earth's first heat-producing nuclear reactor was accomplished without the assistance of humans. It produced 100 kW for 150,000 years, 1.8 billion years ago in Oklo, Gabon. The residue of fission fragments in the soil and the reduction in ^{235}U content from 0.7 to 0.4 % proves the hypothesis. This happened without cooling pumps, fuel-rods and emergency-core-cooling systems. This natural thermal reactor operated merely because a good, rich deposit of uranium was concentrated in a watery estuary. Natural-uranium at that time was 4 % ^{235}U, before it decayed to today's 0.7 %. Water was needed to place uranium into solution, carry it downstream and concentrate it the estuary basin. This multi-step process concentrated the moderately rare uranium into a robust uranium ore. Water also slowed (moderated) high-energy fission neutrons to low velocities, to enhance the probability to fission. A reactor needs only a good concentration of fissionable materials (^{235}U, ^{233}U, ^{239}Pu) in a reasonable geometry, with a moderator (H_2O, D_2O, graphite). No nuclear engineers were needed for the Oklo nuclear reactor (Fig. 3.1).

Electricity generation. When a bar magnet is placed near a cathode-ray-tube television screen or an oscilloscope, the pattern becomes distorted. Electrons moving in a magnetic field experience a force that changes their path. In a similar way, if a copper wire is pushed perpendicular to a magnetic field, the electrons in the wire experience a force along the wire, perpendicular to the velocity of the wire. This force is used to generate an electrical current. Loops of wire spun in a magnetic field develop a voltage that pushes electrons in a wire to drive motors.

D. Hafemeister, *Nuclear Proliferation and Terrorism in the Post-9/11 World*,
DOI 10.1007/978-3-319-25367-1_3

Fig. 3.1 Stagg field pile, 1942. Drawing of the first self-sustaining chain-reaction nuclear pile at Stagg field, University of Chicago on 2 December 1942. The pile contained 400 tons of graphite, 6 tons of uranium metal and 58 tons of uranium oxide. The person in the foreground is withdrawing a control rod (Atomic Heritage Foundation)

Coal, natural gas and nuclear generate electricity: Flames from burning fossil fuels heat closed boilers from the outside, similar to a pressure cooker on the stove. Heat from nuclear fission heats the boiler inside or from heat transferred by pipes from the reactor. The boiler constrains steam pressure to about 2000 psi. The high-pressure steam is released through nozzles, hitting turbine blades to make them spin. The rotational motion spins coils of wire in a magnetic field to generate electricity. One other feature must be mentioned. High efficiency is needed to make a kilowatt-hour of electricity from a pound of coal. This can be done with the aid of cooling water, either from the ocean or cooling towers. It is important to reduce the backpressure on the far side of the turbine, to allow the pushed turbines to spin with gusto. By using heat exchangers to condense steam into water, the back-pressure is reduced.

Nuclear power status. Nuclear power has been producing 20 % of US electricity since the 1990s, but its growth has stopped, as it is slowly declining. An early projection in 1972 suggested there would be 1200 GW_e nuclear capacity by 2000. This fell far short, as capacity is now slightly under 100 GW_e. Global nuclear power capacity has remained fairly constant over the past five years at 350 GW_e. Seventeen nations use a higher percentage of nuclear power than the US, although the US has the largest capacity. French capacity is 63 GW_e, which is 76 % of its grid, while Japanese capacity (at maximum, before Fukushima) was 43 GW_e for 35 % of its grid. Global electricity (2012) was produced 67 % from fossil fuels [coal (40 %), natural gas (23 %), petroleum (4 %)], and from the non-fossil fuels [nuclear (17 %), hydro (17 %), renewables (5 %)]. US electricity (2013) was produced from coal (39 %) natural gas (27 %), nuclear (19 %), hydro (7 %) and renewables (6 %).

Increased cost of nuclear power and the accidents at Three Mile Island (1979), Chernobyl (1986) and Fukushima (2011) halted plans for new plants. Low cost natural gas from fracking with electrical generation from 60 %—efficient combined-cycle gas turbines further removes nuclear power as a competitor. The continuing unrest about radioactive waste disposal has effectively forestalled the addition of new nuclear plants. Some of this threat of leaking wastes has been exaggerated. Cheaper photovoltaic solar energy is a factor, but PV's and wind power will need cost-effective energy storage to penetrate the grid by more than 25 % or so. Solar and wind cannot compete with base-loaded power. I am happy with my PV's on the roof, time of day pricing and only a Leaf electrical car on the driveway.

Nuclear electricity generation. A nuclear power plant is similar to the fossil plant except for major differences:

- Heat is produced inside the boiler (or enters by heat-transfer pipes) since oxygen is not needed for fission, and fission needs water to moderate neutrons for the chain reaction and to transport the heat. The nuclear reactor is the boiler with nuclear fuel rods inside and exiting hot water pipes. The pressurized water reactor (PWR) boiler does not produce steam in the reactor because it is maintained at high pressure.
- PWR Heat exchangers transfer heat from the reactor's core to a second loop of hot water. The second loop's steam is released with nozzles to spin turbines. The heat exchanger prevents radioactivity from being released from the primary loop.
- Nuclear power plants need 25 % more cooling than fossil plants because fossil plants release considerable heat in combustion gases to the atmosphere.
- A typical light water reactor (LWR) has a nuclear core of 100 tonnes of 4.4 % enriched ^{235}U. One-third the core is removed as spent fuel every 1.5 years.
- Spent fuel ponds cool the radioactive spent fuel, or it could burst into flames when packed too closely. It is transferred to on-site surface storage after 5 years, and ultimately, perhaps, to final disposition in an underground repository.

Plutonium production. A typical light water (LWR) reactor uses low-enriched uranium of 4.4 % ^{235}U for fuel and normal water as a coolant and moderator. The LWR produces 220 kg of plutonium per year (1 gigawatt = 1 GW_e = 1000 MW_e). Since fuel rods remain in the LWR for 4 years, there is plenty of opportunity for the newly-produced ^{239}Pu to capture a second neutron to become ^{240}Pu. LWR reactors produce reactor-grade plutonium with over 20 % ^{240}Pu (Fig. 3.2).

CANDU reactor. The Canadian Deuterium Uranium (CANDU), heavy water reactor (HWR) uses heavy water as a coolant and moderator. The CANDU operates with natural uranium fuel of 0.7 % ^{235}U, compared to 3–5 % for LWRs, because D_2O absorbs fewer neutrons than H_2O. CANDUs produce more plutonium per year (350 kg/GW_e-yr). This is because CANDUs are continuously fueled, leaving less residency time to burn the newly produced ^{239}Pu. The shorter residency time also converts less ^{239}Pu to ^{240}Pu. Thus CANDUs produce more plutonium/year and it is weapons-grade plutonium

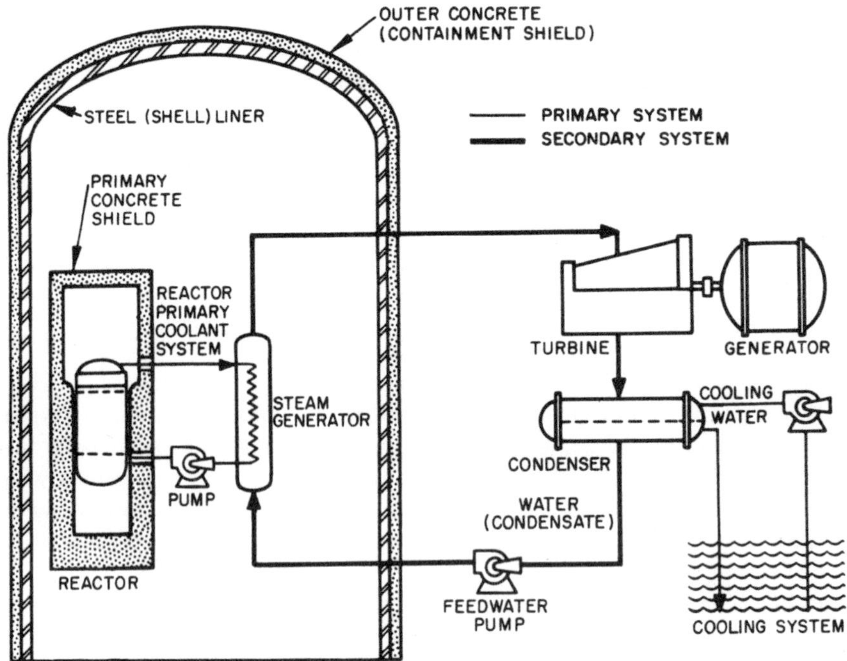

Fig. 3.2 Pressurized water reactor (PWR). There are three water loops in a PWR. The primary coolant system remains within the containment structure, transferring heat to the secondary system in a steam generator. The steam from the secondary loop drives the turbines exterior to the containment structure. The cooling water from the ocean or cooling towers is the third coolant loop, reducing the back pressure by condensing the steam after exiting the turbines (ERDA/DOE)

(6 % ^{240}Pu). However, more uneconomic reprocessing is required to obtain plutonium from CANDU spent fuel since its spent fuel contains only 0.2 % Pu as compared to almost 1 % Pu for LWRs.

Thorium cycle. Thorium contains only one isotope, ^{232}Th, which is fertile, meaning that if it absorbs a neutron it will make an isotope that can fission. When thorium absorbs a neutron, it decays to ^{233}U, which is fissile. It has been long recognized that a nuclear cycle based on thorium could reduce proliferation risks. If one used a Pu seed and blanket approach, with the Pu seed surrounded by a thorium blanket, one could consume the plutonium surplus, extend fuel supplies and reduce plutonium stocks. The main consideration is economics. When the 1977–79 International Nuclear Fuel Cycle Evaluation examined the thorium issue, it was favorable except that thorium's economic competitiveness was immature. The benefits of thorium are listed below. Thorium's main challenge is that it is difficult to work with to obtain the ^{233}U fuel because of ^{232}U's strong gamma-rays.

- thorium is 3–4 times more abundant than uranium,
- thorium produces less radioactive actinides,
- thorium cycle could incinerate excess plutonium,
- thorium is more fertile than ^{238}U,
- ^{233}U gives more neutrons/fission over a wider range of neutron energy,
- thorium dioxide is more chemically stable than uranium dioxide,
- ^{232}U in thorium cycle gives considerable radiation, making warheads difficult.
- Combining molten fluorides of Th, Be and Li lowers melting point to 360 °C, reducing pressure to atmospheric.

Breeder reactors were once thought to be THE path to a world without energy worries, however high costs and other complications have dimmed their future. France was the leader, but now its two Phoenix breeders are shuttered. Japan, Russia and a few others are still trying to the make breeder reactors economically successful, but the British, Germans and Americans have abandoned the chase for the liquid-metal fast breeder reactor (LMFBR).

A ***breeder reactor*** produces more fuel than it burns. A one-GW$_e$ breeder with a breeding ratio of 1.5 produces 1.5 tonnes of Pu per year, with a net growth of 0.5 tonnes in fissile material. Breeder plutonium is weapons-grade Pu. It is mostly produced in the blanket around the reactor where the neutron flux is lower than in the core. In order to have good breeding conditions, it is necessary for the breeder's core to be small, keeping a large flux of neutrons

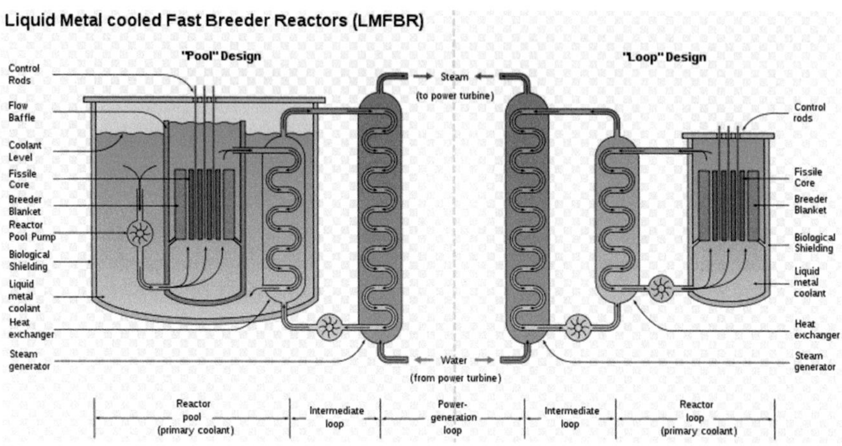

Fig. 3.3 Breeder reactor. The figure to the *left* is the pool configuration, which maintains liquid sodium in the reactor pool structure, including the heat exchanger. The figure to the *right* is the loop configuration, which allows liquid sodium to leave the reactor to enter the external heat exchanger (Atomic Heritage Foundation)

Table 3.1 Power reactor fuel cycles

Type	Pu/GW$_e$-year (kg/yr)	Type of Pu	Comments
LWR	220	Reactor grade	4.4 % U with H_2O
CANDU	350	Weapons grade	Natural U with D_2O
LMFBR	1500	Weapons grade	Liquid–Na cooled

on the blanket for breeding. This geometry requires better cooling with liquid-metal coolants (sodium or potassium), which requires greater engineering expertise (Fig. 3.3 and Table 3.1).

3.2 Nuclear Safety

Extra heat release can melt a reactor core. This extra heat comes from very short-lived beta-decays of fission fragments. In a loss-of-coolant accident, fission stops because the moderating water has run out, but what remains is beta decay heating. During the first minute after fission has stopped with the loss of water, beta decay heat continues at 7 % of the 3 GW of thermal power. The 200 MW of beta-decay heat is considerable and rapidly raises the temperature in one minute of the core to the molten state at 2200 °C. Reactors need emergency core cooling systems to obtain quickly additional cooling water. Extra motor-generator sets are present at the reactor site to do this. Some reactors utilize ponds at higher elevations, to avoid the support the back-up pumps.

Nuclear reactor disasters don't usually happen because of a *single large failure*, but they can happen as a culmination of multiple, smaller operational failures or mistakes. Reactor malfunction is often due to poor maintenance practices or design errors. It is possible for nuclear reactors to suffer a loss-of-coolant accident (LOCA) from a pipe break that is followed by an *emergency core-coolant system* (ECCS) failure. The reactor core would then melt through the steel reactor vessel into the ground below; hence, the nickname *China Syndrome*, as molten fuel headed downwards into the Earth towards China in the Hollywood movie. It should be pointed out that very few (essentially none) Americans died in a US commercial nuclear power accident thus far. The number of deaths per-kilowatt-hour for nuclear power is far smaller than it is for coal or natural gas.

3.2.1 *Five Nuclear Accidents*

- The 1975 Browns Ferry accident was caused by a loss of electrical power in a fire. This caused evaporation of considerable coolant, but power was restored before the core melted.

- The 1979 Three Mile Island accident occurred because cooling water for the core was mistakenly stopped. The core was one-third melted, but only small amounts of radioactivity were released.
- The 1986 Chernobyl accident was caused by stuck control rods with a *positive void coefficient* that lacked a containment structure.
- The 2002 *almost* accident at the Davis–Besse reactor was a result of boric acid consuming a soft-ball size volume in the containment vessel. It could have been detected by monitoring the cooling water.
- The 2011 triple reactor melt-down accident at Fukushima was not caused by the 9.0 earthquake, but rather it was caused by the 14 m (46 ft) foot tsunami, which removed cooling. This $35 billion accident was failed by bad planning.

Radiation has positive uses, such as medical diagnostics, cancer therapy, radio-surgery from collimated beams, nondestructive testing of structures, smoke alarms, and food irradiation. However, the use of radiation must be controlled because of possible health effects. Without taking time to think about possible danger, Louis Slotin used his bare hands to separate a critical mass at Los Alamos on 21 May 1946. He died in nine days, the first victim of the postwar nuclear arms race. Slotin did not die of cancer; rather his body stopped functioning after a dose of over 10 Sv (1000 rem). A one-time dose of 4.5 Sv (450 rem) is lethal to 50 % of victims. The most famous case of radiation murder was that of Alexander Livinenko, who was injected with a dose of 1 μg of ^{210}Po in London subway in 2006. This chapter does not deal with large doses, but rather with doses under 10–100 mSv (1–10 rem), which can cause cancer after a latency period of some 10–30 years.

Chernobyl. The 1986 accident in Ukraine killed considerably more people than the Fukushima accident in 2011. A burning carbon reactor, without a containment structure, propelled radioactivity in large plumes to great distances. The 49,000 living in Pripyat were permanently evacuated. A 2005 UN science panel determined that 56 deaths are directly attributable to the Chernobyl disaster, with the death of 47 emergency workers and 9 young children who developed thyroid cancer from drinking milk contaminated with radioactive iodine. Some 4000 young children were afflicted with thyroid cancer, but prompt medical care successfully reduced these deaths to nine. These results strongly confirm the concept of stockpiling KI pills near reactors. The UN experts predict that there will be 5000 additional deaths from the accident. An exclusion zone with a radius of 25 km reduced long-term effects, but 100,000–200,000 continue to be severely affected mentally and financially. To cover this kind of situation, US reactors have a federal Price-Anderson insurance policy of $300 million, renewed to 2017 (Fig. 3.4).

Fukushima. The 2011 9.0—magnitude earthquake and tsunami in Japan released the second greatest amount of radiation. The cooling generators and pumps were located 6 m above the ocean, but a 14-m tsunami shut down the cooling of the

Fig. 3.4 Chernobyl (Hafemeister, Atomic Heritage Foundation)

three operating reactors and the 6 cooling ponds. Hydrogen explosions exacerbated the situation, and impacted the global nuclear power community. Fukushima mortality was about one-tenth that of Chernobyl. Fukushima was designed for a 8.2 quake but with upper-limit error bars extending to 9.0. The reactors survived the 9.0 earthquake, but not the tsunami. Japan is at a time of transition. How many of its 43 plants will remain closed? A total shutdown would cost Japan $60 billion, in a nation with few energy resources. Japan has 9 tonnes of Pu in Japan, with another 35 tonnes equally divided in UK and France. This plutonium, along with a new reprocessing plant in Japan, was intended to make Pu–U fuel for thermal reactors and breeder reactors.

Nuclear mitigation. Because an accident at a nuclear plant is more dangerous than one at a fossil plant, the following steps are taken to enhance safety:

- Nuclear power plants operate at lower temperatures and pressure. This reduces their efficiency to 33 % as compared to fossil plants that can be as high as 40 or 60 % for a combined cycle gas turbine.
- If a nuclear plant loses cooling water, its fuel can begin to melt with a minute. If cooling water enters at this point, it might release hydrogen from water, which could be explosive. For this reason, emergency core cooling systems (ECCS) inject water quickly, as long as there is not a major pipe break.
- Reactors have a containment dome that would, most likely, contain most of the radioactive materials.
- To reduce the effects of pressure variations, reactors use a large tank, called the accumulator, half-full of water and half-full of nitrogen, connected to the reactor. If pressure suddenly increases, the gas in the accumulator is compressed by the rising water in the accumulator, reducing pressure in the reactor, allowing the reactor to operate at constant pressure.

Mortality estimates for nuclear power. We assume a linear low-dose rate for radiation health effects. This assumes a doubled-dose doubles the health risk. The annual number of deaths from nuclear power is a product of at least eight complex functions, each of which must first be first multiplied for each failure mode and then summed over all failure modes. The eight functions are as follows:

- n, number of nuclear plants
- P, annual probability for a failure mode
- S, amount of released radioactivity
- B, biological function, coupling radiation and mortality
- A, plume area
- ρ, population density in the plume
- W, wind and weather function
- t, time spent in irradiated region.

Estimates of an *extremely serious* reactor accident in a populous location point to numbers that could be devastating: 3000 immediate fatalities, 240,000 thyroid nodules, 45,000 latent cancer fatalities, 30,000 latent genetic effects, and a cost of over $10 billion, contaminating an area of 8000 km^2. Locations vary. The area near the Indian Point reactors, close to New York City, contains 17 million persons within 50 miles. The Three Mile Island accident was fearful but not lethal, with essentially no deaths. U.S. Nuclear reactors have, thus far, not had a serious life-destroying accident. The have obtained a total radiation exposure of 5 person-Sv/yr, causing 0.3 of a death per year. Safety could be enhanced with smaller cores, lower power densities and greater heat capacity to totally prevent loss-of-coolant accidents. The cost of these modular reactors could be high. New reactor designs are not totally passive-safe, but rather they use passive-safe circulation systems.

Fault tree analysis is widely used by industry to analyze accident scenarios and estimate relative and absolute accident rates. It is widely used by aeronautical engineers when designing airplanes. New safety features add weight to planes, requiring judicious choices. Probabilistic risk assessment (PRA) was used to estimate accident rate scenarios for nuclear power plants in the 1975 *Reactor Safety Study* under Norman Rasmussen of MIT. This report was criticized for underestimation of error bars and for not adequately considering common-mode failures, such as earthquakes, which can remove more than one safety system at a time. The study predicted very serious accidents would be infrequent. Such accidents could be caused by overlapping small problems. PRA is useful in determining *relative* risks, as the best approach available, but it does not include some high risk accident modes, such as the risk of fires and of large tsunamis in Japan.

Serial logic. A process of many different tasks, each depending on the success of the previous task, is *serial logic*. To prepare a meal form uncooked ingredients, we need a stove with a working valve, a working pilot light, and available nature gas. Assume all the task failures are $P_i = 10\ \%$. The probability of obtaining a cooked meal is 0.9 cubed, or 73 %, or a probability that cooking dinner fails 27 % of the time. The more tasks in a serial process, the less likely the process will be successful.

Parallel logic. Duplicate back-up systems enhance success by operating in parallel. Reactors have several back-up power options in case there is a loss of line electricity for the water pumps. A back-up motor-diesel generator is available when external power fails, and a second generator is available if the first generator fails. If there are two reactors, then a third generator is available. Some plants have further backup water available from raised ponds. Each branch indicates a chance to recover with a new, parallel system. The overall probability of maintaining electricity to operate a reactor depends on the failure probability for line voltage and the failure probabilities of the two diesel generators (as well as other systems). If each of three power sources has a 10 % failure probability, the *system* failure rate is 0.001, or 1 in 1000, which is much better than a failure rate of 27 % for a meal prepared by three serial tasks, each with a 10 % failure rate.

The ***Rasmussen Report*** estimated a median accident probability of 5×10^{-5}/reactor-year for a core meltdown, with an upper bound of 3×10^{-4}/reactor-year. The 100 US power reactors, operating for 30–40 years, have had more than 3000 reactor-years experience. The pessimistic, *highest probability* estimate, combined with the US experience of about 3000 reactor-years, gives a probability for one core-melt accident in 30 years. This equals the US experience of one serious accident in 30 years. The accident was the partial core melt at Three Mile Island. It has often been assumed that 90 % of core-melt accidents would be contained in the concrete domes, and that 10 % of core melts would cause a steam explosion that would break containment and spread a plume. Some claim this assumption is too pessimistic.

Spent fuel ponds. Three-fourths of US spent fuel is stored in cooling ponds, and another 25 % is stored in dry casks (2011). The age of terrorism raises the issue of the possible attack on spent fuel ponds. After 1 year, spent fuel radioactive heating is 15 kW/tonne and at 10 years it falls to 2 kW/tonne. The spent fuel problem has been exacerbated because the density of spent fuel in ponds was increased in response to the 1977 decision to ban reprocessing of spent fuel to obtain plutonium. Extra fuel rods in ponds give additional heat and reduce convective and radiative cooling by narrowing the spaces between rods. The problem is lessened by moving older, cooler rods to surface storage with natural air-convection cooling.

Yucca Mountain repository. In 1977 the US government offered to accept utility spent fuel for a repository fee of 0.1 ¢/kWh, amounting to $0.75 billion per year. Since the underground repository has not been available over all that time, the electrical utilities charged the government with not fulfilling its agreement. The courts agreed, requiring DOE to pay utilities to keep spent fuel on-site until such time as a repository can accept it, ending the payments to DOE. By 2015, the trust fund collected $30 billion, while spending $7 billion. DOE hopes to have a repository open by 2048.

In 2009, the US Court of Appeals accepted the EPA limit of 15 mrem/year for the first 10,000-year period, and an annual limit of 100 mrem/year between 10,000 and a million years. DOE stated it can easily meet these limits. EPA considered persons living near the repository, eating some local grown food and drinking local water. The nearest community to Yucca Mountain is 30 km away in Beatty, Nevada with a population of 1000.

Since the area is sparsely populated, it would seem preferable to use criteria based on the death rate in persons/year, rather than a radiation rate of mrem/year in a sparsely populated place. Normally environmental regulation uses pollution levels (mrem/year or ppm) for pollution. But this has meaningful limits if population is sparse. What if only one person lived there? Another issue is how to link the risks of each step in a multiplicative process. Should risk parameters be mid-range estimates or extreme values? One should compare the number of projected deaths from Yucca Mountain with the projected number of deaths from other storage choices, since radioactivity must be disposed somewhere.

The political stakes are high for future of nuclear power. In 2009, Secretary of Energy Steven Chu removed Yucca Mountain from consideration. Retrievable surface storage now holds one-fourth of the spent fuel, which will grow until another repository can be found. Ultimately spent fuel should be removed from the surface of Earth. These are some of the choices: (1) Accept the risks at Yucca Mountain, (2) find a new location with a water-saturated medium to prevent oxidation, as Sweden did, or (3) bury spent fuel in a sea trench.[1] Sending it to space has its own set of problems.

Spent fuel at Yucca would be buried 300 m below the surface and 300 m above the water table, with a small water flow of 4 mm/year, entering the water table. But the time to flow from the surface to the water table is faster than expected, with radioactive ^{36}Cl traveling 300 m in less than 50 years. Ten-thousand years and a million years are long times, oxidation will degrade the titanium drip shields and the nickel-based alloy C-22. The media is not water saturated, so oxidation will happen at some rate. It is difficult to estimate underground flow rates in these *heterogeneous media*. Yucca Mountain would have had an initial capacity of 77,000 tons (70,000 tons spent fuel plus military waste). The mass of US spent fuel after 3500 GW_e years of operation in May 2013 is 69,720 tons, filled in 40 years. At that point 49,620 tons were in ponds and 20,100 tons were in air-cooled storage.

Heat loading. In the first several hundred years, spent fuel beta-decay heat comes primarily from fission fragments ^{90}Sr, ^{90}Y, ^{137}Cs, and ^{137}Ba with half-lives of 3–30 years. After 1000 years, heat comes primarily from alpha decay of actinides ^{239}Pu, ^{240}Pu, and ^{241}Am. Spent fuel rods can be placed in geological storage at varying heat loads, ranging from 2 kW_t/ton at 10 years to 1 kW_t/ton after 30 years. Since the Yucca Mountain site is oxidizing, there is concern that metal containers will be breached over a period of 1000–100,000 years. Because water percolation rates to the repository are higher than expected, the design uses engineered barriers to supplement geological barriers. There are very few metals other than gold that can resist corrosion for such long times, but stainless steel/nickel canisters with titanium drip shields are claimed to be robust for thousands of years. Radioactive heat should keep the canisters above 100 °C to keep water away over the initial 1000–2000 years.

Pu migration. The 1978 American Physical Society panel on radioactive wastes concluded that "Pu [is] efficiently confined" to regions close to the 1.8-billion year

[1]Hollister and Nadis (1998).

old Oklo, Gabon, natural reactor. This conclusion was readily accepted because Pu has low solubility in water and it has a tenacious capacity to cling to mineral surfaces. However, in the past 15 years there has been some evidence that Pu adheres to colloids less than a micron in size. On the other hand, some argue that the data are misleading, since the migration might be a result of nuclear weapons explosions that created underground fissures, increasing Pu transport. Should the issue be narrowed to mainly focus on the number of people affected?

3.3 New Reactor Designs

The Nuclear Regulatory Commission received 24 applications for new plants, but only three are under construction. Many are in abeyance when the price of natural gas dropped from $13/MBtu in 2008 to $3/MBtu in 2013. There have been 4 generations of nuclear reactors for commercial production of electricity.[2] We won't consider breeder reactors, led by France but now shuttered.

- ***Generation 1***. Few of the 250-MW_e, gas-cooled reactors still exist in the UK.
- ***Generation 2***. The existing nuclear fleet of light water reactors (LWR). Three of the 440 pressurized water reactors (PWR, Westinghouse) and boiling water reactors (BWR, General Electric) with a capacity of 360 GW_e. One hundred of these are in the US with an average age of 30–40 years, but 63 units have had their licenses extended from 40 to 60 years. Some are now talking of 80-year licenses. France has about 60 and Japan about 43.
- ***Generation 3***. These advanced PWR/BWR reactors and gas-cooled pebble bed modular reactor (PBMR) are designed to be simpler and safer. They have lower costs then the current generation by using passive safety features that rely on gravity, natural circulation and compressed air to provide cooling of both core and containment in case of a severe accident. This allows a smaller system to force coolant into the system. For example, the Westinghouse AP1000 has 50 % fewer valves, 35 % fewer prompts, 80 % less pipe, 48 % less seismic building volume, and 70 % less cable. The US had 22 pending licenses in 2012, with over 50 % for the AP1000. However the cost of the two Vogtle reactors in Georgia is large at $14 billion. Compounding this with the falling price of natural gas, the nuclear industry is on uncertain grounds. The AP1000 extends the ability to passively cool a reactor without electricity for 3 days with a cooling capacity of 17 m^3/MW. It also plans to increase the life of the battery pack to 3 days. Industry claims the AP1000 is ten times safer than its predecessors. There are now 61 global reactors under construction, with 38 in China, which plans to have a total of some 200 GW by 2030, or so. At this point China gets 80 % of its electricity from coal.

[2]Ahearn (2006), Piore (2011).

- ***Generation 4***. Higher natural gas prices in 2002 gave an impetus to nuclear power when the US led the 2002—formation of a 10—nation organization to determine the direction for the next-generation nuclear plants. These are aimed for deployment before 2030. Six types were selected for examination: The very-high temperature reactor (VHTR) was selected by the US with helium cooling and either a graphite core or a pebble bed with ceramic and graphite-coated fuel particles, to produce hydrogen as well as electricity. The hope is to use higher temperatures and simpler systems to raise thermodynamic efficiency from 33 to 45 %. Other designs use lead-cooled or sodium-cooled fast reactors, supercritical water-cooled reactors, gas-cooled fast reactors with either a helium or CO_2 working fluid, and the molten salt reactor.

Small, modular reactors. Section 2.3 showed that dimensional scaling determines the critical mass of nuclear weapons, what birds eat, the size of cow bones, and the minimum wood-fire. Extensive work is now being done by the *NuScale Power Company* to design and build smaller reactors that cannot melt after a loss of coolant accident.[3] Radioactive heat power is proportional to the volume, or radius cubed. Heat is lost through the surface of the volume, proportional to radius squared. The ratio of power heat power gain over loss is proportional to r^3/r^2, or the radius. The smaller the power of the reactor, the safer it will be, producing less heat traveling through a unit of area. NuScale advertising points out that "by the time my son is grown, electric vehicles will need 35,000 MW_e (35 GW_e) of power." This assumes 30 million electric cars, which also might be powered by solar energy.

NuScale is planning an array of twelve 45 MW_e reactor modules underground for a plant size of 0.54 GW_e. The 65 foot by 14-foot diameter units can be carried by barge, truck, or rail. They are cooled by gravitational-convection without pumps, "ensconced in a pool of water. Submerging the reactors adds a safety barrier and increases seismic damping, aiding earthquake resistance." There are two types of *economic scaling*: Make reactors large (1 GW reactors), or mass produce small modules (as in automobile production) and combine them in flexible combinations. DOE supported modular developments by Babcock and Wilcox, working with the Tennessee Valley Authority, and by NuScale.

3.4 Low-Dose Radiation

The increase in the death rate is linear at higher doses as seen in the data on Japanese atomic bomb survivors in Fig. 3.5. There is no doubt about this, but can this linear relationship be extrapolated to risks in the low-dose region? Is there a

---

[3]NuScale Power, Corvallis Oregon.

threshold dose below which biological repair is significant? Hiroshima/Nagasaki data show a linear mortality for radiation dose above 200 mSv (20 rem), but below this threshold the data are more uncertain.[4] The SI radiation units *sievert* and *gray* are defined below, as well as the colloquial *rem* and *rad* units.[5]

Linear dose rate. Data for uranium miners above 100 mSv point to linearity. Can this linear effect be extrapolated into low doses of less than 1 mSv? US citizens receive an average radiation dose of 3.60 mSv/year (360 mrem), most of it—3.0 mSv—from natural sources, including radon in buildings, and an additional 0.6 mSv from manmade sources, such as medical X-rays. Natural radiation damages most of the cells in our bodies. Experiments show that the number DNA breaks in cells is proportional to dose at a rate of 6.3 DNA breaks per human cell per gray (see radiation units in the footnote below). These experiments show that most of the breaks were repaired after a relaxation time of several hours. Figures 3.5 and 3.6 display the 1994 United Nations Scientific Committee on the Effects of Atomic Radiation (UNSCEAR) data on Japanese atomic bomb survivors. Japanese copper samples are currently being measured for the content of ^{63}Ni, which is made from neutron irradiation of copper. It is planned to separate gamma rays doses (the highest source of radiation) from fast neutron doses.

DNA breaks. Below some threshold, it is not clear *from the data* whether DNA breaks are *sufficiently* repaired to prevent the damaged cells from causing cancer. It is *very difficult* for epidemiology to settle this issue at very low doses. Some say that the threshold theory is valid, since the human system repairs DNA after radiation exposures. Others say that a Taylor expansion of a continuous function of the relationship contains a linear term. Most say a double DNA break is needed to cause cancer. This requirement is fulfilled when a natural break and a radiation break occur together before repair, supporting the linear theory. The BEIR-VII committee rejected (2006) the existence of a threshold dose level. It is generally

---

[4]The two atomic bombs dropped on Japan had quite different characteristics.

Hiroshima at 1000 m, gamma ray dose = neutron dose = 3 gray.

Nagasaki at 1000 m, gamma ray dose = 10 gray, neutron dose = 1 gray.

Peterson and Abrahamson (1998).

[5]Rate of decay:

1 curie (radiation from 1 g of radium) = 1 Ci = 3.7×10^{10} decay/s.

1 bequerel (SI) = 1 Bq = 1 decay/s.

Absorbed energy in air:

1 roentgen = 1 R = 87 ergs/g = 0.0087 J/kg.

Physical dose of absorbed energy:

1 rad = 100 erg/g = 0.01 J/kg.

1 gray (SI) = 1 Gy = 1 J/kg = 100 rad.

Biological dose equivalent is absorbed dose times a *relative biological effectiveness* Q.

X-rays, γ-rays, and electrons have $Q = 1$, neutrons have $Q = 5$–20 and alpha particles and fission fragments have $Q = 20$.

1 Roentgen equivalent man (rem) = 0.01 J/kg.

1 sievert (SI) = 1 Sv = 1 J/kg = 100 rem.

believed that double-strand breaks in DNA are harder to self-repair, and are therefore more dangerous. The double breaks could be caused by one energetic particle causing multiple damage, or by higher doses or by combining the background radiation with the source of radiation. The natural cancer rate should be subtracted from the total cancer rate to determine the dependence on radiation.

Dose models are key in discussing regulation of nuclear materials in reactors, wastes, and stored weapons. *Low-dose radiation* is an issue, as it makes a contribution in estimating potential deaths. The Committee on Biological Effectiveness of Ionizing Radiation (BEIR of the National Academy of Sciences) accepted the hypothesis that the rate of additional cancers is linear with dose. However, in 1990 BEIR commented:

> at such low doses and dose rates, it must be acknowledged that the lower limit of the range of uncertainty in the risk estimates extends to zero.

The experts cannot prove the linear theory with threshold is incorrect, but they believe this from examining all the data (2006). A controversial related issue is that of *hormesis*, which theorizes that very small radiation doses *reduce* cancer rates. In Kerala, India, radiation doses are 20 mSv/year, 6 times the US rate including radon, yet people live longer in Kerala than in the rest of India. Of course, their longevity might be due to the other good reasons to live in Kerala. It might be explained with the hormesis theory, which conjectures that small amounts of radiation are beneficial (Figs. 3.5, 3.6 and Table 3.2).

An alternative approach is the *relative risk model*, which uses radiation dose as a multiplicative factor to obtain cancer enhancement. Perhaps the truth lies somewhere between the *absolute risk* and *relative risk* models. Radiation induces cancer that would not have taken place. It is the hot electrons produced by nuclear particles that break DNA bonds. Recent studies show that electron energies as low as 3 eV are sufficient to break these bonds. Clearly, unburned carbon, ultraviolet light and nuclear radiation can do that.

Background radiation. Radiation from natural and human-made sources give an average dose of 3.6 mSv/year (360 mrem/year) in the United States. This gives an absolute death rate of about 1 % for 60 years natural exposure to this level. The 1 % death rate is a 20th, or 5 % of the normal 20 % cancer death rate. Most of the 20 % cancer death rate is primarily due to genetics, what we are born with. This is done under the linear low-dose theory (1990 BEIR-V, with about 2000 rem causing a statistic death). Death rates of rare diseases at small doses are too small to use with scientific credibility, thus deaths are linearly interpolated from the region of 0.1–2 Sv (10–200 rem). The major contributions to background radiation are as follows:

- 3.5 mSv/year (350 mrem/year) US average yearly dose.
- 13 mSv/year (1300 mrem) average dose in Kerala, India.
- 2.0 mSv/year (200 mrem) from radon in buildings.
- 0.31 mSv/year (31 mrem) from cosmic radiation.
- 0.28 mSv/year (28 mrem) from living at sea level.

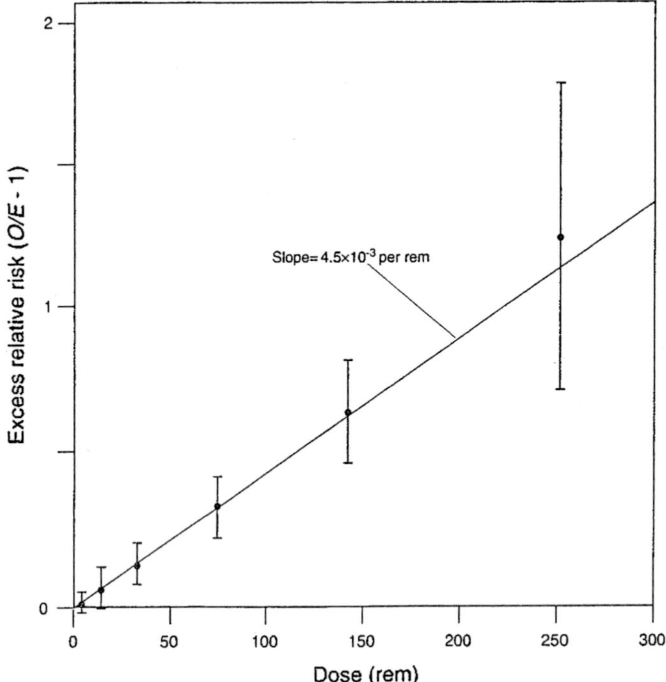

Fig. 3.5 Excess relative risk for solid-tumor mortality versus dose for Japanese atomic bomb survivors. The *error bars* correspond to plus or minus one standard deviation. A *straight-line* fit to the data yields the high-dose rate relative risk factor of 4.5×10^{-1}/Sv (4.5×10^{-3}/rem). The two data points below 20 rem are examined further in Fig. 3.2 (Schillaci 1995)

- 0.81 mSv/year (81 mrem) living in Denver (60 m elevation adds 1 mrem/year).
- 0.39 mSv/year (39 mrem) from natural radioactivity in the body.
- 0.3 mSv (30 mrem) from a mammogram.
- 0.4 mSv (40 mrem) for a full set of dental X-rays.
- 0.1 mSv (10 mrem) from a chest X-ray.
- 0.53 mSv (53 mrem) US average annual yearly exposure from X-rays.
- 14 mSv (1400 mrem) from a gastrointestinal (upper and lower) X-ray.
- 10 mSv (1000 mrem) from a PET scan.
- 2–10 mSv (200–1000 mrem) from a CT scan (higher level for abdomen).
- 2–9 mSv/year (200–900 mrem) for airline flight crews.
- 0.05 mSv (5 mrem) for round trip transcontinental flights.
- 0.001 mSv (0.1 mrem) for an airport full-body scan.
- 56 mSv (5600 mrem) shuttle ride for 18 days.
- 1300 mSv (130,000 mrem) Mars mission.

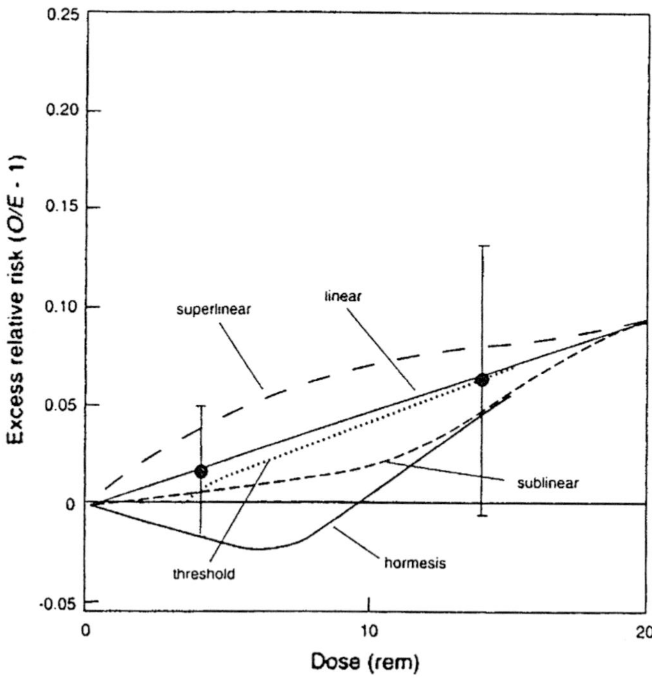

Fig. 3.6 Extrapolation of high-dose Japanese data to doses below 0.2 Sv (20 rem). The low-dose data from Fig. 3.1 are compared to possible curve fits based on superlinear, linear, sublinear, threshold, and hormesis couplings (Schillaci 1995)

Table 3.2 Japanese atomic bomb cancer data

Dose (Sv)	Subjects	Cancer	Expected	Excess
<0.01	42,702	4286	4267	19
0.01–0.1	21,479	2223	2191	32
0.1–0.2	5307	599	574	25
0.2–0.5	5858	759	623	136
0.5–1	2882	418	289	129
1–2	1444	273	140	133
2+	300	55	23	32
Total	79,972	8613	8106	507

Dose to the large intestine (colon) in siverts; comparing subjects, cancers, expected cases, and excess cases of cancer.

3.5 Radiation Standards

If 10,000 people were exposed to an annual low dose of 0.01 Sv (1 rem) over their lifetime, above natural background, there would be a cumulative dose of 100 Sv (10,000 rem). If 20 Sv (2000 rem) caused one statistical death, this cumulative dose would cause 5 deaths. Out of the group of 10,000 about 2000 would normally die from cancer, but not caused by radiation. The radiation dose would raise the death rate from 2000 to 2005 for the group of 10,000.

Occupational exposure. The regulated-limit on annual occupational exposure to radiation dropped from 700 mSv (70 rem) in 1924 to 300 mSv (30 rem) in 1934, to 50 mSv (5 rem) in 1958, and to an integrated dose unit of 10 mSv (1 rem) times a person's age in years in 1993. The annual limit for continual exposure of the public (nonnuclear employees) was set at 5 mSv (500 mrem) in 1960 and was lowered to 1 mSv (100 mrem) in 1990. To set standards *scientifically*, rule-making authorities need to know the low-dose coefficients, threshold values, benefit to society from radiation, value of a lost life, cost to mitigate radiation, and nonradioactive alternatives.

Traveling to Mars in 2030 on NASA's Curiosity Mission would increase radiation dose since Earth partially shields us and Earth's magnetic field deflects low-energy particles. NASA calculates that a passenger on this 6-month voyage would get 0.66 Sv (66 rem). This is 200 times greater the annual natural dose when living on Earth of 0.003 Sv/year (0.3 rem/year). This is three times the dose when living on the Space Station of 0.2 Sv/year (20 rem/year).

Radiation-dose conclusions. The 2006 BEIR-VII report concluded the following on low-dose effects of radiation spread over many persons. Effects from *single incidents of exposure* spread over many people:

- BEIR VII: 480 male or 660 female excess deaths per 10,000 person Sv (1 million person rem), which can be caused by 100,000 persons getting 100 mSv each (10 rem each), or by 10 million persons getting 1 mSv each (100 mrem each). The average of 570 excess deaths is in a background of 20,000 cancer deaths without radiation. The 100 mSv dose raises the cancer *death rate* from 20 to 20.3 %. One statistical death results from 20.8 Sv (2080 rem) for males and 15.2 Sv (1520 rem) for females. The excess cancer rate is double the mortality rate.
- EPA/NRC (2003) used a risk value of 4×10^{-2}/Sv (4×10^{-4}/rem), which inverts to 25 Sv (2500 rem) for a statistical death. This gives a 0.8 % probability of death for an exposure of 0.1 Sv (10 rem).

The 1991 International Commission on Radiological Protection conclusions:

- 500 deaths from 10,000 person Sv (1 million person rem).
- 20 Sv (2000 rem) spread over many people = 1 death.
- US background rate of 3.5 mSv/year gives a dose of 0.28 Sv over 80 years.
- This dose gives a 1 % cancer death rate at 20 Sv/mortality.
- 100 nonfatal cancers per 10,000 person Sv (1 million person rem).
- 130 severe heredity disorders per 10,000 Sv (1 million person rem).

Risk-benefits of CT scans. In 2010 the American Association of Physicists in Medicine called for an open discussion on the radiation hazards from computed tomography (CT) scanning:

Predicting cancer deaths from radiation is not the same as assessing deaths from other causes such as automobile accidents or gun shots, in these latter cases victims can be counted without much ambiguity in the cause of death. Because radiation-induced cancers are exactly the same clinically as normally occurring cancers, there is no way to know who died from a radiation cancer and who died from a naturally occurring cancer. This issue is compounded by the fact that the number of predicted radiation-induced cancers is tiny compared to the very large cancer rate in humans (about 25–30%), making the impact of radiation on cancer rate very hard to measure. Finally, false positives can induce un-needed surgery and other treatment, which is harmful. It is not obvious that mammograms should be continued for all women.

Iodine contamination. Digested radioactive ^{131}I and ^{90}Sr, obtained from a grass-to-cow-to-milk pathway, can be a concern. Iodine collects in the thyroid gland, which in turn enhances iodine concentration by a factor of 7 in adults and a factor of 100 in infants. Most likely, clean milk would be imported to an affected population after an accident for a period of weeks to months. Potassium iodide pills can flood the thyroid gland with iodine, reducing ^{131}I retention. It took time for the Nuclear Regulatory Commission to convince the utilities to distribute KI since the utilities perceived that KI pills on the shelf would be worrisome for residents. The relatively short 8-day half-life of ^{131}I and the avoidance of contaminated milk can greatly remove the threat of iodine contamination. However, KI pills will not protect against bone-seeking ^{90}Sr ($T_{1/2} = 28$ year), nor would KI assuage against the effects of *dirty bombs*, which do not contain ^{131}I.

The 1979 Three Mile Island accident released only 20 curies of the core's 64 million curies of ^{131}I. This is not surprising since containment was not breached, but it has been conjectured that the relatively small release of iodine was a result of iodine bonding to Cs, making nonvolatile CsI. At the other extreme, the 1986 Chernobyl accident released 150 MCi, which was widely dispersed by the burning of the carbon moderator. The estimated number of fatalities from Chernobyl is about 10,000, but this figure is not well documented. Chernobyl's radioactive iodine caused 2000 cases of thyroid cancer in children under age 14, considerably above the normal rate. While thyroid cancer has a high cure rate, the effects of radioactive iodine could have been mitigated if KI pills were readily available.

3.6 Weapon Accidents and Indoor Radon

The threat of plutonium dispersal from a nonnuclear accident with a nuclear warhead is a slight risk, but it has deeper repercussions since it ties into the nuclear test ban debate. First, we discuss the issue of nonnuclear accidents with warheads. A safer weapon design uses *insensitive high explosives* (IHE). It less likely to explode by impact, making it less prone to accidental detonation as compared to *sensitive* high explosives (HE), which has a higher explosive energy density than IHE. For this reason HE is used to implode the size-constrained nuclear weapons on MIRVed submarine-launched ballistic missiles (SLBMs). For the same reason, rocket fuel on SLBMs is also more energetic and more vulnerable. Intercontinental ballistic missiles (ICBMs) are not as volume limited, hence IHE and more resilient rocket fuel are used on ICBMs.

However, this was not always the case. Both ICBMs and bombers were outfitted with more sensitive explosives until the 1960s. The shift to safer warheads with IHE and fire-resistant pits was encouraged because of the Pu dispersal after the 1966 B-52 collision over Palomares, Spain and the 1968 B-52 fire in Thule, Greenland. The Palomares incident resulted in extensive crop damage from the removal of plutonium-contaminated soil to the United States.

The HE on the Trident W-76 and W-88 warheads allowed designers to obtain Trident yields at 0.5 Mton for an 8000-km range. With the cold war over, the Trident SLBMs will be outfitted with only four warheads, instead of the previous eight. The empty space could be filled with larger and safer IHE warheads. But completely replacing W-76 and W-88 warheads is deemed too expensive by DoD, as they continue to support the use of HE on SLBMs. Some opponents of the Comprehensive Test Ban Treaty (CTBT) have called to reestablish nuclear testing to include testing of new IHE warheads for submarines. Analysis of potential deaths from nonnuclear warhead accidents is relevant to the CTBT debate.

Cost of weapons versus value of life. Is the loss of 1000 lives, as projected from a worst-case plutonium dispersal accident, an *acceptable* risk? There are uncertainties in this estimate, particularly in the amount of Pu that attaches to aerosols, but we will assume the figure is correct. A true *risk assessment* should consider alternative possibilities. In this case, the alternative is the additional cost of building *safer* warheads and missiles, as well as the global political cost of renewed nuclear testing. Here we only compare the *value of human life* in contrast to the cost of building new nuclear weapons.

The debate on new IHE warheads arose during debate on nuclear testing prior to consideration of the CTBT. Those opposed to CTBT said that warheads with regular HE were not safe enough and testing of new warheads with IHE was needed. However, the Pentagon maintained that the HE-loaded warheads were safe enough and that it was too expensive to rebuild them. On the other hand, Department of Defense testified in 1992 that a test ban was not a good idea because DoD wished to continue to test weapons for reliability, safety and new designs. Legislation required the government to quantify these trade-offs with a cost versus safety analysis before testing could resume, but this has not been done since testing stopped.

3.6.1 Indoor Radon

An employee at a Pennsylvania nuclear power plant arrived at work in 1984 and surprised himself and others by triggering the plant's radioactivity alarm. It was assumed that he had taken radioactivity home for the night, but it was soon determined that he went home without a trace of radioactivity. Surprisingly, what happened was that he carried *radon daughters* from *home to work*. The radon level in his house was *700 times the EPA-recommended indoor limi*t. Radon exposure is a major issue because the average radon dose of 2 mSv/year (200 mrem/year) is over 50 % of the total background rate of 3.6 mSv/year (360 mrem/year). This is partly due to the fact that people spend 86 % of their time indoors (with the other 6 % in vehicles and 8 % outdoors).

The principal health risk from radon arises not from ^{222}Rn, which, as a rare gas, does not adhere to the lungs, but rather its four radioactive daughters that chemically attach to aerosols and are trapped in the lungs. Radon concentration is increased with increased local radon source strength, with reduced air infiltration through walls/ceilings that traps radon inside, and with increased air coupling between radon ground sources and house interiors. (Increased air infiltration through walls and ceilings reduces radon levels, but it also increases infiltration heat transfer that is about a quarter of cooling/heating energy.) The radon level in your home can be measured with a $20 kit. There are both energy savings and adverse health effects from reduced infiltration. One can both save energy and reduce radon hazards by using air-to-air heat exchangers.

Problems

3.1 **US light water**. Why did the US choose light–water reactors?

3.2 **Canadian heavy water**. Why did Canada choose heavy–water reactors?

3.3 **Proliferation of HWR versus LWR**. Why are heavy water reactors both more and less dangerous for proliferation of nuclear weapons?

3.4 **Breeder commercialization**. What three factors affect commercialization of breeder reactors? What is the plutonium economy?

3.5 **Coal versus nuclear**. What are differences in health, safety and environmental parameters for coal versus nuclear power?

3.6 **Oklo**. How do we know that this reactor existed 1.8 billion years ago?

3.7 **Old uranium**. What was different about the uranium of 1.8 billion years ago? What else was needed for Oklo to happen?

3.8 **Plutonium production**. Describe the cycle to obtain pure Pu.

3.9 **Reactor meltdown**. What creates the heat to melt a reactor in a loss-of-coolant accident?

3.10 **US melt-downs**. Sketch the probability of a big reactor accident from the US record.

3.11 **Fault tree to start a car**. Sketch the diagram to gainfully start your car.

3.12 **rem/year**. How many rem/year do you receive? How many chest X-rays is this at 10 mrem/X-ray?

3.13 **Potassium iodide**. Why put KI in the medicine cabinet?

3.14 **Cost of a life**. How compare added costs to make weapons safer versus cost of a life risked?

3.15 **Spent fuel**. Why is spent fuel placed in a water pond? Could this be a problem?

3.16 **Geological repository**. One design keeps spent fuel close together to keep it hot for 1000 years. Why?

3.17 **Dangerous to transport**? How dangerous is it to ship spent fuel down the highway?

3.18 **Radon and energy**. What is the health versus energy tradeoff for radon in buildings?

3.19 **Radon in basements**. Why do houses without basements have less radon than houses with basements? How mitigate this?

3.20 **Air-to-air heat exchanger**. How do they reduce both radon and energy use?

Seminar: Reactors, Radiation, Radon

AP1000 LWR, back-up power at reactors, BEIR reports, coal versus nuclear power safety and environment record, delayed neutrons and reactor control, CANDU, emergency core cooling system, EPA standard at Yucca Mountain, fault tree analysis (series vs. parallel), Generation III sales, Generation III technology, Generation IV technologies, global status of commercial reprocessing, heavy water production, indoor radon levels by region, liquid metal breeder reactor, LMFBR status in leading countries, mixed oxide fuel (MOX), plutonium economy, potassium iodide pills, Phoenix and Super-Phoenix LMFBR, Price Anderson Law, thorium reactors, Rasmussen Report, spent fuel in ponds versus air-cooled storage, transport of spent fuel, uranium from seawater, WIPP storage, Yucca Mountain.

Bibliography

Ahearn, J. (2006, April). Advanced nuclear reactors: Their use in future energy supply. *Physics and Society, 35*, 2–6.

Ahearne, J., et al. (1997). Nuclear waste issues. *Physics Today, 50*, 22–66.

Alverez, R., et al. (2003). Reducing the hazards from stored spent power reactor fuel in the United States. *Science Global Security, 11*, 1–51.

American Physical Society. (1975, Summer). Light–water reactor safety. *Reviews of Modern Physics 47*(S1), S1–S124.

American Physical Society. (1978, January). Nuclear fuel cycles and waste management. *Reviews of Modern Physics 50*(1), S1–S186.

American Physical Society. (1985, July). Radioactive release from severe accidents at nuclear power plants. *Reviews of Modern Physics 57*(3), S1–S154.

Barkenbus, J., & Forsberg, C. (1995). Internationalizing nuclear safety. *Annual Review of Energy Environment, 20*, 179–212.

Bodansky, D. (1996). *Nuclear energy*. NY: AIP Press.

Boeker, E., & Grondelle, R. (1995). *Environmental physics*. NY: Wiley.

Bupp, I., & Derian, J. (1978). *Light water*. NY: Basic Books.

Craig, P. (1999). High-level nuclear waste: The status of Yucca mountain. *Annual Review of Energy Environment, 24*, 461–486.

Crowley, K., & Ahearne, J. (2002). Managing the legacy of US nuclear-weapons production. *American Scientist, 90*, 514–523.

Fetter, S., & von Hippel, F. (1990). The hazard from plutonium dispersal by nuclear-warhead accidents. *Science Global Security, 2*, 21–41.

Flynn, F., & Slovic, P. (1995). Yucca mountain: Prospects for America's high-level nuclear waste program. *Annual Review of Energy Environment, 20*, 83–118.

Golay, M. (1993). Advanced fission power reactors. *Annual Review of Particle Science, 43*, 297–332.

Hecker, S. (1995). Radiation protection and the human radiation experiments. *Los Alamos Science* 23.

Hollister, C., Anderson, D., & Heath, G. (1981). Sub-seabed disposal of nuclear wastes. *Science, 213*, 1321–1326.

Hollister, C., & Nadis, S. (1998, January). Burial of radioactive waste under the seabed. *Scientific American*, 60–65.

Keeny, S., et al. (1977). *Nuclear power: Issues and choices*. Cambridge, MA: Ballinger.

MacFarlane, A. (2001). Interim storage of spent fuel in the United States. *Annual Review of Energy & Environment, 26*, 201–236.

Marcus, G., & Levin, A. (2002). New designs for the nuclear renaissance. *Physics Today, 55*(4), 54–60.

National Research Council. (2014). *Lessons learned from the Fukushima nuclear accident for improving safety of US nuclear plants*. Washington: National Academy Press.

Nero, A. (1979). *Guidebook to nuclear reactors*. Berkeley, CA: University of California Press.

Nuclear Regulatory Commission. (1975). *Reactor safety study (Rasmussen)*. Washington, DC: Nuclear Regulatory Commission.

Office Technology Assessment. (1991). *Complex cleanup: The environmental legacy of nuclear weapons production*. Washington, DC: OTA.

Peterson, L., & Abrahamson, S. (Eds.). (1998). *Effects of ionizing radiation: Atomic bomb survivors and their children*. Washington, DC: Joseph Henry Press.

Piore, A. (2011, June). Nuclear energy. *Scientific American*, 49–53.

Rasmussen, N. (1981). The application of probabilistic risk assessment techniques to energy technologies. *Annual Review of Energy Environment, 6*, 123–138.

Roberts, R., Shawand, R., & Stahlkopf, K. (1985). Decommissioning of commercial nuclear power plants. *Annual Review of Energy Environment, 10*, 251–284.

Schillaci, M. (1995). Radiation and risk (UNSCEAR94 data). *Los Alamos Science, 23*, 104–105.

Slovic, P. (2000). *The perception of risk*. Sterling, VA: Earthscan Publication.

Wilson, R. (1999). Effects of ionizing radiation at low doses. *American Journal of Physics, 67*, 322–327.

Chapter 4
Missiles and War Games

A nuclear war cannot be won and must never be fought.
[President Ronald Reagan, 7 December 1987]

4.1 Rocket Motion

The German V1 buzz-bomb terrified London in World War II. The V1 was an air-breathing cruise missile, supported by lift forces from the air. The German V2 was the first ballistic missile that flew above the atmosphere without oxygen. The first of 3700 V2 flights took place on 2 October 1942. V2s carried 1000-kg conventional explosives some 300 km, similar to today's Russian Scud B. The accuracy of the V2 was poor, only 35 % landed within 2 km of their targets. At this rate, ICBM accuracy, over its range of almost 10,000 km, would be 60 km. As ICBMs improved, the nuclear arms race shifted from slow, recallable bombers to fast, non-recallable, MIRVed (multiple, independently targetable reentry vehicles) ICBMs. Increased ICBM accuracy led to decreased weapon yields to attack silos, dropping from a megaton to one-third Mton. Huge warheads of 10 Mton (US) and 100 Mton (USSR) were deployed to produce electromagnetic pulses to shut-down military and civilian communication. To enhance leadership attacks, the US developed earth-penetrating warheads.

Rocket motion can be explained in terms of momentum conservation. Put on your roller blades. Have someone hand you a big rock and throw to the forward direction. You quickly notice that you are now going backwards. You obtained this motion by yourself, by exerting a force on the rock that went forwards, but yet you went backwards. Newton's Third Law says that for every force, there is an equal and opposite force. In this case, the rock exerted an equal and opposite force on you, which gave you backwards motion. I consider myself a system made up of the rock and myself. In this case, these forces are *internal* forces because they exist within the system and not from outside the system. I exert a force on the rock, which is cancelled within my system by the force of the rock on me. I gave momentum (mass times velocity) to the rock that is equal and opposite to the momentum that the rock gave to me. Before I threw the rock my system had no

D. Hafemeister, *Nuclear Proliferation and Terrorism in the Post-9/11 World*,
DOI 10.1007/978-3-319-25367-1_4

velocity and no momentum. After I threw the rock, the momentum of the system is still zero as the momentum of the rock and person are equal and opposite.

External forces. Airplanes need the air to fly, sending air backwards with propellers or jet engines to gain forward motion. Airplanes also need air to support the plane using lift force on the wings of the plane. Rockets and missiles do not need air for rocket motion, as they can travel in airless space. If a missile is in air, its motion must be corrected for gravity and air drag and lift *external* forces.[1] Different trajectories need different amounts of energy. A depressed trajectory that travels through the atmosphere needs more energy because of atmospheric drag. The missile launch angle greatly affects the energy needed. If a baseball is thrown vertically or horizontally, it won't move forward to the target. Throwing a baseball at an angle of 45° above the ground gives the maximum range. On the other hand, long-range missiles get their maximum range if launched at 22° above the ground.

Launch velocity. When a missile runs out of fuel it has reached its launch velocity. The final velocity depends on the exhaust velocity of the gas coming out of the rocket. Exhaust velocity depends on the fuel type and the nozzle configuration. Liquid fuels are faster at 3.6 km/s, but solid-fueled rockets, in spite of their smaller 2.7-km/s exhaust-velocity, are preferable for their quick response, longevity, safety and reduced maintenance. The Soviets had a difficult time perfecting solid-fueled rockets, so they continued to use liquid fuels for land-based SS-18s and submarine-launched ballistic missiles (SLBM). Solid-fueled rockets are difficult to control since closing valves does not stop the burning of fuel. Thus, all the solid fuel must be used. Solid-fueled missiles make complicated maneuvers to use the excess fuel, while maintaining excellent accuracy. Another approach is to explosively blow a hole in the missile side-wall to release gas pressure at the proper moment.

Launch velocity depends on the ratio of total initial mass of the entire rocket divided by the final mass of the re-entry vehicles and its bus of warheads. The result from the ratio of initial/final masses is reduced, the launch velocity is proportional to the logarithm of the mass ratio. Thus, the effectiveness of adding extra propellant has limitations. This can be partially overcome by using several rocket stages to reduce the mass of each succeeding stage. The *theoretical launch-weight to throw-weight ratio* for a one-stage strategic rocket with a launch velocity of 7 km/s is 10. In practice this ratio is 20–30 because of the inefficiencies of wasted propulsion, air drag and gravity.

V2. The one-stage V2 delivered its payload to a range of 300 km. The V2 had a launch-weight of 12.8 tonnes (1 tonne = 1000 kg) and carried a 1-tonne warhead as part of its 4-tonne throw-weight. It was fueled with liquid oxygen and ethanol with an exhaust-velocity of 2 km/s. The theoretical terminal velocity was 2.3 km/s, but this is reduced for gravity and air drag.

SS-18. The SS-18 is a huge, two-stage rocket, which delivers 10 warheads 11,000 km with an 8-tonne throw-weight. The SS-18 parameters are as follows:

[1]The drag force is proportional to air density, the velocity squared and the effective cross-sectional area.

Fig. 4.1 Launch of a Minuteman missile (Department of Defense)

First stage mass $m_1 = 146$ tonnes, second stage mass $m_2 = 30$ tonnes, throw-weight $m_{Twt} = 8$ tonnes, exhaust gas velocity $V_{ex} = 3.0$ km/s and the mass of an empty stage is 13 % of its initial mass. Using these values gives a launch velocity of 7 km/s. The two stages contribute equally to the final velocity. The SS-18 launch-weight to throw-weight ratio m_{Lwt}/m_{Twt} is 23. Note that SS-18s with only one stage have a considerably smaller launch velocity of 5.4 km/s (Fig. 4.1).

Shuttle and SpaceX. The shuttle was intended to fly once a week, but it ended up flying once every three months, with 135 trips in 30 years. Elon Musk cut the cost in half to $56 million a launch. He hopes to cut it in half again by reusing boosters and stages, but with some recent accidents. Is there a market for trips at the $100,000 level? Can elegant things be done remotely in space without astronauts, saving money and lives?

4.2 ICBM Accuracy

The reported accuracy of the US Peacekeeper and Trident II is about 100 m at a target 10,000 km away. Much of this error is due to errors in initial parameters of 10 parts per million (10 ppm). The accuracy of a ballistic missile is determined primarily from three errors in the initial conditions of the ballistic trajectory:

- terminal velocity $\Delta v = 0.5 \times 10^{-5}\ v = 0.5 \times 10^{-5}\ (10^4\ \text{m/s}) = 0.05$ m/s,
- range, vertical angular error $10^{-5}\ \theta = (10^{-5})\ (0.5\ \text{rad}) = 0.5 \times 10^{-5}$ rad,
- tracking azimuthal error $\Delta\phi = 10^{-5}$ rad.

Rotating Earth. ICBM inertial guidance systems take into account acceleration, the force of gravity and the moving positions of the launch site and the target site. For instance, during an ICBM flight of time $T = 30$ min, a site at 45° north latitude moves to the east 600 km. An error of 10 parts per million in flight time would increase the aiming error by 5 %. ICBM errors are caused by the following factors:

- initial ballistic velocity and direction,
- timing of trajectory and other close events,
- accelerometer (bias, calibration, misalignment, vibrations),
- gyroscope (initial and acceleration induced drift, vibrations),
- thrust termination,
- energy loss maneuvers,
- gravitational anomalies,
- guidance computation, and
- re-entry buffeting and fusing.

US accuracy improved from 1400 m in 1962 to 90 m in 1988. The Soviets were on average about seven years behind the US trend, improving from 2000 m in 1961 to 230 m in 1986. These errors can be addressed by updating three-dimensional position and velocity vectors during flight via star locations. Since the kill probability (with theoretical reliability of 1) from attacks with hard-target warheads on silos is close to 1.0, there is no great need to increase accuracy for those cases. It would be possible to develop accuracy of 50 m through maneuvering re-entry vehicle (MaRV) technology. The US used MARV on its Pershing II, based in Europe, but this is expensive. Better accuracy could be used to lower weapon yields for attacking fixed points, but such accuracy is already available with cruise missiles.

Nonspherical gravitational bias. US and Soviet/Russian ICBMs are intended to travel near the North Pole. Because the polar radius is 21 km (0.3 %) smaller than its equatorial radius, guidance computers must take into account changes in gravitational forces from the non-spherical Earth. Highly accurate three-dimensional, gravitational potentials were developed for the Earth by observing variations of satellite motion in orbit. When a satellite approaches a concentrated extra mass, the satellite speeds up slightly and it slows after it passes the mass concentration, called a mascon. Corrections for local gravity at launch sites is very important, since slowly rising missiles spend more time near the modified gravitational force. If the oblate-spheroid shape is used to determine gravitational bias error, the difference in range is 15 km. Guidance computers must calculate gravitational bias corrections to better than 1 % accuracy because a 15-km error is 150 times larger than 100-m accuracy. The conventional wisdom is that guidance computers can do this calculation.

GPS accuracy. US Global Positioning Satellites (GPS) and Russian Global Navigation Satellites (GLONAS) accurately determine locations at receivers on and above the Earth. Cruise missiles passively receive GPS signals to determine their locations to a few meters. This can be done accurately using the unclassified channel by comparing the location to a known location nearby. However, if the cruise missile is flying over the ocean or flat plains, this marker is not available.

GPS navigation is not detectable since GPS signals are passively received. This is not the case for cruise missiles that send and receive radar signals to determine location. Since radiated radar from a cruise missile can be detected by the other side, radar-mapping cruise missiles are vulnerable to attack. The absolute location of a cruise missile can be determined to less than a meter by using the unclassified channel and referencing its location to known locations, using *differential GPS*

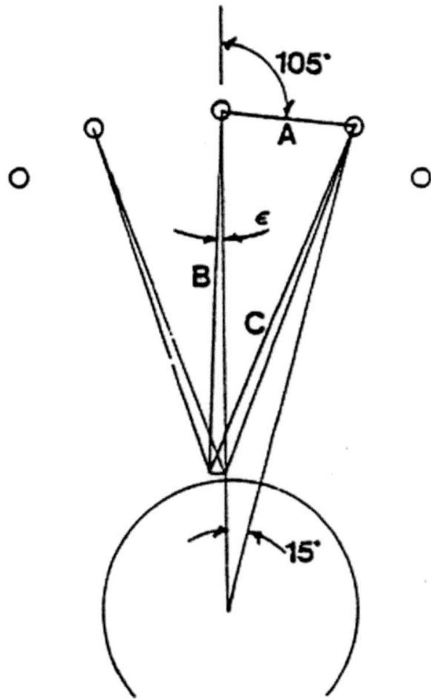

Fig. 4.2 Global positioning satellites. As a missile moves ahead 1 m, side *C* increases by 0.35 m with an angle $\varepsilon = 5.6 \times 10^{-8}$ rad. This increases the transit time by a measurable 2.4 ns (Hafemeister, *AJP* 51, 215 1983)

navigation. Scientists have devised ways to circumvent degraded unclassified GPS signals to achieve results better than expected. GPS is helpful to ground troops, ships, bus drivers, airlines, surveyors, hunters and cruise missiles. GPS is not used, thus far, with strategic missiles. Position is obtained by triangulation from the timed signals from 3 or more of the 24 GPS satellites, spaced at 15° intervals at a distance of 24,000 km from Earth's center (Fig. 4.2).

4.3 Kill Probability

After the two superpowers spent $10 trillion to build 100,000 nuclear weapons and their delivery systems, one might ask the retrospective question, "How much was enough?" The path to an answer begins with determining parameters and performing analysis, and continuing with discussions of political, theological, sociological, psychological and historical implications. The survival of missile silos depends on at least five basic parameters (*CEP*, *H*, *Y, R and n*). These calculations may seem too complicated to be believable, but it is a place to begin the discussion.

1. ***Accuracy of missiles*** is given in the circular error probable (*CEP*) radius of a circle in which 50 % of warheads fall.
2. ***Maximum overpressure*** that a silo survives is called hardness (*H*), which is a pressure (force/area). What actually destroys a silo is the delivered impulse

(pressure × area × Δ time), but this is simplified to a hardness pressure. US silos are hardened to about 2000 psi, while the Soviets built some silos to take larger overpressures. Hardening is increased with better construction, reduced coupling to the Earth with hanging straps to hold ICBMs, and massive springs and shock absorbers.

3. ***Yield*** (*Y*) of weapons, given in kilotons (kton) or megatons (Mton).
4. ***Reliability*** (*R*) of a weapon varies between 0 and 1.
5. ***Number (n) of warheads targeted at a silo***. Each successive warhead gives a smaller additional probability of destroying the target because each previous warhead may have already destroyed the target.

The ***"cookie-cutter" approximation*** assumes that a target is destroyed if overpressure exceeds the hardness of the target but it survives if the overpressure is less than the hardness. Reality expects that the step function, the cookie cutter probability, which is either 0 or 1, should be smoothed to allow some silos to survive at slightly higher pressures and some silos to be destroyed at slightly lower pressures. However, the additional sophistication of a smoothed cookie cutter doesn't change the results significantly because the uncertainties in parameters are larger than analytical gains of adding another parameter. This is often true in public policy discussions, knowing the more important parameters with higher accuracy is more important than including less important theories.

Accuracy versus yield. Assume that a 1-Mton warhead has a single shot skill probability of 90 % against a target (with reliability = 1). How much can yield be reduced if accuracy is improved by a factor of 2, while retaining the same *kill probability*? The answer is that a *CEP* reduced by a factor of 2 allows the yield to be reduced by **a factor of 8**. For our example, this gives a reduced yield of $(1/2)^3$(1 Mton) = 1/8 Mton. Changes in accuracy are more important than changes in yield. US weapon yield was reduced as accuracy improved by a factor of 4 as Minuteman II (370 m) was replaced with Peacekeeper (90 m). The reduction by a factor of 4 in *CEP* implies that yield could be reduced by $4^3 = 64$, but yield was in fact reduced *only* by a factor of 4 from Minuteman-II to Peacekeeper. The difference between ratios of 4 and 64 is that the Peacekeeper was designed for harder silos and in an era when higher kill probabilities were sought.

Senatorial misrepresentation. Soviets always had larger weapons because US accuracy always surpassed Soviet accuracy. Even today, the reported accuracy of the SS-18 (200 m) is 1/2 that of Peacekeeper's 100 m. Senate hearings on SALT made much of the large size of Soviet SS-9s as compared to US Minuteman. Senators misled the public by waving around large models of Soviet missiles and smaller models of Minuteman. They emphasized launch-weight and yield, but they neglected the two most important parameters, accuracy and reliability.

Accuracy versus hardness. As US accuracy increased, the Soviets moved their ICBMs from launch pads to silos with 300 psi hardness, then to silos with 2000 psi hardness and finally to some silos with a hardness of 5000 psi. During this period US accuracy improved from 1300 m in 1962 to 300 m in 1970 to 90 m in 1986. It is generally accepted that accuracy won the race against hardness. Perhaps, super-hardened silos might someday be able to withstand 20,000 psi, but the cost would become very large. In addition, when crater radius is similar to *CEP*, the kill mechanism becomes cratering and not overpressure. US hard-target warheads can produce craters with radii approaching their accuracy.

Relative constancy of hard-target yield. US and Russia maintain warheads of 0.3–0.5 Mton for hard-target weapons. The record was set with the test of the Soviet's 58-Mton weapon in 1962, which was later reported to be but a part of a 100–150 Mton weapon. These parameters are consensus numbers from the International Institute for Strategic Studies.

Reliability. Warhead accuracy requires many tasks be carried out reliably. The total reliability of a ballistic missile is the product of the reliabilities for command-control-communication-intelligence (C^3I) reliability, missile reliability, and warhead reliability. The U.S. Congressional Budget Office quoted a reliability of 85 % for US ICBMs. It is generally believed that warheads have a higher reliability, greater than 95 %. This is higher than the missile reliability that carry the warheads. It is generally believed that Soviet reliability was not as good as that for the US For two hard-target weapons (Peacekeeper, Trident/W88, SS-18, SS-27) attacking a silo, the number of surviving targets is essentially determined by the reliability of the attacking system.

Lethality. A warhead's prowess is often discussed in terms of its main parameters of yield and accuracy by combining it into the lethality (L) parameter ($L = Y^{2/3}/CEP^2$). Note that L is proportional to the ratio of destroyed area (proportional to $Y^{2/3}$) and inversely proportional to missile attacked area (CEP^2). Some debates by prestigious individuals compared total lethality of the two superpowers to determine which side was ahead in the arms race. This method employs poor logic since L doesn't take into account the hardness of targets. Also a missile with a tremendous L value (very accurate with very large yield) could have essentially the same kill probability as a weapon that had only 50 % as much lethality, since the kill probability would be saturated. Lethality is useful as a starting point, but it is only a beginning. It was used unfairly in some debates, claiming very high values had relevance (Table 4.1).

The situation for B is much better than for A. It takes two A-warheads to accomplish what B can do with one. It follows that A improves its one-warhead kill probability more with 10 % improvements than 10 % improvements for B. For A,

Table 4.1 Improvements from enhanced *R*, *Y*, *H* and *CEP*

Attacker	Y (Mt)	H (psi)	CEP (nmi)	R (0–1)	L
A	0.75	2000	0.135	0.85	45
B	0.30	2000	0.05	0.9	252

Attacker	L	P_{k1} (%)	P_{k2} (%)	$\Delta P_{k1}/P_{k1}$ (%):	R (%)	Y or H (%)	CEP (%)
A	45	62	85		10	3.3	9.8
B	252	90	99		10	0.04	0.1

Improvements in one-warhead kill probability, $\Delta P_{\text{kill-1}}/P_{\text{kill-1}}$, from 10 % improved reliability, yield, hardness and accuracy for two situations.

10 % improvements in reliability ($\Delta R/R = 0.1$) and accuracy ($\Delta CEP/CEP = 0.1$) gives 10 % improvements in $P_{\text{kill-1}}$, while a 10 % yield increase ($\Delta Y/Y = 0.1$) raises $P_{\text{kill-1}}$ by only 3.3 %. For B, which has much better accuracy, 10 % improvements increase $P_{\text{kill-1}}$ by 10 % for reliability, 0.04 % for yield and 0.1 % for *CEP*.

Two RV's per Silo from different missiles. Since missile reliability is the most likely reliability failure mode, warheads from different missiles are used to target a silo. A failure of a missile carrying two warheads for one target would cause both the first and second warheads to fail. For the case of a very lethal warhead and $R_{\text{missile}} = 0.8$, 80 % of the silos would be destroyed and 20 % would survive, since second warheads fail with the first failure. If different missiles were used for the two warheads, the kill probability for two warheads would be raised from 80 to 96 %.

2 versus 3 per Silo. Why not a third warhead per silo? This raises kill probability from 96 % for two warheads to 99.92 % for three warheads. This seems very close to total success of 1.0. With two warheads, 40 of 1000 targets survive. With three warheads, only a tenth of one target (statistically) survives but 40 targets are destroyed at an extra cost of 1000 attacking warheads. The argument to attack with 3 warheads becomes more favorable if the targets are MIRVed with 10 warheads. Then an extra 400 warheads are destroyed at an extra cost of 1000 warheads. Where to target the warheads to avoid fratricide (see below) from negating the successive waves of warheads? Are too many targeted warheads a poor trade off, in these macabre terms? Government calculations, using three-warheads targeted per silo by the Soviets, were used in Senate testimony to show vulnerability of US systems, but this scenario was unlikely when one gets into the location of the 3rd warhead.

One high and one low, where goes 3rd warhead? One usually assumes the two warheads are truly independent. But, if both explosions are surface blasts (or both high-altitude air bursts), one warhead might miss the target, but still destroy a brother warhead by fratricide. Kill probability for two warheads can avoid fratricide by using one surface warhead and one high-altitude warhead, but the high altitude blast is less effective on the silo. There are many mechanisms that cause *fratricide*, the killing of one warhead by another:

- blast waves and dust can destroy the second warhead,
- an electromagnetic pulse from the first warhead can destroy the second warhead's electronics, and
- neutrons from the first warhead can damage electronics or pre-initiate the second warhead.

Most of these effects take place in a narrow time window, reducing the problem, but dust from a nearby first surface blast can damage a second surface warhead. The timing separation needed between explosions can be difficult, considering that the two warheads are launched from separate missiles 10,000 km away. A third warhead has to avoid fratricide from two warheads.

Detailed fratricide. We consider three fratricide situations: (1) The first warhead destroys the target with a probability. (2) The first warhead misses the target, but destroys the second warhead with reliability *R*. (3) The first warhead misses the target, but does not destroy the second warhead. For simplicity, we consider *completely*

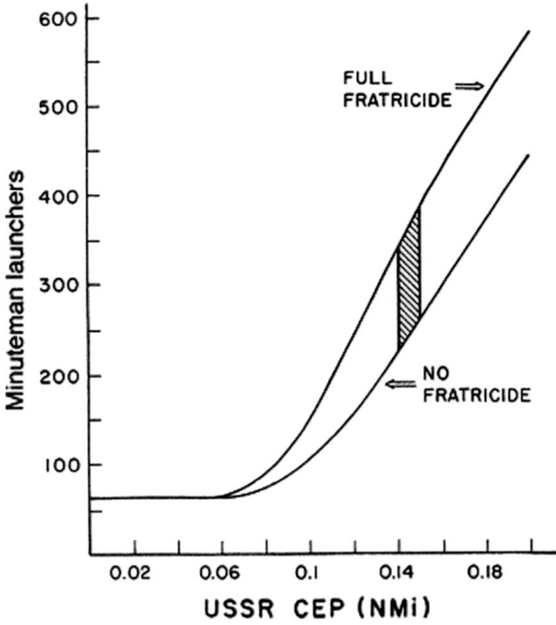

Fig. 4.3 Fratricide. The number of silos that survive calculated as a function of accuracy for two situations: no fratricide and totally effective fratricide when a first warhead misses a target and destroys a second warhead. Totally effective fratricide increases the number of surviving silos by about 10 %. These kinds of discussions should include the survivable submarine fleet (Hafemeister, *AJP* 51, 215 1983)

effective fratricide. For very reliable and lethal weapons ($R = 1$, *Single Shot Kill Probability* = 1), fratricide is irrelevant since it takes only one reliable warhead to destroy a silo. However, if reliability is not 1, but the weapons are very lethal with $SSKP = 1$, then two-shot kill probability with fratricide and without fratricide reduces to the same answer. However, when *SSKP* is less than 1, there is a marked difference. In Fig. 4.3, we plot, as an example, the number of surviving silos as a function of accuracy using the above equations for $P_{\text{kill-2}}$ and $P_{\text{kill-2-fratricide}}$. Accuracy is varied while yield and reliability remain fixed. The curves with and without fratricide coincide for accuracy better than 0.06 nautical mile since *SSKP* approaches 1 at that point, but they separate for larger *CEP*s. The shaded area indicates that, at most, 100 additional silos (10 % of 1000) could survive because of fratricide.

4.4 Launch on Warning

Tight trigger versus loose trigger. The most likely nuclear conflict would probably start as an accident, driven by false information, rather than a sudden "bolt out of the blue" madness. The 1964 movie, *Dr. Strangelove*, set the stage for discussions based on the logic of deterrence, fear, stability issues and morality. Most people believe that the most dangerous moment in the Cold War took place in 1962, during the Cuban Missile Crisis. This was a tense moment, but there have been others. As mere graduate students, Gina and I filled our bathtub with water for future use. That was ridiculously done in Champaign, Illinois, far away from missile fields, Washington and New York. In those days, the recallable bombers dominated US nuclear forces.

Launch-on-warning errors. US and Russia each have had about 1000 long-range strategic, nuclear weapons on alert, as a significant deterrent, letting others know they should not attack us. One problem with this scheme is that this allows the possibility of a miss-informed, accidental, launch-on-warning. Over the past decades there have been at least four cases in which human and technical errors significantly increased the possibility of nuclear war:

- 1979: Missile crews received warnings that a massive nuclear attack from the Soviet Union was under way. Luckily, President Carter decided not to respond with a *launch-on-warning* attack. Later it was discovered that the attack was bogus, the result of a training tape accidentally left in the computer system.
- 1980: A defective computer chip caused a NORAD computer to broadcast warnings of a bogus massive attack by Soviet missiles. Luckily, President Carter ignored the situation.
- 1983: The Russian in-charge, Stanislav Petrov, ignored data from the Soviet early warning satellites. He was not fooled by sunlight reflected from clouds, indicating a US missile attack, but he also was uncertain what it meant. He told his superiors that he did not have the data, while he was staring at seemingly good data. His future in the military suffered because of this. A movie, *The Man Who Saved the World*, was made in 2015 (Fig. 4.4).
- 1995: Russian radars detected a Black–Brant scientific rocket, launched from Norway, which was mistakenly interpreted as a Trident attack. Fortunately, the nuclear response was averted, when Russian President Boris Yeltsin ignored the false alarm.

Fig. 4.4 Stanislav Petrov ignored nuclear attack data that would have called for a launch-on-warning Soviet nuclear strike against the US. The data was later shown to be incorrect, but this heroic act cost Petrov his military career (Atomic Heritage Foundation)

- Requests to Launch. After World War II, the Pentagon has asked for permission to launch nuclear weapons. During the Korean War, Generals MacAuthur and Ridgeway requested 38 nuclear weapons in 1950 and 1951. This was followed by requests during the Cuban crisis in 1962 and during the Viet Nam War (1964). Thus far the US presidents have held firm, denying permission.

4.4.1 Reduced Probability of Launch on Warning

Standard integrated operational plan (1961–2003). The classified SIOP war plan allowed both Launch on Warning (LOW) and Launch under Attack (LUA) scenarios. I hope large fences have been placed around LOW, requiring our leaders to await LUA confirmation of attacks on US territory. Some suggest that remote sound recorders and cameras be located near missile silos, to give accurate information to the leadership of the LUA. Since US submarine nuclear forces are difficult to destroy, there is time remaining to stop and think. As of December 2008, the SIOP is now called the *Operations Plan*. For further details, examine the Congressional mandated *Nuclear Posture Review Report*, which summarizes the current policy and discusses the US relationships with Russia, China, Iran, North Korea and other countries.

Triad to diad choices. If the nuclear triad is vulnerable, there is pressure to launch on warning, to *use it rather than loose it*.

- **Heavy bombers**. The main threat to the safety of nuclear weapons has been caused by bomber crashes. Nuclear weapons were removed from US heavy bombers at the end of the Cold War, but they can be replaced after a delay of a few days. Since the bombers cannot carryout an attack in days, they are not a LOW threat. The three bomber bases could be destroyed with a total of 3–6 nuclear weapons.
- **SLBM**. It is generally agreed that submarine-launched ballistic missiles are essentially invulnerable when on station at sea. SLBM can respond much later, and they are not a LOW threat.
- **ICBM**. ICBM warheads can be attacked, but it is difficult to destroy them all. It is disarming to attack single-warhead ICBM. It takes more than a single, hard-target warhead to be certain of destroying one ICBM warhead. The main LOW target issue is that of the ICBMs, for which there is a concern of use them or lose them. Nothing will protect the cities from a large attack.

Longer alert time. Nuclear systems have hair-triggers, to be able to act quickly. It was human-presidential deciders that prevented the LOW bombs from exploding. It is for this reason that many have called for placing additional strategic weapons on off-alert or delayed-alert status. Nuclear policy is based primarily on *worst-case analysis*, when one side covertly surprises the other side with an all-out attack. The launch on warning scenario is driven by the fear that it is better to use all the weapons on a timely basis, rather than have fewer weapons aimed at empty silos.

The bolt-out-of-the-blue scenario guides policy more than accidents, with too much emphasis on bolt out of the blue. What can be done?

De-alerting. Bombers and submarines are already on de-alert. It was proposed to the Air Force that additional, verifiable physical barriers be inserted to increase the minimum response time of missile silos to a day, or so. It would be best if these actions could be verified to the other side. Time impediments, such as removing ICBM batteries and extra mass on silo doors, were suggested and rejected.

Early warning sensors. A ballistic missile takes 25 min to travel from the Russia to the US. US travel time to Russia is short, some 15 min, because of the closer locations of US submarines. Early warning is provided by satellites and land-based radar.

- ***Satellite detection***. The launch of missiles with their large thermal plumes are detected from space. Burning hydrocarbons produces water and carbon dioxide, which give characteristic optical spectra. By using high quality filters, the visible earth-shine and thermal radiation can be blocked from the sensor, giving good detection of the spectra, night or day. The US shifted from the Defense System Program satellites to the Space Tracking and Surveillance System satellites.
- ***Radar detection***. The US has large-phased-array radars (LPAR) located further from the US than does Russia, giving the US earlier notification.
- ***Improve early warning***? Podvig estimates Russia has 16 min of early warning, while the US has 22 min. At first thought, it would seem that improving early warning systems of both countries would enhance stability. But this is a mixed situation. It is also true that some improvements can convince the owners to have too much faith in their data, and then make a big mistake.
- ***Negotiations***. Russia and the US agreed to de-target their missiles, to avoid accidental launches in 1994. In an emergency, the target list can be changed from target to target, or from no-target to target. Targeting is not verifiable and it is changeable, but de-targeting is generally favorable. There has been discussion of sharing early-warning data, but not much has happened in practice.

4.5 Nuclear Conflict and MAD

Previously we were concerned with attacks on individual targets, such as nuclear silos. The bigger picture is driven by questions of the stability of nuclear forces, and the minimum level of weapons needed to have a nuclear response. The answers to these questions frame the discussion on nuclear arms treaties. Deterrence can be quantified, to some extent, by considering *Mutually Assured Destruction* (*MAD*). The required level of survivable forces for a very damaging second strike is determined by calculation, deterrence considerations, morality, politics and an attempt at common sense. But ultimately it is a political discussion. The range of this discussion is narrower in Washington, DC than in coffee halls and on campuses. When writing sections on national security in the Senate report on the Ratification of START I, I knew the treaty was effectively verifiable. But I felt I had to prove, even *under worst-case analysis*, that sufficient nuclear weapons would survive the worst Soviet attacks reasonably imagined. No one disputed the results. And in fact, the Senate under Senator Jesse Helms followed Senator Pell's staff, using the same *worst-case analysis*, to ratify START II, which never entered into force because of arguments over the Anti-Ballistic Missile (ABM) Treaty.

McNamara's MAD numbers. Many theories were used during the Cold War to determine the number of nuclear weapons needed to deter the Soviet Union. During the Eisenhower administration the doctrine of *massive retaliation* was used, but these goals were not quantified. In the 1960s Secretary of Defense Robert McNamara quantified the discussion with *Mutually Assured Destruction*. MAD has been referred to as placing two scorpions in the same bottle. In the nuclear case, neither side would survive as a modern society. The first criteria of MAD is that a survivable force must be large enough to annihilate 25 % of the other side's population to deter that nation from attacking first. Calculations in 1980 showed it would take 50 one-Mton weapons to destroy 25 % of the Soviet population. A calculation showed that the US could lose 25 % of its population from 130 one-Mton weapons. Soviet population was more concentrated in large cities than the US, making the difference The second criteria of MAD was to determine the force level needed to destroy 50 % of the national industrial base. An attack of 200 one-Mton weapons would destroy 50 % of industrial installations either nation. These calculations were done on an industry-by-industry basis. It would take 10 weapons to destroy 50 % of the aluminum industry, 20 weapons to destroy 50 % of the steel industry, and 50 weapons to destroy 50 % of the petroleum refineries. Since the 1980 US strategic arsenal contained 4000 effective megatons, with only a modest US force needed to survive a first strike to carry out MAD. These numerical discussions continued until the 1991 collapse of the Soviet Union.

Minuteman and SS-18 vulnerability. Concern has been voiced over the years that ground-based ICBMs are vulnerable. In spite of this concern, Peacekeeper (PK) was based in silos that were vulnerable, which is only one of the reasons it was withdrawn. Silo vulnerability has to be discussed within the context of the other two legs of the triad, heavy bombers (B-52 and B-2) and Trident submarines. Silo vulnerability would be less important if MIRVed systems were discarded. With fewer warheads, two-on-one targeting becomes too costly to consider. Mathematics might imply the winner would be the side with the most remaining warheads, but such a deduction ignores the other triad legs and the vulnerability of cities. Minuteman has been configured to carry only one warhead, compared its former level of three. If two warheads attack a silo, each attacking warhead would destroy at most ½ warhead. Such an attack is warhead costly and disarming. An attack of 1000 warheads attacking 500 targets destroys less than 500 warheads. On the other hand, Russia is working to replace the SS-18 with a new liquid-fueled system of 10 warheads. This seems like old Cold War thinking. Because of MIRV, this decision favors a US attack as compared to a Russian attack on our single warhead Minutemen.

MIRV and stability. Peacekeeper was finally based in former Minuteman silos due to a lack of viable options. It was understood that this basing was unstable. It was possible for 10 SS-18s with 100 warheads to attack 50 silos, destroying 90 %

of 500 warheads on 50 ICBMs in a worst-case scenario. The ratio of warheads destroyed/used for the case of the Soviets attacking first would be (US destroyed)/(USSR attack) = 450/100 = 4.5. This situation would be reversed if PK were to attack first, reversing the ratio to 0.22. The ratio of ratios, determined by which side attacked first, is 4.5/0.22 = 20. In spite of this large instability, PK was placed in vulnerable silos, knowing that other US warheads, such as those on SLBMs, would deter attacks. But this approach magnified the potential problem of protecting PK with a launch-on-warning preemption that might be based on incorrect information.

Other basing modes. There are no completely survivable silos. This shortcoming is not too serious for single-warhead ICBMs, since it takes two warheads to reliably destroy a single warhead in a silo, making the silos "sinks" for attacking warheads. On the other hand, SS-18s and PKs have 10 reentry vehicles per missile, which makes them strategically vulnerable. In 1981, the Office of Technology Assessment examined 10 basing modes for PK to overcome this vulnerability, but they failed to recommend any of the following options.

- 200 PK moved into 4600 horizontal/vertical shelters with decoys,
- shelters and silos with ABM defense,
- launch-under-attack, launch-on-warning,
- small submarines or surface ships, each with 2 PK,
- 75 wide-bodied aircraft with dash-on-warning,
- densely-packed silos defended by fratricide, and
- deep underground basing, PK burrows out after the attack.

Nuclear triad. Was the nuclear triad of ICBMs, SLBMs, and heavy bombers vulnerable to attack, requiring the full triad to truly deter?[2] Was the Cold War driven by consideration of "worst-case" analysis and inter-service rivalry for money? The 1992 US General Accounting Office (GAO) report concluded that the Soviet threat to the US triad was overstated, that the performance of existing US systems was understated, and that the performance of new US systems was overstated.

Land-Based ICBM. The GAO concluded the following

> In the case of the land leg, the claimed 'window of vulnerability' caused by improved Soviet missile capability against [US] silo-based ICBMs was overstated on three counts. First, it did not recognize the existence of sea and air leg deterrence—that is, the likelihood that the Soviets would hesitate to launch an all-out attack on the ICBM silos, given their inability to target submerged US submarines or on-alert bombers and their thousands of warheads that could be expected to retaliate. Second, the logic behind the claim assumed the highest estimates for Soviet missile performance dimensions as accuracy, yield and reliability, while at the same time discounting very substantial uncertainties about performance that could not have been resolved short of nuclear conflict. Third, it ignored the ability of US early warning systems to detect a Soviet ICBM attack, and thereby allow a reasonably rapid response. [*The US Nuclear Triad,* US General Accountability Office, 1993]

---

[2]The 2002 Nuclear Posture Review defined the new triad as consisting of nuclear weapons, conventional weapons, and an information-based transformed military.

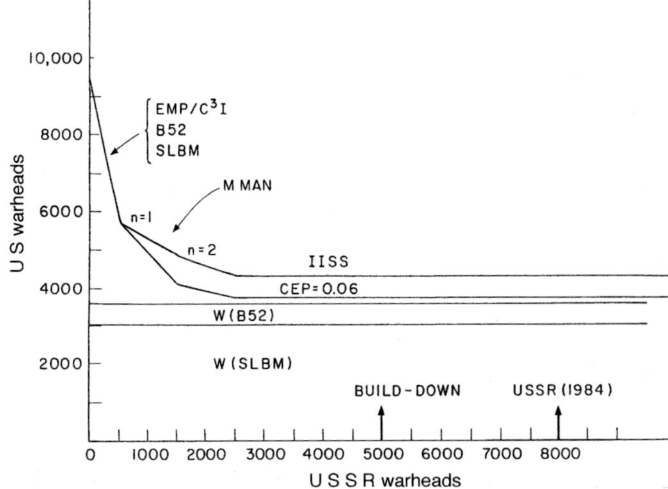

Fig. 4.5 1984 Draw-down curve. The number of US warheads that survive a Soviet first-strike is estimated as a function of the number of attacking Soviet warheads. The *first section of the curve* denotes the reduction in bombers and submarines in port. The *second section* denotes the reduction in silo-based warheads from a single-warhead attack on each silo. The *third section* denotes the reduction in silo-based warheads when two Soviet warheads are targeted on each silo. Note that the marginal return from the use of additional warheads decreases rapidly (Hafemeister, in *Arms Control Verification*, 1986)

The Air Force is experiencing major changes in its transition from the Cold War to the War on Terror. Unattended aerial vehicles (UAV), such as the Predator drone, reduce the need for pilots. The end of the Cold War diminishes the need for heavy bombers. The accuracy of cruise missiles launched from submarines far exceeds the accuracy of bombs from bombers. The high cost of new aircraft has reduced the size of the future air force. The stealthy F-35 Joint Strike Fighter will be able to carry the B-61/12 bomb. The US plans to purchase 2457 F-35s by 2037 (Fig. 4.5).

There are many interesting technologies available for future strategic systems. Most of them have not been used because they are not good enough or they are too expensive. Rather than discuss these options, we list them without discussion:

- *Depressed trajectory re-entry vehicles* to fly under ABM systems.
- *Fast-burn missiles* to rise 80 km in 60 s, complicating the defense.
- *Maneuverable Re-Entry Vehicles* to avoid ABM attacks.
- *Boost glide vehicles*, launched from ICBMs to fly with wings to the target.
- *Earth penetrating warheads* to attack underground targets.
- *ICBM and RV midcourse position updates* from GPS systems.
- *New rocket engines*, using ion drives or Hall-Effect thrusters or using electric and magnetic focusing to avoid thermodynamic efficiency. Exhaust-gas velocity is increased, reducing fuel requirements by a factor of 10.

- *Defense suppression* from radar masking, infrared masking, penetration aids, to attack the defense and super-hard silos.
- *Triad to Diad.* Many think heavy bombers play a small, and expensive role and are too vulnerable and should be shifted to non-nuclear roles. Others think ICBM silos are too vulnerable, encouraging launch on warning. Others think that missiles on the ground provide a vivid red line for others to cross. Others like the recallable, slower threat of bombers.
- *Reduce and Remove Counterforce Roles*. Even though first-strikes seems unlikely, it could happen by mistake. Some say shift to responsive forces only.
- *De-alert Nuclear Forces*. Insert time delays in the process. With a strong SLBM force there is no need to be too prompt.
- *Others say perception is all that matters*. They argue not to soften, as it encourages attacks even when it should not do that.

SLBM. In a similar fashion, GAO concluded that threats to SLBMs had been overstated in "unsubstantiated allegations about likely future breakthroughs in Soviet submarine detection technologies, along with the underestimation of the performance and capabilities of [US] nuclear powered ballistic missile submarines." Threats to SLBMs have been categorized as "non-acoustic anti-submarine warfare," which use radar, laser, or infrared detectors on satellites to search out signatures of nuclear submarines. Two submarine signatures that have been identified are the slightly raised ocean surface above a moving submarine (the Bernoulli hump) and the V-shaped wave above a moving submarine (the Kelvin wave). In principle, these signatures might be observed from submarines near the surface if one were to know where to look, using orbit-based synthetic-aperture radar accompanied by significant computer capabilities, but this would be very difficult. When detecting searching radar, submarines can increase their depth and avoid discovery.

Trident SSBN's are 110 m long and weigh 360 tonnes. The power of the 26-MW motors must be quieted from the noise from vibrations and turbulence. The rotating propeller makes voids on the low-pressure side that grow into bubbles and collapse (cavitation). Special shaping of propellers and cruising at deeper levels constrain cavitation noise. Newer SSBN's use pump jets instead of propellers to create less noise and less cavitation. As of 2015, it is assumed that the US will have 12 Trident submarines, with two under maintenance and five at sea at all times. New subs cost $5 billion each.

Prompt Global Strike. Conventional explosives can be used on distant targets within an hour by 2 of the 24 Trident-D5 SLBMs. But, Prompt Global Strike basing might inadvertently trigger a nuclear war if Russia and others misinterpreted the launch of conventional warheads as beginning a nuclear attack. A National Academy panel encouraged deployment of these systems, but with these comments: Consider other launch methods. Locate these systems away from US nuclear launchers, to prevent the impression that a conventional attack is a nuclear one. Develop notification protocols to avoid misinterpretation by Russia and others. Since conventional warheads have a limited kill radius, Prompt Global Strike weapons must have very good accuracy to be viable. The Air Force is pursuing a prompt response based on Minuteman ICBMs that launch a glider, guided by a laser beam from a satellite, to obtain good accuracy.

ICBM versus SLBM. GAO concluded that the offensive power of the sea leg (SLBM) was essentially equivalent to that of the land leg (ICBM):

> The sea-leg's performance has been understated (or poorly understood) on a number of critical dimensions. Test and operational patrol data show that the speed and reliability of day-to-day communications to submerged, deployed SSBNs [ballistic missile submarines] are far better than widely believed, and about the equal in speed and reliability of communications to ICBM silos. Yet conventional wisdom gives much higher marks to ICBM command and control responsiveness than to that of submarines. In point of fact, SSBNs are in essentially constant communication with national command authorities and, depending on the scenario, SLBMs from submarine platforms would be almost as prompt as ICBMs in hitting enemy targets. Other test data show that the accuracy and reliability of the Navy's D-5 SLBM are about equal to DOD's best estimates for the Peacekeeper. Further, its warhead has a higher yield than the Peacekeeper's. In short we estimate that the D-5 has a hard target kill capability about equal to the Peacekeeper's, while its platforms remain virtually undetectable, unlike easily located silos.

Breakout from strategic treaties. In 1990, the two superpowers declared they each had 12,000 deployed strategic nuclear warheads. The 1991 START-I Treaty lowered this to 6000 warheads each. The US Senate (January 1996) and the Russian Duma (April 2000) ratified START II (3500 warheads, ban on MIRVed ICBMs), but it did not enter into force because President George W. Bush withdrew the US from the ABM Treaty (June 2002). A limit of 2000–2500 warheads was agreed to by Presidents Bill Clinton and Boris Yeltsin in Helsinki on a prospective START III (March 1997), but Congress and Duma blocked progress. On 6 March 2003, the US Senate ratified the Strategic Offensive Reduction Treaty (SORT) with a limit of 1700–2200 "operational" warheads for 2012. This limit is similar to the START III limit since it ignores 240 warheads on two Trident submarines in overhaul. The 2002 Nuclear Posture Review used 2200 operational warheads on Minuteman (450), operational SLBMs (1440), and bombers (300). When added to 1000 warheads on additional bomber positions, plus 3000 more for the hedge, reserves, and tactical missions, the total could rise to about 6000 warheads. Russia protested that they wanted more transparency on warhead reductions because of these large numbers, but in the end they accepted this approach without further verification measures because they could retain 138 SS-18s and a considerable number of SS-19s. The total operational number of warheads was set at 1550 in the New START Treaty.

The main concern on cheating at lower levels of warheads is the possibility that downloaded MIRVed missiles (SS-19, Minuteman) or single reentry vehicle missiles (SS-25/27) could be covertly uploaded with additional warheads. Russia could upload 1500 warheads (100 SS-19 with 5 more warheads = 500, plus 500 SS-25/27 with 2 warheads = 1000), while the United States could upload 3000 warheads (SLBM, 14 subs × 24 SLBM × 3 warheads = 1008, plus 450 Minuteman × 2 warheads = 900, plus more than 1000 on bombers).

The terms of the five START treaties are discussed in Chap. 6. It was expected that President Obama would try to reduce the number of deployed strategic nuclear weapons from the 1550-limit in the New START Treaty to 1000, but President Putin was not interested. Because New START was barely ratified by 71–26, it is unlikely that a *Newer START* treaty could be ratified. There are other options: (1) Executive agreements must be passed by both the Senate and the House with a majority vote (not 2/3) if it is covered by US law. SALT-I was an Executive Agreement. (2) At the end of the Cold War, Presidents Bush-I, Gorbachev and Yeltsin made rapid changes with *reciprocal-unilateral reductions* by stating their plans to reduce and then fulfilling them without monitoring. Congress was not needed for this approach.

Over the last two decades the US decommissioned 10,251 warheads. Between 2011 and 2015, US decommissioned 158 strategic warheads and 88 strategic launchers. A 2012 Department of Defense report, *Modernizing US-Nuclear Strategy, Force Structure and Posture,* concluded that the current U.S. and Russian arsenals "vastly exceed what is needed to satisfy reasonable requirements of deterrence." The Report finds that there is "no conceivable situation" in which nuclear weapons would be used by either side and that "the actual existing threats to our two countries (and the globe) cannot be resolved by using our nuclear arsenals." One of the authors is General James Cartwright (retired), former head of the Strategic Command (2004–2007) and Vice-Chair of the Joint Chiefs of Staff (2007–2011). Cartwright devised a force structure with 900 warheads by 2022, which we describe below. About 50 % of the force would be deployed, with the remainder in reserve. The deployed force would be off-alert, requiring 1–3 days to become launch ready. The reserve forces would take weeks to months to become launch ready. Currently, the ICBM forces are ready to launch within minutes. The JCS in 2013 supported a limit of 1000–1100 warheads, which is 2/3 the 1550 level from the *New Start Treaty*. In 2015, the US and Russia both had about 5000 active warheads:

US: 2080 strategic deployed, 2680 in-storage, 2340 for dismantlement
RF: 1550 strategic deployed, 1500 non-deployed, 2000 tactical

Shortly after these results were published, Secretary of State John Kerry confirmed them by stating that the US had 4717 warheads, close to the 4760 listed above.

- **ICBM: 450** Minuteman in silos with 200 W78 (335 kton) and 250 W87 (300 kton). This will be reduced to 400, while extending its lifetime to 2030. General Cartwright would dismantle the Minuteman-III force, which now carries 450 warheads. Air Force Chief-of-Staff General Norton Schwartz disagrees in that he wants to force an adversary to attack the US homeland, raising the deterrence and risk levels. Sid Drell and Jim Goodby would keep 100 Minutemen missiles, and have fewer warheads on submarines.
- **SLBM**: The Ohio-class Trident II—5D submarine fleet consists of 12 active SSBNs, 8 in the Pacific and 6 in the Atlantic. In 2015 these SSBNs carried 24 SLBMs with 4 warheads each (down from 8), for a total of 1152 warheads ($12 \times 24 \times 4$) on 288 SLBMs. In addition, two SSBNs are non-deployed, in port for maintenance. The operational warheads consist of 768 W-76 (100 kton) and

384 W-88 (455 kton). The SLCM W-80/0 are being dismantled. Pursuant to the New START treaty, the number of launch tubes is being reduced from 24 to 20 on each sub by 2018, for a total of 240 SLBMs. Twelve replacement SSBN(X) are being designed with 16 tubes. The number of annual patrols dropped from 64 in 1999 to 30 in 2014. The Navy has been encouraged to use 20 % enriched reactor fuel rather than the 90 % fuel it uses now. This would reduce the demand for weapon grade materials, as called for in the Fissile Material Cut-off Treaty.

- **Heavy bomber**: The B2 and B52/H bombers are in off-alert status, without carrying the B2 allotment of 16 bombs (B61, B83) and the B52 allotment of 20 ALCMs. The total number of bomber warheads is 1000, including 528 on ALCMs, with some 750 stored at Kirtland AFB, New Mexico. In 2015 there were 16 B2 and 44 B-52/H planned for nuclear reactivation, if needed. The total bomber fleet consists of 18 B2 and 93 B52. These will be replaced by 2025 with 30–50 new bombers. The bombers are very vulnerable to attack. This is somewhat balanced as bombers can be put on petrol to show that the US is very concerned about the issue at hand. The bombers are located at air force bases in Barksdale, LA, Minot, ND and Whiteman, MO. The Air Force plans to acquire 1000 cruise missiles that are capable of nuclear and conventional roles (Fig. 4.6).

CBO cost estimates: The Congressional Budget Office estimated a cost of $348 billion in 2015 for US nuclear forces for the 2015–2024 decade: SLBM ($83 B), Bombers ($40 B), ICBM ($26 B), Tactical and others ($19 B), Command, Control and communications ($52 B), Laboratories and support ($79 B), projected growth in costs ($49 B).

Tactical non-strategic: 180 B61 bombs are deployed in five European countries (Belgium, Germany, Italy, Netherlands and Turkey). The F-35 fighter can carry the B61/12 bomb.

How many nuclear weapons is enough? These estimates are usually based on worst-case-analysis. One usually assumes that the very complicated attacks go as planned. Two-warhead targeting of hardened silos is based on the use of accurate

Fig. 4.6 Nuclear Airplane. In the 1950s the US modified a B36 bomber to carry a 3 MW nuclear reactor. The reactor operated for 36 h, without actually flying the plane (Department of Defense)

timing to avoid fratricide and other issues. The complete systematic attack cannot be tested, thus it would be used without testing. Political debates don't allow arguments that nay-say foreign prowess by saying the other side is not competent. But even with worst-case analysis, it is easily shown that there is sufficient second-strike prowess remaining after a *full-throated* attack by the other side.

Senate ratification: The size of the future US arsenal was an issue for the ratification of the Strategic Arms Reduction Treaties. In 1992, the staff of the Foreign Relations Committee estimated the minimum number of surviving warheads after a brutal, out-of-the-blue attack.[3] In 1992, it was first assumed that U.S. nuclear forces under START-1 forces would deploy the following strategic forces:

- 1400 warheads on ICBMs,
- 3456 warheads on SLBMs and
- 3700 warheads on heavy bombers, for a total of 8500.

The total number of *strategic* warheads of 8500 is gargantuan, much larger than today's number of about 1500. The Senate START-1 report estimated the surviving warheads, based on the warhead number times kill probability for each of the legs of the triad:

$$\text{ICBM } (1400 \times 0.1 - 0.2) + \text{SLBM } (3456 \times 0.65) + \text{Bombers } (3700 \times 0.3) = 3500\,\text{W-head.}$$

Exchange models lay down attacking warheads to maximize destruction of warheads by choosing which warhead types to attack specific targets. This approach was used by Steinbruner, Bing and May to size nuclear forces after START-I was adopted.[4] On 26 June 1992, General Colin Powell, Chairman of the Joint Chiefs of Staff, concluded the following before the Senate Foreign Relations Committee hearings on START:

> We are not dueling with each other, my warhead against your warhead. The question is, does the Untied States force structure give us enough capability to deliver a devastating blow against any nuclear State that may choose to attack us? If it does, then that is a deterrent to that nuclear State ever contemplating such an action.

4.6 Conventional Conflict

In today's world, tank battles between major nations seem less likely as we have nuclear weapons and smart conventional weapons. But tanks were used in the Iraq and Afghanistan wars and the military forces from Islamic State in Iraq and Syria (ISIS) have obtained tanks from their opponents. The US defense budget is about

[3]Senate Foreign Relations Committee (1992)

[4]Steinbruner et al. (1988).

35 % of the global arms budget of $1600 billion (2014). The International Institute for Strategic Studies obtained and modified the 15 largest military budgets for 2014: U.S. ($581B), China ($129B), Saudi Arabia ($81B), Russia ($70B), United Kingdom ($62B), France ($53B), Japan ($48B), India ($45B), Germany ($44B), South Korea ($34B), Brazil ($32B), Italy ($24B), Israel ($23B), Australia ($23B), Iraq ($19B), plus rest of world ($330B).

Lanchester equations. Modeling conventional wars is more difficult than modeling nuclear wars. Local geography varies considerably, supply lines matter more, psychological state of military personnel matters more, and so forth. The Pentagon predictions for the 1991 Gulf War were overly pessimistic. In 1914, Frederich Lanchester developed equations to describe destruction of military forces without considering their force structure and political will to fight. The Lanchester equations assume the x and y forces are diminished at a rate proportional to the opponent's strength, which assumes that all the troops on one side shoot at all the other troops on the other side. Parameters are used to describe initial *force strengths* and *fighting effectiveness*. These are interesting pedagogical calculations but not very accurate.

Richardson model. Lewis Richardson was moved by his World War I experience to examine what happens when a nation acquires weapons in increasing numbers, which encourages their neighbors to do the same. The two world wars started very differently. This was the case before World War I, when European nations increased their armaments in response to each other, and then stumbled into war with the *Guns of August*. A counter example occurred when nations failed to respond to Hitler as he built his war machine. This allowed him to annex his neighbors, resulting in World War II.

There is no simple analysis for all situations, but Richardson showed that military spending by one side begets military spending on the other side in his study of 315 wars from 1810 to 1953. Richardson developed log–log plots of numbers of conflicts versus numbers of fatalities. (World Wars I and II had direct military casualties as high as 3 million per year, but civilian deaths in World War II greatly outstripped this number.) An extension of Richardson's graph gives a probability of wars with over 10 million fatalities at 1 % per year, or one such war in 100 years. World War II was this war in Richardson's model. An all-out nuclear war could have killed hundreds of millions, with a Richardson probability of about 0.1 % per year (one in 1000 years). At the end of the Cold War, most experts would down-rate the probability of nuclear war, but recent events are more pessimistic. However, mathematics can not describe the chances of an accidental nuclear war caused by launch-on-warning errors. Compounding this with the failure to agree on lower numbers of warheads and nuclear testing, de-alerting, additional verification measures, the Iraq and Afghan conflicts, and the ISIS war gives one pause.

Richardson's model gives the action-reaction response of one nation to the threat of increased military spending by its adversary. The two-nation model uses coupled equations to describe armament levels of two nations. The first effect is "threat and response" of a nation to a neighbor's military spending. These terms give arms races that can be exponentially unstable.

Global arms trade. Conventional arms stocks and sales can affect nuclear weapons policy. This was true in the Cold War when the Soviets had a numerical superiority in conventional arms in Europe, even though their overall quality was lower. Global sales of conventional arms for the years 2010 and 2011 were quite different. The US 2010 total for exports was $21.4 billion. In 2011 it rose to $66.3 billion. This anomaly was caused by a very large sale of $33.4 billion to Saudi Arabia for 84 new F-15 fighter aircraft and upgrades to 70 existing Saudi planes.

This sale raised the US total to $56.3B for sales to *developing countries*, raising the US share of *this* market to 79 % as compared to 44 % in 2010. With the demise of the Soviet Union, the US became the clear leader of conventional arms exports. Russia, the 2nd ranking supplier, fell from $8.9 billion in 2010 to $4.8 billion in 2011. Congressional Research Service data for global arms sales in 2011 are as follows; (1) United States, $66.3 billion, (2) Russia, $4.8B, (3) France, $4.4B, (4) China, $2.1B, (5) South Korea, $1.5B, (6) Italy, $1.2B (7) Ukraine, $1.1B, (8) Turkey. $0.8B, (9) Spain, $0.5B, (10) United Kingdom, $0.4B and (11) Israel, $0.4B. US global sales of $66.3 B was four times that of the next ten states at $17.2B.

Carrots and sticks. Conventional arms sales have long been used as sticks and carrots in international affairs. They can be carrots for our allies, as a sign that we want to continue to work with them and make them dependent on our spare parts and help. On the other hand, the threat of a denial of conventional arms sales can be used as a stick to coax them to stay in line with our policies. Conventional arms sales also represent considerable jobs and profits for the seller nations. Conventional arms sales also help balance trade deficits. The issues often get clouded. I witnessed the following in the Senate Foreign Relations Committee hearings:

- Difficult issues can be summarized easily with ratios, such as a 7 to 4 ratio of military sales to Turkey, compared to Greece, or a two to one ratio of gifts of weapons to Israel, compared to Egypt in the 1979 Camp David accords.
- Levels of sophistication vary between neighboring nations. Who will get AWACS reconnaissance planes in the Middle-East?
- Approval of military-equipment transfers to third-party nations is retained by the exporter. Are they fulfilled with adequate inspections?
- Will US tanks be used on both sides in a war between Pakistan and India?
- Will equipment exported to Syrian rebels land in the hands of the Asad's Syrian army or the religious ISIS army?
- Can conventional arms agreements and treaties be monitored to track the number and flow of weapons? The Conventional Forces in Europe Treaty (CFE) considered tagging individual tanks and other limited equipment. This was rejected as too complicated, as fly-over aerial counting was deemed sufficient. CFE was for large forces, what about smaller forces?
- Will other nations produce our weapon designs without our permission?

Problems

4.1 **Rocket motion**. Rockets get motion from the reaction force of pushing gas from the back-end. Use this to explain how you walk, and how you do flip-turns in the swimming pool.

4.2 **No air for missiles**. Why do planes need air, but ballistic missiles don't?

4.3 **Terminal velocity**. Falling bombs have an almost-constant final velocity. What parameters determine this? Why don't re-entry vehicles reach terminal velocity?

4.4 **Stopping thrust**. A liquid-fueled missile ends burning fuel to avoid more velocity by turning off a valve. What is done in the case of solid-fueled missiles?

4.5 **Optimal launch angle**. Without air drag, a baseball thrown at 45° above the ground obtains maximum range. Qualitatively, why are long-range missiles launched at half-this angle for maximum range?

4.6 **Best accuracy**. How do missiles obtain the best accuracy?

4.7 **GPS relative accuracy**. How can GPS signals be used to give cruise missiles very good accuracy without using the best GPS channel?

4.8 **Five kill probability parameters**. What are these parameters, how have they improved?

4.9 **Fratricide**. What are fratricide mechanisms? What are three possible outcomes?

4.10 **Improved accuracy**. If CEP is lowered by a factor of three, how much can yield be reduced for the same effect?

4.11 **Diminishing effect of lethality**. Which is better to attack hard-targets, two 0.5 Mton warheads or one 1 Mton warhead? Why is just keeping track of yield nonsensical?

4.12 **Minuteman vulnerability**. What limits vulnerability claims on Minuteman ICBMs?

4.13 **Earth penetrating warheads**. What was their purpose in the Cold War and after 9/11? What limits these weapons?

4.14 **Breakout from START/SORT**. Design a force structure that is insensitive to attack by many covert nuclear warheads. Compare to the present US and Russian forces.

4.15 **Nuclear posture review**. What are current defense plans?

4.16 **Lanchester conventional forces**. General Lee was disappointed when his extra forces did not arrive at Gettysburg. Explain his concerns in terms of Lanchester's results.

4.17 **Richardson's conventional forces**. Why was Richardson correct in his analysis of the causes of World War I and incorrect for World War II? How did these wars differ?

4.18 **Conventional Forces in Europe treaty**. What did the CFE Treaty do? Who cut more, NATO or Warsaw Pact? What is CFE status today; what are the pressures on CFE?

4.19 **ISIS**. Where did ISIS get troops and equipment?
4.20 **Numbers**. What are the force numbers for the *P5* military forces?

Seminar: Missiles and War Games
Chinese submarine progress, cookie cutter approximation, Earth penetrating warheads, enduring stockpile, fratricide, intelligence community, German scientists in the US, GPS guidance, LOW versus LUA, MARV versus MIRV, mascons, Minuteman vulnerability, National Ignition Facility, Nuclear posture review (2002, 2009), Peacekeeper basing, reliable replacement warhead, Prompt Global Strike, Russian submarines progress, Single Integrated Operating Plan (SIOP), stealth airplanes, triad analysis, US considered using nuclear weapons, World War I dreadnaught numbers, WW-I versus WW-II causes.

Bibliography

Blair, B. (1991). *Strategic command and control*. Washington, DC: Brookings.
Blair, B., et al. (2011). One-hundred nuclear wars. *Science Global Security, 19*(3), 167–194.
Carter, A., Steinbruner, J., & Zraket, C. (1987). *Managing nuclear operations*. Washington, DC: Brookings.
Collina, T. (2012, June). Former STRATCOM head call for cuts. *Arms Control Today*, 27–28.
Congressional Commission on the Strategic Posture of US. (2009). Washington, DC: US Institute of Peace Press.
Department of Defense. (2015). *Nuclear posture review report*. Washington, DC: US Congress.
Drell, S., & Goodby, J. (2012, June). Nuclear deterrence in a changed world. *Arms Control Today*, 8–13.
Feiveson, H. (Ed.). (1999). *The nuclear turning point*. Washington, DC: Brookings.
Hafemeister, D. (1997). Reflections on the GAO report on the nuclear triad. *Science and Global Security, 6*(3), 383–393.
Hafemeister, D. (2014). *Physics of societal issues*. NY: Springer.
Herring, T. (1996). The global positioning system. *Scientific American, 274*(2), 44–50.
Hobson, A. (1989). ICBM vulnerability: Calculations, predictions and error bars. *American Journal of Physics, 56*, 829–836.
Hobson, A. (1991). The ICBM basing question. *Science Global Security, 2*, 153–198.
Hoffman, D. (2009). *The dead hand*. NY: Doubleday.
International Institute of Strategic Studies. (2015). *The military balance*. Oxford, UK: Oxford University Press.
Kristensen, H., & Norris, R. S. (2015). US nuclear forces, 2015. *Bulletin of Atomic Scientists, 69*, 107–119.
Levi, B., Sakitt, M., & Hobson, A. (1989). *The future of the land-based missile*. NY: AIP Press.
Madow, R. (2012). *Drift: The unmooring of America's military power*. NY: Random House.
May, M., Bing, G., & Steinbruner, J. (1988). Strategic arms after START. *International Sec, 13*, 90–113.
National Academy of Sciences. (1997). *The future of US nuclear weapons policy*. Washington, DC: NAS Press.
Office Technology Assessment. (1981). *MX missile basing*. Washington, DC: OTA.
Padvig, P. (2006). Reducing the risk of an accidental launch. *Science Global Security, 14*, 75–115.

Schields, J., & Potter, W. (Eds.). (1997). *Dismantling the cold war*. Cambridge, MA: MIT Press.

Schultz, G., Andreasen, S., Drell, S., & Goodby, J. (2008). *Reykjavik revisited*. Stanford, CA: Hoover.

Schwartz, S. (1998). *Atomic audit*. Washington, DC: Brookings.

Senate Foreign Relations Committee. (1992). The START treaty. *Executive Report*, *102–5*, 52.

Snyder, R. (1987). Approximations for the range of ballistic missiles. *American Journal of Physics, 55*, 432–437.

Steinbruner, J., Bing, G., & May, M. (1988). Strategic arsenals after START-I. *International Security, 13*, 90–133.

Stockholm International Peace Research Institute. (2015). *SIPRI yearbook*. Oxford, UK: Oxford University Press.

Tannenwald, N. (2007). *Nuclear taboo: US and non-use of nuclear weapons since 1945*. Cambridge, UK: Cambridge University Press.

Woolf, A. (2012, January). Modernizing the triad on a tight budget. *Arms Control Today*, 8–13.

Chapter 5
Ballistic Missile Defense

... to make nuclear weapons impotent and obsolete.
[President Ronald Reagan, March 23, 1983]

"Then we can gang up twenty of them, and we can send out 100-mJ pulse, 20 times per second, and hit any target we want. The impact energy then will be on the order of 20–30 kg of explosives."
"And that will kill any missile anybody can make."
"Yes sir." Major Gregory smiled.
"What you're telling me is, the thing—tea clipper—works."
"We validated the system architecture," the general corrected Ryan. "It's been a long haul since we started looking at this system. Five years ago there were eleven hurdles. There are three technical hurdles left. Five years from now there won't be any. Then we can start building it."
"The strategic implications..." Ryan said, and stopped.
"Jesus."
"It's going to change the world," the General agreed.
[Tom Clancy, The Cardinal of the Kremlin, 1989]

5.1 ABM History

Every military weapon, it seems, has an anti-weapon defense. Swords versus shields; tanks versus tank traps; poison gas versus gas masks; and so on. Anti-ballistic missile (ABM) defensive weapons are important for several reasons. The connection between offensive missiles and defensive technologies played a role in the nuclear-weapons debate with the former Soviet Union. As long as the US has some ABM prowess, Russia has been less cooperative on strategic arms treaties. The US spends considerable funds, some $10 billion per year on ABM research and deployment. American ABM systems might be launched to defend against North Korea, Iran, China, Russia or other nations. It is a hope that anti-ballistic missiles could destroy accidental launches of a few ballistic missiles. These are serious risks for the proliferated 21st century. But at the same time, we should understand what

D. Hafemeister, *Nuclear Proliferation and Terrorism in the Post-9/11 World*,
DOI 10.1007/978-3-319-25367-1_5

ABM systems are and their short-comings, and how countermeasures can nullify them.

ABM. The US has spent over $100 billion on ABM research, development and deployment since President Reagan's 1983 speech. Defenses have gone from ABM, to SDI (Strategic Defense Initiative), to BMD (ballistic missile defense), to NMD (national missile defense) and to theater missile defense. The Soviets started much of this with their deployment of the Galosh ABM system around Moscow in 1962. Galosh used 2-Mton nuclear warheads with a kill radius of 300 km at four sites, located 135 miles north-west of Moscow. The Soviets started building the accompanying radar system in 1964, which began service in 1967. President Johnson knew these facts when he met in Glassboro, NJ with Soviet Premier Kosygin in 1967. In 1968, the US considered an attack on 17 Galosh facilities with 100 Minuteman missiles and 6 Polaris SLBMs. Sixteen 1-Mt warheads were aimed at each of the four interceptor sites. The Galosh defense was intended to protect Moscow, but it made Moscow much more vulnerable because of the large US response. This has always been true, if the defense looks promising, the offense will build more offense, which is usually cheaper to do.

The Johnson administration proposed building the *Sentinel* defense to protect US cities, but it is difficult to protect soft buildings, extended over large urban areas. If cities could be completely defended, a nation might (unwisely) attack first without fear of retaliation. It is the consensus view that ABM systems cannot defend against a small or modest first strike. If the attacking nation had a robust ABM system, it might convince itself it was able to defeat a weakened, responsive, second strike. This is the famous ABM strategic instability. Defensive weapons could make offensive weapons more powerful. It is doubtful that any ABM would ever be strong enough to contain Russia, or visa versa. Deployment of ABM systems can be counterproductive since the existence of Galosh caused the United States to increase targeting of Moscow. For these reasons, Johnson proposed the ABM and SALT treaties to restrain both the defense and the offense. Constraints on defense systems were originally rejected by Soviet leader Alexi Kosygin in 1967, he stated that defensive weapons were "moral." Upon further thought he agreed with Johnson to complete the ABM Treaty.

The Nixon administration negotiated the 1972 ABM Treaty, limiting each side to 100 ABM launchers at two sites (later one site) and banning nationwide defenses. This paved the way for the deployment of the *Safeguard* system to defend US ICBM silos from Soviet attacks. Safeguard gave up the idea of defending US cities. Safeguard used megaton-sized Spartan missiles to attack reentry vehicles (RVs) in their midcourse phase, 1000 km above the Earth. If Spartan failed, then kiloton-sized Sprint missiles would attack RVs in their reentry phase in the atmosphere. The exoatmospheric Spartan was similar to the Russian ABM-1B system and the high acceleration endoatmospheric Sprint was similar to the Russian Gazelle system. Safeguard was deployed at a cost of $7 billion (in 1975 dollars), but it was decommissioned after 6 months in 1976 because it was ineffective against countermeasures. Safeguard was also vulnerable to losing its sensors, its eyes and ears, in an attack against its radars.

SDI. On 23 March 1983, President Ronald Reagan made his *Star Wars* speech, stating that the United States needed

> a comprehensive and intensive effort to define a long-term research and development program… on measures that are defensive… [to destroy Soviet missiles] before they reach [US] soil or that of our allies.

The new concept of SDI was to attack ballistic missiles in their *boost phase*, before they released their reentry vehicles. *Boost phase attacks* on SS-18s could reduce the number of targets by a factor of 10 since ten RVs remained on the boost-phase SS-18 bus. A boost phase attack is potentially much better than an attack on the midcourse and reentry phases since the 10 RVs would have become separate targets. Such a strategy *requires space-based weapons* to be close to the SS-18 launch positions, some 10,000 km away. Even if X-ray weapons had a gigantic range, the Earth's matter is between US and USSR to absorb X-rays. The fate of the SS-18 would be more viable if SS-18s had countermeasures to protect it.

Directed Energy Weapons. The initial SDI program was based on directed energy weapon systems (DEWS) with space-based lasers, particle beam weapons, rail guns, and other approaches. SDI was shifted in 1987 from beam weapons to kinetic kill vehicle (KKV) weapons, which would collide with RVs. One version, "brilliant pebbles," was to remain in orbit for over a decade, always ready to accelerate toward an incoming RV.

SDI and Gorbachev. Why did Gorbachev seem to respect SDI when he had advice from Roald Sagdeev (Director of the Soviet Space Institute) and Evgeny Velikhov (Vice President, Soviet Academy of Sciences and Deputy Director, Kurchatov Institute) that SDI beam weapons were over-promised. Some referred to SDI as modern-day Lysenkoism, created by a Soviet biologist who used his position to attack Darwinism for political power. Sagdeev summarized the early Soviet approach: "If Americans oversold SDI, we Russians overbought it."

Most scientists and engineers who examined SDI technologies concluded that SDI was several orders of magnitude away from being a viable system. A simple proof of this is to ask a question. Can a successful US X-ray laser, on a pop-up missile on a submarine in southern Arctic Ocean, have sufficient time to destroy missiles launched 3000 km south in Kazakhstan? The answer is that there is not sufficient time to destroy the boost phase of these ICBMs. In addition, Soviet fast-burn boosters could circumvent the X-ray laser by unloading its RVs quickly enough to survive the X-ray laser.

Beyond these complications, SDI is further compromised by the difficulty of managing a competent battle management structure to direct thousands of DEWSs against a massive nuclear attack. SDI management could only be partially tested because it is too expensive to test the whole system against actual targets. This pessimism is further compounded by the relative ease of countermeasures to foil SDI. The best technical analysis of SDI came from the *American Physical Society* (*Reviews of Modern Physics*, July, 1987). The APS DEWS panel of 17 scientists had a strong contingency of insiders, including four from government weapon labs and three from industrial laboratories. These insiders were well versed on government SDI research. In 1987, SDI changed direction from directed energy weapons to kinetic-kill vehicles. Why didn't the DoD Defense Science Board point out SDI's deficiencies in 1983, rather than launching such a major effort?

SDI shortens or extends Cold War? Scientists debate *facts*, while politicians often debate *perceptions*. Why did Gorbachev try so hard to kill SDI? Roald Sagdeev comments (quotes from Nigel Hey's *Star Wars Enigma*):

> I think the moment Gorbachev understood that SDI wouldn't work, he decided that those who were trying to push a futile system must have some kind of hidden agenda. Even some military spokesmen thought… [the US was] planning to use the cover of SDI to deliver nuclear weapons from orbit.

Much has been made of the possible agreement between Reagan and Gorbachev to ban *all* nuclear weapons at Reykjavik in October 1986. (Would the Pentagon and Congress have concurred?) The ban on nuclear weapons failed because Reagan did not agree to Gorbachev's request to ban SDI tests in space. Both Reagan and Gorbachev were saddened by their failure to conclude a weapons ban, but they soon agreed on many other matters.

Gorbachev responded: "I cannot agree that the SDI initiative had this much importance." Velikhov comments: "The idea that it accelerated the collapse of the Soviet Union is nonsense." Sagdeev replies: "The reasons were completely internal… Many in the West were persuaded that SDI intimidated the Soviets so much that they decided to dismantle the communist system." Edward Teller replied that SDI did not shorten the Cold War with further comments on the Soviet collapse: "The obvious reasons for the failure… were, first, misgovernment, and second, failure to acquire military superiority beyond Eastern Europe." The impact of SDI on Gorbachev was much less than that of economics and nuclear instability. SDI could have shortened or extended the Cold War by, at most, 6 months.[1] Six months after Reykjavik, the U.S. abandoned beam weapons in favor of kinetic-kill vehicles. At that point SDI became irrelevant and did not prevent the Soviets from accepting the INF, TTTBT, CFE, START and CTBT treaties. One could argue that SDI slowed the end of the Cold War, preventing these agreements at Reykjavik. One could also argue that the extra push from SDI, on top of his other problems, made Gorbachev more compliant, not out of fear but out of system-overload. But, either way, the effect was not large.

Science Court on SDI. Arthur Kantrowitz invented the *Science Court,* with quasi-judicial standards for due process and entering data into evidence. An application of this was the *Science Advisory Procedure*, held at Dartmouth College on 23 May 1985, which was a giant step forward to learn scientific truths. The *conclusions were advisory* since there was no legal authority and there was no up or down vote (see Masters and Kantrowitz). Richard Garwin and Ed Gerry came from opposing sides on ABM defense issues, but yet they came up with an impressive list of agreed statements. For example, Garwin and Gerry agreed to these

[1]D. Hafemeister, "Review of *Star Wars Enigma*" by N. Hey, *J. Cold War Studies* 10(2), 141–143 (2008).

statements: “No viable defensive system can allow space mines to be placed within lethal range of space assets… The utility of pop-up for boost phase intercept can be negated by fast-burn boosters.” It is very useful to remove issues by agreement, so time can be spent on areas of disagreement.

BMD/NMD. Patriot missiles were ineffective against Iraq’s short-range Scuds in the first Gulf War of 1990, partly because the Patriots were originally designed to attack planes and not missiles. However, their use served a timely political purpose by calming Israel not to respond to Iraq’s missile attacks. In 1993, the Clinton administration changed the SDI program to the BMD program, which was intended to defend against theater missiles with ranges up to 1000 km, and not strategic missiles. One BMD weapon is the Theater High Altitude Air Defense (THAAD) missile. Russia and China have been concerned that US defenses might ultimately protect the entire United States, undercutting their “second strike” deterrent. The coupling between defense and offense is scenario dependent, making mixed offense-defense agreements difficult to negotiate.

Abrogation of ABM Treaty. Beginning in 2001, President George W. Bush gave strong support for protection of the entire United States through a National Missile Defense with increased budgets. In his role as president he had the power to withdraw the United States from the ABM Treaty, which he formally did in June 2002. It is ironic that senators are needed to ratify treaties, but presidents can abrogate them without Congressional approval. The Defense Science Board considered a return to the nuclear-armed ABMs of the 1970s, but this was rejected.

The Bush II administration abrogated the ABM Treaty to perform *tests* that would violate the Treaty because *deployment* of ground-based interceptors (GBI) for NMD violates Article I, “Each party undertakes not to deploy ABM systems for a defense of the territory of its country.” To attack the midcourse phase the Bush administration proposed to base GBIs on land and on sea. To attack the boost phase, they proposed space-based lasers, space-based kinetic kill vehicles (KKV), and the airborne laser (ABL). The 2003 American Physical Society report goes into detail on this topic. The cap-stone deployment by Bush-II was to deploy ground-based KKV interceptors at Fort Greely, Alaska and Vandenberg Air Force Base, California to attack mid-course phase missiles from North Korea. By 2010 there were 30 GBIs available to defend against a hypothetical missile attack by a dozen missiles from Asia. As of 2015, it is planned to have 40 GBIs at Ft. Greely, a half-dozen at Vandenberg and some on the US East Coast. Ultimately there might be 100 GBIs in Alaska and California, with another 125 in North Dakota. The Government Accountability Office was not optimistic, stating in April 2014 that GBI was “capable of intercepting a simple threat in a limited way.”

European-based defenses. The second part of the Bush-II plan was to base ten GBI’s in Poland, with radars in the Czech Republic to defend again missiles from Iran. The Obama administration shifted the Poland-Czech deployment to base GBIs on destroyers, closer to the threatening nations. This is more effective, but more expensive. It shifts the mission to attack the boost phase and the mid-course phase, rather than just the mid-course phase. The 2003 APS report analyzed sea-based interceptors attacking the boost phase. It concluded a better approach to use "existing US Navy Aegis Missiles to engage short- or medium-range ballistic missiles launched from sea platforms without significant modification, provided that the Aegis ship is within a few tens of kilometers of the launch platform." In other words, the Aegis *must be deployed* near known launch platforms, but its reach would be less effective against sites further inland. Some of these systems were purchased on the basis of "capability-based acquisition," avoiding expensive testing.

The 2003 APS report analyzed the amount of mass needed in orbit for *space-based* kinetic-kill vehicles: "we find that a thousand or more inceptors would be needed for a system having the lowest possible mass and providing a realistic decision time. Even so, the total mass that would have to be orbited would require at least five- to ten-fold increase over current US space-launch rates, making such a system impractical."

Target discrimination. Discrimination between targets and decoys is a key issue. Defenses can be spoofed in several ways: infrared (IR) detectors can be avoided if attacking RVs are cooled; RVs can be placed in large aluminized balloons to obscure their location from radar and IR detectors; RV-shaped balloons can be used as decoys; and radar can be confused by releasing small wires, called chaff. In addition, many small bomblets of anthrax could be released in the boost phase, overwhelming the defense. It is conceivable that missiles could be attacked in the boost phase before they released the bomblets, but the time window for realizing this is very narrow. All offense and defense systems have their vulnerabilities.

US withdrawal from the ABM treaty and US BMD deployment in Europe concern Russia. These BMD systems don't threaten the Russian ICBMs, but there is a Russian concern that BMD could be upgraded to offensive weapons aimed at Russia. Many believe that a cooperative defense with Russia, of some sort, is needed for progress in strategic arms control.[2] In 2013 it seems that some kind of accommodation is needed on BMD with Russia in order to obtain deeper strategic cuts. In 2013, North Korea threatened Japan and the US with nuclear weapons. The US responded by adding 14 additional ground-based interceptors in Alaska. To pay for this and get political accommodation with Russia, the US cancelled phase IV of its European BMD system, which was to develop boosters with 5 km/s to attack ICBMs, a plan that disturbed Russia.

ABM technologies. This chapter examines many types of lasers (chemical, excimer, free-electron, ganged fiber or slab solids, and nuclear-pumped), a variety of basings (space, air, land, and sea), neutral particle beams (NPBs), electromagnetic rail guns, and KKVs. Countermeasures are described that can overcome these defenses. The American Physical Societies studies on DEWSs (1987) and on boost-phase interception (2003) are good references for those who want details. A recent concern is the internationalization of the ABM issue. To increase technical cooperation on missile defense with other countries, there have been calls to

---

[2]Wilkening (2012).

weaken the Missile Technology Control Regime. The MTCR regime is intended to slow proliferation by preventing missile technology transfer to nations that might threaten other nations.

5.2 Nuclear ABMs

Exoatmospheric ABM. The offense can spoof the defense with decoys or with cooled RVs to reduce infrared signatures. The offense can also spoof with RVs in large balloons or by releasing small bio-weapon bomblets in the boost phase. Because of this weakness, there was a call for the Bush-II administration to implement nuclear weapons on ABMs to attack the midcourse phase above the atmosphere. Such a weapon would circumvent some countermeasures by attacking all weapons within its kill radius. This nuclear ABM would be similar to Spartan, which was deployed in Grand Forks, North Dakota, for 6 months in 1975. At altitudes of hundreds of kilometers, the lack of air raises the fraction of X-ray energy to 75 % of the yield. The fluence delivered from a 5-Mton, W-71 warhead, 1 km from the target above the atmosphere, is enough to destroy a hardened RV. The kill range is increased somewhat because X-rays also deliver an ablative shock impulse.

Endoatmospheric ABM. Low-altitude interceptors equipped with nuclear weapons were devised to attack RVs in their reentry phase. The 1970s Sprint interceptors had an acceleration of over 100 g's, rising to 10 km in seconds. Because some atmosphere exists at this height, this location is called *endoatmospheric*. The incoming system is attacked with lethal neutron fluence at 1 km. At sea level, the neutron fluence is reduced by a factor of 100, but at 15-km altitude it is reduced by only a factor of 2. Fast neutrons damage semiconductor circuits, fissile materials and high explosives. Neutrons also can pre-initiate the nuclear detonation, lowering yield by direct absorption at a critical moment.

5.3 Particle Beam Weapons

Neutral particle beam weapons might defend a target in near *real time*, since beams travel at almost the speed of light. The Earth's magnetic field deflects charged-particle beams, but neutral-charged beams have straight-line trajectories. Neutral beams are formed by passing an energetic hydrogen ion beam (H^-) through a stripper gas to remove one of its two electrons, forming neutral hydrogen. NPBs would attack ICBMs and RVs above the atmosphere to avoid beam degradation by the air. SDI proposed to base NPB accelerators in orbit, ready to attack boost-phase missiles before their RVs are released.

A 1-m diameter beam fixed on an ICBM at a distance of 1000 km must be tracked to small angles of 0.0001 of a degree. The NPB would destroy a missile by heating its explosives to ignite at 500 °C. The particle beam also raises aluminum-body temperatures, causing internal stress and misalignment. Small NPB doses can confuse semiconductor logic, while larger doses destroys these components. To reduce beam current, it could be trained on ICBMs for a longer period, perhaps 10 s, but this is difficult. Doing this requires continuous tracking, since a 1-m diameter RV moves out of a 1-m beam diameter in 0.0001 s.

Building, deploying, and maintaining accurate and reliable neutral particle beam weapons would be a formidable task. The American Physical Society study concluded that high electrical power of a GW_e peak power would be needed for NPB weapons and electromagnetic rail guns. Nuclear power is the only known way to obtain these power levels in orbit. For this reason the Strategic Defense Initiative Organization (SDIO) planned the 100-kW_e SP-100 space reactor with a thermal power of 2.5-MW_t. The APS panel concluded that a hundred or more space-based nuclear power plants would be needed for SDI.

Burning a hole in the atmosphere. If space-basing fails, ground-based particle beams might be considered. The energy to make a 10-km path through the atmosphere is 10 % of the initial energy of the beam. Stability issues and degraded angular resolution are further issues. In addition, there are beam losses in the beam hole. Temperature must be raised from 300 to 3000 K to reduce air density by a factor of 10.

NPB roles. It is unlikely that NPBs would be effective enough to attack ICBMs, but they have been considered for other roles. NPBs in orbit could destroy satellites more easily than attacking ICBMs since satellites are soft targets in predictable orbits. However, there are easier ways to attack satellites. NPBs might be used to determine if midcourse-phase objects are decoys or warheads. A warhead could be distinguished from a decoy after being hit with energetic protons, releasing characteristic gamma rays. SDIO planned to use this information to enhance defensive targeting. The most sophisticated idea for targeting warheads, called *adaptive preferential targeting*, works in the following way: Accurate and early RV tracking by radar and optical sensors are used to identify the particular targeted silo for that RV. If a *single* RV was targeting a silo, defense weapons would attack the RV. If *two or more* RVs were targeting a silo, defense weapons would not attack these RVs. In this way, effectiveness of a limited defense would be maximized to save the maximum number of silos. For this approach to succeed, the defense must be able to accurately separate warheads from decoys and the RV trajectories must be known very accurately, early in the trajectory. Acquiring this information on a timely basis is difficult.

5.4 Laser Weapons

Aircraft can be destroyed with a 25-kJ laser pulse, acting at short distances. In 2013, the Navy announced it would base lasers on ships to shoot down drones. This is easier since drones are soft and move slowly, allowing time to absorb more energy.

The Navy acknowledged that it would not be effective against planes and missiles. Because clouds and rain absorb laser beams, ground-based lasers are not effective against ICBMs, limiting them to good weather. The atmosphere also broadens laser beams, a drawback that, to some extent, might be overcome with adaptive optics. Space-based lasers could avoid these problems. In addition, space-based lasers might reduce the number of targets if MIRVed ICBMs were attacked in the boost phase, before their RVs were released. The broadened resolution of a beam is proportional to the laser wavelength divided by the focusing-mirror diameter (λ/D).

Orbital chemical lasers. Radiation from CO_2 lasers is relatively long, giving considerable broadening, precluding its use as a space-based laser. To reduce broadening, the ground-based, 2-MW Mid-Infrared Advanced Chemical Laser (MIRACL) in New Mexico uses deuterium fluoride (DF) with shorter wavelengths, readily passing through the atmosphere. The hydrogen fluoride (HF) version is not useable for ground basing since its photons are absorbed by the atmosphere, but it is preferred for space-basing since it suffers less broadening. An HF beam used with a 3-m diameter mirror gives a beam diameter of 2 m at a distance of 1000 km. But the American Physical Society concluded that the HF laser

> output powers at acceptable beam quality need to be increased by at least two orders of magnitude for hydrogen-fluoride and deuterium-fluoride lasers for use as an effective kill weapon in the boost phase.

Fuel in orbit. The amount of fuel needed in orbit to combat 2000 ballistic missiles of the former Soviet Union is very scenario dependent. *If all the lasers were in the correct location at the correct time* to attack 2000 Soviet ICBMs with two shots per booster, the lower-bound estimate would be (2000 ICBM) (3 ton/ICBM) = 6000 ton of HF. Since the lasers are mostly not in the correct locations, it would require more than 10 times this amount.

Laser availability. Only a fraction of orbiting lasers are available for use at one time, since at any point in time the other parts of their orbit are mostly unavailable. The number of lasers needed for defensive action depends on the location of targeted silos and submarines. The fraction of time on duty to attack, the *duty factor*, is less than 25 % since only one-fourth of the satellites are in the former Soviet one-half hemisphere at a time. But even this estimate is too optimistic since lasers have a limited range and silos are based in clumps, but one has to also consider SLBMs based at sea as targets. On this basis it is reasonable to suggest a duty factor of about 10 %. Yet this could be too optimistic since SLBMs could be gathered in small regions, saturating the defense at that location by monopolizing available lasers and then allowing 100 % transmission. If the duty factor were 10 %, the fuel required would be 10 (6000 ton) = 60,000 ton of HF in orbit. To put this in perspective, the payload of 308 SS-18s (at maximum strength), at 8 ton each, is 2500 ton, which is 4 % of the orbital 60,000 ton of HF needed. This estimate does not take into account the mass of lasers, missiles and other equipment. Thus, the defense payload greatly exceeds the offense payload. Since launch-weight is 20 times the throw-weight, total launch-weight is over 1 million tons.

National Ignition Facility. NIF was completed at LLNL in 2009. It is a research tool, and not a defensive weapon. It has some distance to go before obtaining energy break-even with ignition of *deuterium and tritium*. NIF was designed to obtain a pulse of 1.8 mJ of ultra-violet energy on target from 192 beams. The thermal pulse on target is 10^{15} W over a nanosecond, a thousand times larger than the US electrical grid of 10^{12} W. NIF is being used to answer weapons questions, such as the equation of state of plutonium, uranium and hydrogen fusion by bombarding cylindrical hohlarium of 1 cm by 0.5 cm (heat with X-rays from the inner surface). When heating directly a 0.2 cm spherical capsule, NIF obtained 50 million degrees K and 150 billion atmospheres of pressure. On 4 March 2014, NIF obtained 9.6×10^{15} neutrons with 27 kJ of energy.

5.5 Earth-Based Lasers

Fighting mirrors. Because of the difficulty of orbiting many chemical lasers, SDIO explored the notion of Earth-based lasers to attack ICBMs. Lasers based on the ground cannot attack boost-phase missiles because the curvature of the Earth blocks laser beams. The only way for ground-based lasers to attack the boost phase is to use *relay mirrors* to reflect beams to *fighting mirrors* that focus beams onto boost-phase missiles. If the relay mirrors were located in geosynchronous orbit, only a few mirrors would be needed because of their fixed relative position. However, relay mirrors would have to be very large to reflect the large beam size at geosynchronous orbit. If relay mirrors were located at lower altitudes of 5000–10,000 km, sizes of relay mirrors would be smaller, but more mirrors would be needed since they would not remain above Russia. The APS estimated that energy from large, land-based excimer lasers and free-electron lasers (FELs) was 3–4 orders of magnitude too small to attack the ICBM boost phase.

Adaptive optics. Astronomers use adaptive optics (AO) to remove the broadening effects on stellar images caused by atmospheric turbulence, by improving resolution of land-based telescopes to the natural diffraction limit. AO can reduce effects of turbulence on laser beams, going upward through the atmosphere from ground-based mirrors. AO flexes the main mirror to geometrically shape laser pulses to overcome the just-measured effects of atmospheric turbulence. Here is how it is done: Looking upward, AO observes light stimulated by a second laser to measure turbulence in the high atmosphere. Sensors determine the slope of distortions in the incoming wave front, which gives information to deform the main mirror by a submicron (less than a millionth of a meter) to modify the upward-traveling wave front. Turbulent air cells in the troposphere are 50 cm in size at 10 km altitudes, limiting angular resolution of ground-based telescopes to an arc-second (1/3600 of a degree). This limit is 25 times larger than the diffraction limit of a large mirror. Actuators deform flexible mirrors quickly, in less than 0.001 s, much faster than atmospheric turbulence.

Free-electron lasers. FELs have attained peak power of a megawatt at wavelengths of 1 μm. FELs result from the motion of a beam of electrons, passing over a periodic array of magnets, alternating up and down. More than 1 GW is needed to power FEL weapons because Earth-based laser beams would have to travel to geosynchronous orbit, as compared to space-based lasers, with 1000–km ranges, requiring 20–200 MW. One cannot simply use ratios of distances squared when configuring FEL weapons because scenarios depend on battle-mirror distances to boost-phase ICBMs. The 1987 APS study concluded that peak powers of 100–1000 GW would be necessary for FELs. Another possibility is the excimer laser, which uses krypton fluoride, but its power level is too small by several orders of magnitude.

X-ray laser pumped with a nuclear explosion. SDI primarily planed for space-based weapons in low-Earth orbit, but Edward Teller sought another approach because he realized that space-based weapons were vulnerable to attack. He envisioned *pop-up missiles* based on submarines in the Arctic Ocean, quickly leaping into space to attack Soviet boost-phase missiles. Teller and Lowell Wood contemplated powerful X-ray lasers, pumped with energy from a nuclear weapon. The limited tests of the X-ray laser were not successful, a fact that was kept from many policy makers. The next paragraph shows that basing a perfect X-ray laser has serious timing problems.

Pop-up, missiles. Basing submarines in the Arctic Ocean, close to some former Soviet ICBM silos, is necessary for the X-ray laser. Pop-up missiles must be ready to attack the boost phase in a timely manner and with considerable energy on target. This entails a race in time between boost-phase duration and the time for the X-ray laser (Excalibur) to rise above the atmosphere, sufficiently to attack the boost phase. If the response time and Excalibur's acceleration were good, the laser could look down on the rising ICBM. However, if Soviets used fast-burn boosters to release RVs in one minute, the pop-up laser would be too late. There is a downside to being closer, since close-in submarines are vulnerable to attack. Certainly the X-ray laser could not have defended against Soviet missiles launched in Kazakhstan. Secondly, did nuclear weapon tests show that it was possible that X-ray laser could be pumped with a nuclear weapon? I have been told that laser multiplication was marginally achieved, far below what was needed to make a weapon. To obtain sub arc-second targeting accuracy on an accelerating pop-up missile at a distance of 3000 km is daunting. The cartoon X-ray laser on magazine covers showed many beam tubes emerging from the weapon, each independently, targeting different objects, which is daunting (Fig. 5.1).

X-ray laser fluence. A distant missile would be destroyed if X-ray beams could focus a significant fraction of the nuclear weapon yield. In the 1980s, X-ray lasers were developed by pumping input energy from optical lasers. Very intense visible laser light is focused on a thin wire, vaporizing it so quickly that it implodes. A selenium wire was used for the wire, ionizing its atoms 24 times and lazing it. In the 1990s, pulses from the Nova laser created tabletop X-ray lasers that lazed in a single pass since there were no credible X-ray mirrors for this task. It has been claimed that some X-ray lazing took place after the Livermore nuclear explosive

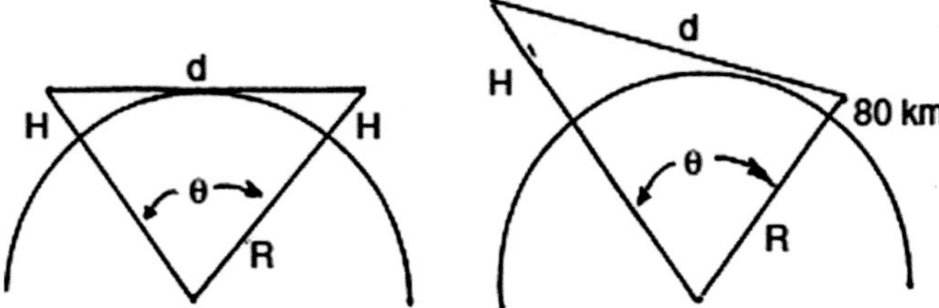

Fig. 5.1 Laser trajectories. For the case of a slow-burn booster, Excalibur rises to altitude *H* (above the Earth's radius *R*) to shoot a beam a distance *d* to a missile that is preparing to launch RVs at altitude *H*. The minimum distance *d* is the case of equal altitudes of missile and Excalibur, as shown on the *left*. For the case of a fast-burn booster, the RVs are released at lower altitude. Excalibur must climb to higher altitudes and the beam must travel further, as shown on the *right*. Atmospheric absorption of X-ray beams protects missiles up to an altitude of 80 km

test of Excalibur, but the US General Accounting Office deemed this as overstated. Geometric broadening gives a sharper beam angle from smaller diameter rods, but thinner rods produce wider beams because of diffraction-broadening. Obtaining the combined requirements for (1) popup basing, (2) a high-efficiency Excalibur and (3) accurately aiming is a tremendous task, without considering possible defensive counter-measures.

5.6 Kinetic Kill Vehicles

From DEWS to KKVs. SDIO shifted its emphasis from directed-energy beam weapons to space-based KKV in 1987. The first KKV was the space-based interceptor (SBI), which was to be based in groups of 10 in satellite garages. The KKVs were expected to attack the boost phase at a distance of 300 km from higher orbits in 50 s with a relative velocity of 6 km/s. Timing requirements can be relaxed when attacking the midcourse phase. After SBI stalled, *brilliant pebbles* were proposed to collide with midcourse RVs. These were 1-m long, self-propelled rockets of 50 kg. They were to be based in space, awaiting a call to attack boost-phase and midcourse-phase missiles. Thousands of pebbles would be needed to cover all situations; they would function without maintenance for at least a decade.

As SDI made little progress toward deployment, changes in missile defense were encouraged. Patriot missiles, designed to attack airplanes and not missiles, were used to attack Iraqi Scuds in the 1991 Gulf War. The success of Patriot was exaggerated; its overstated success reenergized the quest to develop BMD theater missiles with ranges of 300–600 km and to create a National Missile Defense. Radar and IR sensors obtain trajectories for the defense, guiding for kinetic kill or explosion. The kinetic energy density of KKV projectiles is much larger than that of explosives. Even a 30-g pebble has a considerable energy. The Ground Based Interceptor is a kinetic kill weapon, to attack the midcourse phase at 7–8 km/s.

KKV technologies have not been fully tested in combat to know if they work, except for the case the Israeli *Iron-Dome* theater missile defense. During the 2012 Israel-Gaza hostilities, hundreds of missiles from Gaza were intercepted by the *Iron-Dome* system, reporting 80–90 % effectiveness. Iron Dome can protect an area of 150 km^2 for missiles with ranges of 4–70 km. The US supported this program with a $1 billion.

5.7 Airborne Laser

The Airborne Laser was developed to destroy the theater-based boost phase. The ABL planned to use 3-MW, oxygen-iodine lasers, located on a fleet of seven Boeing 747 s by 2011. Ultimately only one ABL aircraft was built; ABL was canceled by the Obama Administration after 16 years and $5 billion. ABL was needed to fly continually above suspicious areas, which is costly. ABL employs a 1.5-m diameter mirror with adaptive optics, giving a beam area of 0.5 m^2 at 300 km. The energy on target is degraded since the air is warmed by beam absorption over it 100–300 km path, resulting in thermal blooming that degrades energy on target. Timing is difficult since Scud or fast-burn missiles take 30 s to clear the clouds, leaving ABL only 30 s to target. ICBMs would allow more time to target after observation. The 2003 APS report stated that 2 min is available to disable a solid-fueled ICBM *after launching* and 3 min would be available for liquid-fueled ICBMs. It is unlikely that the beam would be on target for more than 1 s.

ABL problems. Clouds can hide rising missiles and confound ABL. Because of the short reaction time needed to attack the boost phase, ABLs have to be on station to be relevant. And they are vulnerable. On October 4, 2001, a misfired Ukrainian surface-to-air-missile destroyed Siberian Airlines flight 1812 from a distance of 250 km. The 2003 APS report concluded the following:

> Because solid-propellant ICBMs are more heat-resistant, the Airborne Laser's ground range against them would be only about 300 km, too short to defend against solid propellant ICBMs from either Iran or North Korea… Countermeasures against the ABL could include applying ablative coatings or rotating the ICBM to reduce the amount of heat the missiles absorbs, launching multiple missiles to overwhelm the ABL's capabilities or attacking the aircraft carrying the laser.

5.8 Anti-Satellite Weapons

Satellites are a mainstay of US national security, watching for missile attacks and other military matters. (A drawback in their role is that a nation could believe a massive attack was imminent if its satellites were attacked, perhaps responding too quickly in time of crisis.) Satellites are also the source of national technical means

(NTM) of verification, monitoring weapons on foreign soil and helping select locations for arms control inspections. National security and arms control applications of NTM are a stabilizing factor of global security *when* properly used. It might be understandable for a country to destroy enemy satellites in time of conflict, but to threaten satellites can cause war from overreactions. For these reasons, the US Congress imposed a ban on anti-satellite (ASAT) tests in the 1980s.

In the 1980s, the United States deployed an ASAT missile on F15 fighters, lifting KKVs into orbit, while the Soviets deployed ASAT missiles with fragmentation explosives. Ground-based lasers can easily destroy satellites that are soft objects in predicable orbits. The MIRACL laser at White Sands, New Mexico attacked a retired US satellite in 1997. A land-based ASAT beam can track a satellite through a 60° arc in the sky, covering 200 km of flight. This gives an engagement time of 30 s, long enough to destroy the satellite, but clearly not enough to destroy an incoming ICBM.

Problems

5.1 **Nuclear ABM**. What were the 1970 Sprint and Spartan ABM systems?

5.2 **Destabilizing ABM**? Under what conditions are ABM systems destabilizing?

5.3 **ABM Treaty**. What actions did the ABM Treaty forbid and what did it allow?

5.4 **Radar power limit**. Why did the ABM Treaty limit radar power and radar area at national boundaries?

5.5 **Boost phase *sine qua non***. What are two reasons the boost phase is the best place to attack missiles?

5.6 **Space basing**. Why is space basing necessary to attack the boost phase?

5.7 **Layered defense**. If an ABM layer is defeated by countermeasures or overwhelmed by numbers, how might warheads penetrate the defense?

5.8 **Punched hole**. Why would ground-based lasers punch a hole in the atmosphere?

5.9 **Countermeasures**. What countermeasures by the offense could defeat the defensive on the boost phase, on the midcourse phase and on the re-entry phase?

5.10 **Laser beam width**. What factors determine the sharpness of the laser beam?

5.11 **Duty factor**. What affects the fraction of time that a laser in is useful in orbit?

5.12 **Neutral Particle Beams**. Why is it necessary for particle beams to have no electric charge?

5.13 **NPB midcourse discrimination**. How can neutral particles determine if an object is decoy or warhead?

5.14 **GEO battle mirrors**. Locate a battle mirror in orbit with big lasers based on US ground.

5.15 **Distance to target**. Does it matter how far silos are from Excalibur on pop-up missiles on submarines in the Arctic Ocean?

5.16 **Adaptive preferential targeting**. What is this?

5.17 **Boost-phase timeline**. KKVs might be based on ships near North Korea. Why do this?

5.18 **Airborne Laser**. What are the difficulties with the ABL system?

5.19 **Anti-satellite weapons**. Why are satellites easier to attack than ICBMs?

5.20 **ASAT attack**. How would you make satellites less vulnerable to attack?

Seminar ABM, BMD, ASAT

Adaptive optics, Aegis destroyer BMD deployments, Air Borne Laser, brilliant pebbles, broad versus narrow interpretation of the ABM treaty, BMD success rate in Gulf-I war and over Gaza, countermeasures, decoys, defense can be provocative, electrical power in orbit, European deployment of US BMD, Excalibur X-ray laser pumped with a nuclear weapon, free-electron laser, Ground-Based Interceptor deployments and technology, Israeli Arrow BMD system, Nitze (Paul) requirements for ABM deployment, Soviet Galosh ABM system near Moscow, Space-Based Interceptor technologies and deployment, Soviet views of SDI, targeting accuracy in orbit, testing SDI in space, US Nike-Zeus defensive system.

Bibliography

American Physical Society. (1987). Science and technology of directed energy weapons. *Review of Modern Physics 59*, S1–S202 and *Physics Today 40*(3), S1–16.

American Physical Society. (2003). *Boost-phase intercept systems for national missile defense*. College Park, MD: APS.

Bridger, S. (2015). *Scientists at war*. Harvard University Press: Cambridge, MA.

Broad, W. (1985). *Star warriors*. NY: Simon and Schuster.

Carter, A., & Schwartz, D. (Eds.). (1984). *Ballistic missile defense*. Washington, DC: Brookings.

Coyle, P. (2013, January). Back to the drawing board: The need for sound science in US missile defense. *Arms Control Today*, 8–14.

Coyle, P. (2014). Science, technology and politics of ballistic missile defense. In *AIP Conference Proceedings on Nuclear Weapon Issues in 21st Century* (Vol. 1596, pp. 135–142).

FAS and Soviet Scientists. (1989). Space reactor arms control. *Science and Global Security, 1*, 59–164.

Fitzgerald, F. (2000). *Way out in the blue*. NY: Simon and Schuster.

Forden, G. (1999). The airborne laser. *IEEE Spectrum, 36*(3), 40–49.

Garwin, R. (1985). How many orbiting lasers for boost-phase intercept. *Nature, 315*, 286–290.

Gronlund, L., et al. (2000). Continuing debate on national missile defense. *Physics Today, 53*(12), 36–43.

Hey, N. (2006). *The star wars enigma: Behind the scenes of the cold war race for missile defense*. Washington, DC: Potomac Books.

Masters, R., & Kantrowitz, A. (1988). *Scientific adversary procedure: The SDI experiments at Dartmouth, technology and politics*. Durham, NC: Duke University Press.

National Research Council. (2012). *Making sense of ballistic missile defense*. Washington, DC: National Academy Press.

Office Technology Assessment. (1985). *Ballistic missile defense technologies*. Washington, DC: OTA.

Office Technology Assessment. (1987). *SDI technology, survivability and software*. Washington, DC: OTA.

Sessler, A., et al. (2000). *Countermeasures*. Cambridge, MA: Union Concerned Scientists.

Stupl, J., & Neuneck, G. (2010). Assessment of long range laser weapon engagements. *Science Global Security, 18*(1), 1–60.

Taylor, T. (1987). Third-generation nuclear weapons. *Scientific American, 256*(4), 30–38.

Wilkening, D. (2004). Airborne boost-phase ballistic missile defense. *Science Global Security, 12* (1), 1–67.

Wilkening, D. (2012). Cooperating with Russia on missile defense. *Arms Control Today*, 8–12.

Wilkening, D. (2014). NRC study: Making sense of ballistic missile defense. In *AIP Conference Proceedings on Nuclear weapon issues in 21st century* (Vol. 1596, pp. 123–134).

Wright, D., Grego, L., & Gronlund, L. (2005). *The physics of space security*. Cambridge, MA: American Academy of Arts and Sciences.

Chapter 6
Verification and Arms Control Treaties

Trust, but verify.
[President Ronald Reagan to General-Secretary Mikhail Gorbachev, 8 Dec. 1987]

Super-power leaders, Regan and Gorbachev, had just signed the 1987 *Intermediate-Range Nuclear Forces* (INF) Treaty in Washington. Signing *INF* marked the beginning of the end of the Cold War, two years before the November 1989 fall of the Berlin Wall, 3 years before the *Conventional Armed Forces in Europe Treaty* (*CFE*) was signed in 1990 and 5 years before the *Strategic Arms Reduction Treaty* (*START*) was signed.

6.1 Verification Context

Nations remain in arms control treaties because they judge that the disadvantage of controls on weapons to be far less dangerous than a system with no controls. The failure of the United States to ratify the Comprehensive-Test-Ban Treaty (CTBT) and its subsequent withdrawal from the Anti-Ballistic Missile (ABM) Treaty are exceptions to a global consensus; the world waits for viable, meaningful arms control, where possible.

Monitoring. It is important to distinguish between *monitoring* and *verification*. Monitoring is data collecting through satellites, aircraft, movable and stationary sites and other means. The monitored data is collected from photographic, infrared, neutron, X-ray, gamma ray, electromagnetic pulse, radar, seismic, hydroacoustic, radionuclide, infrasound, and analog/digital communications data. Optical resolution by US satellites improved from 12 m in 1960 to 0.7 m in 1980 to better than 0.1 m at present; the United States now sells photos with resolution of 0.5 m. The KH satellite series began with images on film that were jettisoned to Earth by parachute. The advent of charge-coupled devices now allows direct electronic transmission of digital images. By 1988 satellites began to transmit radar images. In the atmosphere, the Air Force uses RC-135 planes, the Navy uses EP-3 planes, and

D. Hafemeister, *Nuclear Proliferation and Terrorism in the Post-9/11 World*,
DOI 10.1007/978-3-319-25367-1_6

the Army uses Predator/Raptor drones to gather further information. These technologies are an interesting scientific tours-de-force, as well as being politically important. To avoid being labeled "secret," the information is described as obtained by National Technical Means (NTM) of verification. NTM is buttressed with on-site inspections (OSIs) that take data at declared and other sites. Interagency groups analyze the data and come to preliminary judgments to determine if decision makers should be informed about compliance problems.

Verification is the quasi-judicial act that determines whether monitoring data are sufficient to prove that a treaty party has or has not complied with the terms of an agreement. The level of confidence in the data determines the adjectives applied to the word *violation,* such as, *possible* violation, *likely* violation, or *probable* violation, or just plain *violation.* The charge of *likely* violation is close to a *preponderance-of-the-data* standard but less than a standard of *beyond all reasonable doubt.* The verification process has been occasionally politicized, but it has mostly been accurate. The Soviet Krasnoyarsk radar was an *unambiguous* violation of the ABM Treaty. An example that cuts the other way was the falseness of the US charge of a *likely* violation by the Soviets of the Threshold Test Ban Treaty (TTBT) in the 1980s, an issue with which the author was intimately involved as the State Department technical lead on testing issues.

Effective verification. During the ratification of the INF Treaty, Ambassador Paul Nitze defined *effective verification:* "… if the other side moves beyond the limits of the treaty in any militarily significant way, we would be able to detect such violation in time to respond effectively and thereby deny the other side the benefit of the violation." Thus, any cheating must be detected in a timely manner before it can threaten national security. Since nations are already threatened with a plethora of *legal* nuclear weapons, the "effective" criterion concerns the degree of marginal threat due to cheating beyond the legal level of threat agreed to by the treaty. During the ratification of START, Secretary of State James Baker repeated this definition, but added a new criterion: "Additionally, the verification regime should enable us to detect patterns of marginal violations that do not present immediate risk to the US security." During the negotiations, the United States and Russia decided how much verification was enough: there are wise verification measures and there are superfluous ones. An excellent measure from START is the reentry-vehicle on-site inspection (RVOSI) that counts physical RVs (warheads) on a missile. RVOSIs are an extremely important tool if an agreement limiting numbers of actual warheads is achieved.

6.2 Arms Control Treaties

The US executive branch selects policy options in interagency groups (IG) led by the National Security Council (NSC), working with the State Department, the Defense Department, the Joint Chiefs of Staff (JCS), the Central Intelligence Agency (CIA), and in some cases the National Nuclear Security Agency (NNSA).

Formerly, the Arms Control and Disarmament Agency (ACDA 1961–99) was in this process. If a dispute exists, the issue goes to the full NSC, led by the president and composed of the secretaries of State and Defense, the chairman of the JCS and the Director of National Intelligence. Finally, the president makes the decision to endorse, amend, or reject the IG agreement. In this section we list the central provisions of key arms control treaties that have passed through this process.

6.2.1 Nuclear Testing Treaties

LTBT. The Limited Test Ban Treaty bans nuclear tests in the atmosphere, outer space, and under water. (Signed and entered into force (EIF) in 1963.)
TTBT. The Threshold Test Ban Treaty bans underground nuclear tests of over 150 ktons. Its 1988 protocol added On-Site Inspections. (Signed 1974, EIF 1990.)
PNET. The Peaceful Nuclear Explosions Treaty limits PNEs to underground explosions with a maximum of 150 kton for individual PNEs and 1500 kton for group explosions. The U.S tested PNEs 27 times, while the Soviets tested PNEs 124 times. (Signed 1976, EIF 1990.)
CTBT. The Comprehensive-Nuclear-Test Ban Treaty bans nuclear tests of any yield at any place and any time. Three nuclear weapon states (NWSs) ratified CTBT (France, Russia, and United Kingdom) and two have not (China and United States). Ratification failed in the US Senate by 51–48 in 1999. By May 2015, 183 nations signed CTBT, 164 nations ratified, and 36 of 44 nuclear capable nations ratified. Of the 321 proposed IMS detection stations; 281 were certified, 19 were installed, 19 were under construction and 18 were planned. (Signed 1996, not EIF.)

6.2.2 Strategic Nuclear Arms Treaties

SALT I. The Strategic Arms Limitation Talks (1972) led to a 5-year *executive agreement* (not a treaty, but with the force of law) that passed both houses of Congress with a majority vote. SALT I capped the number of allowed intercontinental ballistic missiles (ICBMs) and submarine-launched ballistic missiles (SLBMs) at their 1972 levels, but it did not cover bombers.
ABM treaty. The Anti-ballistic Missile Treaty (1972) constrained each side (US and USSR) to 100 ABM launchers located at no more than one site (1974). The ABM Treaty forbade national area defense and radar construction within a nation. It created the Standing Consultative Commission (SCC) to implement the SALT and ABM treaties. SDI beam weapons were constrained by a provision that certain other physical principles (such as beam weapons) "would be subject to discussion." President George W. Bush withdrew the United States from the ABM treaty in June 2002. Russia continues to be concerned about US BMD deployments in Europe.

SALT II. The unratified SALT II limits (1979–1986):

- 2400 ICBMs, SLBMs, and heavy bombers,
- 1320 MIRVed ICBMs, SLBMs, and heavy bombers with cruise missiles,
- 1200 MIRVed ICBMs and SLBMs,
- 820 MIRVed ICBMs and
- 1 new ICBM type during the 5-year duration.

INF. The Intermediate-Range Nuclear Forces Treaty (1987, EIF 1988–2001) bans all ballistic and cruise missile systems everywhere between ranges of 500 km and 5500 km and established notifications and OSIs of historic proportions. By 1991 the United States eliminated 283 launchers and 846 missiles capable of carrying 859 warheads, and the Soviets eliminated 825 launchers and 1846 missiles capable of carrying 3154 warheads. In July 2014, the US State Department charged Russia with a violation of INF with a test of a banned ground-launched cruise missile. Russia responded that the launcher was legal since it had been previously tested. Russia claimed that the US Mark-41 launcher, needed for BMD deployment in Romania and Poland, was illegal. These systems have not yet been deployed.

START I. The first Strategic Arms Reduction Treaty (1991, EIF 1994) set the following 15-year limits:

- 1600 ICBMs, SLBMs, and heavy bombers,
- 6000 accountable warheads (about 8000 actual warheads),
- 3600 tons throw-weight (46 % reduction, cutting 50 % of 308 SS-18s), and
- 13 types of OSIs, many declarations, and perimeter portals at mobile-missile production facilities.

START II. Signed in 1993, ratified by US Senate in 1996 and by Russian Duma in 2000, but it did not enter into force because of US withdrawal from the ABM treaty. The US wanted more missile defense and Russia did not mind the demise of START-II since they wanted to keep 138 SS-18s and many SS-19s. *START II* set the following limits:

- 3000–3500 warheads on ICBMs, SLBMs, and heavy bombers (fully counted),
- elimination of all heavy SS-18 missiles,
- elimination of MIRVed ICBMs, except for 90 SS-18 silos converted for SS-25/27,
- 105 SS-19s, which must be reduced from 6 warheads to 1.

START III. Its framework was agreed to in 1997 by Presidents Clinton and Yeltsin in Helsinki. It called for a limit of 2000–2500 warheads and enhanced transparency on warhead inventories and destruction. It collapsed when START II failed. It is not clear how far warhead transparency would have progressed.

SORT (START IV). The Strategic Offense Reduction Treaty (SORT, Treaty of Moscow) was signed in May 2002 by Presidents Bush and Putin, and ratified on 6 March 2003 by a vote of 95–0. It limits *operational* warheads to 1700–2200 by

2012. This is the START III limit, when adding the 240 warheads on 2 Trident submarines in overhaul. SORT expires on the day the limits come into affect and it does not include intermediate compliance dates. SORT does not include verification measures.

New START (START V). Signed on 8 April 2010 and ratified on 22 December 2011 by a vote of 71–26. The treaty only removed 30 silos, 34 heavy bombers and 56 SLBM tubes, but it re-established START verification process, which lapsed after 15-years of START-I. The 10-year agreement limits each side's deployed operational strategic warheads to 1550. The strategic delivery systems, deployed and non–deployed, such as submarines in overhaul are limited to 800, with a sublimit of 700 deployed systems. The number of Russian facilities inspected for START was 70 and for NEW START it is now 35. The number of on-site inspections was reduced from 28 for START to 18 for NEW START. The NEW START warhead limit is 25 % lower than the 2200 for SORT, depending on how you do the math on bomber weapons. The US retrains 420 Minuteman silos, 60 nuclear-capable bombers (42 B52s and 18 B-2s), and 14 submarines with 20 SLBM tubes each (down from 24), with 240 operational SLBM warheads. Heavy bombers carry up to 20 nuclear weapons, but they only count as 1 strategic warhead because of their slow response and mostly off-alert status. The U.S. will spend $180 billion to modernize the nuclear weapons complex and delivery systems over 10 years. In 2010 the United States had 2200 operational warheads, plus 240 warheads on 2 Tridents in overhaul, plus an *active reserve, responsive force* of 2000 warheads, which could be deployed in "weeks, months or years" depending on the delivery platform. This is being reduced to the 1550 number, with a total of 5000, which is about the same for Russia. In addition, the United States will have additional warheads in the *inactive reserve* stockpile, plus a stockpile of primary and secondary stages. "Russia would not be able to achieve a military significant advantage from its strategic nuclear forces" (US Strategic Command 2013).

What next? (START VI). The annexation of Crimea and the invasion of Eastern Ukraine complicate matters with Russia in 2015. As the two super-power stockpiles are reduced to about 5000 warheads each, we should consider the stockpiles of the other new nuclear weapon states. According to Hans Kristensen (FAS) and Phillip Schell (SIPRI), the other warheads are divided among UK (225), France (300), China (250), India (90–110), Pakistan (100–120) and Israel (80). This totals to 1075 warheads, which is 10 % of the 10,000 total for US and Russia.

Obama has called for a limit of about 1000 for strategic warheads, plus tactical warheads and non-deployed warheads, for a total of 2000 warheads. This would be done by reciprocal, unilateral measures as done by Bush-I and Gorbachev, and not by treaty. Putin stated that he is not interested in deeper cuts under a follow on START Treaty. Putin wants more concessions on BMD before he would do this. The 2013 US total is 4650 (including those waiting dismantlement) and the Russians total is 5000.[1] Verification might be two-tired, stronger verification for

[1]Pifer (2010).

strategic operational warheads, and weaker verification (declarations and spot checks) for tactical and nondeployed warheads. If the number of delivery vehicles was lowered to 600, this could give the U.S. a force of 192 SLBMs (12 SSBNs with 16 SLBMs each, with 2 more SSBNs in maintenance), 368 ICBMs and 40 heavy bombers. The U.S. is retiring its 400 submarine-launched cruise missiles (W-80/0 on SLCM). The U.S. would retain some 400 B61 bombs, half deployed in Europe. Will it be a treaty, an executive agreement, or reciprocal-unilateral agreement in word only?

General James Cartwright, former Commander-in-Chief of the Strategic Command and Deputy-Chair of the Joint Chiefs of Staff, presented a plan to Congress to reduce by 80 % the total number of US warheads, from 5000 to 900, with half deployed and half in storage.[2] Ten Tridents would carry 720 warheads. Warheads would be carried on eighteen B-2s and thirty new strategic bombers, removing the B-52 from this role. All land-based ICBMs would be retired. The enduring stockpile would be reduced from six warhead types to four. Tactical warheads would be eliminated. This would require the production of 18 new plutonium pits per year at Los Alamos.

Western NATO wants to remove US tactical weapons from Europe, while Eastern NATO wants to maintain them. The US will resist restraining conventional forces greatly in Europe. Russia wants to constrain missile defense in Europe and keep tactical weapons for its southern borders. Russian fears of NATO will have to be overcome. Some say the main hope for an agreement is on bilateral cooperative ballistic missile defenses to start the process.

Global zero? An interesting development has come from Stanford's Hoover Institute. Former Secretaries of State Henry Kissinger and George Schultz, former Secretary of Defense Bill Perry and former Chair of the Senate Armed Services Committee Sam Nunn. They committed to work towards the goal of removing all nuclear weapons from the Earth. This goal was considered at the 1986 Reykjavik summit by President Reagan and Soviet leader Gorbachev, both pushed hard to do this, but to be undone by the ABM issue. Four op-ed pieces have appeared in the *Wall Street Journal* by the *Gang of Four* (4 January 2007, 15 January 2008, 7 March 2011 and 5 March 2013), laying out steps that move us towards zero. They readily admit they do not have all the answers in terms of verification and nuclear stability, but they also realize that the recommended steps are useful even if we do not get to zero. The four statesmen concluded the following:

> But the risks from continuing to go down the mountain or standing pat are too real to ignore. We must chart a course to higher ground where the mountaintop becomes more visible.

An interesting question is "why would these leaders risk political flack from their political parties to recommend the goal of zero nuclear weapons?" My guess is that they see the problems of the world as so severe that they are willing to take political

---

[2]Collina (2012).

smears to make some progress, with zero as a very distant goal. See references by Drell and Schultz. Some are concerned about the "blow-back" from talking about zero too much, namely that the nations under the US nuclear umbrella will feel uneasy about the US deterrent and move towards nuclear weapons. At the highest levels of government, this fear has not been realized, but it is a possibility that must be addressed. Foreign policy perceptions can out-weigh reality.

6.2.3 Other Treaties and Agreements

Hague convention (1899, 1907). The Hague Convention contains criteria on the laws of war and war crimes. Germany's use of chemical weapons in World War I violated the Hague Convention.

Washington naval conference agreement (1922). This curtailed a battleship arms race, but it ultimately failed because there were no verification measures, and Japan withdrew in 1936. The nine nations (US/UK/Japan/others) had to follow a ratio of four numbers (5/5/3/1.7) on battleships and aircraft carriers, with constrained gun sizes (up to 16 inches). The London Naval Treaty (1930) expanded this to cruisers and destroyers.

Geneva convention (1925). The Convention is an agreement on the no-first use of chemical weapons. It was negotiated in response to chemical weapons use in World War I. It was ratified by most nations, except for US, Japan and Soviet Union. Chemical weapons have since been used since by Japan, Iraq and Syria.

NPT (1970). The Nuclear Non-Proliferation Treaty (1970) prevents the spread of nuclear weapons, while promoting peaceful uses of nuclear energy. NPT creates two classes of nations, the Nuclear Weapon States (NWS: US, RF, UK, France and PRC) and 180 nonnuclear weapons states (NNWS) without India, Israel, and Pakistan and North Korea (withdrew). NPT was extended without a time limit in 1996. The key agreements in the NPT are as follows:

- NWS not to transfer/assist nuclear weapon technology to NNWS,
- NNWS not to receive/acquire nuclear weapons technology,
- NNWS accept IAEA inspections on nuclear facilities and materials,
- NWS assist NNWS with their peaceful nuclear power programs,
- NWS volunteer inspections at a few sites, and
- NWS undertake negotiations on nuclear disarmament.

BWC (1975). The Biological Weapons Convention bans production or acquisition of biological agents for weapons. BWC does not have a verification protocol. BWC has 171 members as of January 2015. Egypt, Syria and 7 African nations signed but did not ratify, while Israel and 15 African nations did not sign.

MTCR (1987). The Missile Technology Control Regime bans export of ballistic and cruise missiles with a payload over 500 kg, with a range of over 300 km. MTCR has 34 members. In 2002, the International Code of Conduct Against

Ballistic Missile Proliferation was signed by 134 states, but not China, India, Iraq, Iran, North Korea, and Pakistan.

CWC (1997). The Chemical Weapons Convention bans production, acquisition, stockpiling, transfer and use of chemical weapons and essential chemicals with extensive verification. As of January 2015, CWC had a strong consensus of 190 members, lacking only Angola, Egypt, Israel, North Korea and South Sudan. (Angola and Myanmar are slated to do so.)

Nuclear weapons-free zones. Antarctica (1961), sea beds (1972), outer space and moon (1967), South America (Treaty of Tlatelolco, partial EIF in 1968, more complete with 23 nations by 1989), South Pacific (Treaty of Rarotonga, signed 1996), Africa (Treaty of Pelindaba, signed 1996), and Central Asia (6 former Soviet Republics, signed 2002, but not EIF).

CFE (1992–2007). Conventional Armed Forces in Europe Treaty ended the Cold War in Europe in 1992, allotting equal forces to NATO and the Warsaw Treaty Organization. WTO reduced from 200,000 to 80,000 tanks, artillery, personnel carriers, aircraft and helicopters, while NATO had essentially no reductions. Russian President Vladimir Putin suspended CFE observance on 14 July 2007. US suspended CFE implementation on 22 November 2011 because Russian troops were in Georgia and Moldova. Russia is threatened by NATO expansion into Warsaw Pact nations, responding aggressively.

Hot line (1963, 1984). Hot line agreements assure quick and reliable communications between US and Russia. The Nuclear Risk Reduction Centers Agreement (1987) is the *cool line* to transmit large amounts of arms control information between Washington and Moscow.

Mine Ban treaty (1999). The Ottawa Treaty outlaws the use, stockpiling, production, and transfer of anti-personnel mines. It has 162 members as of January 2015. Thirty-four states signed but did not ratify, including the United States, Russia and China. There were 3508 mine casualties during 2013, when 185 km^2 were cleared, containing 275,000 anti-personnel mines.

Open skies (2002) commits 34 Eurasian and North American nations to open airspace for over-flights for photography and radar reconnaissance. US and Russia are allotted 42 inspections per year. In 2009, US used 13 inspections, 12 over Russia and 1 over Belarus. Open Skies allows parties to examine their neighbors military forces. It could assist monitoring the Comprehensive Test Ban Treaty by involving nations without reconnaissance satellites.

Convention on cyber crime (2004). Forty-five members, mostly Europe, US and Japan, but not Russia and China. The convention is the first international treaty on crimes committed via the Internet and other computer networks, dealing particularly with infringements of copyright, computer-related fraud, child pornography and violations of network security. Its main objective, set out in the preamble, is to pursue a common criminal policy aimed at the protection of society against cyber crime, especially by adopting appropriate legislation and fostering international cooperation.

Arms trade treaty (2014). About 400,000 persons die each year from illegal transfer of small and light weapons. The UN voted 154–3 in favor (2013). ATT regulates conventional arms sales of tanks, aircraft and small arms. The increased transparency helps to prevent diversions of these weapons. US signed ATT in 2013, but it has not been submitted to the Congress for ratification. ATT entered into Force 24 December 2014 with 66 nations ratifying, while 130 signed. Only Iran, North Korea and Syria opposed it at the UN in 2013.

Iran-P5+1 joint plan of action (signed 2015). The *P5 plus Germany* reduced economic sanctions in trade for a reduction in Iran's enrichment capability, extending breakout time from a few months to a year for ten years.

6.2.4 Global Organizations

- *United Nations* (1945, 193 members) passed UN Security Council Resolution 1540 criminalizes misbehavior on weapons of mass destruction with non-sate actors,
- *Zangger Committee* (nuclear export trigger list, 1971),
- *Nuclear Suppliers Group* (nuclear export criteria, 46 members, 1975),
- *United Nations Conference on Disarmament* (1979),
- *Australia Group* (chemical weapons, 1985),
- *Wassenaar Arrangement* (conventional arms technologies, 41 members, 1996),
- *Ottawa Group* (1997, landmines). US hasn't signed, but restrains 3 million anti-personnel landmines on the Korean Peninsula,
- *International Criminal Court* (2002) has 122 members, but not US and Russia.

6.3 Optical Reconnaissance

Reconnaissance optical systems have considerably improved in many ways:

- Aerial survey films are available with high contrast of 1000 lines/mm (1 μ between lines).
- Charge-coupled devices (CCD, 1980s) allow real-time photography at excellent resolution.
- Chromatic aberration is reduced with computer-aided lens design and machining.
- Blurring by satellite motion is removed by moving camera with precision servo-controls.
- Similar-appearing objects reflect and emit differently in the visible and IR, measured with multi-lens, multispectral cameras at many wavelengths.
- Two views of the same terrain at different angles give stereoscopic images to obtain heights of objects. Cartographic cameras cover 25,000 km^2 in stereo with 10-m resolution, using only two pictures.

- Fourier transform infrared spectroscopy rapidly and accurately determines trace impurities in air.

Resolution of film versus CCD. The Hubble Space Telescope was launched in 1990 and may last until 2020. It is generally assumed that reconnaissance satellites are similar to the Hubble. The resolution obtained from film is limited by the size of film grains, the distance from the object to the satellite and the diameter and focal length of the satellite mirror. CCD resolution is similarly limited, except it is the pixel size in the semiconducting material that matters, instead of the size of film grains. Film can give better resolution, but not significantly better. CCDs have replaced film because of other advantages.

Diffraction broadening can be ignored since geometrical spreading from 10-μm pixels is much bigger in a CCD array of 1600 × 1600 pixels. With 5–10 cm resolution, it would not be possible to read *Pravda's* masthead from orbit, but it would be possible to distinguish a VW Beetle from a Saturn. CCDs have become the verification sensor of choice, with a small resolution sacrifice, because of the rather dramatic advantages listed below:

- CCDs are used in real time with computers, while film requires time delays and scanning.
- CCDs are reusable, removing considerable weight in orbit.
- CCDs are sensitive to a broader range of wavelengths.
- CCDs have a much greater black/white dynamic range.
- CCD response is extremely linear, compared to film's great nonlinearity.
- CCDs are 70 times more light-efficient than film.

Adaptive optics. Point-like stars twinkle at night because atmospheric turbulence continually changes refraction angles. Large disk planets mask the twinkling. Luckily for reconnaissance photography, angular, *seeing* resolution for looking downward from a satellite is considerably better than for the case of looking upward from a telescope, which is about 1 arcsecond. Since seeing is a small problem for downward-looking satellites, it is not necessary to enhance reconnaissance cameras with adaptive optics. However, adaptive optics is important for observing reentry vehicles or satellites with ground-based telescopes. Adaptive optics is used at the Air Force Maui Optical Station (AMOS) in Hawaii, which is at 3-km altitude to improve seeing and reduce atmospheric and cloud absorption. AMOS's location at 20°N latitude allows it to observe most satellites, with 5-cm resolution for objects at an altitude of 125 km. AMOS telescopes track RVs with a velocity of 5° per second and accelerations of 4° per second squared. Adaptive optics is used with the 1.6-m AMOS telescope to reduce visible seeing to the diffraction limit. However, better photos can be obtained with cameras in space on Hubble or other nearby satellites.

Digital image processing. The development of electro-optic CCDs and very large integrated circuits enhances the ability to monitor military activities and arms control treaties. Digital image processing restores and enhances photographs that are blurred by many causes. Some approaches use the subtraction of one picture from another to observe changes. Other approaches enhance edges and lines, remove high-spatial-frequency noise, enhance contrasts by removing clouds and search for patterns of missile silos or mobile missile launchers. These digital processing techniques are used on images obtained by photography, radar, sonar, infrared, and X-rays. A photo of a silo can be hidden by lack of contrast and photographic noise, but analysis (Fig. 6.1) can extract the silo image from the noise. The known circular shape of a silo can be used to further extract the signal from the noise.

Infrared reconnaissance. Infrared detectors are made with gallium impurities in silicon. IR cameras obtain thermal images of objects from orbit, which are useful to find covert facilities and weapons, as well as missile and weapon tests. IR reconnaissance detects temperature differences of 0.1 °C. IR spatial resolution is not as sharp as in the visible since IR's longer wavelengths increase diffraction-broadening. The continuous distribution of heat radiation provides less information than visible spectra, but IR resolution is sufficient to detect re-entry vehicles, silos, vehicles and human activity. Two atmospheric wavelength windows transmit IR. Space-based photography prefers the 10-μ longer-wave length window because Earth emits at that frequency and the window is far from the reflected near-infrared rays from the sun. The longer IR wavelengths gives poorer resolution since diffraction-broadening is 20 times greater than that of light.

IR sensors on satellites are naturally cooled in space. It is expensive to cool IR sensors with liquid nitrogen on the Earth in our cars, boats, drones or remote locations. Without cooling, the thermal noise signal swamps the IR image. However, new bolometric infrared detectors have solved this problem. A weak IR signal marginally warms a small pixel, raising the temperature by milli-Kelvins. This temperature rise increases the electrical resistance, which can be measured.

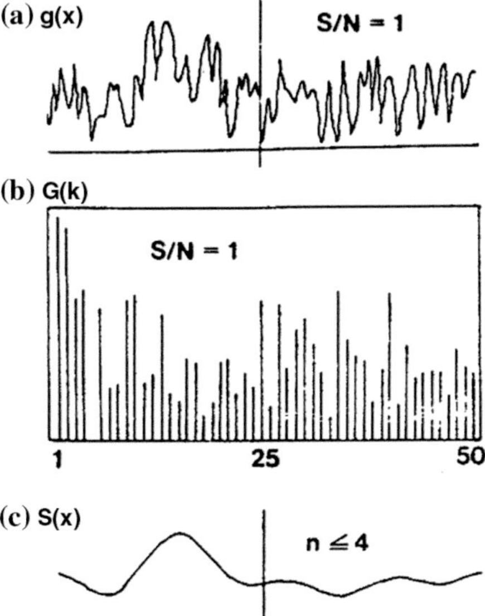

Fig. 6.1 Fourier-transformed image. **a** The image of the silo g(x) is severely hidden by a large signal-to-noise ratio *S/N* = 1 (NN = 2 times the silo height). The silo is 20 pixels wide in a field of 100 pixels. **b** The spatial data is Fourier-transformed to obtain its frequency spectrum: *F* [g (x)] = *G(k)*. **c** Most of the noise is removed from the photo by Fourier-transforming the four lowest frequency components back to position space to make the silo more apparent

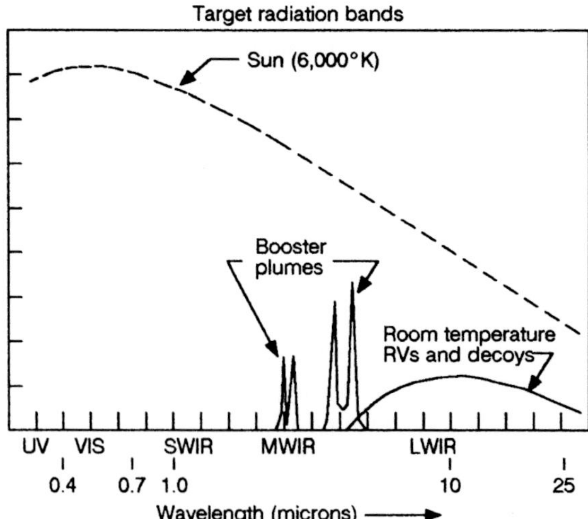

Fig. 6.2 Electromagnetic radiation at different temperatures. Radiation from the sun at 6000 K extends from the ultraviolet (UV) to the infrared (IR), peaking in the visible (VIS) region. Missile plumes with CO_2 and H_2O combustion products are readily detectable at geosynchronous orbit in the short- and mid-wave infrared (SWIR and MWIR). Colder bodies, such as reentry vehicles, decoys, booster bodies, satellites, and the Earth radiate in the long wave infrared (LWIR). Different detectors are required to detect the various objects (*SDI Technology,* Office of Technology Assessment, 1988)

The small electrical current raises the pixel's temperature. Background noise can be eliminated by statistics. The bolometric sensors are used on Predator drones, on expensive cars and by energy-conserving house doctors (Fig. 6.2).

6.4 Radar Monitoring

SAR and InSAR radar. Synthetic aperture radar (SAR) systems on satellites obtain images to monitor arms control treaties. Earth-based phased-array radar monitors ballistic missile tests and obtains images of missiles and RVs. Because radar has long wavelength, SAR resolution cannot compete with optical systems. However, radar is used at night and it locates objects through clouds and rain, as well as inside wooden buildings, and it detects reflections from metal objects. Radar's long wavelength requires a SAR antenna several miles in length to obtain good resolution. SAR antennae are created by using the satellite's orbital motion, increasing the effective size of the antenna with time. Resolution of a meter is obtained at 10 GHz at lower altitudes.

Subsidence of the Earth's surface is measured with an accuracy of 2 mm, using inter-ferometric synthetic aperture radar (InSAR) that combines two SAR images. In some situations it is possible to detect nuclear explosions of one kton with

InSAR. The most important result from an InSAR observation is to reduce the error in the location of the test to 100 m, much better than seismology. Tests observed with seismology, could be accurately located with InSAR. This is important for CTBT on-site inspections (Fig. 6.3).

Large phased-array radar. Large phased-array radars (LPAR) are located at the *Cobra Dane Radar Station* in the Aleutian Islands and the *Cobra Judy* on the *Observation Island* ship and elsewhere. Cobra Dane operates at 200 MHz, giving a resolution of 1 m. The 96-array pattern has a diameter of 10 arrays. Reentry vehicle size is measured with ground-based radar using *inverse synthetic aperture radar* (ISAR), in which the moving and rotating RV supplies antenna motion. The Doppler shift of a tumbling RV gives size resolution of 10 cm.

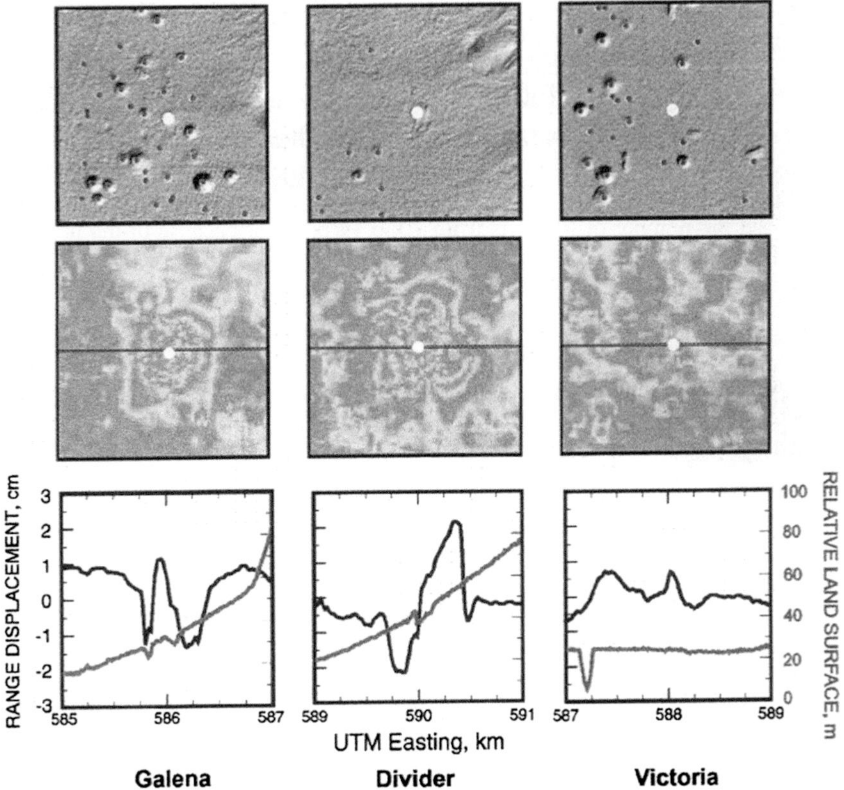

Fig. 6.3 Subsidence after three 1992 nuclear tests at NTS. (*Top*) Digital elevation maps with 10-m resolution locate sites of underground tests (*dots*). (*Center*) Interference pattern displays a cycle of 2.8-cm vertical displacement, showing subsidence during the time interval between two SAR pictures. (*Bottom*) Profile plots of near vertical-displacement (*left scale*) and surface topography (*right scale*). Galena (*left*, 3.9 M_L) and Divider (*middle*, 4.4 M_L) show readily observable subsidence, whereas Victoria (*right*, 2.7 M_L) was not observable (Vincent et al. 2003)

Ballistic missile coefficient. START limits change in throw-weight on *new types* of missiles because extra throw weight increases the capacity to add more reentry vehicles. RV mass and range accuracy are measured with radar Doppler shift measurements that observe RV motion, including the effects of atmospheric drag. Small drag implies a good aerodynamic shape for good range accuracy.

6.5 How Much Verification Is Enough?

Nations need to quantify national security threats to determine if an arms control treaty can indeed be "effectively" verified. The "effective" standard requires that the United States has sufficient, timely warning to respond to an attack from covert warheads that could significantly damage US national security beyond what could be done by the already legal strategic forces. Verification would be carried out through a variety of monitoring inspections. An important one for START is the *re-entry vehicle onsite inspection* (RVOSI), which determines if more than the allowed number of RVs is present on ICBMs, SLBMs and bombers (Sect. 6.2). The decision is often made on the basis of observing whether a bump on the RV bus is filled with an RV. Since it is difficult to have more than one nuclear warhead in an RV slot, the observation of a filled-bump is deemed to be due to a warhead. START II would have raised the number of RVOSIs from 10 to 14 per year because of the possible danger, for example, of adding one or two small warheads to the single-warhead SS-27 or by up-loading formerly-MIRVed Minuteman IIIs and SS-19s.

Problems

6.1 **Unratified treaties**. A country has signed but not ratified a treaty. Is this country responsible for compliance to the treaty terms? What are examples of this, even after failed ratification?

6.2. **Vienna convention on treaties**. What is this? What is the US status?

6.3. **When wise to join?** Under what conditions should US join arms control treaties, and when should the US not do so. Give some examples. Was something gained in the process?

6.4. **League of Nations**. What was the main sticking point? Is that issue still with us?

6.5. **Peaceful technologies**. Which treaties protect dual-use technologies for international commerce. How is this done?

6.6. **BWC issue**. Why is BWC monitoring difficult?

6.7. **Evolution of SALT to New START**. Briefly sketch how the strategic arms treaties evolved.

6.8. **SALT I and ABM**. How did the offense and defense mix in these treaties?

6.9. **Optical resolution**. What determines the optical resolution of satellite images?

6.10. **IR photos**. How do IR and visible resolution compare? What can IR do that visible cannot?
6.11. **Moving satellite film**. Why must satellite film be moved when pictures are taken?
6.12. **CCD reconnaissance**. Why might objects, such as silos or railroads, be easier to find?
6.13. **Digital image processing**. What approaches digitally improve photos?
6.14. **Optical *seeing***. You are inside a shower with your hand placed just inside a translucent shower curtain. Look over the top of the curtain to observe your hand in a distant mirror. How does this image of your hand compare to the image of your hand when it is placed just outside the curtain while you remain in the shower? What can you conclude?
6.15. **Radar resolution**. What determines the size of SAR antennas? What affects resolution?
6.16. **Ballistic missile coefficient**. What are the drop times and terminal velocities for 1, 4 and 9 coffee *Mr. Coffee* filters released from 2-meter height?
6.17. **High frequency components**. Why would you expect a 5-m silo might give low frequency photographic signals, while pixel noise gives a higher frequency?
6.18. **Test ban evolution**. Trace the evolution of nuclear test bans from the LTB to the CTBT.
6.19. **Regional seismic stations**. What are advantage of close-in seismographs?
6.20. **Threshold bias**. Why does geological substrata matter when determining absolute yield?
6.21. **Reliability of nuclear warheads**. What is the most critical element to maintain reliability of nuclear warheads? What tests tell the most about warheads?
6.22. **Verification standard: START versus CTBT**. What does the *effective* verification standard require of START and CTBT? Discuss congressional acceptance of START versus CTBT in terms of the effective verification standard.
6.23. **Militarily significant violations**. Assume the US has 1500 New START warheads comprised of 50 % SLBMs (1/2 at sea and 1/2 in port), 25 % 1-warhead ICBMs in silos, and 25 % bombers. Assume 1500 Russian warheads with 0.8 kill probability against the silos. (a) How many US warheads survive a worst-case Russian attack within treaty limits? (b) Determine US survivable warheads with Russian violations of 200 and 1000 warheads. Are these violations militarily significant?

Seminar: Treaties and Verification

Annual Compliance Report, bolometric infrared detectors, confidence building measures, Corrtex measurement of weapon yield, Environmental Modification Treaty, hot and cold line communication with Russia, infrared detector material, infrasound detection of nuclear tests, INF Treaty status, interferometric synthetic aperture radar detection of nuclear tests, National Environmental Policy Act, NPT renewal in 1995, nuclear testing data, optical bhang meter, peaceful nuclear tests, seismic arrays, Vienna Convention on Treaties, Washington Naval Agreement of 1922, World Court.

Bibliography

Cahn, A. (1998). *Killing detente: The right attacks the CIA*. University Park, PA: Penn State University Press.

Collina, T. (2012, September). Cartwright urges nuclear spending cuts. *Arms Control Today*, 36–37.

Dahlman, J., et al. (2011). *Detect and deter: Can countries verify the nuclear test ban*. NY: Springer.

Drell, S., & Purifoy, R. (1994). Technical issues of a nuclear test ban. *Annual Review of Nuclear and Particle Science, 44*, 285–327.

Drell, S., & Schultz, G. (2007). *Implications of the Reykjavik summit on its 20th anniversary*. Palo Alto, CA: Hoover Press.

Dunn, L. (1980). *Arms control verification and on-site inspection*. Lexington, MA: Lexington Press.

Elachi, C. (1987). *Introduction to the physics and techniques of remote sensing*. NY: Wiley.

Feiveson, H., et al. (2015). *Unmaking the bomb*. Cambridge, MA: MIT Press.

Fetter, S. (1988). *Towards a comprehensive test ban*. Cambridge, MA: Ballinger.

Graham, T. (2002). *Disarmament sketches: Three decades of arms control international law*. Seattle, WA: U. Washington.

Hafemeister, D. (1997). Reflections on the GAO report on the nuclear triad. *SAGS, 6*, 383–393.

Hafemeister, D. (2005). Presidential report: Net benefit analysis of US/Soviet arms control. *SAGS, 13*, 209–217.

Hafemeister, D. (2007). Progress in CTBT monitoring. *Science Global Security, 15*(3), 151–183.

Hafemeister, D. (2014). *Physics of societal issues*. NY: Springer.

Hersh, S. (1991). *The Sampson option*. NY: Random House.

Jeanloz, R. (2000). Science-based stockpile stewardship. *Physics Today, 53*(12), 44–50.

Jones, P. (2014). *Open skies*. Palo Alto, CA: Stanford University Press.

Knight, A. (2012 April 5). The mysterious end of the Soviet Union. *NY Review of Books*, 74–78.

Krass, A. (1997). *The United States and arms control*. New York: Praeger.

Krepon, M., & Caldwell, D. (1991). *Politics of arms control treaty ratification*. NY: St. Martin's Press.

Krepon, M., & Umberger, M. (1988). *Verification and compliance*. Westport, CT: Ballinger.

Lynch, H., Meunier, R., & Ritson, D. (1989). Some technical issues in arms control. *Annual Review of Nuclear and Particle Science, 39*, 151–182.

Moynihan, M. (2000). The scientific community and intelligence collection. *Physics Today, 53* (12), 51–56.

National Academy Sciences. (2002). *Technical issues related to the comprehensive nuclear test ban treaty*. Washington, DC: NAS Press.

National Academy Sciences. (2005). *Monitoring nuclear weapons and nuclear-explosive materials*. Washington, DC: NAS Press.

National Academy Sciences. (2012). *The comprehensive nuclear test ban treaty: Technical issues for the United States*. Washington, DC: National Academy Press.

Nelson, R. (2002). Low-yield earth-penetrating nuclear weapons. *Science Global Security, 10*, 1–20.

Office Technology Assessment. (1991). *Verification technologies*. Washington, DC: OTA.

Pifer, S. (2010). *The next round: The US and nuclear arms reductions after new START* (pp. 8–14). Washington, DC: Brookings.

Reed, T., & Stillman, D. (2009). *The nuclear express*. Minneapolis, MN: Zenith.

Richelson, J. (1998). Scientists in black. *Scientific American, 278*(2), 48–55.

Sabbins, F. (2000). *Remote sensing*. San Francisco, CA: Freeman.

Schaff, D., Kim, W., & Richards, P. (2012). Seismological constraints on low-yield nuclear tests. *SAGS, 20*, 155–171.

Schelling, T. (1960). *The strategy of conflict*. Oxford, UK.

Schultz, G., et al. (2008). *Reykjavik revisited: Steps toward a world free of nuclear weapons*. Stanford, CA: Hoover.

Schultz, G., et al. (2011). *Deterrence: Its past and future*. Palo Alto, CA: Hoover.

Smith, G. (1980). *Doubletalk: The story of SALT I*. NY: Doubleday.

Tsipis, K., Hafemeister, D., & Janeway, P. (Eds.). (1986). *Arms control verification*. Washington, DC: Pergamon.

US Arms Control and Disarmament Agency. (1996). *Arms control and disarmament agreements*. Washington, DC: ACDA.

Vincent, P., et al. (2003). New signatures of underground nuclear tests revealed by satellite radar interferometry. *Geophysical Research Letters 30*, 2141–2145

von Hippel, F., & Sagdeev, R. (Eds.). (1990). *Reversing the arms race*. New York: Gordon and Breach.

Wallace, R., & Melton, H. (2008). *Spycraft, the secret history of the CIA's spytechs*. NY: Penguin.

Zellner, W. (2012 March). Conventional arms control in Europe: Is there a last chance? *Arms Control Today*, 14–18.

Chapter 7
Winding Down the Cold War

Three kilograms of ninety percent enriched weapons-grade uranium were stolen in Moscow.
[US Central Intelligence Agency, 1994]

Although not independently confirmed, a 1998 theft in Chelyabinsk Oblast had 'quite sufficient material to produce an atomic bomb,' the only nuclear theft that has been so described.
[Viktor Yerastov, Chief of Minatom's Nuclear Materials Accounting and Control]

7.1 Then and Now

This chapter discusses progress in winding down the four-decades of Cold War (1945–1992). We didn't obtain a lasting "New World Order" as was discussed at the time. The future is uncertain with the 2014 Russian annexation of Crimea and the 2015 Russian-assisted rebellion/invasion in the Luhansk and Donetsk People's Republics. We are now in an era of tension, compounded by disagreements. There are two sides to these stories. Was Russia over-aggressive in Ukraine, breaking agreements on the boundaries of its former republics? Does Russia feel entrapped and lonely by NATO and EU expansions, which turned Warsaw Pact countries into NATO countries? The October 2014 election showed that Ukraine wanted to move westward towards EU markets, away from the Russian marketplace. This is a major blow to Russia, to loose its main Slavic companion. EU/US won't go to war over these distant changes of national boundaries, but the trade and economic embargos hampered Russia at a time of falling petroleum prices. Will this spat with Russia spill over into other areas where we have common interests? Some suggest the West should refrain from recruiting Eastern Europe, including Ukraine, to its side, but keep it as a neutral buffer. But this is difficult to do and it seems too late for this option. There was discussion on this topic between Secretary of State James Baker and Soviet leader Mikhail Gorbachev, but no strong agreements.

D. Hafemeister, *Nuclear Proliferation and Terrorism in the Post-9/11 World*,
DOI 10.1007/978-3-319-25367-1_7

During the darkest days of the Cold War, the Soviet Union and the US worked closely together on nonproliferation issues, keeping nuclear weapons in the hands of the nuclear-P5 and out of the hands of the non-nuclear-P183. Proliferation policies were kept separate from other policies, as they both built more strategic weapons. Without Russian and Chinese support, it will be difficult to maintain the nonproliferation regime of haves and have-nots in an era of religious expansion.

This chapter reviews the efforts of Russia and US to conclude agreements on the control and reduction in nuclear warheads and military weapon-usable materials. Some negotiations attempted to codify arms limitation and reduction measures, others were aimed at constraining the spread of fissile materials and technologies to the non-nuclear weapon states. A number of the negotiations had elements of both arms control and non-proliferation. Many of the provisions in nuclear agreements between Russia and the US are *transparency measures*—those that give confidence that a state is fulfilling its obligations. Some transparency measures are unilateral and are intended to enhance confidence or goodwill. *Verification measures*, on the other hand, usually require more intrusive monitoring, enough to ensure a high likelihood that the parties are in compliance with a treaty. These measures apply to the nuclear weapon complexes, with the exception of warhead facilities.

7.1.1 Belarus, Kazakhstan, Ukraine

On 1 December 1991, Ukraine voted to separate from the Soviet Union. This was followed by the secessions of the other 13 Soviet republics and by Russia. On 23 May 1992, the United States and the four nuclear capable Soviet republics (Russia, Ukraine, Belarusian, Kazakhstan) signed the Lisbon Protocol, which made these five nations State-Parties to the START Treaty. In addition, all Parties agreed to join the NPT. In Table 7.1, the strategic weapons of the four nuclear former Soviet Republics are listed.

When Ukraine voted to leave the Soviet Union, it stated it would give up all the Soviet nuclear weapons located on its territory. However, it became apparent that money could be obtained from these weapons, as the US bought 500 tonnes of high-enriched uranium, as 4 % reactor fuel, from Russia. On 20 December 1991,

Table 7.1 Location of nuclear weapons in four former Soviet republics

	Kazakhstan	Ukraine	Belarus	Russia	USSR
ICBM/WH	104/1040	176/1240	54/54	1067/4278	1401/6612
SLBM/WH	0/0	0/0	0/0	940/2804	940/2804
Bomber/WH	40/370	34/416	0/0	88/800	162/1600

The numbers are in the ratio of launchers over warheads. Soviets had 62 SSBN submarines in 1992 (*The START Treaty*, Senate Foreign Relations Committee, 18 September 1992)

Ukraine changed its mind, stating it wanted compensation for having helped produce this windfall in uranium economics. When the Ukrainian Parliament, or *the Rada*, ratified START I on 18 November 1993, it declared a number of stiff conditions, which Russia would not accept. After extensive negotiations, Ukrainian President Leonid Kravchuk, US President Bill Clinton and Russian President Boris Yeltsin signed the Trilateral Statement in which they agreed that all warheads would be withdrawn from Ukraine to Russia. For agreeing to return the warheads, Ukraine was paid economic benefits and given guarantees that its territory would not be violated by the other signatories. This agreement was accepted by the Ukrainian Rada on 16 November 1994. The Belarusian parliament ratified START I on 4 February 1993 and Belarus acceded to the NPT on 22 July 1993. Kazakhstan ratified START I on July 1992 and formally acceded to the NPT on 14 February 1994. With these actions, three former (partial) nuclear weapon states returned to the status of nonnuclear weapon states.

Conventional Forces 1991. Before nuclear weapons could be reduced in Europe, it was necessary to constrain conventional military weapons controlled by the Warsaw Pact and NATO. The *Conventional Forces in Europe Treaty* (CFE) was ratified by the Senate on 27 November 1991. The CFE Treaty limited 5 types of treaty-limited equipment (TLE) made up of tanks, armored carrier vehicles (ACV), artillery pieces, combat aircraft and attack helicopters. The controlled area extended from the Atlantic to the Urals. NATO and the Warsaw Pact could each field up to 20,000 tanks, 30,000 armed combat vehicles, 20,000 artillery pieces, 6800 combat aircraft and 2000 attack helicopters, for a total of 78,800 TLE. When CFE was first considered in 1989, NATO had 76,000 TLE, the Warsaw Pact had 197,000 TLE. The Soviet Union owned 151,576 (77 %) of the Pact's 197,000 TLE. Thus, the Warsaw Pact had to destroy 120,000 TLE. However some of the WTO equipment was older and should have been scrapped. NATO had to destroy essentially nothing, but it did cascade transfer TLE from NATO North to NATO South. The ratio of equipment held by the two blocks and the Soviet Union changed swiftly with these major events:

1. CFE Treaty,
2. Collapse of the Warsaw Pact, and
3. Collapse of the Soviet Union,

as shown in Table 7.2. The ratio of TLE between the Warsaw Pact and NATO was initially 2.7. The final TLE ratio between the Russian Federation and the other 21 CFE parties was 0.51. The Soviet–Russia ratio changed from 2.7 to 0.51, a factor of 5.3; this is a tsunami-sized change.

Table 7.2 TLE belonging to CFE Treaty parties

	January 1989	November 1990	EIF + 40 months
USSR	151,576	72,645	52,975
B/CZ/H/P/R	35,000	35,684	25,825
GDR	11,000	10,674	0
Warsaw Pact	197,000	119,003	78,800
United States	15,000	15,246	13,172
NATO minus US	60,000	60,505	65,628
NATO	76,000	75,751	78,800
Ratios			
USSR/NATO	2.0	0.96	0.67
WTO/NATO	2.7	1.57	1.00
SU/21-CFE	1.2	0.59	0.51

The CFE Treaty, Senate Foreign Relations Committee, 19 November 1991

7.2 Control on Warheads and Fissile Materials

Proposals for controlling and accounting for warheads and fissile materials have a long history, dating back to the first meeting of the United Nations General Assembly, in January 1946. The first General Assembly resolution established the UN Atomic Energy Commission, with the mandate to

> make specific proposals… for the elimination from national armaments of atomic weapons and of all other major weapons adaptable to mass destruction.

At the first meeting of the UN commission in June 1946, US Representative Bernard Baruch put forward a proposal for international control, with a call to create an International Atomic Development Authority that would own or manage all nuclear activities for military applications. The proposal also called for the dismantlement of nuclear warheads. Ultimately, the Baruch Plan failed because of the irreconcilable differences between the Soviets and the US. The Soviet Union would not accept the provision for sanctions against violations without the right of a veto by the five permanent members of the UN Security Council. It also wanted a prohibition on nuclear weapons before a verification system was put in place, which the United States would not accept. Comprehensive nuclear disarmament remained on the UN agenda, but reliance on nuclear weapons during the Cold War blocked attempts to achieve even modest measures. With the entry into force of the Treaty on the Non-proliferation of Nuclear Weapons (1970, NPT), the role of the International Atomic Energy Agency (IAEA) was expanded to include monitoring non-nuclear weapon states (NNWS). The five formally recognized nuclear weapon states (NWS) were not required to accept IAEA safeguards on their nuclear facilities. The international experience with implementing IAEA safeguards for non-proliferation purposes is relevant for tasks that would be part of a comprehensive nuclear arms reduction regime.

Fissile materials and the 1967–69 dismantlement demonstration. In 1956 President Dwight Eisenhower proposed a ban on the production of fissile material for weapon purposes. In the following decade, the Warsaw Treaty Organization, the Western countries and many non-aligned states made similar proposals, but no serious negotiations took place. In 1966, the US made a proposal to the Conference on Disarmament (CD) that was limited but remarkable for the times. Under this proposal the US and the USSR would transfer high-enriched uranium (HEU) from weapons to peaceful uses under international safeguards. The US offered to transfer 60 tonnes of HEU under the condition that the USSR would transfer 40 tonnes to peaceful uses. Both states were expected to demonstrate

> the destruction of nuclear weapons to make HEU available for transfer to peaceful nuclear energy under international safeguards, and to halt the production of weapon-usable nuclear materials.

As part of the US Government's assessment of the verifiability of this proposal, the Arms Control and Disarmament Agency (ACDA), working with the US Atomic Energy Commission and the Department of Defense (DOD), created *Project Cloud Gap* for demonstration dismantlement inspections. The experiments were carried out at the Pantex (Texas), Rocky Flats (Colorado), Paducah (Kentucky) and Oak Ridge (Tennessee) facilities. Inspectors were given extensive access to the Pantex facility for close observation and monitoring of weapon dismantlement. At Rocky Flats they monitored the disassembly of warhead pits and separation of materials into plutonium, uranium and other residue. At Paducah they monitored the separation of materials into salvageable categories and the disposal of classified residue. At the Y-12 plant at Oak Ridge, they monitored the disassembly of HEU parts and the melting and casting of HEU into ingots. The inspectors carried minimal equipment, such as cameras, scales, Geiger counters, portable neutron counters and gamma-ray spectrometers. They collected samples for mass spectrometers to measure isotopic concentrations. The experiment monitored 40 warheads undergoing scheduled disassembly, along with 32 fake warheads. The principle behind the experiments was to provide unrestricted visual access to the dismantlement process in order to ensure that warhead dismantlement was taking place. *There was no attempt to conceal classified information.* With this degree of open access, the ACDA report conclusion was no surprise that classified information would be revealed. However, the report also concluded that information "could be protected by redesign of facilities and equipment." This project highlighted the tension between obtaining the needed degree of confidence that weapons were being destroyed and protecting sensitive information—a tension that is still central to efforts to design effective monitoring arrangements.

The Slava experiment. On 5 July 1989 a team of Soviet and US scientists measured gamma-ray spectra with a high purity germanium detector from a Soviet warhead mounted on a SS-N-12 cruise missile on the *Slava* cruiser. The most detectable gamma transitions showed the presence of uranium-235, plutonium-239 and uranium-232. The presence of uranium-232 indicated that the uranium in the Soviet warhead had resided in a nuclear reactor before being used as feedstock for an enrichment plant. The US government was very interested in the results obtained by the Federation of Atomic Scientists and the Natural Resources Defense Council, both non-governmental organizations.

The Soviet–US team also monitored neutron emissions from the warhead located on the *Slava*. A helicopter, carrying neutron detectors, flew at distances of 30–80 m from the warhead. The detectors were designed to observe warheads at distances of 100–150 m with the requirement that the signal must be more than three times the standard deviation (σ) of the background. The neutron data from the passage of the helicopter at 30 m were two to three times greater than the 3σ-background level.

7.3 Warhead Monitoring in INF and START

The major agreements to limit or reduce offensive nuclear arms were negotiated by the two superpowers during and immediately after the cold war, namely the SALT-I and SALT-II agreements, the INF Treaty, the START-I and START-II Treaties, followed by SORT and New START. These agreements focused on delivery vehicles and launchers. Warheads were dealt with mainly through *counting rules* that attributed a certain number of deployed warheads to a particular delivery vehicle, for the following reasons:

1. Ballistic missiles are the major delivery vehicle for nuclear warheads.
2. Ballistic missiles, silos, submarines and bombers are much larger and easier to count than nuclear warheads. They are much more difficult to hide than warheads or their fissile material components. Technologies such as templates, attributes and information barriers were not available at that time to properly verify warhead dismantlement without the risk of revealing sensitive design information. National technical means (NTM) could assess delivery vehicle inventories. Neither the US or USSR were willing to accept the intrusiveness required to verify limits on warheads.
3. The number and characteristics of the Soviet and US deployed strategic delivery vehicles and launchers provide better measures of strategic significance of nuclear arsenals than the size of warhead or fissile-material stockpiles. The traditional concerns of both states were with warheads that can be delivered rapidly and accurately over long distances, although delivery by aircraft, ships and trucks was also a concern.
4. Modern strategic delivery vehicles are expensive, typically costing ten times more to develop, produce and maintain than the nuclear warheads they carry. The elimination of delivery vehicles creates a large barrier to reconstituting deployable nuclear weapons.

INF and START both contain provisions relating to warheads. The INF Treaty preceded the wind-down of the Cold War. INF was the first Soviet–US agreement to eliminate an entire class of nuclear weapons, banning possession and deployment of ground-launched missiles with ranges of 500–5500 km. In carrying out its INF obligations, the USSR destroyed 1800 missiles, capable of carrying over 3000 warheads, and the US destroyed 850 single-warhead missiles.

Certainly INF was a success, but Russian tests of ground-based cruise missiles (GLCM) have caused complications for the INF Treaty in 2014. One of the most important verification issues was the need to determine that the banned SS-20 missile was no longer deployed by the Soviet Union. This was difficult, because the first stage of the permitted long-range (strategic) SS-25 intercontinental ballistic missile (ICBM) and the prohibited intermediate-range (theater) SS-20 missile are very similar. In addition, because SS-25 canisters are larger than SS-20 canisters, an SS-25 canister could contain an SS-20 missile.

The problem was further complicated by Soviet deployment of SS-25s at former SS-20 bases. SS-25s carried one warhead, while SS-20s carried three. This asymmetry makes the geometrical pattern of neutron and plutonium gamma-ray emissions from the two-missile types very different. Under INF, radiation detection equipment measures the flux of neutrons while the missile is in its canister. This gives no critical information on warhead design. Under INF, parties had the right to establish permanent continuous monitoring systems. The US built a perimeter-portal continuous monitoring (PPCM) system at the Soviet Votkinsk Machine Building Plant, 500 km east of Moscow, and the USSR monitored the US Hercules Plant at Magna, Utah, where Pershing II rocket engines were produced. The PPCM monitoring facility operated at Votkinsk from 1988 to 2001. Under INF, treaty inspectors measured the length and weight of all objects entering and leaving the missile factory. All road and rail shipping-containers, large enough to hold a SS-20 missile, were made available for inspection at Votkinsk. X-ray imaging with a modified version of the *CargoScan* commercial scanner was used. The X-ray images showed the length and diameter of the first stage of the missiles to ensure that they were not SS-20 first stages. In addition, inspectors visually inspected and measured missiles inside their canister eight times a year. This random inspection of canisters provided a great deterrent to cheating. US inspectors also patrolled the 5-km perimeter fences around Votkinsk.

7.4 Post Cold-War Initiatives

The end of the Cold War offered both hope and danger. Soviet/Russian and US leaders saw an opportunity to transform their relationship from a hostile to a cooperative one, reducing the risks that the two nuclear arsenals posed to both military forces and homeland, and more broadly, to international security. The collapse of the Soviet Union raised fears for the security of thousands of Soviet nuclear warheads, their proliferation as *loose nukes* to other countries and as tonnes of weapons-usable Pu and HEU. In responding to these risks and opportunities, Russian and US leaders undertook remarkable initiatives that provided basic foundations for much more comprehensive arrangements to control nuclear warheads and materials. The START Treaties are discussed in Chap. 6.

Reciprocal-unilateral initiatives on nuclear weapons. In 1991 President George H.W. Bush announced the withdrawal of all US ground- and sea-launched tactical nuclear weapons to the US. All of the ground-launched and about half of the sea-launched weapons would be destroyed. Soviet President Mikhail Gorbachev responded with the announcement that all Soviet tactical nuclear weapons would be withdrawn to the Russian Federation, and that nuclear artillery, ground-launched missile warheads and nuclear mines would be destroyed. In 1992 Russian President Boris Yeltsin confirmed and extended Gorbachev's pledges. In addition to destroying all ground-launched tactical warheads, he announced that Russia would destroy half of its air-launched tactical warheads, half of its nuclear warheads for anti-aircraft missiles and one-third of its tactical sea-launched nuclear warheads. Full implementation of the pledges in the 1991–92 Presidential Nuclear Initiatives (PNIs) stated that 5000 US tactical warheads would be destroyed. The number of Russian warheads scheduled for destruction is more difficult to judge; the US Central Intelligence Agency gave an estimate of 5000–15,000 warheads. Under the PNIs, the unilateral reductions were not subject to monitoring, nor were there meaningful transparency measures. It is therefore not known whether the reductions were carried out completely.

Fissile materials. The end of the cold war left Russia and the US with large stockpiles of plutonium and HEU, far more than they could possibly need for nuclear weapons production or maintenance of stockpiles. The 2013 Annual Report of the International Panel on Fissile Materials Concluded the following:

> The global stockpile of high enriched uranium (HEU) as of the end of 2012 is estimated to be about 1380 ± 125 tons. This is sufficient for more than 55,000 simple, first-generation implosion fission weapons. About 98 % of this material is held by the nuclear weapon states, mostly by Russia and the United States…. The global HEU stockpile has been shrinking. Over the past two decades, about 630 tons of HEU has been blended down, mostly by Russia, which has eliminated a total of 488 tons as of the end of 2012. This includes 473 tons of excess weapon-grade material. The United States, which has eliminated about 141 tons of mostly non-weapon-grade HEU, has chosen to set aside 152 tons of excess weapons HEU for a naval fuel reserve. The global stockpile of separated plutonium in 2012 was about 495 ± 10 tons. Almost half of this stockpile was produced for weapons, while most of the rest has been produced in civilian programs in nuclear weapon states. As a result, about 98 % of all separated plutonium is in the nuclear weapon states. Most of the uncertainty is due to a lack of official information about Russia's plutonium production history.

Both governments gradually came to the conclusion that continued production of fissile material was unnecessary. They took unilateral action during the late 1980s and early 1990s to close down the fissile material manufacturing facilities that were still in operation. After the accident at the Chernobyl nuclear power facility in 1986, there was widespread public concern, particularly in the US, about the environmental hazards associated with nuclear energy in general and plutonium production in particular. The resulting public pressure added further impetus to the decision to stop the production of fissile material.

The process of closing US production facilities extended over more than two decades. Production of HEU for weapons ceased in 1964, although production of HEU continued for naval and research reactors until 1988. The US Government announced in November 1991 that all HEU production would be suspended. Plutonium production reactors were closed beginning in 1964 as new reactor designs went on-line and as the need for plutonium diminished. The last two operating production reactors, located at Savannah River, South Carolina, were closed in 1988 because of safety concerns. The House of Representatives passed an amendment to the Defense Department budget in July of the following year urging the president to negotiate with the Soviet Union a bilateral ban on fissile material production for warheads. Finally, in July 1992 President Bush announced that, as part of a non-proliferation initiative, the US would no longer produce fissile material.

The Soviet Union stopped the production of weapon-grade uranium in 1988 and of plutonium in 1994 (except at three power reactors, which are now being closed). President Yeltsin, reiterating an offer made earlier by Gorbachev, suggested in January 1992 that Russia and the US negotiate a bilateral fissile material production cut-off treaty. An announcement was made that same month that Russia would stop all production of weapon-grade plutonium by 2000 regardless of whether an agreement was reached. However, the three production reactors are still operating, to provide heat and power for local residents. The Russian and US governments tried for decades to form a plan to replace the reactors with an alternative source of energy.

Biden amendment. The 1992 US Senate debate on ratification of the START I Treaty raised concerns about Russia's ability to rapidly redeploy warheads that have been removed from their delivery vehicles. There was also great concern about the security of nuclear weapons and materials. To address these concerns, an amendment to the START resolution of ratification was proposed by Senator Joe Biden to require the President to seek an appropriate arrangement

> to monitor the numbers of nuclear stockpile weapons on the territory of the parties to this Treaty; and the location and inventory of facilities on the territories of the parties to this treaty capable of producing or processing significant quantities of fissile materials.

The Biden amendment provided a major impetus for the US to explore technical and policy approaches to monitor warheads, including production and dismantlement.

Cooperative Threat Reduction. The collapse of the Soviet Union raised fears of a loss of control over thousands of deployed strategic and non-strategic nuclear weapons and hundreds of tonnes of fissile material—the scenario for a proliferation nightmare. In December 1991, a bipartisan effort led by Senators Sam Nunn and Richard Lugar addressed these dangers with a proposal that became law. The legislation authorized the president to transfer up to $400 million from the 1992 defense budget, making DOD the first major agency engaged in what became known as the Cooperative Threat Reduction (CTR) program. After 20 years, US assistance for CTR and other programs totaled some $10 billion.

In the early years, CTR focused on assisting Belarus, Kazakhstan and Ukraine in their efforts to return all former—Soviet warheads on their territories to Russia and to dismantle or destroy the associated strategic nuclear delivery vehicles and silos.

CTR also provided assistance to Russia to eliminate excess strategic nuclear arms on its territory. Altogether, CTR facilitated the dismantlement of over 2000 former Soviet strategic missiles and launchers. CTR also contributed to funding construction of the nuclear materials storage facility at Mayak. The CTR funded diverse activities as the provision of nuclear material containers, the refurbishment of Russian railway wagons for the transport of nuclear materials and the acquisition of nuclear accident response equipment.

The Russian and US governments recognized the risk of theft or diversion of fissile material posed *a clear and present danger to national and international security*. Russian–US programs improved fissile *materials protection, control and accounting* (MPC&A) in the former Soviet Union. These programs were shifted from DOD to the DOE to identify more accurately facilities for MPC&A upgrades and to define responsibilities for participating organizations.

CTR was a remarkable initiative, undertaken in response to extraordinary circumstances. Engaging directly to ensure security of nuclear warheads and fissile materials gave the US unprecedented access to Russian facilities. Despite the difficulties involved in the implementation of many of its programs, CTR represents an essential part of the foundation for more comprehensive limits.

Laboratory-to-laboratory programs. Not surprisingly, the implementation of new programs was initially slow, given the long tradition of secrecy in the Russian nuclear complex. To circumvent these difficulties and to take advantage of the potential to build trust through direct contacts between scientists, the DOE national laboratories and their Russian counterparts initiated a wide variety of contracts for joint research on technologies for monitoring, physical security and accountancy of nuclear weapons and materials. Established in 2000, the National Nuclear Security Administration (NNSA), a semi-autonomous agency within the DOE, now has responsibility for DOE cooperative security programs, including MPC&A. The laboratory-to-laboratory contracts are intended to transfer successful technologies between the parties in order to enhance transparency and arrive at the best monitoring options. The activities are wide-ranging and include:

a. physical security and containment of facilities;
b. radiation detection techniques;
c. fissile material accounting;
d. plutonium disposition in general;
e. plutonium storage at Mayak;
f. purchase of Russian HEU; and
g. monitoring warhead dismantlement.

The laboratory-to-laboratory exchanges helped technical experts of both states to become familiar and confident with monitoring techniques and information barriers. For example, cooperative gamma-ray measurements of classified objects were carried out without releasing classified information. Many believe that such programs progressed successfully because they were developed away from the political spotlight. They engaged technical experts who shared knowledge and an appreciation of the issues at the technical level.

7.5 Initiatives to Limit Fissile Materials

To carry out broad initiatives to control and reduce nuclear weapons and materials stockpiles, specific programs for Russian–US activities were implemented in the 1990s. Important programs that sought support and alternative employment for Russian nuclear scientists and commercial activities for the former closed nuclear cities are mentioned briefly. The varied programs are listed under four headings:

- diplomatic frame-work;
- production and disposition of fissile material;
- improvement of fissile material MPC&A; and
- monitoring of warheads.

Although Russia and the US appeared to be moving towards an initial regime for warhead and fissile material reductions, Russia broke off the talks in late 1995 and they were not resumed. Some US experts believe that the agenda was simply too broad and ambitious for the time and circumstances. Matthew Bunn cites three reasons for "the transparency that never happened."

1. The historical legacy of tsarist and communist secrecy made the Russian Government *extraordinarily reluctant to open nuclear secrets.*
2. Many in the US Government were equally unwilling to make US facilities accessible to Russia.
3. The US Government never offered significant strategic or financial incentives to overcome Russian reluctance.

7.5.1 Fissile Material Cut-off Treaty

Four of the five nuclear weapon states have officially declared they have stopped production of HEU and plutonium for nuclear weapon purposes. In a major initiative, the 1992 Russian–US informal agreement to ban the production of fissile materials was expanded to create the concept of a multilateral Fissile Material Cut-off Treaty (FMCT). On 27 September 1993, President Clinton proposed at the United Nations a multilateral agreement to halt the production of HEU and plutonium for nuclear explosives. In December 1993 the General Assembly adopted by consensus a resolution calling for the initiation of negotiations. The January 1994 the Clinton–Yeltsin summit meeting produced a joint statement calling for "the most rapid conclusion" of the FMCT.

The FMCT concept focuses primarily on the five NPT-recognized NWS and the four de facto NWS (North Korea, India, Israel and Pakistan), but all other states would be invited to join the regime. In 1995 the CD agreed by consensus to establish an ad hoc committee to negotiate a treaty, but progress stalled over a number of issues. For example, India and a few other states have declared that they would not sign an FMCT unless a strict deadline was set for the NWS to fulfill their

NPT Article VI obligations to eliminate their nuclear weapons. Issues of ballistic missile defense, weaponization of outer space (raised by China) and no-first-use of nuclear weapons have blocked progress. Since the CD operates on a consensus basis, a deadlock can easily be created, as happened in this case.

The cost of verifying an FMCT would vary greatly depending on the approach adopted. It is unlikely that the treaty's monitoring provisions would apply to stockpiles of fissile material produced in the NWS before it entered into force. The FMCT could establish safeguards at all the power plants in the NWS, which would raise costs since there are as many nuclear power plants in the NWS as there are in the NNWS. However, safeguarding all reactors worldwide would not double the IAEA's burden since the IAEA also performs other tasks. It is envisaged that the IAEA would conduct routine FMCT inspections at plutonium and HEU production and storage sites in the NWS.

Precedents relevant to an FMCT. A number of international arrangements offer precedents and experience that could be useful for an FMCT. A ban on the production of HEU is monitored under the 1989 Hexapartite Enrichment Project, in which six states (Australia, Germany, Japan, the Netherlands, the UK and the USA) placed all their civil centrifuge plants under IAEA safeguards. Monitoring to distinguish between HEU and low-enriched uranium (LEU) is an integral part of this arrangement. This monitoring could be extended to all types of enrichment plants. States which have nuclear-powered submarines have asked for an exemption for HEU fuels for naval propulsion. This issue could be avoided by designing the next generation of naval power plants to operate at levels well below 90 % uranium-235 enrichment, which several states have done.

7.5.2 Plutonium Management

Most NWSs have sufficient weapon-grade plutonium, so they no longer reprocess military spent fuel. A ban on reprocessing is easy to monitor on a permanent basis when plants are closed, but it is more complicated if the plants continue to be used to reprocess civilian spent fuel to obtain separated plutonium for fabrication into mixed oxide (MOX) fuel. The reprocessing plants in the NNWS were originally designed to accommodate IAEA material accounting measurements, but plants in the NWS were not. The monitoring of plutonium under an FMCT would have to ensure that new plutonium remained inside the civilian nuclear fuel cycle and not in weapons. To obtain accurate material balances and track material throughout its use, it would be necessary to measure flow rates at predetermined key measurement points in the plant.

In order to be confident that clandestine production of HEU or plutonium is not taking place in the NNWS, the IAEA has instituted the Strengthened Safeguards System (INFCIRC/540), by which states are required to make declarations about their research and development for enrichment and reprocessing technologies. INFCIRC/540 also establishes environmental monitoring to detect clandestine reprocessing and enrichment plants. Special inspections already allow further inspection of a declared site to confirm declarations. Special inspections can also be applied at undeclared sites. (The IAEA had requested such inspections in North Korea.) The inspection regime will allow managed access to undeclared facilities in order to confirm the absence of undeclared production.

Disposition of excess plutonium. Recognizing the greater proliferation risks from excess weapon-grade plutonium, President George H.W. Bush's National Security Advisor, General Brent Scowcroft, asked the National Academy of Sciences (NAS) in 1992 to study options for plutonium management and disposition. In a two-volume study released in 1994 and 1995, the Academy's Committee on International Security and Arms Control (CISAC) recommended that Russian and US excess weapon plutonium be converted into a form that is at least as inaccessible for weapon use as the plutonium in spent-fuel rods from civilian nuclear power production. This would put weapon plutonium in the category of risks posed by spent fuel, which the CISAC also strongly recommended addressing. CISAC determined two approaches were acceptable to fulfill the *spent-fuel standard*:

- Encapsulation of diluted plutonium in a radioactive matrix (immobilization) for eventual geological disposal with other high-level nuclear waste; and
- Use of plutonium as MOX fuel in existing reactors without subsequent reprocessing.

Russia and the US signed the Plutonium Management and Disposition Agreement (1 September 2000): Each party must remove 34 tonnes of plutonium from its nuclear weapons program and convert it into forms that will be irreversibly removed from military purposes. The agreement is to remain in effect until the plutonium is irradiated to a specified level or is immobilized for geological storage. Fifteen years later, progress on converting plutonium into MOX fuel has been slow and expensive. There are now calls to adopt the formerly rejected approach of direct burial.

Mayak storage facility agreement. In 1991 Minatom Minister Viktor Mikhailov requested financial help to build a large facility near Tomsk to store excess weapon-usable materials under secure conditions. In January 1996 US Secretary of Defense William Perry and Mikhailov agreed on the construction of a storage facility for excess weapon-usable fissile materials at Mayak (Chelyabinsk-65). The Mayak facility was designed in 1996 to accommodate 50,000 canisters filled with 66 tonnes of plutonium and 536 tonnes of HEU. The first wing cost about \$500 million, with the US paying 90 %. It is designed to store 25,000 containers, holding a total of 50 tonnes of plutonium and perhaps 200 tonnes of HEU. The two states agreed on "joint accountability and transparency measures" that permits the US to confirm Mayak's holdings. The US Congress expects some confirmation that the materials are weapon-usable, but it is difficult to verify the plutonium originated from dismantled warheads. This requires measuring attributes of the plutonium pits when they are brought to the Russian pit processing and packaging facility for conversion into spheres or hockey-puck shapes. The monitoring arrangement grants the some US access to Mayak, but with modest monitoring.

Russian spent-fuel repository. Another approach is to build a global spent fuel repository in Russia. A repository that held 10,000–20,000 tonnes of spent fuel could raise $20 billion for Russia. The availability of such a repository could reduce the pressure to reprocess for many countries. Russia appears to be planning to store its spent fuel for 10 years before reprocessing it to make MOX fuel. In addition, a geological repository is needed for 40,000 tonnes of foreign-owned, US-origin spent fuel. There is some pressure to reprocess the spent fuel to reduce the need for storage. This would obtain 400 tonnes of reactor-grade plutonium. In July 2001 President Vladimir Putin signed a law that allowed Russia to import spent fuel for storage and reprocessing. This presents many legal and political difficulties. The US Nuclear Non-proliferation Act of 1978 constrains transfer and reprocessing of US-origin spent fuel and requires that safeguards will be maintained. Russia's civil nuclear exports to India and Iran could also complicate an approval from the US.

HEU management. HEU poses a more serious proliferation danger than plutonium since it is easier to manufacture into nuclear warheads. HEU is not a significant spontaneous neutron emitter and can be fabricated into a nuclear warhead with the simple gun-type design. At the same time, HEU has the great advantage that it can be easily converted into LEU fuels that have considerable commercial value. By contrast, the use of plutonium in MOX fuels is costly. For these basic economic reasons, significant progress has been made in reducing the Russian and US excess HEU stockpiles, while little progress has been made in disposing of excess plutonium. Under the 1993 HEU Agreement, the US purchased 500 tonnes of Russian HEU as down-blended to LEU. Roughly speaking, the original price of $12 billion was based on $8 billion for enrichment services (separative work units, SWU) and $4 billion for the natural uranium feed.

7.6 Warhead Monitoring After START

The Joint Statement issued at the conclusion of the March 1997 Helsinki summit meeting called for a START III agreement to include

> measures relating to the transparency of strategic nuclear warhead inventories and the destruction of strategic nuclear warheads… to promote the irreversibility of deep reductions including prevention of a rapid increase in the number of warheads… explore, as separate issues, possible measures relating to… tactical nuclear systems, to include appropriate confidence-building and transparency measures, and [to] consider the issues related to transparency in nuclear materials.

Monitoring warheads is difficult because warheads are much smaller than missiles and bombers, and their design and construction details are state secrets. It would be necessary to monitor the birth and death of warheads and their status in stockpile. The US dismantled 12,000 warheads at Pantex between 1990 and 1999, and now dismantles about 300 per year, with 2500 awaiting dismantlement (2014). With the exception of the monitoring provisions specified in INF and START,

Table 7.3 Global nuclear weapon estimates

Russian federation	8000 (4300 operational, 3700 excess)
United States	7300 (1980 deployed, 2540 await dismantlement)
France	300
China	250
United Kingdom	225
Israel	80
Pakistan	100–120
India	90–100
North Korea	<10

Kristensen and Norris (2014)

described above, warheads remain outside the bounds of transparency or verification measures. In the years to come, this will be the greatest challenge in the development of a comprehensive regime to control warheads and fissile materials. In the 1990's the US and Russia researched a variety of technical ways to monitor the presence and counting of nuclear warheads, but things are different in 2015.

7.6.1 NAS on Monitoring Warheads

The 2005 National Academy of Sciences study, *Monitoring Nuclear Weapons and Nuclear Explosive Materials* concluded that "a system could be designed such that inspection of a relatively small sample could be sufficient to prove a high degree of confidence in the authenticity of all weapons and components." Two basic approaches are as follows:

Attribute identification. If it quacks like a duck, it is a duck. Attributes cannot determine a specific warhead type, but templates can. Attributes give less information, but they are less risk to national security, as they do not store sensitive weapon information on foreign soil. A 1 kg spherical piece of plutonium may not be a pit, but to produce it is almost the cost of producing a pit. The authenticity of a plutonium pit is determined with measurements of six unclassified parameters:

- Is plutonium present?
- Is it weapons-grade plutonium? (240/239 ratio less than 0.1)
- What is the age of plutonium since separation?
- Is plutonium mass greater than half a kilogram?
- What is the symmetry of the plutonium mass?
- Is the plutonium in the form of plutonium oxide?

Template identification. Each primary and secondary of a weapon emits unique gamma radiation patterns. To gain experience, DOE technical teams carried out monitoring experiments on 30 warheads and their pits and secondaries. See Figs. 7.1

Fig. 7.1 Negotiations in Moscow and Kiev, December 16–22, 1991. Moscow in December 1991 (*left*), Soviet Deputy Foreign Minister Victor Karpov and US delegation (*center*), Tom Cochran (*right*), Natural Resources Defense Council, Harold Agnew, former Los Alamos Director, and William Jack Howard, former Vice-President of Sandia National Laboratory (D. Hafemeister)

Fig. 7.2 Press conference in Kiev in which Ukraine takes back its promise to give up Soviet weapons, unless paid their fair share (which took two more years) (*right*). Discussing INF/START verification implementation details with General Vladimir Medvedev, Director of Soviet On-Site Inspection Agency, Ministry of Defense (*center and right*) (D. Hafemeister)

and 7.2 for plutonium gamma-ray spectra. A gamma ray spectrum from each warhead is obtained during an inspection. The intensity of the various peaks depends on the isotopics, material composition, geometry and configuration of materials and components. The ratios of the peaks are the same for weapons of the same type. Templates are the most accurate way to do warhead monitoring, as they can identify specific warhead types. Arms-control advocates generally prefer templates, while those more concerned with national security secrets favor attributes.

Information barriers: Reverse engineering of classified gamma-ray spectra can lead to design information, but this information can be protected with information barriers and a closed process. It is imperative that sensitive weapon design information, obtained in the measurements, be kept secret. In reality, Russia probably would gain little new knowledge from the data to affect national security, but that argument won't convince those in charge of keeping the secrets. Information barriers can prevent electromagnetic transmissions through barriers. The information inside the barrier must not have memory when the power is removed (Figs. 7.3 and 7.4).

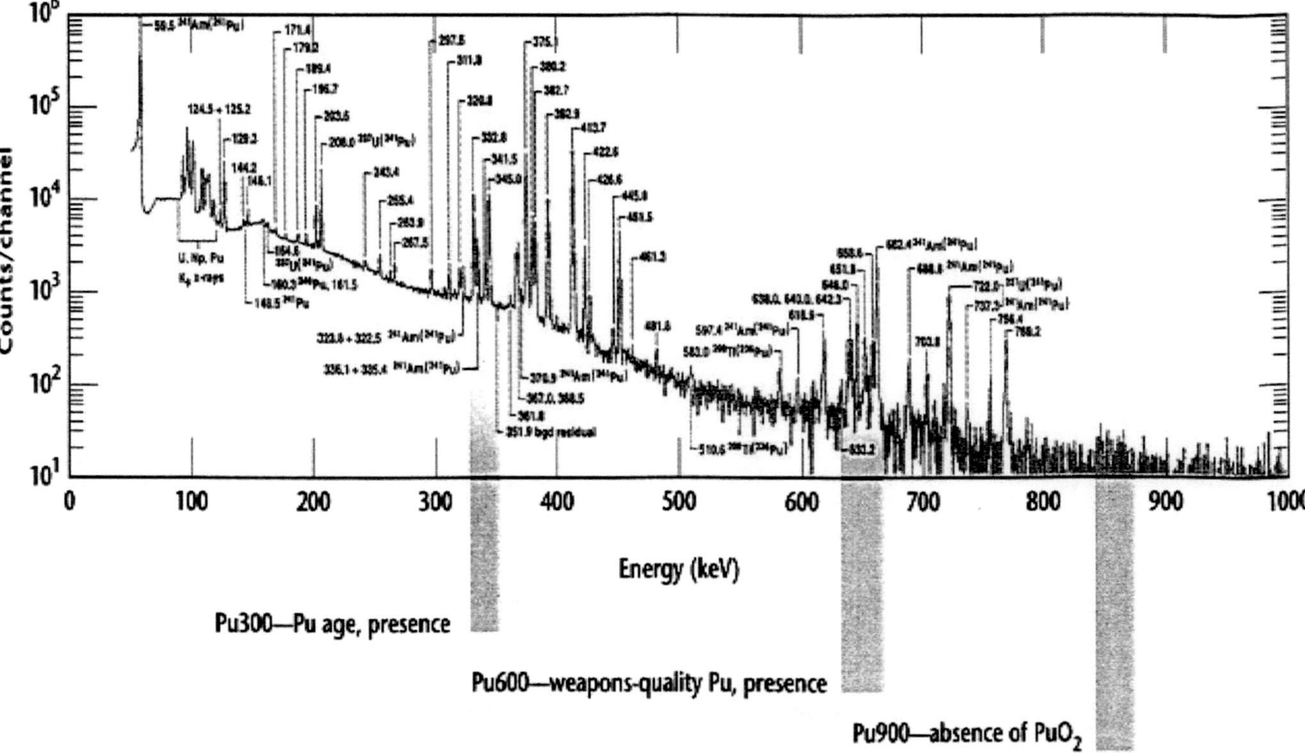

Fig. 7.3 $Pu^{300}/Pu^{600}/Pu^{900}$ monitoring. By monitoring gamma-ray windows near 300, 600, and 900 keV it is possible to determine (1) Pu presence, (2) Pu age since reprocessing, (3) Pu content to determine if Pu is weapons-grade, and (4) absence of plutonium oxide (along with other measurements). In addition, gamma-ray spectra can give a minimum Pu mass estimate. However, an estimate for Pu mass is more accurately obtained through neutron counting (*Technology R&D for Arms Control*, Department of Energy, 2001)

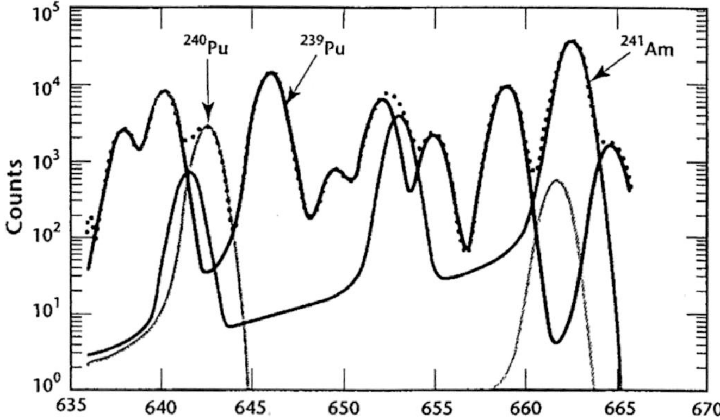

Fig. 7.4 Pu quality. The gamma-ray window between 635 and 665 keV displays transitions from both ^{239}Pu and ^{240}Pu. Weapons-grade Pu contains 6 % ^{240}Pu, while reactor-grade contains more than 20 % ^{240}Pu. Monitoring at the Mayak storage facility near Ozersk will use a ratio of $0.1 = {}^{240}\mathrm{Pu}/{}^{239}\mathrm{Pu}$ to separate the two materials (*Technology R&D for Arms Control*, Department of Energy, 2001)

7.7 Quo Vadis After Cold War?

Negotiations and initiatives to reduce Cold War nuclear arsenals and to strengthen transparency have led to cooperative programs and measures that would have been inconceivable a decade earlier. Much has been accomplished in enhancing controls over fissile materials and establishing mutual monitoring rights. Security at Russian nuclear facilities is enhanced through the MPC&A program, and excess fissile materials are being constrained with the construction of the Mayak storage facility and the HEU Agreement. In 2015, we find such progress at a standstill, with little hope.

The National Research Council's 1999 study concluded that there had been significant progress with Russian materials control, but that there was much more work to be done. It also concluded that the Russian MPC&A program would be a "high-priority national security imperative for the United States." The MPC&A program addressed the following deficiencies in Russia:

- lack of unified physical protection standards and inadequate defenses within sites;
- lack of perimeter-portal monitors to detect nuclear materials leaving sites;
- inadequate central alarms and assessment and display capabilities;
- inadequate protection of guards from weapons and an inadequate guard force;
- lack of material accounting procedures to detect and localize nuclear losses;
- inadequate measurements of waste and scrap nuclear materials during reprocessing, manufacture and transport; and
- antiquated tamper-indicating seals and tags that fail to provide timely detection.

The NRC study recommended long-term indigenization of MPC&A activities and stressed the importance of nurturing Russian ownership of technical solutions resulting from the Russian–US programs. Former Soviet nuclear weapon scientists were faced with the stark choices of unemployment, work in a non-nuclear government job, emigration to another country or conversion of their skills for work in the civilian sector. The Initiatives for Proliferation Prevention program was a cooperative Minatom–DOE program to assist scientists with nuclear weapon expertise to apply their skills to development and product manufacturing in the commercial sector. NNSA claimed the program provided alternative, peaceful employment to roughly 8000 former Soviet specialists on weapons of mass destruction. Another program was the Nuclear Cities Initiative, which improved the commercial sector in Russia's 10 formerly secret nuclear cities. The International Science and Technology Centers in Moscow and in Kiev provided former Soviet nuclear weapon scientists with opportunities in non-military research.

The 2001 *Russian Advisory Task Force*, chaired by former Senator Howard Baker and former Presidential Counselor Lloyd Cutler concluded the following:

- The most urgent national security threat to the United States is the danger that weapons of mass destruction, or weapon-usable material in Russia, could be stolen and sold to terrorists or hostile states and used against US troops abroad or citizens at home.
- The current non-proliferation programs of the DOE, DOD and related agencies have achieved impressive results, but their limited mandate and funding fall short of what is required to adequately address the threat.
- The president and congress face the urgent national security challenge of devising an enhanced response proportionate to the threat. The panel recommended $30 billion in additional funding over the next decade, which would be 1 % of the projected US defense budget for this period.

7.7.1 Lessons Learned

Over the past few years, progress has waned as competing pressures in each state caused delays. The following problems must be resolved if progress is to be renewed:

Access rights and reciprocity. The lack of access to critical facilities in both states adversely affected the ability to win consensus on monitoring regimes. US officials have made many more visits to Russia than Russian officials have to the US since Russia has more excess material, some of which is not adequately guarded, with more MPC&A measures needed. US is asking feels entitled to access rights because it is purchasing Russian HEU and funding the Mayak storage complex. For the overall benefit of their cooperation, it is clear that the US should ensure the development of as much symmetry as possible between the two sides.

Degree of monitoring. The level of monitoring can rise with increased experience and trust, as in the case of monitoring the HEU Agreement.

Secrecy. The former Soviet Union was often obsessed with secrecy, but the US also exhibits this tendency. Segments of the US Government are negative towards the constraints of mutual monitoring. The concerns about the loss of secret information from Los Alamos sparked the creation of the National Nuclear Security Administration and the introduction of lie-detector tests at the national laboratories, which affected staff morale. While there are legitimate reasons for keeping national security information secret, relatively harmless facts are also kept secret, which impedes Russia–US progress in reducing the nuclear threat. The US favors *transparency measures*, but in Russia there is a fear they allow the stronger party to spy on the weaker one. The US has learned a great deal about the Russian nuclear complex. Although this knowledge may not be of great assistance to the US military, it is often hard to convince Russian officials of this.

Entanglement with other issues. Cooperation and progress were slowed by unrelated issues; such as the US involvement in wars in Bosnia/Herzegovina and Kosovo, the enlargement of NATO, the abandonment of the ABM Treaty and the planned deployment of missile defenses. Russian annexation of Crimea and the wars with Syria and the Islamic State in Iraq and Syria have magnified this issue. Will Russia and China work together to prevent further cooperation on international proliferation. These issues adversely affect Russian–US relations, they made the task of improving controls over nuclear weapons and materials harder to achieve.

Diplomatic strategy. The eagerness of the US to move forward on a large and complex agenda in 1995 may have frightened Russia into pulling out of negotiations on the Agreement for Cooperation and taking a more hesitant position concerning transparency and irreversibility. The US has more personnel available to conduct negotiations, thus causing Russia to suffer from 'negotiation fatigue'. This may be one of the reasons why Russia prefers a slow, 'step-by-step' approach. Ultimately, both states will act only when they view a particular arrangement as beneficial to their national security. Diplomacy is difficult in 2015.

Incentives. The financial assistance, which the US provided Russia in exchange for monitoring rights, has created an incentive that has sometimes helped move the agenda forward. Once the US completed its purchase of the Russian HEU, paid for the Mayak facility and helped with MPC&A, it is less clear what type of financial arrangements can promote mutual monitoring. National pride, the fear of revealing secret information and the declining price of Russian oil all contribute. Funding alone will not be enough to determine the best approach to devising the best arrangements, and there are reasons to believe this approach should be gradual, with a negotiation strategy based on unilateral measures and executive agreements.

Securing the Bomb. The Nuclear Threat Initiative commissioned Matthew Bunn at Harvard's *Managing the Atom* program to oversight US efforts to secure nuclear weapons materials. Bunn and his colleagues worked a decade on the analysis: This is what the Securing the Bomb Project viewed as the highest risks in 2010:

- Pakistan, where a small and heavily guarded nuclear stockpile faces immense threats, both from insiders who may be corrupt or sympathetic to terrorists and from large-scale attacks by outsiders;
- Russia, which has the world's largest nuclear stockpile in the world's largest number of buildings and bunkers; security measures that have improved dramatically but still includes important vulnerabilities (and need to be sustained for the long haul); and substantial threats, particularly from insiders, giving endemic corruption in Russia; and
- HEU-fueled research reactors, which usually (though not always) use stocks of HEU, in forms that would require some chemical processing before they could be used in a bomb, but which often have only the most minimal security measures in place—in some cases little more than a night watchman and a chain-link fence.

7.7.2 Describing Progress

- During *FY2009*, security and accounting upgrades were completed at 29 additional weapons-usable nuclear material buildings in Russia, bringing the total for such buildings upgraded in Russia and the Eurasian states to 210, only 19 short of the target of 229 buildings to be completed through *FY2012*.
- Since President Obama launched the four-year nuclear security effort, four countries have eliminated all the weapons-usable nuclear material on their soil, with US help. To date, the United States has helped remove all HEU from more than 47 facilities in countries around the world.
- Discussions about eliminating all HEU in several developing or transition non-nuclear-weapon states with the largest HEU stocks are well advanced.
- Cooperation to improve nuclear security is continuing in Pakistan, though the specifics are classified; the United States and Russia have greatly broadened their exchanges of best practices, efforts to strengthen the security culture, and cooperation to ensure effective nuclear security will be sustained for the long haul; and detailed dialogue with China on improving nuclear security and accounting is continuing.

7.7.3 Steps to Control Fissile Materials

- Build a sense of urgency and commitment worldwide.
- Broadened consolidation and security upgrade efforts.
- Get the rules and incentives right.
- Take a partnership-based approach.
- Broadened best practices exchanges and security culture efforts.
- Create mechanisms to follow-up and to build confidence in progress.

- Build a multilayered defense.
- Provide the needed leadership, planning, and resources.

***Nuclear threat initiative* grades** nuclear materials protection and accountability programs in 50 nations to determine the viability of these programs. Below are listed the criteria for the NTI Nuclear Materials Security Index, along with 18 sub-criteria, to grade these programs. This is a useful approach as we move from the end of the Cold War to proliferation and terrorism situations.

1. Quantities and sites:
    - Quantities of nuclear materials,
    - Sites and transportation,
    - Material production and elimination trends.
2. Security and control measures:
    - On-site physical protection,
    - Control and accounting procedures,
    - Security personnel measures,
    - Physical security during transport,
    - Response capabilities.
3. Global norms:
    - International legal commitments,
    - Voluntary commitments,
    - Nuclear security and materials transparency.
4. Domestic commitments and capacity:
    - UNSCR 1540 Implementation,
    - Domestic nuclear materials security legislation,
    - Safeguards adoption and compliance,
    - Independent regulatory agency.
5. Societal factors:
    - Political stability,
    - Pervasiveness of corruption,
    - Group(s) interested in illicitly acquiring materials.

Nuclear warheads in 2014. The *Nuclear Notebook Project* has produced the best open literature estimates for global nuclear weapons arsenals. Over the past three decades these results have been published in the Bulletin of Atomic Scientists, which are useful for policy discussions. Table 7.3 gives a breakout for the 5 Nuclear Weapon States and the 4 de facto Nuclear Weapon States, with a total of 16,300 nuclear weapons.

Problems

7.1 **Cold War highs and lows**. When were 3 good and 3 bad times between US and SU?

7.2 **Almost to war**. When were these three times? When did an person make a big difference?

7.3 **Future cooperation**. What 3 issues, beyond nuclear weapons, are impacted by relations with Russia?

7.4 **Transparency**. What are 3 transparency measures the US has with Russia?

7.5 **Reciprocal-unilateral measures**. What issues could be improved by using reciprocal-unilateral measures with Russia? How could Congress/Parliament constrict this option?

7.6 **Project Cloud Gap**. What were the experiments to monitor warhead dismantlement in 1967–69? What was the main debate about?

7.7 **Slava in the Black Sea, 1989**. Why was this NGO monitoring a breakthrough? Why did Grobachev allow this to go ahead?

7.8 **Warhead monitoring, neutrons versus gamma-rays**? How is this done? Which reveals more secrets?

7.9 **Enrichment and reprocessing bans**. What is the status on enrichment and reprocessing in the NWS(P5), defacto-NWS(4), NNWS(180)?

7.10 **Fissile Material Ban**. What are the arguments for and against the FMCT? Constraints?

7.11 **Storage at Mayak**. What is the current status at Mayak? What access does the US have?

7.12 **Ten secret Soviet cities**. How has the transition gone at these 10 nuclear laboratories?

7.13 **No HEU for SSBNs**. What nations switched naval reactor fuel from 90 % HEU to 20 %?

7.14 **MOX in US and RF**. What is the status on using Pu in reactors as MOX fuel? What role is France playing?

7.15 **INFCIRC/540**. What is monitored pursuant to this agreement? Results?

7.16 **US-Origin spent fuel abroad**. How much is there? What is its disposition, reprocessing, storage, and pressure for US to take it back? What is the status on the Russian site?

7.17 **Information barriers**. Measuring gamma ray emissions can reveal secrets. How is this protected with information barriers? Can these be subverted?

7.18 **Wars**. What is the status of the ISIS, Syria, Afghanistan, Ukraine wars? What kind of direction and policy would you chart in this area?

7.19 **NTI Nuclear Materials Security Index**. What are the results on this Nuclear Threat Initiative? Where are HEU and Pu safe and not so safe, and why?

7.20 **Nuclear Weapon Totals**. How have these numbers changed over the past 10 and 30 years in the P5 NWS and the defacto-4 NWS?

Seminar: Wind-down of Cold War

Bernard Baruch, Biden Amendment, Cooperative Threat Reduction results, current status (Ukraine, ISIS, INF, CFE), Gorbachev agrees to CFE cuts, Mayak storage progress, Middlebury College graduate program in nuclear proliferation prevention in Monterey, nuclear weapon monitoring (attributes, templates, information barriers), Project Cloud Gap, Slava experiment in 1989, Soviet Republic status (Belarus, Kazakhstan, Ukraine), Soviet secret cities progress, START prefers missile controls compared to warhead controls, transparency with Russia, US present uranium reactor fuel purchases, Votkinsk perimeter-portal monitoring.

Bibliography

Bunn, G. (1992). *Arms control by committee: Managing negotiations with the Russians*. Stanford, CA: Stanford University Press.

Bunn, M. (2010). *Securing the bomb 2010*. Cambridge, MA: Belfer Center, Harvard University.

Cliff, D., Elbahtimy, H., & Persbo, A. (2010). *Verifying warhead dismantlement*. London, UK: Vertic.

Cooperative Threat Reduction. (2013). *Annual report to Congress*. Washington, DC: Department of Defense.

Feiveson, H., et al. (2015). *Unmaking the bomb*. Cambridge, MA: MIT Press.

Feshbach, M., & Friendly, A. (1991). *Fratricide in the USSR*. NY: Basic Books.

Fetter, S., et al. (1990a). Detecting nuclear warheads. *Science Global Security, 1*(3), 225–327.

Fetter, S., et al. (1990b). Gamma ray measurements of a Soviet cruise missile warhead. *Science, 248*, 828–834.

Gorbachev, M. (1987). *Perestrokia: New Thinking for Our Country and the World*. NY: Harper-Row.

Graham, T. (2002). *Disarmament sketches: Three decades of arms control international law*. Seattle, WA: University of Washington Press.

International Panel Fissile Materials. (2010). *Country perspective on challenges to fissile material (cutoff) treaty*. Princeton, NJ: Princeton University.

International Panel Fissile Materials. (2011). *Managing spent fuel from nuclear power reactors*. Princeton, NJ: Princeton University.

International Panel Fissile Materials. (2013a). *Global fissile material report*. Princeton, NJ: Princeton University.

International Panel Fissile Materials. (2013b). *Nuclear warhead stockpiles and transparency*. Princeton, NJ: Princeton University.

Knight, A. (2012). The mysterious end of the Soviet Union. In *NY Review of Books* (pp. 74–78). April 5, 2012.

Kristensen, H., & Norris, R. S. (2014). Worldwide deployments of nuclear weapons. *Bulletin of the Atomic Sciences, 70*(5), 2014.

Matlock, J. (1995). *Autopsy on an empire*. NY: Random House.

Matlock, J. (2005). *Reagan and Gorbachev: How the Cold War ended*. NY: Random House.

Mearsheimer, J. (2014). Why the Ukraine crisis is the West's fault. *Foreign Affairs 93*(5), 77–89.

National Academy of Sciences. (2002). *Technical issues related to the CTBT*. Washington, DC: NAS Press.

National Academy of Sciences. (2005). *Monitoring nuclear weapons and nuclear-explosive materials*. Washington, DC: NAS Press.

National Academy of Sciences. (2012). *The comprehensive nuclear test ban treaty: Technical issues for the United States*. Washington, DC: National Academy Press.

Nuclear Threat Initiative. (2014a). *Annual report*. Washington, DC.

Nuclear Threat Initiative. (2014b). *Nuclear materials security index*. Washington, DC.

Plowshares Fund. (2013). *Making an impact, annual report*. Washington, DC.

Sarotte, M. (2014). A broken promise? What the West really told Moscow about NATO expansion. *Foreign Affairs 93*(5), 90–97.

Woolf, A. (2013). The evolution of cooperative threat reduction. In *Congressional Research Service*. Washington, DC.

Zarimpas, N. (2003). *Transparency in nuclear warheads and materials, SIPRI*. Oxford, UK: Oxford University.

Chapter 8
Nuclear Proliferation

There are indications because of new inventions, that 10, 15, or 20 nations will have a nuclear capacity, including Red China, by the end of the presidential office in 1964. This is extremely serious. There have been many wars in the history of mankind and to take a chance now and not make every effort that we could make to provide for some control over these weapons, I think, would be a great mistake.

[John Kennedy, Presidential Debates with Richard Nixon, 13 October 1960]

We are here to make a choice between the quick and the dead. That is our business. Behind the black portent of the new atomic age lies a hope which, seized upon with faith, can work our salvation. If we fail, then we have damned every man to be the slave of fear. Let us not deceive ourselves: We must elect world peace or world destruction.

Science has torn from nature a secret so vast in its potentialities that our minds cower from the terror it creates. Yet terror is not enough to inhibit the use of the atomic bomb. The terror created by weapons has never stopped man from employing them. For each new weapon a defense has been produced, in time. But now we face a condition in which adequate defense does not exist.

[Bernard Baruch, US Representative to UN Atomic Energy Commission, 14 June 1946]

8.1 Proliferation History

In a dramatic moment before the United Nations, Bernard Baruch described the American plan to internationalize and control nuclear weapons. In biblical fashion, Baruch described the making of the "choice between the quick and the dead" (*Apostle's Creed*) caused by the forthcoming global spread of nuclear weapons. This prediction came to fruition 15 years later when presidential candidate John F. Kennedy gave the above warning in the Third Presidential Debate with Vice President Richard M. Nixon.

D. Hafemeister, *Nuclear Proliferation and Terrorism in the Post-9/11 World*, DOI 10.1007/978-3-319-25367-1_8

Kennedy's projection of 20 states *seeking the bomb* was correct, but it took a few decades to arrive at the number 25 or more. There are nine nuclear weapon states by possession (not the NPT definition). They are, in order of appearance, United States, Soviet Union, United Kingdom, France, China, Israel, India, Pakistan, and North Korea. South Africa, the 10th nuclear weapon state, gave up its six nuclear weapons in 1991, but did not reveal this until 1993. The former Soviet republics of Belarus, Kazakhstan, and Ukraine had *legal control* over their borders and their nuclear weapons for five years after 1991, becoming the eleventh, twelfth and thirteenth (almost) nuclear weapon states. They legally gave up their weapons in the mid-1990s by negotiations with the Russian Federation. When the striving nations of Iraq, Libya, and Iran are included, the total number of nuclear aspirant states rises to 16, exceeding Kennedy's middle bound of 15. The total information in Table 8.1, which includes the other past-striving nuclear powers, gives a total of 26 states having worked on nuclear weapons, which exceeds Kennedy's upper bound of 20 states. And there are more nations that could be put in this list. Chapter 10 gives additional information on the *proliferated states* (Fig. 8.1).

Table 8.1 Nuclear weapon state status

5 NWS: US (fission 1945/fusion 1952), FSU/Russia (1949/1953), UK (1952/1957), France (1960/1966), China (1964/1967)
4 defacto NWS: India (1974/1998-claimed), Pakistan (1998), Israel (1968?), North Korea (2006)
4 former defacto NWS: South Africa (1979–1993); Belarus (1991–6), Kazakhstan (1991–95), Ukraine (1991–96)
12 former nuclear weapon programs: Algeria, Argentina, Brazil, Germany (1939–1945), Iraq (1975–91), Japan (1945), Libya, South Korea, Sweden, Switzerland, Syria, Taiwan
1+ current nuclear weapon programs: Iran…

Twenty-four nations are categorized in terms of their progress to nuclear weapons with the dates of the first (fission/fusion) tests listed for the five nuclear weapon states (NWS as defined in the NPT) and for the four emerged, defacto NWS. Dates for the other states indicate the years of their first nuclear tests

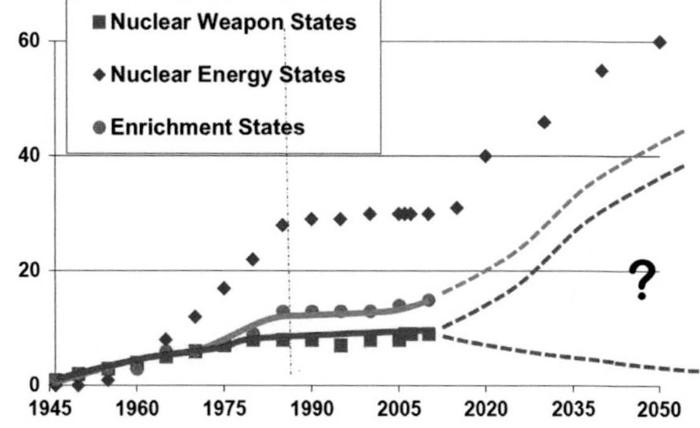

Fig. 8.1 Evolution of nuclear weapon states (*squares*), nuclear energy states (*diamonds*) and enrichment states (*circles*). Will the NPT be able to constrain the growth of nuclear weapon states, as nuclear energy and enrichment states expand? (A. Pregenzer)

Nuclear power connection. Commercialization of nuclear power raises three basic issues:

Nuclear proliferation from the spread of nuclear technology and materials to nations that want nuclear weapons;
nuclear safety resulting from the release of large amounts of radioactivity from reactor accidents (including spent fuel fires); and
disposal of nuclear waste to underground or surface storage sites.

This chapter deals only with nuclear proliferation, as nuclear safety and wastes were covered in Chap. 3. In my view, the severity of these issues is ranked as follows: Nuclear proliferation is the highest concern, followed by reactor safety of medium concern, followed by waste disposal as least dangerous (but more political). The safety and waste issues have to be considered in comparison to the dangers from the replacement power, as they all do damage.

8.1.1 Nuclear Danger: Weapons > Safety > Waste

Connection between peaceful nucleis and military nuclei. Does the development of commercial nuclear power contribute to the spread of nuclear weapons? The answer is clearly yes, but more analysis is needed to address the severity of this issue. At some point one has to deepen the question with a second question. How much of the spread of nuclear weapons was caused by commercial nuclear power and how much was caused by the political and military leaders that wanted nuclear weapons in their arsenal. But certainly, commercial nuclear power was used as a political cover by some nations when heading towards the bomb. Iran and India would be two prime examples, but there are others. We will raise this issue again when discussing the Nuclear Non-Proliferation Treaty. If commercial nuclear power is not available, clearly there would be less proliferation. But nuclear power is widely used, many nations want nuclear power because they lack domestic energy sources and they are too impatient to bet on renewable power, which is now looking stronger. There is concern about nuclear weapons in the Middle East, but the Middle East is not a bastion of commercial nuclear power. The biggest nuclear power programs are those of Iran and Pakistan (barely in the Middle East). The evolution of Germany's power grid should be watched, as it diminishes coal and nuclear and enhances alternative power. Nuclear power in the international sector is growing slowly (mainly China), and shrinking slowly in the Western sector. In twenty years, many of these plants will be retired, what will replace them? We can't answer that question here.

Proliferation policy is more complicated than superpower strategic nuclear weapons policy. This is because nuclear power supplies commercial energy to 30 nations, which could grow to 50. On the other hand, the SALT/START/SORT/New-START treaties involve only two nations (United States and Russia) and the treaties do not impact commercial energy supplies and commerce. In 2012, 30 nations produced 16 % of the world's electricity with nuclear power for a total capacity of 374 GW_e. The United States had 102 GW_e of nuclear power in 2012, producing 19 % of its electricity, but nuclear power is slowly getting smaller. The US load factor rose to 90 %, as higher-enriched uranium fuel remains longer in reactors, giving less frequent shutdowns. Many nations depend heavily on nuclear energy; these include France (63 GW_e, 75 % nuclear), Japan (Fukushima, 3–44 GW_e, 2–20 %), Russia (24 GW_e, 18 %), South Korea (21 GW_e, 30 %), Germany (12 GW_e, 16 %), United Kingdom (9 GW_e, 18 %), and Belgium (6 GW_e, 51 %). Spent fuel from France, UK, Japan, Belgium are reprocessed in France and UK, but these foreign contracts are coming to an end. Most countries plan to use Earth-burial, but with few specifics at this time. Japan has almost completed its Rokkasho reprocessing plant, capable of producing 8 tonnes of plutonium a year from 800 tonnes of spent fuel. The Monju breeder reactor was shut down in 2010 so the fall-back position produces mixed-oxide (MOX) fuel for 15 light water reactors. But, much is uncertain after the Fukushima accident. Other East-Asian nations (China, South Korea and Taiwan) are thinking about reprocessing, while they store spent fuel, waiting for a geological repository.

Incentives for Nuclear Weapons. Nations obtain nuclear weapons mostly from mutual mistrust with a neighboring state, rather than from fear of a distant superpower. Such a *nearest neighbor interaction* is exemplified by these national duos: US/USSR, USSR/China, China/India, India/Pakistan, North-Korea/South-Korea, Israel/Arab-states, Argentina/Brazil, China/Taiwan, Iran/Israel and Sunni/Shiite/ISIS Islam. Proliferation started with Klaus Fuchs and Ted Hall passing American nuclear secrets to help Soviet designers. Friendly cooperation in the US Manhattan Project helped Britain on its way, and it has been stated that the United States gave indirect assistance to the French. The Soviets gave major assistance to China to help it become a nuclear weapons state. The French helped Israel with the sale of the Dimona reactor and associated reprocessing technology. Canada assisted India with the sale of the Cirus reactor, to which the United States supplied heavy water and reprocessing technology. China helped Pakistan with designs, materials, and missiles. Pakistan's A.Q. Kahn sold centrifuges and weapon designs to North Korea, Iran and Libya. Kahn also approach Iraq, but Saddam Hussein didn't trust him. And so the story goes.

Such events prompted Tom Lehrer to write his song, "*Who's Next?*"

First we got the Bomb, and that was good,
Cause we love Peace and motherhood.
Then Russia got the bomb, but that's OK.
The balance of power's maintained that way.

Who's next? [France, Egypt, Israel]
…………
Luxembourg is next to go.
Then who knows, maybe Monaco.
We'll try to stay serene and calm
When Alabama gets the Bomb.
Who's Next?
[Tom Lehrer, Reprise Records, 1965]

NPT cover-up? Nuclear weapons programs usually grow out of dedicated national military programs, rather than civil nuclear power programs. But there often is a connection between the peaceful and military fuel cycles that use reactors, making plutonium in reactors, or high-enriched uranium (HEU) with centrifuges. India, Pakistan, Israel, Iraq, Iran, and other countries used research programs to hide military programs. Proliferation watchers are rightly concerned when nations with small electrical power grids move towards nuclear power. If a small nation risks destabilizing its grid by connecting it to large nuclear power stations, there could be another reason. Similarly, small nations that plan to build an enrichment or reprocessing plant may want to establish a weapons program since it is not cost effective to enrich or reprocess on a small scale.

Reactor-grade Pu. As we discuss later, reactor-grade plutonium (RgPu) with 20 % ^{240}Pu from civil power plants could be used to make lesser-quality nuclear weapons. The US and Russia could make significant 20-kton weapons from RgPu, but that takes sophistication and experience. Still, reactor-grade Pu can be made into smaller weapons of, perhaps, 0.5 kton. This is why commercial production and storage of RgPu is monitored by the International Atomic Energy Agency (IAEA). Nations that practice with reactor grade plutonium gain knowledge that would help future weapon programs. Thus far, the nine nuclear weapon states with plutonium weapons have used *only weapons-grade plutonium* (WgPu) from dedicated, non-power reactors.

Literature search to make weapons. In 1964 Lawrence Livermore National Laboratory asked three recent physics-PhD graduates to design a nuclear weapon using only un-classified literature. They opted for the more difficult implosion design for plutonium weapons.[1] After a year, the *Nth Country Experiment* was successful, sending a message to Washington that was ignored until the Indian bomb of 1974. Hans Kristensen and Stan Norris of the Federation of American Scientists have compiled estimates of the global stockpile of nuclear weapons. Their results are published in the *Nuclear Notebook* in the *Bulletin of Atomic Scientists* and as an appendix in the *SIPRI Yearbook*. Their estimates for the number of deployed nuclear weapons as of 2014 are as follows: Russia (4500), US (4760), France (300), China (240), UK (225), Israel (80), Pakistan (90–110), India (80–100) North Korea (<10) for a total of 10,200. The US and Russia have about 7000 additional weapons in line for dismantlement (Fig. 8.2).

[1]D. Stober, "No Experience Necessary," *Bulletin of Atomic Scientists*, March 2003, 57–63.

Fig. 8.2 Iraq's supergun. The 350-mm diameter "supergun" that Iraq tested at Jabal Hamrayn, 200 km north of Baghdad to shoot at Israel (United Nations Inspection Team 1991)

Atoms for Peace. President Dwight Eisenhower gave his "Atoms for Peace" speech to the United Nations (8 December 1953) in which he proposed that the IAEA should impound, store, and protect fissile materials. Eisenhower claimed that nuclear fuel could be *proliferation-proof*: "The ingenuity of our scientists will provide special safe conditions under which such a bank of fissionable material can be made essentially immune to surprise seizure." Did Eisenhower mean plutonium would be immune to surprise attack because of excellent physical protection, accounting and safeguards? Or did he mean that the presence of the spontaneous neutron-emitting ^{240}Pu prevents the use of plutonium in nuclear bombs? At that time, scientists incorrectly believed that 20 % ^{240}Pu gave enough early neutrons to preinitiate a chain reaction, greatly reducing the yield for all weapons designs. This belief has been proven false, but with qualifications (Fig. 8.3).

Fig. 8.3 Iraq's EMIS enrichment plant. View of control room under construction at the Tarmiyah Industrial Enrichment Plant for uranium enrichment with electromagnetic isotope separation (EMIS) (United Nations 1991)

8.2 The NPT

NPT, helps or hurts? An interesting question to discuss is whether actions subsequent to the *Atoms for Peace* speech reduced or increased global proliferation of nuclear weapons? I agree whole-heartedly with Len Weiss: "Did the 50-year-old Atoms for Peace program accelerate nuclear weapons proliferation? The jury has been in for some time on this question, and the answer is yes.[2]" There is no simple answer, to the broader question of "was the Atoms for Peace program wise?" I take a long-range point of view. Eisenhower did the right thing because his policies fostered the establishment of the IAEA in 1957 and the Nuclear Nonproliferation Treaty in 1970. Alone, the United States has never had the leverage to dictate its wishes to all nations. With all nations agreeing on a common framework of rules and inspections, the world can act in concert, at a certain level. Without the IAEA and NPT, which is the core of the international nuclear regime, the world community would not effectively exercise control over NNWS weapons programs and plutonium/uranium stockpiles. With the passage of time, the technical barrier to make nuclear weapons become smaller, as secrets leak and technologies improve. Without a global, political commitment to nonproliferation, the march of technology would be unstoppable. There is no simple

[2] L. Weiss, "Atoms for Peace," *Bulletin of Atomic Scientists*, November 2003, 34–44.

Fig. 8.4 Signing of the NPT by president Lyndon Johnson in 1968. The NPT entered into force under president Richard Nixon in 1970 (Atomic Heritage Foundation)

answer to this question because it is very, very difficult to obtain a stronger treaty than the NPT. I think it is NPT or nothing. I wish I was wrong, but I await the improved NPT regime (Fig. 8.4).

It is clear that proliferation accelerated under *Atoms for Peace*, as plutonium reprocessing was declassified and taught to foreign scientists at the National Laboratories. The Indian scientists who reprocessed plutonium at Argonne and Oak Ridge went home to help India produce its bombs. The NPT also gave cover to weapons programs that might not have existed otherwise. The Iraqi, Iranian and North Korean violations took place, however, at undeclared locations that were not inspected. If IAEA members had allowed the organization to act more vigorously, it could have taken stronger actions, such as challenge inspections. South Africa developed gaseous nozzle separation and nuclear weapons with some help. They succeeded with an expenditure of only hundreds of millions of dollars, much less than what Saddam Hussein spent on his covert, unsuccessful program. There is no doubt that openness on nuclear matters contributed to proliferation.

IAEA's burden. The world had 435 nuclear reactors with 370 GW_e of power in 2012, with 71 additional reactors under construction. The accumulation of military HEU and plutonium are shown in Table 8.2. The conflicting IAEA mission of promoting nuclear power and control of nuclear materials has grown substantially as IAEA membership grew to 140 states. The 2013 verification and safeguards budget of $170 million covered 699 facilities, 565 material balance areas. These facilities contain some 188,500 significant quantities (8 kg Pu-239, and 25 kg U-235) of nuclear materials in stockpiles, reactors, spent fuel rods, separated Pu, mixed oxide and HEU. The IAEA carried out 11,859 person-days of inspections with 250 inspectors on 2613 inspections. The IAEA appears to have a good record on monitoring *declared* sites with no major mistakes. The famous Iraqi, Iran, Libya and North Korea violations involved *undeclared* locations. After the Gulf War, the IAEA changed its procedures to

Table 8.2 Fissile Material stocks (2012)

	HEU (tonne)	Weapon Pu	Civil Pu
Russia	695	128	50.1
US	604	83.2	0
France	31	6	57.5
China	16	1.8	0.014
UK	21.2	3.5	91.2
Pakistan	3	0.2	0
India	0.8	5.2	0.24
Israel	0.3	0.084	–
North Korea	–	0.03	–
Others	15	–	61
Total	1390	230	269

The global stockpile of highly enriched uranium (HEU) is about 1390 tonnes. The global stockpile of separated plutonium is about 490 tonnes, of which 260 tonnes is in civilian custody. (International Panel on Fissile Materials, Princeton, 2013)

- increase use of intelligence information from large nations;
- take environmental samples to search for clandestine enrichment/reprocessing plants;
- establish special inspection procedures for undeclared sites.

It is not clear how effective these additional measures will turn out since they require more funding, but it is clear that a multilateral approach is essential for managing proliferation in a world of 190 nations. The well-respected *Office of Technology Assessment* 1977 report on *Nuclear Proliferation and Safeguards* concluded the following:

> In the long run two general rules apply: (a) Solutions to the proliferation problem will have to be found primarily, though not exclusively, through multilateral actions, and (b) the extent of US influence will vary from country to country.
> [*Nuclear Proliferation and Safeguards*, OTA, Washington, DC, 1977]

NPT details and disparities. There are obligations and disparities in the NPT. The Nuclear Nonproliferation Treaty created a discriminatory regime with two classes of nations. The *have nations,* the *big five* of World War II, are defined in the NPT as nuclear weapon states (NWS: United States, Russia, UK, France, and China). NWS are defined in Article IX as having exploded a nuclear device *before 1 January 1967*. NWS have both nuclear weapons and UN veto power in the Security Council. The *have-not nations* (non-nuclear weapon states, or NNWS) are the remaining 180 parties to the NPT. The main NPT holdouts are India, Israel, and Pakistan, while Iran and North Korea are NPT parties, "not in good standing."

Article I. NWS can own nuclear weapons, but cannot transfer to NWS or NNWS.

> Each [NWS] undertakes not to transfer to any recipient whatsoever nuclear weapons or other nuclear devices or control over such weapons or explosive devices directly or indirectly; and not in any way to assist, encourage, or induce any NNWS to manufacture or otherwise acquire nuclear weapons or other nuclear explosive devices, or control over such weapons or explosive devices.

Article II. NNWS reject nuclear weapons in many ways.

> Each [NNWS] undertakes not to receive the transfer from any transferor whatsoever of nuclear weapons or other nuclear explosive devices or of control over such weapons or explosive devices directly or indirectly; not to manufacture or otherwise acquire nuclear weapons or other nuclear explosive devices; and not to seek or receive any assistance in the manufacture of nuclear weapons or other nuclear explosive devices….

Article III. NPT requires safeguard inspections on all NNWS nuclear facilities, but inspections are *not* required on NWS nuclear facilities. Some NWS have volunteered a few of their nuclear facilities for exemplary inspections, but the IAEA does not have the funds to carry this out on all NWS facilities and the NWS don't want it in the first place.

> Each [NNWS] undertakes to accept safeguards, as set forth in an to be negotiated and concluded with the [IAEA] in accordance with the statute of the [IAEA] and the Agency's safeguard system,…. With a view to preventing diversion of nuclear energy from peaceful uses to nuclear weapons or other nuclear explosive devices… The safeguards required by this article shall be applied to all source or special fissionable material in all peaceful nuclear activities within the territory of each State….

Article IV. NPT acknowledges that NNWS have the inalienable right to develop, research and produce nuclear energy for peaceful purposes without discrimination. NPT strongly encourages NWS to assist NNWS with their peaceful nuclear power and research programs, as part of the overall bargain of the NPT. This includes the production of plutonium and weapons-grade uranium fuel. Such cooperation has taken place, but the Carter administration constrained commercial plutonium reprocessing, which was strongly condemned abroad, but is now widely accepted.

1. Nothing in this Treaty shall be interpreted as affecting the inalienable right of all parties to the Treaty to develop research, production and use of nuclear energy for peaceful purposes without discrimination and in conformity with Articles I and II of this Treaty.
2. All the parties to the Treaty undertake to facilitate, and have the right to participate in, the fullest possible exchange of equipment, materials and scientific and technological information for the peaceful uses of nuclear energy…. contributing…. to the further development of the applications of nuclear energy for peaceful purposes, especially in the territories of [NNWS] with due consideration for the needs of the developing areas of the world.

Article V. Peaceful nuclear tests for NNWS can be arranged by NWS; a relic provision, never used. Its removal would be heralded by most nations, but the fact it remains in the NPT is evidence of the difficulty of modifying the nuclear language contained in the NPT.

> .... potential benefits from any peaceful applications of nuclear explosions will be made available to [NNWS] on a nondiscriminatory basis and that the charge to such Parties for the explosive devices used will be as low as possible and exclude any charge for research and development....

Article VI. NPT calls upon the NWS to end the nuclear race and move towards a treaty on *general and complete disarmament* under strict and effective international control. The 1994 START, 2003 SORT and the 2012 New-START treaties are steps in that direction, but nonnuclear weapon states want further progress. The 1999 defeat of the Comprehensive Nuclear-Test-Ban Treaty in the Senate was a step back from the US promise to adopt a permanent test ban treaty at the same time the US was calling for NPT extension without a time limit. The CTBT is regarded by NPT nonnuclear states as the litmus test on NWS intentions, more than the numerical limits of the START treaties. The defeat of the CTBT by the US Senate, the Indian–Pakistani nuclear tests of 1998, the Iraqi, North Korean, and Iranian nuclear programs, the modest progress on strategic offense weapons and the demise of the ABM treaty are indicators of problems that the NPT regime cannot ignore.

> Each of the Parties to the Treaty undertakes to pursue negotiations in good faith on effective measures relating to cessation of the nuclear arms race at an early date and to nuclear disarmament on a Treaty on general and complete disarmament under strict and effective international control.

Review Conference. An NPT Review Conference is held every 5 years. The latest Review Conference failed to adopt a final document on 26 May 2015. The US, UK and Canada blocked the final document, which called for a negotiation conference on 1 March 2016 on a nuclear free zone in the Middle East, sponsored by Egypt and the Arab League. These meetings tend to show the disparity of views between the *nuclear haves* and *have-nots*. There is often a lack of consensus on a final statement, however the 2000 conference concluded with the *Thirteen Steps* to improve NPT practices, listed below in Table 8.3.

The 2015 NPT Review Conference did not obtain agreement on a final document. Egypt, acting "on behalf of the Arab League" called to convene a conference, by 1 March 2016, to launch negotiations on a zone free of nuclear and other weapons of mass destruction in the Middle East. The US, UK and Canada objected to this conference since it would place extra stress upon Israel. The future of the NPT regime is under stress for multiple reasons.

Table 8.3 Thirteen Steps of improve NPT practices from the 2000 NPT review conference

1. Achieve the early entry into force of the Comprehensive Test Ban Treaty (CTBT).
2. Maintain a moratorium on nuclear testing pending entry into force of the CTBT.
3. Work toward moving the Conference on Disarmament (CD) to begin a program of work leading to a verifiable negotiated Fissile Material Cutoff Treaty.
4. Work toward establishing a subsidiary body within the CD dealing with nuclear disarmament.
5. Apply the principle of irreversibility to nuclear disarmament, nuclear and related arms control, and weapon reduction measures.
6. Establish an "unequivocal undertaking" by the nuclear-weapons states to accomplish the total elimination of their nuclear arsenals.
7. Achieve the early entry into force and full implementation of START II, conclude the START-III Treaty, and strengthen the Anti-Ballistic Missile Treaty.
8. Complete and implement the trilateral initiative between the United States, Russia, and the IAEA, whereby the IAEA will take custody of surplus fissile materials arising from the dismantlement of nuclear weapons.
9. Engage in: a. further unilateral reductions of nuclear weapons b. increase transparency of nuclear weapon capabilities c. further reductions of non-strategic nuclear weapons d. measures to further reduce operational status of nuclear weapons e. diminish the role of nuclear weapons in security policies; and f. start the process, "as soon as appropriate," leading to the total elimination of nuclear weapons
10. Create arrangements by which fissile material no longer needed for military purposes will be placed within IAEA or other relevant international verification arrangement such that the material remains permanently outside military programs.
11. Reaffirm commitment to general and complete disarmament under international control.
12. Issue regular reports on implementation of Article VI
13. Develop further verification capabilities needed to assure compliance with disarmament agreements.

L. Weiss, *Arms Control Today*, p. 21–26 (December 2003)

8.3 Non-Proliferation Policy

Incentives and disincentives for nuclear weapons. The 1977 report by the Congressional Office of Technology Assessment, *Nuclear Proliferation and Safeguards*, continues to be useful today. The decision to go nuclear is complex.

But, fear or the madman dictatorship can force the decision quickly to choose bigger weapons with more prowess. National leaders don't want to look weak, as they will be ridiculed as weak with little power. There is more to discuss on how these decisions are made. There can be blowback against going to the bomb from the nation's corporate elites, from religious, legal and intellectual leaders and others that point out there are incentives and disincentives to the bomb. There are unintended consequences that must be considered. It's useful to list the incentives and disincentives as discussed by the OTA panel.

8.3.1 Incentives to Acquire Nuclear Weapons

Deterrence. Possession of nuclear weapons can deter other countries from conventional attacks. The evidence for this is weak, there have been many attacks and wars since 1945 without reference to American nuclear weapons. The P5 nations don't want to use nuclear weapons because it lowers the threshold for their use. The P5 are concerned about unintended consequences, there has been no use for nuclear weapons in a military situation since 1945. China did not have nuclear weapons during the Korean War and was not dissuaded in attacking American troops. The Viet Kong did not consider American nuclear weapons during the Vietnamese War.

Increased international status. It has been thought that possession of nuclear weapons gives extra international status and benefits. This is not clear when we see that the US and the Soviet Union were often ignored.

Domestic political requirements. This cuts both ways. By building extra nuclear delivery systems, jobs are created to make political happiness, but this takes money out of the marketplace that could have enhanced the gross domestic product. But then, both Democrats and Republicans called for more nuclear weapons.

Increased strategic autonomy. An example of this is the French going alone without NATO. Now, many years after de Gaulle, the French are working with the British and NATO, even though NATO headquarters has moved to Brussels.

Strategic hedge against military and political uncertainty. If you live in an unhealthy neighborhood, you may want more protection by hedging with a nuclear weapons program. The NPT/IAEA regime is intended for neighboring nations to ban nuclear weapons, then Norway doesn't fear Denmark.

Possession of a weapon of last resort. Israel wants its neighbors to know it will respond mightily if Tel Aviv or Jerusalem is destroyed by chemical weapons. But then, doesn't that prompt Iran, Egypt and Saudi Arabia to obtain nuclear weapons?

Leverage over the industrialized countries. Some third world countries hope to get better economic bargains if richer countries know the weaker countries are vulnerable. This incentive for weaker nations to stay away from nuclear weapons has to compete with its neighborhood nuclear situation.

8.3.2 Disincentives for Nuclear Weapons

Resource diversion. The US spent some \$6 trillion (\$6000 billion) on the nuclear arms race. The countries that didn't do this saved financial resources to compete in manufacturing cars and producing steel. In 1950, the US GDP was 50 % of the world total and in 2014 it is 25 % of the total. These are apples and bananas questions, economic well-being and national security are different issues, but yet they are coupled. Military manufacturing can make considerable export revenue.

Adverse public opinion. In many countries, public opinion opposes crossing the nuclear threshold. This is true for Japan, but their constitution will probably be amended to strengthen their military forces.

Disruption of established security guarantees. Other than the US, UK and France, the other NATO members do not have nuclear weapons. If a NATO nonnuclear nation moved towards nuclear weapons it could mean their removal from the NATO pact, making them more vulnerable than they were before.

Infeasibility of a desired nuclear strategy. Many nations started working on nuclear weapons, but years later they floundered because of a lack of resources or a change of heart. This is evidenced by events in Argentina, Brazil, Iraq, Iran, Libya, and North Korea.

Adverse international relations. A state can draw economic sanctions and other slights if they start a nuclear weapons program.

Advocacy of neutralist aims. Countries like Australia, Sweden and Switzerland did not obtain nuclear weapons since this can undercut a main tenant of being a neutral nation.

8.3.3 Other NPT-Policy Issues

Plutonium economy. The U.S. abandoned the plutonium economy with President Carter's 1978 cancellations of the

- Clinch River Liquid Metal Fast Breeder Reactor (LMFBR) in Oak Ridge, Tennessee with 350 MW_e power and 1000 MW_t heat, and the
- Barnwell Nuclear Fuel Plant, built by Allied General Nuclear Services, to reprocess plutonium from spent fuel in Barnwell, South Carolina.

The down-grading of plutonium emanated from the US State Department, not the Department of Energy, because of the international implications of a plutonium economy. This change was economically wise, as the breeder reactor did not live up to expectations. France's Phoenix and Super-Phoenix reactors have been shuttered. Russia, Japan and India maintain active plans for the breeder but at a great cost.

Reduction in weapons-grade HEU and Pu. At the end of the cold war, the new Russian Federation had 1200 tonnes of high-enriched uranium and 135 tonnns of separated plutonium. The protection of Russian SNM was greatly enhanced by US programs, including the megatons-to-megawatts program in which Russia converted 500 tonnes of HEU to reactor fuel for US purchase (1993–2013). In addition, the US donated half the construction funds to build the nuclear-materials storage facility at Mayak. Senators Sam Nunn and Dick Lugar created the Cooperative Threat Reduction Program in 1991 "to secure and dismantle weapons of mass distraction and their associated infrastructure in the former Soviet Union states." This program is housed in DOD's Defense Threat Reduction Agency (DTRA). The bilateral Materials Protection, Control, and Accounting (MPCA) Program has been successful in this regard. The Congress refused the Pell-Cranston amendment to expand purchasing of uranium to include surplus Russian plutonium in December 1991, primarily because plutonium is more difficult. Congress was correct in ranking HEU purchase first in importance since it is easier to make HEU into weapons and HEU is harder to detect, but it would have been helpful to extend this to the purchase of the more difficult plutonium. Fifteen states possess weapons usable materials, requiring sophisticated protection and control programs on these materials.

The United States and Russia agreed to each dispose of 34 tons of weapons grade plutonium, as well as 500 tons of Russian HEU and 174 tons of US HEU. The down blending of Russian HEU has been successful, but the production and use of mixed-oxide (MOX, plutonium and uranium) fuels lags.

Assured supply of uranium and controls on plutonium. To reduce interest in plutonium recycling of commercial-reactor spent fuel, countries need to be *assured* that they will receive assured supplies of LEU reactor fuel. The United States maintains control with its legal power to decide if spent fuel, enriched in the US or irradiated in US-supplied reactors, can be reprocessed. Uranium supplies have been helped by the dramatically-slowed growth in nuclear power, the increased uranium efficiency of reactors, and the conversion of Russian weapons-grade uranium into reactor fuel.

Fuel bank. The fuel bank was first proposed in the Eisenhower regime and it continues to resurface as the world searches for proliferation solutions. Some $150 million has been committed by the European Union, Russia, US and Warren Buffett ($50 million) to help nations obtain reactor fuel if their commercial contracts are threatened for other reasons. After its 1979 revolution, Iran lost its German reactor contracts and its US nuclear fuel contracts. The existence of the fuel bank removes the main argument for a nation to want enrichment and reprocessing. The current negotiations with Iran have been aided by Russian offers to guarantee reactor fuel to Iran, as part of Iran's reduction in centrifuges. This is a strong argument, but nationalism can be a stronger one.

Nuclear suppliers group. Nuclear-supplier nations agreed to ban exports of sensitive nuclear fuel facilities (enrichment and reprocessing) and require all of the importer's nuclear facilities be under safeguards. The NSG acts outside the IAEA regime, but it is generally supported by the IAEA nations.

Export criteria. The Indian bomb of 1974 created the climate for the passage of the Nuclear Nonproliferation Act (NNPA) of 1978, requiring safeguards on all nuclear facilities before US exports are allowed. NNPA also established criteria for allowing reprocessing of US-origin spent fuel with a demand for enough timely warning to respond to acts of proliferation before they became militarily significant.

Sanctions. When a NNWS moves toward the bomb, it will be denied nuclear power exports, military equipment, and other items of commerce. Enhanced sanctions were applied to India and Pakistan after their 1998 nuclear tests, but they were removed two years later, mainly because of the lack of cohesion among key nuclear supplier nations. Economic and military sanctions can be a powerful technique, but only if there is a strong cohesion among nations to work together.

Russian weapons usable materials. It is imperative that Russian warheads and weapons-grade nuclear materials remain under firm Russian accounting and control. The US Cooperative Threat Reduction programs significantly improved this situation, which is now in abeyance because of disagreements over Ukraine.

Spent fuel storage. The 35,000 tons of US-origin spent fuel at foreign storage sites do not have a final destination. The glut of spent fuel has overcrowded spent fuel ponds, shifting older spent fuel to air-cooled surface-storage. It is important not to overcrowd the ponds because of potential thermal problems if cooling water is lost. The US does not want to accept US-origin foreign spent fuel rods, nor does it want them reprocessed since this creates more separated plutonium. Russia passed a law in 2001 that allows it to establish a large storage site, which could be a useful constraint on plutonium if it is not reprocessed. As of 2009, global spent fuel was distributed as follows: Canada (38,400 tonnes), Finland (1600), France (13,500), Germany (5850), Japan (19,000), Russia (13,000), South Korea (10,900), Sweden (5400), UK (5850) and US (61,000).

Fissile Material Cutoff Treaty. France, Russia, UK and US declared a halt to production of HEU and Pu, primarily because they have much more than needed. The fifth nuclear weapon state, China, is believed to have halted production, and does not have large stocks of these materials. But its Asian rivals, India and Pakistan, continue to produce these materials. A global ban on production of weapons-usable materials would lessen this regional proliferation concern. The IAEA *Additional Protocol* requires declarations on production and stockpiling of these materials. It allows inspections to confirm the declarations of these materials. Brazil and Argentina have not signed the *advanced protocol,* but monitor without it. FMCT verification is not yet agreed, but if it did not include monitoring all P-5 weapons materials it would be much less expensive. It would be very difficult to establish the IAEA base-line levels for these materials. The US uses about 2 tons of HEU a year for its naval fleet. A conversion for naval fuel to 20 % enriched might be required for FMCT. If FMCT verification is constrained to new production, it would be an easier to monitor.

Anti ballistic missiles. After spending $100 billion, missile defenses have faltered. But has the Israeli Iron Dome system changed the scene for tactical missiles? Will the present missile defense network constrain missiles delivered from Iran, North Korea, and others? Can Russia be convinced that these defensive systems don't threaten their strategic forces?

Vertical and horizontal proliferation. Entrance into force of CTBT and a Newer START at 1000 strategic warheads would lower the political temperature of nuclear weapons and assist non-proliferation efforts. Vertical walls and horizontal floors do cross.

Zero nuclear weapons. In the near-term, ratifying CTBT and going to 1000 nuclear weapons would not diminish the ability to deter attacks on Russia and the US. The *Gang of Four* (Schultz, Kissinger, Perry, Nunn) talks of climbing the mountain through the clouds to the unknown summit. All admit they don't know how to get to zero nuclear weapons in a stable, verifiable manner. Domestic and international politics are often led more by *perception than by facts*. Does discussion of zero nuclear weapons guide policy choices towards a more stable nuclear arsenal? Does talk of nuclear abolition feed the angst of nations under the nuclear umbrella? Will it be necessary to have B-2 over-flights to reduce this angst?

Preemptive counter-proliferation. The multilateral NPT/IAEA regime is not perfect, but it is all we have. Or is it? The June 1981 Israeli destruction of Iraq's Osirak reactor was decried by most UN members since Iraq was a member "in good standing." Today there is little hand wringing over that unilateral attack since Osirak could have produced plutonium for Saddam's weapons. This preemptive attack rolled back proliferation by *counter-proliferation*. Such attacks may become more likely in the future. As with all policies, there are upsides and downsides. Preemptive rollbacks get the job done quickly, while international diplomats argue and postpone. But preemptive attacks shake the foundations of international processes and create new events we haven't considered. *Proliferation due process* should consider international organizations as part of decision making.

Intentions. An examination of motives for nuclear weapon programs and preemptive attacks points to many inconsistencies that consider the *degree of threat* from proliferation. An attacker has to have *both capabilities and intentions*. Preemptive counter-proliferation can be driven by *perceptions* of the intentions of nations, since many states certainly have capabilities and we call them allies. The 2002 US *National Security Strategy* states it clearly: "We will not hesitate to act alone, if necessary, to exercise our right of self-defense by acting preemptively." But in soured relationships it is often difficult to separate worst-case interpretations from lesser interpretations. It is easy to exaggerate the threat, as politics, the press, government, academics, and non-governmental organization can't look weak.

Assassination. Ultimately, counter-proliferation can lead to assassinations of leaders and scientists. The US and others are not blameless in this area. Presidential assassinations are banned by Presidential Executive Order 12,333, which, nonetheless, can be quickly nullified without the approval of congress. One might ask if less stringent criteria to do assassinations could rebound and put our leaders in harm's way. As the third millennium begins, there are ominous signs from North Korea, which withdrew from NPT (11 January 2002) and from Iran, which was charged by the IAEA with violating the NPT for building clandestine centrifuges (6 June 2003). We are hopeful that these situations can be resolved with diplomacy that has sticks and carrots. The outcomes will depend strongly on China (North Korea) and on economic sanctions (Iran) and *a little bit of luck*. Unintended consequences? What is the historical evidence for success and failure?

IAEA safeguards loophole. NPT nations gain nuclear materials under IAEA safeguards. Assume an NPT nation leaves the IAEA in a huff. Are IAEA safeguards still applied to these nuclear materials under the NPT, now owned by a non-member? See article by Goldschmidt.

Problems

8.1 **Baruch Plan**. What were the proposals in Baruch's plan? Why was the plan not acceptable to the Soviets? What were its advantages and disadvantages? Whom did it favor?

8.2 **Eisenhower's Atoms for Peace Speech**. What were the advantages and disadvantages of the actions that resulted from President Eisenhower's speech and policies? How were the results affected by advances in technology? In what ways did it promote civil nuclear power, proliferation and nonproliferation? Did Ike have other motives in making this speech?

8.3 **Next 10 proliferant states?** Who will be the next ten states, beyond the 24 nuclear weapon nations in Table 8.1? What nations have nuclear power, beyond the list of 24 nations?

8.4 **Nuclear Nonproliferation Treaty**. What are the trade-offs and responsibilities for the NWS and NNWS in Articles 1–6? Under what conditions are the NWS committed to be assured suppliers of nuclear fuel?

8.5 **NPT details**. What is the waiting period between a nation's statement of leaving the NPT and its exit? How can the NPT be amended? Is consensus needed for amending and extending the NPT?

8.6 **Commercial power first?** Of the 24 nations (Table 8.1), which ones developed nuclear power first and then worked on nuclear weapons? Which ones went directly to nuclear weapons?

8.7 **Rank 3 problems?** How would you rank the dangers of proliferation, safety and waste?

8.8 **Proliferation verus strategic?** Compare and contrast the issues involved with confronting proliferation and those with strategic nuclear weapons?

8.9 **A.Q. Khan's travels?** Where did he visit? Did the Pakistani government known of his travels?

8.10 **US knowledge of AQ Kahn?** Why did the US have look the other way on Kahn?

8.11 **NPT cover-up**. What nations used the NPT to cover up their weapon plans?

8.12 **Nth country study**. What did the Livermore study prove? What is on the www on these topics?

8.13 **Ike's farewell**. What did Eisenhower say about the Military Industrial Complex on 17 January 1961?

8.14 **NPT Review Conferences**. What happened at the 2015 NPT Review conference? How many of these conferences obtained a consensus report, and how many did not? What are the issues for the 2020 conference?

8.15 **Positive and negative security assurances**. How do positive and negative security assurances differ?

8.16 **CW hedge**. Does the US retain the right to attack a nation that uses chemical or biological weapons against it? What is the history of this situation?

8.17 **Definition of a nuclear weapon**. The NPT uses the phrase "explosive devise" instead of defining a nuclear weapon. Why was this done? How would you define a "nuclear weapon?"

8.18 **Challenge inspections?** What is the IAEA definition of challenge inspection? How many have occurred in the past 10 and 40 years?

8.19 **IAEA split identity?** What two roles does the IAEA have, are they in conflict?

8.20 **OTA main conclusion**. OTA warned that solutions to proliferation will primarily have to be multilateral, and US leverage will vary widely, depending on the situation. Explain this.

8.21 **India a NWS?** Why is this not true, but it is true?

8.22 **Peaceful nuclear explosions**. How many took place? Who did them. What results?

8.23 **Why France left NATO?** When did this happen, and why? Status now?

8.24 **Joint Committee on Atomic Energy**. How JCAE circumvent the Constitution? What were its strengths and weaknesses?

8.25 **Timely-warning criteria?** To obtain US nuclear materials, it must be shown that the US will have sufficient knowledge and time to act before substantial damage could happen to the US. Explain with examples.

8.26 **Russian spent fuel storage?** What is the current status of this?

8.27 **Other sanctions?** In what 3 situations have US applied sanctions on other nations? Did it achieve its policy goal?

8.28 **Nuclear umbrella**. If the US stated it would no longer retaliate if its client states were attacked with nuclear weapons, how would the client states react? Examples?

8.29 **Counter-proliferation**. What are the risks for an aggressive counter-proliferation policy?

Seminar: Proliferation Issues

ABM and NPT intersection, Baruch speech and policy, coal versus nuclear power, future scenarios which include larger issues, incentives and disincentives for the nuclear weapons, India exception with GW Bush, NPT Articles I–VI, NPT Review conference results, NPT 13 steps from 2000 realities, NPT success stories, NPT failure stories, nth country physics results at LLNL, Nuclear Suppliers Group (members, rules, results), peaceful/military power connections, plutonium economy projections, preemptive counter-proliferation examples and results, proliferation complications, RgPu versus WgPu, reprocessing status in the world, reprocessing economics, sanctions after 1974 India test, sanctions in 1980s on Pakistan, sanction after India/Pakistan tests of 1998, spent fuel in Russia, spent fuel in Yucca Mountain, weapon attempted states 25–30.

Bibliography

Cirincione, J. (2000). *Repairing the regime: Preventing spread of weapons of mass destruction*. NY: Routledge.

Cirincione, J. (2002). *Deadly arsenals: Tracking weapons of mass destruction*. Washington, DC : Carnegie Endowment for International Peace.

Cirincione, J. (2013). *Nuclear nightmares: Securing the world before it is too late*. NY: Columbia University.

Cohen, A. (1998). *Israel and the bomb*. NY: Columbia University.

Commission on intelligence capabilities of US regarding weapons mass destruction. (2005). Washingdpon, DC.

Craig, P., & Jungerman, J. (1990). *Nuclear arms race technology and society*. NY: McGraw Hill.

Dunn, L. (2015 April). Finding a way out of the NPT disarmament stalemate. *Arms Control Today*, 8–14.

El Baradei, M. (2011). *Age of deception: Nuclear diplomacy in treacherous times*. NY: Metropolitan.

Feiveson, H., Glaser, A., Mian, Z., & von Hippel, F. (2014). *Unmasking the bomb: A fissile materials approach to disarmament and nonproliferation*. Cambridge, MA: MIT.

Garwin, R., & Charpak, G. (2001). *Megatons to megawatts*. NY: Alfred A. Knopf.

Goldschmidt, P. (2015 March). Securing irreversible IAEA safeguards to close the next NPT loophole. *Arms Control Today*, 15–19.

Hecker, S. (2011 July). Adventures in scientific nuclear diplomacy. *Physics Today*, 31–37.

Hersh, S. (1991). *The Sampson option*. NY: Random House.

International Panel on Fissile Materials. (2012). Global fissile material report, Princeton, NJ.

Meeburg, A., & von Hippel, F. (2009 March). Complete cutoff. *Arms Control Today*, 16–23.

Office Technology Assessment. (1977). *Nuclear proliferation and safeguards*. NY: Praeger.

Office Technology Assessment. (1993). *Proliferation of weapons of mass destruction*. Washington, DC: OTA.

Office Technology Assessment. (1995). *Nuclear safeguards and the IAEA*. Washington, DC: OTA.

Scheinman, L. (1987). *The international atomic energy agency*. Washington, DC: Resources for Future.

Weiss, L. (2003 December). Nuclear-weapon states and the grand bargain. *Arms Control Today*, *33*, 21–26.

Chapter 9
Proliferation Technologies

Pakistan's motivation for nuclear weapons arose from a need to prevent 'nuclear blackmail' by India. Had Iraq and Libya been nuclear powers, they wouldn't have been destroyed in the way we have seen recently.... If (Pakistan) had an (atomic) capability before 1971, we [Pakistan] would not have lost half of our country after a disgraceful defeat.

[Abdul Qadeer Kahn, 16 May 2011]

One of the highest priorities of successive administrations is counter proliferation. But there's little coherence on what agencies do to move that interest forward,

[Ash Carter, January 2014]

9.1 Special Nuclear Material

The IAEA uses the unit *significant quantity* to determine weapons capabilities of weapons-grade nuclear material, the higher-quality special nuclear material (SNM). A significant quantity is the very approximate amount of material needed to make an explosive device. A significant quantity is defined as 25 kg for 90 % enriched ^{235}U of HEU and 8 kg for ^{239}Pu. These values are close to the bare sphere critical masses, in which the solid ball of HEU or Pu is not compressed and has no reflective back-scatter, or tampers to enhance compression and no boosting extra neutrons. Nuclear weapons are made with less than a significant quantity. For instance, US plutonium pits are made with about 50 % of a significant quantity. The 4 kg plutonium pit mass was reduced from 8 kg by enhancing fission efficiency:

1. Neutrons are reflected back into the device.
2. Hollow spheres of fissile material are compressed with explosives to higher densities.
3. Neutron generations are skipped to enhance yield with boosting neutrons from boosting with tritium and deuterium gas in the hollow center.

D. Hafemeister, *Nuclear Proliferation and Terrorism in the Post-9/11 World*,
DOI 10.1007/978-3-319-25367-1_9

Uranium is enriched in ^{235}U by the laws of physics, while plutonium is separated from spent fuel rods in reprocessing plants by the laws of chemistry. The US, UK, China, Pakistan and South Africa (and Iran on U weapons) began nuclear weapons programs with uranium weapons and moved up to plutonium weapons as the US has done. However Russia, France, India, Israel, and North Korea began with plutonium.

Plutonium economy. The *Carter Administration* (1977–81) was the first US administration to lean on foreign governments to constrain their production of weapon's useable SNM. This effort was not appreciated by the US's friends and allies, as they gave testy behavior at international and bilateral meetings. These nations perceived the Carter efforts would diminish their energy plans based on the breeder reactor. Forty years later, the breeder is long delayed, very uncertain and too expensive. It appeared at that time that the world was entering a plutonium economy, with widespread use of plutonium for mixed oxide (MOX) fuels in light water reactors and breeder reactors. At that time it was generally believed that the plutonium route to weapons was easier to accomplish than through uranium enrichment. The commercial plutonium economy did not happen

AQ Kahn's defection in 1978 from URENCO in the Netherlands to Pakistan showed the weakness of the international system. Kahn took centrifuge designs and the associated experience from URENCO, paving the proliferation path for HEU nuclear weapons. Not satisfied with helping Pakistan, Kahn expanded nuclear sales to Iran, Libya, and North Korea. Kahn also tried to sell centrifuges to Iraq's Saddam Hussain, but Saddam didn't trust Kahn. The shift to HEU allows easier weapons designs compared to plutonium-implosion weapons. This was proven when South Africa successfully used gaseous nozzles to obtain HEU for its six nuclear weapons, spending only a few hundred million dollars. On the other hand, Iraq failed with its electromagnetic isotope separation (EMIS), but Pakistan, Iran, and North Korea succeeded with gas centrifuges. Table 9.1 summarizes the global facilities that produce SNM (Table 9.2).

9.2 Uranium Enrichment

Uranium enrichment can increase natural uranium (0.7 % ^{235}U) to reactor fuel (4 % ^{235}U) and nuclear weapon grade (>90 % ^{235}U). The various classes of uranium are defined in Table 9.3.

Enrichment depends on the masses of the two uranium isotopes, ^{235}U and ^{238}U. The lighter the isotope, the greater the molecular velocity. Since uranium is a heavy element, it requires high temperatures to produce gaseous uranium atoms, which is difficult. But when uranium is combined with fluorine to make uranium–hexafluoride molecules, it becomes a manageable gas at lower temperatures. The slight mass difference gives $^{235}UF_6$ slightly more velocity, some 0.4 % faster, than $^{238}UF_6$.

Table 9.1 Enrichment and reprocessing production plants

Brazil (enrichment 1/reprocessing 0)
China (3/1)
France (1/2)
Germany (1/0)
India (1/3)
Iran (2/0)
Israel (0/1)
Japan (1/2)
Netherlands (1/0)
North Korea (1/1)
Pakistan (2/2)
Russia (4/1)
UK (1/2)
US (2/1)

These national facilities are operating or under construction. In 2013, 13 states had 21 enrichment plants, and 10 states had 18 reprocessing plants. The list does not include facilities that have been closed or are in planning. (International Panel on Fissile Materials, Princeton, 2013)

Table 9.2 Significant quantities of isotopes (IAEA)

Isotopic content	Sig. quantity	Inspection times
Pu (<80 % ^{239}Pu)	8 kg Pu	3 months
^{233}U	8 kg	3 months
HEU (>20 % ^{235}U)	25 kg ^{235}U	3 months
LEU (<20 % ^{235}U)	75 kg ^{235}U	1 year
Natural U	10 tonne	
Depleted U	20 tonne	
Thorium	20 tonne	

Table 9.3 Enriched uranium isotopic composition

	^{234}U	^{235}U	^{238}U
Weapons-grade U (%)	1	93.3	5.5
HEU (%)	<1	>20	<80
Natural U (%)	0.0054	0.711	99.3
Depleted U (%)	–	0.2	99.8

Gaseous Diffusion. For decades the main US uranium enrichment technology was gaseous diffusion. The US had three plants in Oak Ridge, Tennessee; Paducah, Kentucky; and Portsmouth, Ohio. These very large plants consumed 6 GW_e, which is 1 % of US total electrical power. The diffusion plants are now shuttered for economic reasons since enrichment technology shifted from gaseous diffusion to gaseous centrifuges. Centrifuges use only 2 % of the electricity needed by gaseous diffusion.

The lighter, faster $^{235}UF_6$ molecules hit the barriers more often than the slower $^{238}UF_6$ molecules, passing through the small pores in the membranes at higher rates. The physical size of the two molecular types are the essentially the same and does not matter. Diffusion is not a size selector but a mass selector. The lighter molecules bump into the membrane more often, with the result they go through the filter more often, as compared to the heavier molecule.

The diffusion barrier is made of small, 1-μ (10^{-6} m) diameter nickel spheres with pore diameters of 0.03 μ between spheres. The nickel spheres are sintered when pressed together at high temperatures and pressures. This gives a gain in the uranium isotope ratio of 0.4 % per stage, with a ratio of 1.004/stage. It takes 1000 gaseous diffusion stages to make 3.2 % enriched fuel and 3500 stages to make 90 % enriched weapons uranium. The diffusion plants are large and clunky, easily visible from satellites, airplanes, and the highway. Gaseous diffusion is less effective for higher-mass UF_6 molecules because *diffusion rates depend on the* mass *ratio*, which in this case is 1.0086. Gaseous diffusion is much more effective for light molecules, such as the separation of ^{6}Li and ^{7}Li with a mass ratio of 1.17. Lithium-6 is used in hydrogen-bomb secondaries and to make tritium for boosted primaries and weapon initiators. In contrast to gaseous diffusion's mass ratio dependence, *gaseous centrifuge and gravitational separation depend on the mass difference*.

Gravitational Separation. The Earth's atmosphere is an example of gravitational separation. The heavier oxygen isotopes of ^{17}O and ^{18}O are enhanced near the Earth's surface. On the other hand, the lighter, more common isotope ^{16}O is enhanced at upper altitudes compared to the Earth's surface. The Manhattan Project used gravitational separation of isotopes to produce slightly-enriched uranium feedstock for gaseous diffusion separators, which was then fed into electromagnetic isotope separation (EMIS), shown symbolically by the arrows below:

$$\textit{gravitational diffusion} \Rightarrow \textit{gaseous diffusion} \Rightarrow \textit{EMIS}.$$

Gravitational separation combined a gravitational-force difference over its 10-m vertical tubes, with moderate heating to 57 °C (135 °F). Gravitational diffusion enriched only by 0.01 % per stage as compared to gaseous diffusion's 0.4 % per stage (forty times greater).

Gaseous Centrifuges. Two years before the *Manhattan Project* in 1940, Paul Hartek (Germany) began centrifuge experiments with force fields greatly exceeding gravity. Centrifuges can produce considerable enrichment per stage, but the German centrifuges self-destructed at higher rotational velocities. Molecules are spun in a cylindrical shape, close to the inside surface of the centrifuge tubular body. At ultrahigh speeds, the two components of the gas separate into two thin layers near the tube wall with heavier molecules to the outside. Two molecular currents are formed, the lighter $^{235}UF_6$ gas rises to the top and is collected and the heavier $^{238}UF_6$ gas falls to the bottom, where it is collected. Today's centrifuges use maraging steel (low C, Ni/Fe alloy), carbon-fiber or glass-fiber tubes. The tensile strengths of aluminum and other metals are smaller under rotational stress.

Centrifuges are, thus far, the 21st century enrichment-technology of choice, playing a large role in A.Q. Kahn's nuclear Wal-Mart. For nations operating a covert nuclear weapons program, the centrifuges can be hidden and protected underground easily. They can be placed in smaller cascades that can be re-configured to make HEU. Centrifuges use much less electrical power, and do not need much chemistry.

Centrifuges of 10 cm radius, operating at 48,000 RPM, give a large gain of 15 % per stage, which is 30 times larger than diffusion. Modern centrifuges, running at 90,000 RPM can obtain 20–50 % gain per stage. Centrifuges need only a dozen stages to obtain 3.2 % reactor fuel compared to a thousand stages for diffusion. A smaller centrifuge plant could produce 25-kg weapons-grade uranium in 2 months. Such a plant can be built clandestinely in a space of 60 m by 60 m and needs only tens of MW as compared to a GW for gaseous diffusion plants. The equilibrium time for a centrifuge plant is only minutes, allowing plant operators to shift quickly from LEU production piping to HEU-production piping, but with great difficulty if monitored. To make sure this is not happening, the IAEA Hexapartite Agreement allowed inspections with 2-h notice to intrusively monitor isotope ratios and sealed valves, and to perform remote monitoring at critical locations.

Electromagnetic Separation. Electromagnetic isotope separators (EMIS) were developed by Ernest Lawrence as the final stage of three separation-types to produce HEU at Oak Ridge in WW-II. These EMIS were called *Calutrons* since Lawrence was from the University of California. They produced enough HEU for the Hiroshima weapon, and not much more by August 1945. Iraq surprised the world by choosing an improved EMIS technology, but its construction was not complete before it was discovered and destroyed by UN inspectors. Iraq was building 140 EMIS separators using 300 MW to produce 30 kg/year of weapons-grade uranium.

A charged particle moving in a magnetic field, experiences a force that confines it to a circular orbit perpendicular to the magnetic field. A radius of 1.2 m was obtained in Calutrons with a 35-keV uranium ion beam in a magnetic field of 3500 Gauss. The beam radii for ^{235}U and ^{238}U ions differed by 0.8 cm, which is easily separated in one pass through EMIS. After the war, the US replaced Calutron EMIS with gaseous diffusion since Calutrons lost considerable feedstock during ionization.

Laser isotope separation. The slightly smaller volume of ^{235}U nuclei provides enough difference in electrostatic energy to produce slightly different energy levels in ^{235}U as compared to ^{238}U. Tunable dye lasers and other lasers can selectively ionize ^{235}U in uranium vapor or UF_6 molecules. The ionized ^{235}U is separated with from neutral ^{238}U with electric fields. Laser isotope separation (LIS) can make weapons-usable HEU but with varying degrees of difficulty. LIS allows proliferators to avoid making plutonium in reactors. High enrichment HEU with LIS is complicated by exchange-charge reactions that transfer electrons from neutrally charged ^{238}U to make ionized ^{238}U, while converting ionized ^{235}U to its neutral state. These collisions add charged ^{238}U-ions to the charged ^{235}U-ion stream, diminishing the enrichment level.

A review panel for the SILEX laser separation plant in North Carolina concluded the following: "Laser-based enrichment process have always been of concern from the perspective of nuclear proliferation…a laser enrichment facility might be easier to build without detection and could be a more efficient producer of high enriched uranium for a nuclear weapons program." Others have stated that LIS is not a significant proliferation problem. At this point, progress on LIS has been slow. Separation of Isotopes by Laser Excitation (*SILEX*) is projected to be sixteen times more efficient than centrifuges to obtain reactor fuel. Francis Slakey of Georgetown University objected to the proposal because carbon dioxide lasers are readily available, creating a proliferation risk. SILEX responded that the entire system needs sophisticated components. SILEX projected it will produce 20 SWU/MW-h, for an electrical cost of $5/SWU.

Aerodynamic Nozzles and Helicons. UF_6 molecular beams turn very tight corners in a Becker nozzle enrichment system invented in Germany. The heavier $^{238}UF_6$ is deflected less than a $^{235}UF_6$ because of its larger mass. Isotopes are separated with a knife blade into two streams after they pass the corner. Gas at 400 m/s in a radial turn of 0.1 mm experiences a tremendous centripetal acceleration of 160 million grams.

South Africa chose a similar approach, but instead they projected UF_6 at right angles in a tight cone, the spiraling $^{238}UF_6$ shifting more to the larger side and $^{235}UF_6$ shifted more to the smaller side. This is similar to a centrifuge with a nonrotating casing. Again, million's of grams are developed in a tight geometry. South Africa's weapons program was amazing, it needed a smaller cost of hundreds of millions of dollars, working mostly in isolation.

Table 9.4 Summary of the enrichment technologies

Gaseous diffusion	With UF_6, $v_{235}/v_{238} = 1.043$, giving ^{235}U more chances to penetrate the membrane of sintered nickel spheres. Thousands of stages needed. The US three facilities consumed 7 GW_e, easy to detect, hard to modify cascades to change from LEU to HEU.
Gravitational diffusion	Manhattan project used this to enrich to a few percent, not used again.
Electromagnetic	Calutrons lost considerable uranium during ionization, so abandoned. Iraq in 1990s improved EMIS, but this was dismantled.
Gas centrifuge	Two concentrate, thin layered cylinders, ^{235}U to the inside and rises, ^{238}U to the outside and falls. Uses much less electricity, easy to hide. Some 10–100 stages to HEU.
Laser enrichment	Tunable laser ionizes ^{235}U, but leaves ^{238}U neutral. Electric field separates charged ions. Charge exchange collisions diminish yield. Some feel this will be economically viable.
Nozzle	S. Africa, Germany, and Brazil. Very small radii give dramatic accelerations, separating ^{235}U and ^{238}U.
Chemical	France and Japan. Lighter isotope preferentially binds to more *loosely bound* compounds, while heavier isotope binds to the more *strongly bound* compounds. Not yet economic.

Chemical ion exchange. France and Japan have developed pilot plants to take advantage of small isotopic-mass differences that affect chemical reaction rates. Using catalysts, they observed that lighter isotopes tend to preferentially bind to more *loosely bound* compounds, while heavier isotopes tend to bind to the more *strongly bound* compounds. Mixing the two types of compounds causes ^{235}U to flow from the more tightly bound compound to the more loosely bound compound. The two compounds are separated chemically, giving enriched uranium (Table 9.4).

9.3 Uranium Details

Separated work units. *This section is intended only for the dedicated.* Separation of isotopes cannot be 100 % complete, except in very small samples. The enrichment process changes the isotopic ratio of the feedstock by increasing the ratio of the desired isotope in the product and decreasing its ratio in the waste (the tails). Mixing separated isotopes increases chaos, raising the system's entropy, and, conversely, isotope separation creates order and lowers entropy. Separation of isotopes lowers chaos and lowers the entropy of feed (F) to the sum of the entropies of product (P) and waste (W). A SWU is the difference in the values V (similar to entropy) needed to convert feed to product plus waste (tails). A SWU costs about $100. Separative work is the difference between the values of the output and input. The total separative value of a sample is its value times its mass, hence its unit is in kg-SWU or tonne-SWU. Enrichment plant sizes are given in units of tonne-SWU/year.

The enrichment of uranium in separative work units is a nonlinear process. A common light water reactor (LWR), producing about a gigawatt of electric power, consumes each year about 25 tons of 3.8 %-enriched fuel, obtained from 210 tons of natural uranium, and requires 120,000 SWU. Figure 9.1 shows that separative work diminishes as uranium is enriched from its natural form of 0.71 % ^{235}U, to its high (90 %) enriched form. It requires about 800 SWU per ton of uranium to make 130 kg of 4 %-enriched reactor fuel. An additional 300 SWU yields 26 kg of 20 % enriched uranium, and a further 200 SWU yields 5.6 kg of 90 % HEU. Put another way, it takes about 60 % of the total work to produce low-enriched LWR fuel, 25 % of the total to enrich the fuel to 20 %, and another 15 % of the total to raise that amount to 90 %. Thus, a stockpile of 20 % enriched material can significantly reduce the break-out time a country needs to make a nuclear weapon. (This was a large issue with the 2015 Iran/P5+1 agreement.)

Uranium countries. Argentina, Brazil, China, France, Germany, India, Iran, Japan, Netherlands, North Korea, Pakistan, Russia, UK and US currently enrich uranium. Commercially enriched stock is primarily supplied by Europe's URENCO (10 million SWU/year), Russia's Rosatom (26 million), and France's AREVA (10 million). Iran's various centrifuges produce 0.8–4 SWU/year. Iran claims its right to pursue uranium enrichment in line with Article IV of the NPT. But its claim is subject to dispute, considering that Iran did not report past enrichment activities

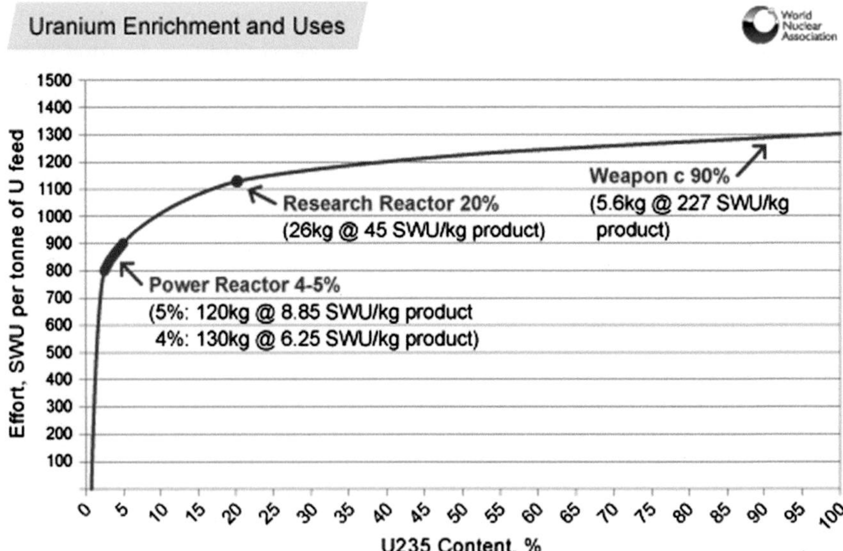

Fig. 9.1 SWUs for reactor fuels and nuclear weapons. The effort required to enrich uranium and increase the concentration of its fissile-isotope ^{235}U diminishes exponentially as the concentration rises. The effort is measured in separative work units. Starting with 1 ton of natural uranium, it takes 800 SWU to produce 4 %-enriched reactor fuel, an additional 300 SWU to produce 20 %-enriched material for research reactors, and 200 more SWU to produce 90 %-enriched material (World Nuclear Association)

to the IAEA. These developments prompted economic sanctions in UN Security Council resolutions. Other non-weapon states that enrich uranium have negotiated acceptable reporting and inspection agreements with the IAEA.

9.3.1 Three Situations for 1000 tonne-SWU/year

3.2 % Fuel. It takes 4.7 kg-SWU to obtain 1 kg of 3.2 % enriched fuel (1970s) from natural uranium with tails of 0.2 %. It takes 5.8 kg of natural uranium feed to make 1 kg of 3.2 % fuel. A 1000-tonne SWU/year plant could produce 210 tonne/year of 3.2 % product from a feed of 1230 tonne/year, while rejecting 0.2 % tails into wastes of 1020 tonne/year.

4.4 % Fuel. Since the 1970s, the time between reactor shutdowns for refueling increased from 12 to 18 months by using higher-enriched fuel of 4.4 % ^{235}U. This change helped increase nuclear power plant load factors to 90 %, making the plants more profitable, while using less uranium and producing less plutonium and less spent fuel. The ratio of enrichments (4.4 % vs. 3.2 %) is roughly the ratio of times between shutdowns. New fuels require stronger materials to remain viable for longer times in the reactor. It takes 8.2 tonne of uranium and 7 tonne-SWU to make 1 tonne of 4.4 % fuel from natural uranium with 0.2 % tails. The cost of uranium and separative work to make 1 kg of 4.4 % fuel is about $1000.

90 % HEU. It takes 226 kg-SWU and 174 kg of natural uranium to make 1 kg of 90 % HEU with 0.2 % tails. A 1000 tonne SWU/year plant can make 4.4 tonne/year of HEU from 765 tonne/year of natural uranium, rejecting 761 tonne/year of 0.2 % tails. The cost of uranium and separative work to make 1 kg of HEU is about $25,000. The HEU in a warhead of 20 kg is worth about $500,000.

Uranium supplies. The US tried to dissuade other counties from entering the plutonium economy of MOX and breeders by stretching uranium supplies. Let's examine some ways this is being done. If 4.4 %-enriched fuels are used in 1 GW_e reactors, 22 tonnes of spent fuel are annually removed. This amounts to a requirement of 1 ton of ^{235}U nuclei per year for a GW_e reactor. The United States has roughly 3 million tons of reasonably assured reserves and speculative resources, which could accommodate 400 GW-lifetimes, enough to run 400 large nuclear plants for their lifetime.

In the 1970s, these reserves were considered minimal because US nuclear power was expected to rise to more than 1000 GW by the year 2000. At the end of the century, the US nuclear power peaked at 100 GW, which will consume 25 % of the 3-Mton reserves. There is additional uranium available in low-grade ores, as well as in "mining" the enrichment tails to 0.05 %, in dismantled nuclear weapons, in coal fly-ash, and in ocean water, but at a higher cost. The shift from the once-through cycle to the breeder ended because of many factors: Uranium surplus, curtailed nuclear orders and inexpensive combined-cycle gas turbines powered by natural gas. Nuclear power might have a larger role in the future because of climate change problems, but this will only happen if governments enter the marketplace.

Megatons to Megawatts. As the cold-war world subsided from 70,000 total warheads to a world of 3000 operational warheads under New START (plus others), there was a concern about Russia's ability to manage its 130 tonnes of plutonium and 700 tonnes of HEU (after the sale to the US). As an encouragement to Russia, the US placed 12 tonnes of plutonium and HEU under IAEA safeguards in hopes that Russia would follow suit. The major barrier to weapons production is the availability of weapons-grade fissile materials, rather than the design and fabrication of weapons. To reduce dangerous proliferation, the US agreed to pay $12 billion over 20 years to purchase 500 tonnes of HEU, blended to low-enriched reactor fuel. This arrangement gave needed funds to Russia at a critical time, preventing its bankrupt nuclear facilities from collapsing. The idea for this exchange came from Tom Neff of MIT. He originally suggested in the New York Times that it should be a swap, US food for Russian enriched uranium, but it ended up being US money for Russian enriched uranium. The US clearly needed the reactor fuel and Russia needed the money.

9.4 Plutonium Details

Reactor-Grade and Weapons-Grade Pu. A typical LWR creates 0.5 ^{239}Pu nuclei for every fission event. Some ^{239}Pu remain as ^{239}Pu in the reactor, some ^{239}Pu nuclei fission, while other ^{239}Pu capture a neutron and become ^{240}Pu. Weapon-grade plutonium (WgPu) contains less than 6 % ^{240}Pu, which usually means the fuel rods remained in a reactor for months not years. However, fuel rods that remain in a reactor for a few years produce reactor-grade plutonium (RgPu) containing over 20 % ^{240}Pu. This is extremely important because of the following:

> Spontaneous neutrons from ^{240}Pu decays can pre-initiate a nuclear explosion, giving a burst of neutrons before the nuclear volume has been assembled or strongly compressed by explosives. These early neutrons can greatly reduce its yield. It is analogous to pre-ignition in a car; when the spark fires too early (advanced too much) combustion suffers. The early explosion of gasoline fumes pushes against the piston, making it more difficult to rise. Similarly, pre-initiation can reduce the yield of nuclear weapons by a factor of ten or more. However, Russia and the US, using sophisticated designs, unavailable to first-time programs, can overcome the pre-initiation problem from RgPu.

Thus far, all nine nations with plutonium weapons chose WgPu and rejected RgPu for their nuclear weapons. Nevertheless, poor weapons of 1 kton or good sophisticated weapons can be made from RgPu. For this reason, the United States constrains reprocessing of US-origin spent fuel and the IAEA maintains safeguards over both types of plutonium with equal authority.

Longer burnup fuels. Longer fuel-time in reactors is referred to as *higher burn-up fuels,* as more fissile isotopes are burned (fission). The extra time for extra burn-up reduces the value to proliferators; the ^{239}Pu isotope-ratio decreases and the *relative* plutonium production rate decreases with greater burnup. To reduce weapons usability, it is imperative to have long-lived fuel rods that are metallurgically robust. Uranium requirements can be reduced by 10 % if fuel remains in reactors for a longer residency. Longer time in the reactor allows a greater fraction of ^{235}U nuclei to fission and it allows greater in situ ^{239}Pu production to fission in the reactor, which increasing the ^{240}Pu contaminant. The properties of the plutonium isotopes are given in Tables 9.5, 9.6, 9.7 and 9.8.

US Ban on commercial plutonium. Plutonium is obtained from spent fuel rods by *plutonium-uranium recovery by extraction* (PUREX), which is a liquid-liquid extraction method. If only uranium is extracted, it is called UREX. Plutonium can also be separated by an electrolytic approach called pyro-reprocessing. During the 1976 presidential campaign, President Ford called for an indefinite deferral of commercial spent-fuel reprocessing. This was followed by President Carter, who called for closing of the Barnwell reprocessing plant in South Carolina and the cancellation of the Clinch River Breeder Reactor in 1977. These steps were largely in response to India's 1974 nuclear test with a Pu bomb. For decades, European nations and Japan have had large plutonium programs to establish MOX fuels and breeders. The end of commercial reprocessing precluded the use of plutonium in MOX fuels for thermal and breeder reactors. The Ford/Carter decision was

Table 9.5 Uranium Enrichment Plants

Facility	Type	Operational	Safeguard	Capacity (tSWU/yr)
Argentina				
Pilcaniyeu	Civilian	Uncertain	Yes	TBD
Brazil				
Resende	Civilian	Being commissioned	Yes	115–120
China				
Shaanxi	Civilian	Operating	Yes	500–1000
Lanzhou II	Civilian	Operating	Offered	500
Lanzou (new)	Civilian	Operating	No	1000
France				
George Besse II	Civilian	Operating	Yes	7500–11,000
Germany				
Gronau	Civilian	Operating	Yes	2200–4500
India				
Ratehalli	Military	Operating	No	15–30
Iran				
Nantanz	Civilian	Under construction	Yes	120
Qom	Civilian	Under construction	Yes	5–10
Japan				
Rokkasho	Civilian	Resuming operation	Yes	50–1500
Netherlands				
Almelo	Civilian	Operating	Yes	5000–6200
North Korea				
Yongbyon	?	?	No	8
Pakistan				
Kahuta	Military	Operating	No	15–45
Gadwal	Military	Operating	No	Unknown
Russia				
Angarsk	Civilian	Operating	Offered	2200–5000
Novouralsk	Civilian	Operating	No	13,300
Zelenogorsk	Civilian	Operating	No	7900
Seversk	Civilian	Operating	No	3800
United Kingdom				
Capenhurst	Civilian	Operating	Yes	5000
United States				
Paducah, KY	Civilian	Shutdown in 2013	Offered	11,300
Pikerton, OH	Civilian	Planned	Offered	3800
Eunice, NM	Civilian	Operating	Offered	5900
Aveva E. Rock, ID	Civilian	Planned	Offered	3300–6600
GLE, Wilmington, NC	Civilian	Planned	?	3500–6000

When a range of capacity is shown, the facility is expanding its capacity—except for Pakistan where the range denotes uncertainty in estimating capacities (International Panel on Fissile Materials, 2013)

Table 9.6 $^{239}Pu/Pu$ ratio and plutonium production rate

Burnup	$^{239}Pu/Pu$ ratio (%)	Pu heavy metal (kg/tonne)
0 MWd/t	100	0
20	78	6
40	58	11
60	47	14
80	43	16

The fractional population of $^{239}Pu/Pu$ decreases with burnup (MW-days/tonne). The Pu-kg/tonne heavy metal ratio rises with burn-up, rising above 1 % Pu/heavy metal at 40 MW-days/tonne.

Table 9.7 Isotopic composition of five grades of plutonium (Mark 1993)

	^{238}Pu	^{239}Pu	^{240}Pu	^{241}Pu	^{242}Pu
Super-grade (%)	–	98	2	–	–
Weapons-grade (%)	0.012	93.8	5.8	0.35	0.022
Reactor-grade (%)	1.3	60.3	24.3	9.1	5.0
MOX-grade (%)	1.9	40.4	32.1	17.8	7.8
Breeder blanket (%)	–	96	4	–	–

Table 9.8 Key properties of Pu and Am isotopes

Isotope	Bare-sphere Crit. mass (kg)	Half life (years)	Decay heat (W/kg)	Neutrons (n/g-s)
^{238}Pu	10	88	560	2600
^{239}Pu	10	24,000	1.9	0.02
^{240}Pu	40	6,600	6.8	900
^{241}Pu	13	14	4.2	0.05
^{242}Pu	80	380,000	0.1	1700
^{241}Am	60	430	110	1.2
WgPu (94 % ^{239}Pu)	10.7		2.3	50
RePu (55 % ^{239}Pu)	14.4		20	460

Other bare-sphere critical masses: ^{235}U (48 kg), ^{233}U (16 kg), ^{237}Np (73 kg) (International Panel Fissile Materials).

prompted by events in several smaller countries, where governments tried to obtain reprocessing plants, often with an eye to obtaining nuclear weapons.

***PUREX** process* uses solvent extraction to separate 98 % of fission products from the uranium and plutonium stream. The fuel is first dissolved in nitric acid. An organic solvent containing tributyl phosphate in a hydrocarbon solvent, such as kerosene, is used to extract uranium and plutonium. Plutonium is separated from uranium by feeding the kerosene solution with aqueous ferrous sulphamate. The uranium is stripped from the kerosene solution with nitric acid. By adjusting the acidity of the solvent, the streams are separated between solvent and aqueous streams. To avoid separating plutonium for safeguard reasons, plutonium nitrate is mixed with uranium nitrate for conversion to the mixed-oxide. Safeguards measurements take place in four basic areas: accountability of input samples, assay of the waste stream, assay of the final product, and assay of hold-up tanks.

Reprocessing, materials unaccounted for. There is concern that nations will use their nuclear power programs to cover other motives, which may be to make nuclear weapons. The Carter policy of a *once-through* fuel cycle without reprocessing gained adherents over time, as nations readjusted their views on the plutonium economy, and as the US curtailed its reprocessing plant in Barnwell, South Carolina. But reprocessing continues in China, UK, France, Russia, India, Israel, Japan and Pakistan. In the past, reprocessing plants had Pu material unaccounted for (MUF) of about 1 % per year, but they hope to reduce this to 0.1–0.2 %. The Japanese Rokkasho plant can reprocess 800-ton/year. This is enough to make 1000 warheads a year with reactor-grade plutonium. A 1 % loss of Pu is (0.01 loss) (0.9 % Pu) (8×10^5 kg spent fuel) = 70 kg/year and a 0.1 % loss is 7 kg/year. A significant quantify of plutonium is 8 kg, but an imploded primary contains about 4 kg.

Rokkasho reprocessing plant. The Rokkasho reprocessing plant (RRP) was intended to produce plutonium for Japan's breeder program, but now it is more likely to produce plutonium for mixed oxide fuels for light water reactors, or perhaps not used at all. RRP was designed to reprocess 800 tonnes/year of spent reactor fuel from 30 GW_e of LWR. RRP serves as a test-bed to examine advanced safeguard technologies, integrated through a collaborative process between IAEA, Japan and US. IAEA will have inspectors present at all times.

Plutonium produced in HEU research reactors. Research reactors provide two paths to nuclear weapons. The first path is the diversion of HEU reactor fuel. The US worked to close this path by reducing ^{235}U content in the research reactor fuel from 90 % to 20 %. The same reactor power can be maintained in most cases by increasing uranium density while reducing the ^{235}U-enrichment level, which has the effect of maintaining the same volume density of ^{235}U, but reducing the enrichment level considerably below 90 %.

Natural uranium research reactors. At the other extreme, Israel's Dimona and India's Cirus research reactors use natural uranium fuel and a heavy water moderator to produce plutonium at a rate of 0.3 kg/MW_t-year. It takes about six months to obtain 5 kg for a Pu warhead from a 40-MW_t reactor, or about two warheads a year. The US urged conversion of natural uranium research reactors to a *higher* enrichment level of 20 % to substantially reduce plutonium production. The rate of plutonium production depends on the ratio of fertile ^{238}U to fissile ^{235}U. By switching from natural U fuel to 20 % fuel, the Pu production rate is reduced from two warheads per year by a factor of 35 to one warhead in 15 years. This greatly increases the time to obtain the material to make a warhead. Since Israel and India obtained their first warheads in 1968 and 1974, respectively, they could each have produced 80 warheads with natural uranium fuel.

CIVEX. Walter Marshall and Chauncey Starr proposed a fuel cycle in 1978 that provides a radiation barrier to prevent terrorists from working with stolen Pu. A high concentration of ^{238}Pu in plutonium increases damaging radiation, increases unwanted thermal power, and increases spontaneous neutron emission. The practical complications of CIVEX precluded its adoption, but it raised interesting questions. The use of the 88-year half-life of ^{238}Pu is long enough to allow Pu to be protected, but short enough to produce considerable radioactivity and heat. If 1 % ^{238}Pu is added to 5 kg of Pu, this gives a heating rate of 40 W. This heat source raises the temperature of a bare sphere by 100–200 °C, which can damage warhead explosives. Clever designs can insert heat bridges to remove heat from warheads of reactor grade plutonium or CIVEX plutonium.

Global nuclear energy partnership. President George W. Bush introduced the Global Nuclear Energy Partnership (GNEP) to encourage a global nuclear renaissance, as well as to assist with nuclear waste and proliferation matters.[1] On 6 February 2006, Secretary of Energy Samuel Bodman introduced reprocessing as part of the policy: "GNEP will leverage new technology to effectively and safely recycle spent nuclear fuel without separating plutonium." PUREX reprocessing was to be modified to the UREX+ process, keeping U and Pu together in a recycled fuel, or Pu with neptunium. This approach had the advantage that the heat load in a repository would be greatly reduced, allowing more underground storage. The US stopped commercial reprocessing in 1977, while other countries did not while they spent considerable funds with little progress. The US did not return to any form of reprocessing. Politically, enhanced relations with India followed agreement on nuclear issues, as India received almost nuclear weapon state status. This was supported by most nations. The rationale was to cement Indian good behavior on nonproliferation matters, but it did create a large exception in NPT matters.

9.5 Missile Technologies

MTCR and Scuds. The first ballistic missile, the German V2, had a payload of 900 kg and a range of 300 km. Fifty years later, Scud-type missiles, similar to the V2, proliferated to 25 nations, similar to the type launched by Iraq in the 1991 Gulf War. The issue of missile proliferation encouraged establishment of the Missile Technology Control Regime (MTCR) in 1987. MTCR constrains exports of missiles and subsystems to deliver *no more than* 500 kg, the size of a crude nuclear warhead at a Scud range of 300 km. In 1993 the throw-weight criteria was removed since biological weapons are much lighter and can be as dangerous. Similar to the NPT, there are exceptions for *peaceful missile uses* of dangerous technologies. This

[1] von Hippel (2008), Lindemyer (2009).

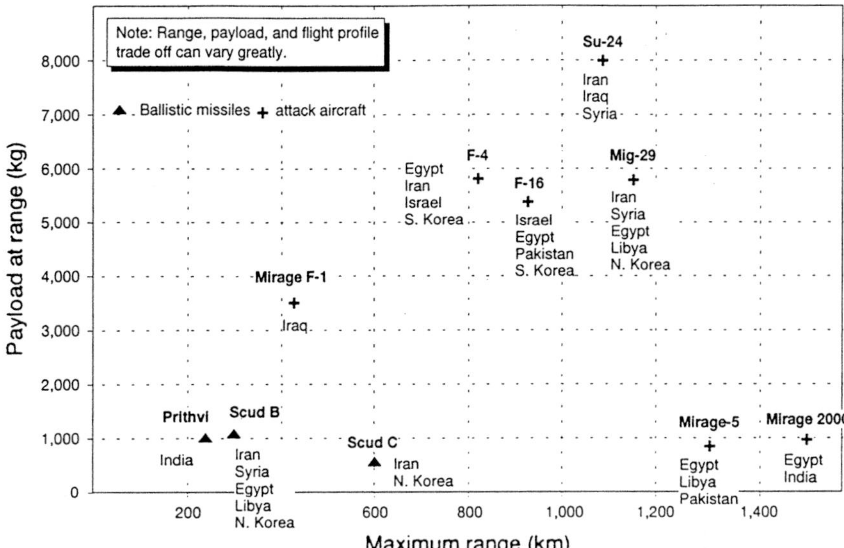

Fig. 9.2 Payload and range of aircraft and missiles. Aircraft can carry larger payloads than Scud missiles at theater ranges of 1000 km. As indicated in the graph, aircraft can be more lethal, but they can be more vulnerable to attack, and they respond less quickly (Office of Technology Assessment, 1993)

can be helpful or it can be a loophole. There are exemptions in MTCR for the transfer of space launch vehicles (SLV) for nonmilitary uses. These exemptions complicate MTCR compliance. Since MTCR is a quasi-executive agreement and not a signed treaty, it has led to misunderstandings, such as the Chinese export of M-11 missiles to Pakistan. Payload and range of aircraft and missiles are displayed in Fig. 9.2.

Throw-weight and range are linked. A reduction in throw-weight allows an increase in velocity v and range R. A 1 % reduction in mass increases the range by 1–2 %, depending on the situation. A miniaturized warhead can have a longer range of attack. By reducing the number of warheads on SLBM's, there is extra propulsion to throw fewer warheads a greater distance.

Scuds. The MTCR Annex lists export constraints. Scud velocity is reported to obtain a fly-out velocity of 2 km/s, to obtain a 300 km range over a 5-min flight time. Its 500 kg warhead obtains an altitude of 100 km. MTCR exports of rocket engines must not exceed a threshold impulse (force times time, or mass times velocity), which is obtained for Scud-like systems with a 500-kg mass at an initial velocity of 2 km/s.

9.6 Safeguards Technologies

The absence of nuclear safeguards, materials accounting and physical security in the Soviet Union was obvious at the end of the Cold War. The former Soviets relied on the robust KGB to protect their nuclear materials, without doing much materials accounting and physical security. The former Soviet KGB relied on fear to make it work. In this section we briefly describe a number of the techniques that are used to monitor plutonium and HEU. Two excellent resources are Glenn Knoll's *Radiation Detection and Measurement* and Jim Doyle's *Nuclear Safeguards, Security and Nonproliferation.*

IAEA safeguards. The key IAEA safeguard criteria (INFCIRC/153) is as follows: "The timely detection of diversion of significant quantities of nuclear material from peaceful nuclear activities to the manufacture of nuclear weapons or other nuclear explosive devices or for purposes unknown, and deterrence of such diversion by the risk of early detection." The mass for implosion weapons is about 50 % of a significant quantity. The IAEA takes a cautious view on plutonium by considering WpPu to be the same as RePu.

IAEA safeguards are based on national quantitative declarations of their special nuclear material stocks and equipment, which are inspected to confirm or deny the declarations. The material-unaccounted-for (MUF) is the amount of uranium or plutonium that has been lost in the system. It is difficult for operators of large enrichment and reprocessing plants to obtain an annual MUF less than a significant quantity. Old fuel facilities show that Pu and HEU are often trapped in obscure pipes. In such cases, the operators must show that they have strong physical security measures in place to prevent clandestine material from exiting the plant. IAEA inspectors have 3 activities to verify operator declarations: The inspector

1. checks the nuclear material accountancy,
2. verifies the nuclear material with visual and nondestructive assay techniques, and
3. uses containment and surveillance to check that nuclear material are not diverted or misused.

Inspections are expensive for the IAEA and intrusive for facility operators. For these reasons the IAEA relies heavily on unattended monitoring systems to determine safeguards violations by measuring radioactivity, heat, pressure, temperature, material flow, vibrations, infrared, optical and EM fields. IAEA installs about 10 unattended monitoring systems a year with some 200 in place by 2013. Inspectors use a variety of measures that are given in Tables 9.9 and 9.10.

The amount of IAEA effort needed is determined by examining the proliferation pathways from acquiring raw material to obtaining separated plutonium and HEU. This approach leads to a unified safeguard system for each facility.

Terrorist nuclear bombs are discussed in Chap. 13. It is of interest to estimate the probability of a terrorist nuclear weapon attack. Matt Bunn (Chap. 13) obtains an attack rate of 29 % in the decade between 2006 and 2016, which is almost

Table 9.9 International reprocessing plants in tonnes heavy metal/year (International Panel on Fissile Materials, 2013)

Facility	Type	Status	Safeguard status	Capacity (tHM/year)
China				
Pilot plant	Civilian	Operating	No	50–100
France				
UP2	Civilian	Operating	Yes	1000
UP3	Civilian	Operating	Yes	1000
India				
Trombay	Military	Operating	No	50
Tarapur-I	Dual	Operating	No	100
Tarapur-II	Dual	Operating	No	100
Kalpakkam	Dual	Operating	No	100
Israel				
Dimona	Military	Operating	No	40–100
Japan				
Rokkasho	Civilian	Starting up	Yes	800
Tokai	Civilian	Temp. down	Yes	200
North Korea				
Yongbyon	Military	Stand by	No	100–150
Pakistan				
Niore	Military	Operating	No	20–40
Chashma	Military	Under construction	No	50–100
Russia				
RT-1	Dual	Operating	No	200–400
Seversk	Dual	Shutdown	No	6000
Zheleznogorsk	Dual	Shutdown	No	3500
United Kingdom				
B205	Civilian	To be shutdown	Yes	1500
THORP	Civilian	To be shutdown	Yes	1200
United States				
H-canyon, SRP	Converted	Special operations	No	15

Table 9.10 IAEA inspection safeguard technologies

• Encrypted data transmission
• Uninterruptible power supplies for their installations
• Unique data signatures
• Cross-correlations with other safeguard measures
• Visual inspection of components and cables
• Tamper indicating tags, seals and enclosures
• Active and passive tags and seals
• Radio-frequency identification (RFID)
• Surveillance of optical sensors
• Supervision of maintenance

completed at this writing. Obtaining significant amounts of nuclear weapon useable materials is the first, and largest, hurdle to make a bomb. The results are speculative, mainly done for policy makers to focus on four types of attack: (1) outsider theft, (2) insider theft, (3) the black market, and (4) provision by a state. Ten parameters are involved in determining the probability of successfully obtaining materials, the number of rogue states or groups that would attack, the ability to transform the material into a warhead, the ability to deliver the warhead, the ability to penetrate the homeland security defenses, and other issues. Len Weiss comments in Chap. 13 that the probability of a terrorist nuclear bomb is exaggerated.

9.6.1 Advanced Nuclear Monitoring Technologies

Improved detector materials are in demand since improved technologies can lead to better IAEA safeguards for difficult tasks. The Department of Homeland Security must vouch for the authenticity of 7 million containers arriving each year by sea and 9 million/year by land. Better detectors means fewer false positives, which can be, perhaps, 100 per day, complicating inspections. This is a tall order.

Gamma ray detectors. Traditional NaI detectors have a resolution of 6 % at 662 keV, while germanium and CdZnTe detectors have better resolutions of 0.13 and 1.7 %, respectively. CdZnTe is preferable to Ge for handheld detectors because it does not require liquid-nitrogen cooling, but it is expensive. There has been progress in high-pressure xenon detectors, in nano-composite scintillators and quantum-dot detectors.

Neutron detectors. An excellent way to detect neutrons is with a ^{3}He-gas tube that costs about $1000, plus electronics. Neutrons and helium-3 combine to give tritium and a proton, with considerable energy for detection. To detect fast neutrons from fission, the gas tube is surrounded with polyethylene to thermalize neutrons, as tritium's thermal absorption is 5000 times greater than its fast absorption. Neutrons are detected more cheaply with plastic scintillators, loaded with boron or lithium. Sensitive neutron detection is done with sophisticated neutron-multiplicity counters. The most sensitive approach is, surprisingly enough, that of calorimetry because it does not depend on the substrate that contains the radioactivity. The heat output is combined with isotopic analysis, either by gamma rays or mass spectrometry, to obtain plutonium mass. A calorimeter consists of (1) a sample chamber, (2) a well-defined thermal resistance, (3) a temperature sensor and (4) a constant temperature environment.

HEU detection. On two occasions, the ABC-TV network successfully smuggled depleted uranium samples through airport portal detectors. HEU is more radioactive than depleted uranium, but fewer neutrons and gamma rays are emitted from HEU than from Pu. It is preferable to use an active approach to locate fissile materials. External neutrons fission uranium and plutonium nuclei, leaving a fission signature of neutrons, fission fragments and gamma rays.

Radiography. The Department of Homeland Security has a massive job to prevent weapons of mass destruction from entering the country. Can it be done? X-ray machines produce 10–100-keV X-rays that are detected by film or electronic media. Gamma-ray radiography uses 662-keV gamma rays from a ^{60}Co source to project onto NaI detectors. Los Alamos National Laboratory developed an approach using cosmic-ray muons, which are deflected electrostatically by heavy-Z nuclei, U and Pu. The muon path, before entry to the chamber, is compared to the exiting muon path. This comparison determines if heavy-Z nuclei were present to deflect muons. Either drift tubes or a gas electron multiplier tubes are used. However, if Pu or HEU was brought in small parts, this approach can be spoofed. If the US is to scan 16 million containers/year it is projected to cost over a billion dollars a year.

Antineutrino detectors. After the first giggle, this approach looks better, but don't forget economics. The decay of ^{239}Pu produces substantially fewer antineutrinos than ^{235}U. The 27 m^3 Livermore detector determines the reactor's fissile inventory and its thermal power as a function of time. Inverse-beta decay is used to detect the anti-neutrinos. As the uranium fuel is consumed, the antineutrino rate drops. If uranium is removed, the system can detect it with the detection of 20–60 anti-neutrinos per day, with an error of about 25 %. Livermore estimates that it can verify the operation of a 5–10 MW_t reactor at a distance of 10 km. However, detection of neutrinos alone, does not locate the nuclear source. Theoretically this can be done by a scheme called Watchman, which examines the resulting Cherenkov radiation. This gives the direction to the source, but it is difficult.

It is much more difficult to monitor a nuclear explosion, with a source of 9×10^{23} anti-neutrinos/kton. Perhaps, 1 kton can be detected at a distance of 10–100 km, or 10 kton at 250 km.

Nuclear forensics. If a foreign nuclear weapon is exploded on an American city, would we be able to determine where the weapon originated? If 1 kg of HEU or Pu was found in a bus terminal, would we be able to trace it to its origin? Forensics is the identification and analysis of these events, using traditional forensics down to femto-gram size samples, fingerprints, grain size and microstructure and cloth fibers. Nuclear forensics determines isotope ratios with radioactivity measurements. For instance, the ratio of ^{18}O/^{16}O and the lead isotopes depend on the location of weapon production sites. Resonance–ionization mass–spectrometry measures the isotope ratio for different elements with the same mass, such as ^{238}U/^{238}Pu. By measuring the ratios of uranium, plutonium and reactor products, it is possible to determine the type of reactor used, the method of enrichment, the age of the sample since reprocessing, the burn-up of fuel and more. Appendix E of the APS/AAAS study, *Nuclear Forensics*, describes the techniques.

Nuclear archaeology. Cumulative plutonium production can be determined by observing the ratio of isotopes that are transmuted over the life of the reactor. Un-irradiated boron has an isotopic ratio, ^{11}B/^{10}B of about 4. This boron ratio increases with irradiation, rising from 4 to as much as 15 for large plutonium production. The Graphite Isotope-Ratio Method (GRIM) was initially tested at a graphite reactor in Hanford. The cumulative plutonium production at North Korea's Yongbyon reactor could be determined using this method to compare to their plutonium declarations to the IAEA. Similarly, ^{235}U enrichment levels can be estimated from the $^{234}U/U_{natural}$ ratio in depleted uranium tails.

Nuclear emergency search team. NEST is staffed with nuclear weapon experts from the three weapon labs at Los Alamos, Livermore and Sandia. If a suspicious weapon is located, portable counting and X-ray equipment can be used to determine the best way to dismantle it. Since NEST staff understands nuclear weapons designs, it should be able to interpret the X-ray data and recommend action for dismantlement.

Problems

9.1 **A.Q. Kahn**. How did A.Q. Khan get his technical training? How did he steal secrets and bring them to Pakistan?

9.2 **Brazilian centrifuges**. Describe Brazil's enrichment program. Why does Brazil get this program while Iran does not?

9.3 **China's weapons choice**. Why did China choose the uranium bomb in the 1950s?

9.4 **Iran enriches**. Why did Iran pick gas centrifuges over gas diffusion?

9.5 **Diffusion membranes**. Take a coin and draw circles next to circles to show how sintered nickel spheres can make a membrane.

9.6. **Laser separation**. How do charge exchange reactions make LIS less effective?

9.7 **South Africa's bomb**. What was South Africa's motive to build a bomb?

9.8 **SWU**. What is a separate of work unit? Give us some feel for SWUs.

9.9 **Gaseous centrifuge versus gaseous diffusion**. What are advantages and disadvantages of these two technologies?

9.10 **Enrich to 20 %**. Why is a stockpile of 20 % enriched uranium dangerous for proliferation?

9.11 **RgPu versus WgPu**. How are these two kinds of plutonium produced? Why have nations chosen not to make weapons with reactor grade plutonium?

9.12 **Pre-initiation**. What causes pre-initiation of a nuclear weapon? How can you design around this?

9.13 **Burnup of fuel**. Is there a nonproliferation advantage by using high-burnup fuels? What does one have to do to make high burnup fuels feasible?

9.14 **SWU total**. Add up the global enrichment capability in Table 9.4. How many warheads per year could this enterprise produce? How many power plants could this capacity operate at the rate of 120,000 SWU per reactor year.

9.15 **US and global uranium resources**. Look up current values for these parameters. How many reactor years of nuclear power would this support.

9.16 **Reprocessing**. What is the difference between PUREX reprocessing and pyro-reprocessing?

9.17 **Need for reprocessing**. A small nation with one power plant decides to build a large reprocessing plant. What is its motive to do this? How should the US try to coax this country away from this path? What if they do not choose to follow this path?

9.18 **Barnwell and Clinch River.** The Carter administration canceled these two projects. Was this wise? Please use some numbers in your response.

9.19 **F-15 and F-16**. Describe these platforms and how they might be used for nuclear weapons.

9.20 **Infrared detection**. Why has IR detection become easier?

Seminar: Proliferation Technology

AQ Kahn biography, danger of 20 % stockpile, entropy reduced in enrichment and increased when combine (chaos and order), EMIS rejected, engineer smaller warhead sizes, gaseous diffusion rejected, muon detection of warheads at Los Alamos, nickel pores in barriers, nuclear archeology, Phoenix status, reactor fuel raised to 4.5 % gains what, Rokkasho status, Russian enrichment sales to US and others, Scud vs. aircraft bomb basing, SILEX status, SWU pricing, uranium resources and reserves, US reprocessing history.

Bibliography

Ahlswede, J., & Kalinowski, M. (2012). Global plutonium production with civilian research reactors. *Science and Global Security, 20*(2), 69–96.

Albright, D., Berkhout, F., & Walker, W. (1997). *Plutonium and highly-enriched uranium: 1996 world inventories, capabilities and policies*. Oxford, UK: Oxford University.

Berstein, A., et al. (2010). Nuclear security applications of anti-neutrino detectors. *Science and Global Security, 18*, 127–192.

Bodansky, D. (2004). *Nuclear energy*. New York, NY: American Institute of Physics Press.

Bukharin, O. (1996a). Analysis of the size and quality of uranium inventories in Russia. *Science and Global Security, 6*, 59–77.

Bukharin, O. (1996b). Security of fissile materials in Russia. *Annual Review of Energy and the Environment, 21*, 467–498.

Bunn, M., & Holdren, J. (1997). Managing military uranium and plutonium in the United States and the former Soviet Union. *Annual Review of Energy and the Environment, 22*, 403–486.

Bunn, M., Fetter, S., Holdren, J., & vander Zwaan, B. (2005). Economics of reprocessing versus direct disposal of spent nuclear fuel. *Nuclear Technology, 150*(3), 209–230.

Craig, P., & Jungerman, J. (1990). *Nuclear arms race technology and society*. New York, NY: McGraw Hill.

Doyle, J. (Ed.). (2008). *Nuclear safeguards, security, and nonproliferation*. New York, NY: Elsevier.

Fuller, J. (2010, December). Verification to the road to zero. *Arms Control Today*, 19–27.

Glaser, A. (2008). Characteristics of the gaseous centrifuge for uranium enrichment and relevance for nuclear proliferation. *Science and Global Security, 16*(2), 1–25.

Glasner, A., & Glaser, A. (2011). Graphite isotope-ratio method extended to heavy-water, plutonium-production reactors. *Science and Global Security, 19*(3), 223–233.

Glaser, A., & Mian, Z. (2008). Fissile material stockpile and production. *Science and Global Security, 16*(3), 55–73.

IAEA. (2005). *Thorium cycle: Potential benefits and challenges*. Vienna, Austria: IAEA.
Johnston, R. (2001). Tamper-indicating seals for nuclear disarmament and hazardous waste management. *Science and Global Security, 9*, 107.
Kemp, R. (2009). Gaseous centrifuge theory and development. *Science and Global Security, 17* (1), 1–19.
Knoll, G. (2010). *Radiation detection and measurement*. New York, NY: Wiley.
Krass, A., et al. (1983). *Uranium enrichment and nuclear weapons proliferation*. London, UK: Taylor-Francis.
Kuperman, A. (2015). *Nuclear terrorism and national security: The challenge and phasing out of HEU*. New York, NY: Routledge.
Lamarsh, J. (1977). *Introduction to nuclear engineering*. Reading, MA: Addison Wesley.
Lindemyer, J. (2009). The global nuclear energy partnership. *Nonproliferation Review, 16*, 79–93.
Mark, J. C. (1993). Explosive properties of reactor-grade plutonium. *Science and Global Security, 4*, 111–128.
National Academy of Sciences. (1994). *Management and disposition of excess weapons plutonium*. Washington, DC: National Academy Press.
National Academy of Sciences. (2005). *Monitoring nuclear weapons and nuclear-explosive materials*. Washington, DC: NAS Press.
National Commission on Terrorist Attacks Upon the U.S. (2003). The 9/11 Comm. Report, Norton, New York.
Nero, A. (1979). *A guidebook of nuclear reactors*. Berkeley, CA: University of California.
Office Technology Assessment. (1993a). *Technologies underlying weapons of mass destruction*. Washington, DC: OTA.
Office Technology Assessment. (1993b). *Dismantling the bomb and managing the nuclear materials*. Washington, DC: OTA.
Parker, A. (2014, October). Promoting international nuclear forensics. Science and Technology Research, 12–19.
Richelson, J. (2009). *Inside NEST*. Norton, NY: American's Secret Nuclear Bomb Squad.
von Hippel, F. (2008, May). Rethinking nuclear fuel recycling. *Scientific American*, 88–93.
Wood, H., Glaser, A., & Kemp, R. (2008). The gas centrifuge and nuclear weapons proliferation. *Physics Today, 2008*, 40–45.

Chapter 10
Proliferated States

Pakistan will fight, fight for 1000 years. If…. India builds the (atom) bomb…. (Pakistan) will eat grass, or, … even go hungry but we (Pakistan) will get one of our own (atom bomb)…. We (Pakistan) have no other choice.
[Pakistan President Zulfikar Ali Bhutto, founder of Pakistan's nuclear weapons program]

I have much to answer for…. The investigations have established that many of the reported activities did occur and these were inevitably, initiated at my behest. In my interviews with the concerned government officials I was confronted with the evidence and findings, and I voluntarily admitted that much of it is true and accurate.
[Abdul Quadeer Kahn, 4 February 2004, Pakistan state television]

10.1 Technology Transfer

Many books have been written on the details of nuclear transfers from one nation to another. There has been lots of high dramas that make interesting films. My favorite books for details on nuclear transfers are these three references:

- *The Nuclear Express* by Thomas Reed and Danny Stillman,
- *In Mortal Hands*: *A Cautionary History of the Nuclear Age* by Stephanie Cooke,
- *Deadly Arsenals*: *Nuclear*, *Biological and Chemical Threats* by Joe Cirincioni, Jon Wolfsthal and Miriam Rajkumar.

This chapter discusses 27 nation's *covert and overt actions* to obtain nuclear weapons. This is fascinating and serious reading. What did Klaus Fuchs and Ted Hall take from Los Alamos? How did A.Q. Kahn take secrets from URENCO's centrifuge plant in the Netherlands to Pakistan and beyond? Why was the West slow to take action on Kahn when they had solid intelligence that he was taking secrets? Did the Pakistani government observe Kahn when he took sales trips to

D. Hafemeister, *Nuclear Proliferation and Terrorism in the Post-9/11 World*,
DOI 10.1007/978-3-319-25367-1_10

Lybia, Iran, Iraq and North Korea? Three of these nations made nuclear purchases, while suspicious Saddam Hussein stood on the sidelines. Why did Pakistan's Prime Minister Pervez Musharraf give Kahn a mild house-arrest, which is now released? Why did he make his statement in English and not Urdu? And so on. Our goal is to discuss the proliferation facts, but don't get overwhelmed in details. We want to focus on what must be mastered if nuclear proliferation and terrorism are to be controlled. We want to prepare you to become a Foreign Service Officer working on these issues in the State Department or a member of a non-governmental organization, looking over the government's shoulders. We need both right (creative) and left (linear-logical) brains, but in this case, the left-brain holds both history and physics. Let the professor assign the details for the seminar portion of your course.

In lecturing on these topics since 1972, I have used two approaches to cover the important historical events in a shorter time. The first approach is to study from the summaries and chronologies in Chap. 1, even though this is drinking from a fire hose. We ruffle some students with this rapid march, but we have no choice if we want to retain time for policy and technical topics. The second approach is to add Chaps. 7 and 10 and appendices A and B with more history for the dedicated. The seminar portion of the course further involves students, getting them up to speed by verbalizing on the proliferation and terrorism landscape.

Proliferation Plots. One-way to gain fast entry to these topics is to prepare a schematic, displaying the multi-dimensioned aspects of proliferation policy. This I have done in Fig. 10.1. Bill Broad independently presented a similar, but different schematic in the December 9, 2008 issue of the *New York Times*. Broad's approach has the advantage that events are presented as a function of time. My approach lists more details and pathways, listing critical times instead of plotting with time.

Many technology transfers stemmed from the US, since it obtained nuclear weapons first, four years before the Soviets. The US mostly constrained itself on sensitive nuclear exports of enrichment and reprocessing, but, nevertheless, these technologies managed to spread. Our wartime partners, UK and Canada folded into the initial plans, but the atomic energy acts established separateness. There is no doubt that nuclear exports played a significant role in proliferation, by increasing the competency of some scientists, by importing technologies and by giving a political cover for covert programs. The sensitive technologies became the *sweetener* in nuclear transfers, coupled with reactor orders and nuclear fuel contracts. When Germany and France made progress on reactor designs, the US monopoly was broken. Germany and France also had to overcome the US lead on long-term contracts for enriched nuclear fuel. The Germans developed the Becker nozzle enrichment process, which was a sweetener to aid Brazil and South Africa. Similarly, France worked with Taiwan, Pakistan, South Korea and Japan to market reprocessing plants in conjunction with nuclear reactor orders. This was followed in the 1970s with URENCO (UK, Germany, Netherlands) developing centrifuges that found their way to Pakistan and later to North Korea, Iran and Libya. Good news followed in 1995, when Argentina agreed to the NPT, joining Brazil in mutual inspections, an example for other nations. After the 1960s, Israel managed to get various materials from the West, stating it "would not be the first to introduce

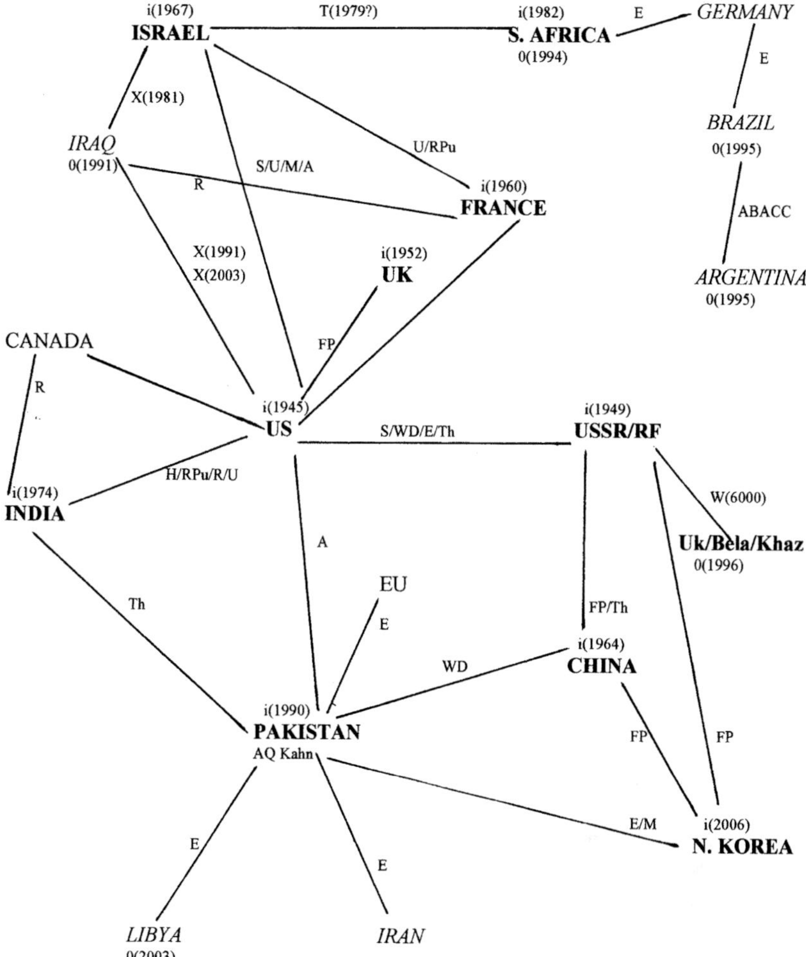

This diagram only covers the major nuclear–proliferation, technology transfers. Nuclear weapon states (past and present) are in **bold**. Aspiring NWS (past and present) are in *italics*. Symbols: A(aircraft), E(enrichment), FP (full–package), H(heavy water), M(missile), R(reactor), RPu(reprocessing), S(spy), T(test), Th(threat), W(warhead), WD(warhead design), X(attack), 0(ending program), i(initial nuclear warhead).

Fig. 10.1 Nuclear technology—transfer flow diagram. Attacks (*X*) to roll back proliferation (counter-proliferation) have taken place twice by the US on Iraq and once by Israel on Iraq's Osirak reactor. Attacks have been considered for Iran, North Korea and others

nuclear weapons to the Middle-East." But, one assumes, Israel is a *screw-driver turn* away from having nuclear weapons, ready to launch. The US fulfills its obligations to the NPT on Israeli matters by not transferring weapons-useable materials or weapons designs. But, the US is involved in other ways, supporting Israel with aircraft, as nuclear-weapon launch platforms, that undercut the spirit of

Table 10.1 Estimated global nuclear weapons inventories, 2014

Russia	5000
United States	4760
France	300
China	250
Britain	225
Israel	80
Pakistan	100–120
India	90–110
North Korea	<10

Russia has 3700 warheads awaiting dismantlement. The US has 2540 awaiting dismantlement, with 1980 deployed and a stockpile of 4760 (H. Kristensen and R.S. Norris, 2014–15)

the NPT, but not the law. In addition, the US exports Patriot and Arrow missile defense systems to fight against a variety of missile types.

The Soviets exported the full package of nuclear technologies to China and somewhat to North Korea. When the Soviets and China split around 1960, China became the protector state for North Korea. The Soviets often forced their client states to tighter standards than the US, requiring the return of spent fuel to the Soviet Union. But the Soviets did not do well when it exported the full-package exports to China. But in general, the Soviets were constrained with its Warsaw Pact nations. Will China now constrain North Korea? Will Iran develop the Shiite bomb that frightens Sunni nations? Will the NPT regime survive to the year 2100? Table 10.1 lists Kristensen and Norris's estimates for the number of nuclear weapons held by nuclear weapon states, the P5 and the de facto-four.

27 Nuclear plans: In the sections below we briefly discuss the proliferation storylines of 27 Nations. Each country has its own individual story, which depends on its technical prowess, its configuration of neighboring nations, its level of fear, its economic situation and, of course, its political situation. Each country initially ignored the proliferation issue. But, in due time the political process broadened, as these countries realized that, without rules, proliferation will continue to threaten the Earth. We choose not to summarize these details, but rather give a summary of a couple pages per country to attain a minimal level of detail. We present the historical data in Sects. 10.2–10.6.

10.2 Five P-5 NWS (US, Russia, China, UK, France)

10.2.1 United States

Proliferation links. The story begins with the US (after Germany). Figure 10.1 displays the United States as the center of the technology transfer pattern. This is to be expected because the US developed nuclear weapon first. The first spoke from the US hub extends to USSR/Russia. This spoke was driven by competitive fear, as well as the spying of Klaus Fuchs and others. The threat of US dominance made Stalin uneasy. The second spoke is drawn towards the UK because UK and Canadian scientists were intimately involved in the Manhattan project. Klaus Fuchs (Germany, UK) returned to Britain as a professor, then landed in jail and, after release, moved to East Germany. The current British submarines and their SLBMs were produced in the US. The Canadian connection stimulated their CANDU reactors and heavy-water technology. Another spoke is drawn to India, which participated in Eisenhower's Atoms for Peace Research at Argonne and Oak Ridge, learning reactors, reprocessing and plutonium-foundry practices. The spoke to Pakistan is primarily to sell F-16 nuclear-weapons platforms to Pakistan. The spoke to Israel is somewhat murky. It is reported that high-enriched uranium was covertly stolen and shipped from Pennsylvania to Israel. And, US-produced F-16 could serve as Israel's nuclear-weapons platforms. The spoke between US and Iraq symbolizes the two Gulf wars, which belatedly discovered the Iraqi nuclear program in the first 1991 war, but mistakenly accursed and invaded Iraq in the second 2003 war.

Weapons complex. According to Kristensen and Norris (2015), the United States has 2080 warheads on-alert, available for immediate delivery with 1150 on SLBMs and 470 on ICBMs. These warheads are based on 800 ballistic missiles and aircraft, with another 300 are off-alert, located at three bomber bases in the US, with another 200 deployed in Europe. An additional 2680 warheads are in storage. In addition, 2340 warheads are retired and await dismantlement at Pantex in Texas. See Sect. 4.5 for details on the evolving triad.

The New START Treaty reduces strategic-deployed warheads to 1550. Obama hinted that 1000 is sufficient for him, but Putin is not interested in this lower number. Coupling the reduction in warheads with the confirmed plutonium-pit lifetime of over 100 years, the size of pit production facility in Los Alamos can be reduced to 20 pits per year. Los Alamos, Livermore and Sandia laboratories are maintaining their historic roles. The US stores its nuclear weapons at 12 sites in the US and at another 6 sites in the European Union. The US no longer stores nuclear weapons at Nellis Air Force Base, Nevada or Barksdale Air Force Base, Louisiana.

Unilateral reciprocal measures. The end of the Cold War allowed President George H.W. Bush to use reciprocal unilateral measures with President Gorbachev in September and October of 1991 to reduce the nuclear threat. The US promised to withdraw 5500 tactical weapons and take off alert 450 Minuteman warheads and 1600 SLBM warheads. He also promised to cancel the production of rail-mobile Peacekeeper missiles, rail-mobile Minuteman missiles, and new short-range attack missiles (SRAM-II). The Soviets promised to remove 12,000 tactical warheads, reduce START to 5000 warheads, take off alert heavy bombers and 500 ICBMs, freeze rail-mobile SS-24 launchers at 36, stop SS-24 out-of-garrison movements, cancel new SS-24 and SS-25 missiles, cut Soviet troop strength from 3.7 to 3.0 million, and stop nuclear tests for one year. A few months later in January 1992, Bush further promised to reduce B2 bombers from 75 to 20, cancel Midgetman mobile-ICBM, convert many bombers from nuclear to conventional weapons, stop production of SLBM W-88 warheads, and stop production of the advanced cruise missile. President Yeltsin responded with promises to take off alert 600 ICBMs, stop production of Bear and Blackjack bombers, and stop production of ACLMs. And there was more, but the conclusion is that these were very large reductions done without diplomacy and done quickly. A rare event indeed.

Modernization plans. The US went through its last large modernization in the 1970s and 80s with the construction of Peacekeeper ICBM, submarine launched cruise missiles (SLCM), Trident submarines, SLBMs with W-88 warheads, B-1 and B-2 bombers, and improvements for the B-52 and Minuteman. The Congressional Budget Office estimated (2013) US plans to spend $355 billion in a decade on these nuclear efforts, up from $213 billion in 2011. The modernization of the triad is in flux, but it looks like this:

ICBM: To comply with the New START Treaty, the US will reduce the number of silos from 450 to 400. This will be followed with modernizations to extend service life of Minuteman to 2034.

SLBM: The number of 70-day submarine patrols has been reduced from 3.5 to 2.5/year. The 400 nuclear SLCMs (W-80 warheads) have been removed. By 2018, the number missile tubes on each submarine will be reduced from 24 to 20. The life-extended version of the Trident II D5 SLBM will be deployed by 2017. The Navy is planning a replacement submarine with 16 missile tubes, but 2000 tons larger than the Ohio-class submarine. The size of the SSBN fleet will be reduced from 14 to 12.

Heavy bomber: The Air Force has begun life extension programs for B-2 and B-52 bombers and their ALCMs. The Air Force now operates a fleet of 20 B-2, 63 B1, 76 B-52 bombers. Of these 16 B-2 and 44 B-52 are nuclear capable. The B-52 no longer carry gravity bombs but they can carry 20 ALCMs. In the mid-2020s the Air Force plans to start replacing B-52 and B-1 bombers with a new long-range bomber B3 with a procurement of 80. The current ALCM is scheduled to remain operational through the 2020s. The Air Force plans to acquire 1000 cruise missiles, capable of nuclear and conventional roles and 2457 F35 stealth fighters.

Warheads: The *3 + 2 strategy* envisions *interoperable warheads* between ICBMs and SLBMs with three warheads for ICBMs/SLBMs and two warheads for bombers.

B-61/12 bomb: Scheduled for deployment around 2020, with greater accuracy from new tail kits. They will be deployed on F-15, F-16 and Tornado NATO aircraft, then in 2024 they will be deployed on F-35A stealth fighters. About 50 % of the B-61s will be on aircraft owned by EU countries.

10.2.2 Russia

Proliferation links. The theft of US nuclear weapon plans by Klaus Fuchs and Ted Hall to the Soviet Union was the first covert act in the nuclear proliferation drama. This shortened research time to four years for the Joe-I test explosion in 1949. The

Soviets initially transferred weapons designs and enrichment technology to China, but then stopped since it was helping a rival in Asia. Since the signing of the NPT in 1968, the US and USSR have cooperated many times over NPT issues, with the Soviets being firmer with its Warsaw-Pact client states, as compared to the US with its NATO and other friends. An example of that is that the Soviet Union required return of spent-fuel, while the US left its US-origin spent-fuel scattered in Europe and Asia. Even now, when times are tense over Ukraine, Russia is working with the P5 to constrain the Iranian nuclear program. The Russians played a key role in offering to supply nuclear fuel to Iran's two existing reactors and perhaps more in the future.

Weapons complex. At the end of the Cold War, Minatom had 1 million employees in a very bloated, unprotected nuclear empire. There were 3 great concerns at the time:

- Nuclear weapons deployed in Belarus, Kazakhstan, Ukraine and elsewhere might not return to Russian control. This was particularly true for Ukraine since it felt it had not received fair compensation for its part in the megatons to megawatts reactor fuel sale.
- Russia might lose control of its nuclear weapons, especially the tactical weapons.
- Russian nuclear materials and expertise might disappear to other countries or to terrorist groups. This threat was summarized by former Sen. Sam Nunn:

> The old threats we faced during the Cold War, a Soviet strike or an invasion of Europe, where threats made dangerous by Soviet strength. The new threats we face today—increased Russian reliance on early launch and first use and increased reliance on tactical-battlefield nuclear weapons—are threats made dangerous by Russia's weakness. The threats of today go beyond nuclear forces and include terrorist groups. Much of Russia's nuclear, biological, and chemical weapons and materials are poorly secured; its weapons scientists and guards are poorly paid. We can't risk a world where a Russian scientist can take care of his children only by endangering ours. [Sam Nunn]

US Cooperative Threat Reduction funding exceeded $10 billion over the years, it was a successful program without major incidents. Table 10.2 displays the number of Russian strategic and tactical warheads and launchers for the time frames of 1990, 2005 and 2014.

Modernization plans. Russia is moving beyond the nuclear structure of the Soviet Union of 2014 (46 SS-18, 30 SS-19, 117 SS-25) to a force that is smaller, more modern and cheaper to maintain. Russia's SS-18 missiles are very large, but

Table 10.2 Russian strategic and tactical nuclear forces (launchers/warheads)

Type	1990	2005	2014
ICBM	1064/6612	611/2436	304/967
SLBM	940/2804	292/1672	160/528
Bombers	162/1600	78/624	72/810
Total Strategic L/WH	2166/11,016	981/4732	536/2305
Tactical WH	21,700	3400	2000

Kristensen and Norris (2014a, b, c)

much less accurate than the US, by a factor of two. This is extremely important when we recall the following: A warhead has the same ability to attack if its accuracy is increased by a factor of two (from 200 to 100 m) while the yield is reduced by a factor of 8. Within the next decade or two, new warheads are planned for its new Bulava SLBM on a fleet of 8 new Borei submarines. Putin promised in 2012 to produce 400 land- and sea-based ballistic missiles. The Blackjack, Bear and Backfire bombers are being upgraded, along with a new subsonic bomber with a new cruise missile. This may all be difficult to do because of Putin's perceived greater need for conventional forces and because of the bad Russian economy from reduced petroleum revenues. As mentioned in Chap. 7, the conflict in Ukraine can further exacerbate matters by hurting the Russian economy, giving Putin a need for something to look strong.

10.2.3 China

Proliferation links. Beginning in 1955, the Soviet Union gave considerable support to China's nuclear weapons program by giving them designs of older weapons, uranium enrichment, nuclear reactors, and other items. By 1960 the Soviets ended their nuclear support, as China became a rival. China's first nuclear test took place in 1964, followed with its first hydrogen test in 1967. China's views on nonproliferation have changed *mightily* over the decades. Initially China favored nuclear weapons proliferation to gain foreign support by helping other nations. Then China shifted into quiet acquiescence in the 1970s by not opposing nuclear proliferation but not supporting it either, which it saw as limiting US and Soviet power. Then in the 1990s China made commitments against proliferation, but it's state owned corporations continued with illicit nuclear or missile transfers to Pakistan, Iran, Libya, North Korea, Syria, Saudi Arabia and Algeria. Now China is mostly on board with support for NPT and CTBT (when the US ratifies) and by membership in the Zanger Committee and the Nuclear Suppliers Group. It is widely hoped China will lean on North Korea to end their nuclear weapons program, but China fears a toppled North Korean government would cause massive immigration to China.

Weapons complex. China originally chose the uranium cycle since China felt the plutonium cycle was too complex without Soviet help, and the Soviets had already given them gaseous diffusion enrichment. Now China has both plutonium and uranium weapons. China continues to make advances in the number and quality of its nuclear arsenal, but what is surprising is that this is done at a modest pace. It is of interest to know why the pace is modest. Perhaps China realizes that an abundant Chinese arsenal would bring threats and excessive nuclear targeting from the US and Russia. And perhaps China realized that trillions of dollars saved by doing less can help their economic engines.

Modernization plans. China's economic clout could easily pay for long-term modernization plans. China has approximately 250 deployed warheads, with 150 deployed on land-based missiles, aircraft and a slowly emerging submarine fleet.

ICBM. Beginning in 1981, China deployed twenty long-range DF-5A ICBMs with a considerable range of 13,000 km, sufficient to hit all the US. These liquid fueled missiles were first placed on launch pads and then in silos. They carry a single large warhead of 4 or 5 Mton. In 2006 China started deploying eight DF-31 solid-fuel missiles. These had a range of 7000 km and carried a single warhead of 250 kton. This was followed in 2007 with twenty road-mobile DF-31A with a range of 11,000 km. China has 40 ICBMs that can reach the United States, and perhaps 60 can reach Moscow. The US national intelligence estimate concluded that China could field some 75–200 MIRVed warheads by 2020, being the last of the P5 to do that. In 2015 China deployed 3-warhead MIRVs on the DF-5, while the DF-31 MIRV lags.

SLBM. China is putting more emphasis on mobile systems on both land and sea. It is shifting its SLBMs from JL-1 (1986, 1700 km) on Xia SSBN to the new JL-2 (2013, 7000 km) on 3 new Jin SSBNs. China's leaders don't like a freewheeling Navy, but they like their ability to hide. China is building ground- and air-launched cruise missiles, with 1500 km range, but it is not updating its old Soviet Badger bomber fleet.

10.2.4 United Kingdom

Proliferation links. Both the Germans and the British began work on nuclear weapons before the US. In January 1941, the Germans were doing reactor experiments with impure graphite in Berlin at the *Virus House*, while the *British Maud Committee* determined in July 1941 that a weapon could be made with 10 kg of ^{235}U. The US National Academy of Sciences endorsed the British results. British, Canadian and other scientists worked on the Manhattan project. On 27 January 1950, Klaus Fuchs, then a British citizen, confessed that he transmitted atomic secrets to the Soviets. The US Atomic Energy Act of 1946 prevented further collaboration with the British and other nations. The UK has been a strong advocate of stemming proliferation, trying to make the NPT work, while maintaining a nuclear force.

Weapons complex. The UK produced 834 warheads and conducted 44 nuclear tests between 1952 and 1991. From the mid-1960s to the mid-1990s the UK stockpile was estimated to contain 300 warheads, while today it is somewhat smaller at 225 warheads. The THORP reprocessing plant in the Lake District has accepted contracts for reprocessing spent fuel.

Modernization plans. Britain eliminated all its air- and sea-based tactical nuclear weapons. There is now a debate whether Britain needs the remainder of its arsenal. The recent special election to remain in the UK damped the issue, but the winning of all but three seats by the nationalists in the Scottish Parliament cuts the other way. Of the P3, Britain has done the most to reduce its nuclear stockpile. The British Trident SSBNs and SLBMs were obtained from the US.

10.2.5 France

Proliferation links. The French Atomic Energy Commission was created in 1945, but it didn't focus on nuclear weapons until the mid-1950s. Uranium ore was discovered in France in 1948. Initially there were exploratory talks with Germany and Italy about military cooperation but these were stopped at an early stage when General de Gaulle returned to office in 1958. The decision to test a nuclear weapon in 1960 was not made until 1958. France used 9 reactors for plutonium production and created its own enrichment facility. At its maximum, France deployed some 538 nuclear warheads, peaking during 1992, but it now is down to about 300 warheads. France has a plant in Normandy for reprocessing spent fuel from Japan and other countries.

Modernization plans. France is in final stages of modernizing its *force de frappe* for 2050. It removed its missile-based forces in 1996. France is replacing its M45 warhead on its two le-Triumphant-class and two le-Redoubtable-class submarines with a new M51 warhead, which has a greater range, payload and accuracy. In 2011 France replaced its air launch cruise missile with a 500 km range version. The Charles de Gaulle aircraft carrier is capable of deploying nuclear weapons but they are not deployed in peacetime. France has 60 nuclear capable aircraft in its Air Force and some 24 carrier-based nuclear capable fighters in its Navy.

10.3 Four Defacto NWS (India, Israel, North Korea, Pakistan)

10.3.1 India

Proliferation links. India has not signed either the NPT or the CTBT. India has long been concerned about its nuclear neighbors, China and Pakistan. The world was shocked by India's 1974 nuclear test, as this was the first nuclear test beyond the P5. Many hoped only rich nations could test a nuclear device, but that was clearly wrong. The US State Department responded with the creation of the Nuclear Suppliers Group in 1975. The Congress responded with the Nuclear Non-Proliferation Act of 1978 and Glenn-Symington sanctions, which cut off nuclear fuel contracts to countries that did not have all their facilities under safeguards and banned reprocessing without US permission on US-origin fuels. Yet, the US and others quietly helped India find suppliers for fuel for its Tarapur reactors, first the French from 1980 to 1992, then China from 1994 to 2004 and finally Russia from 2006 to the present. India learned reprocessing technology from the US, as part of the Atoms For Peace Program. As of 2013, India had 550 kg of weapons-grade plutonium, 4700 kg of reactor-grade plutonium and 2400 kg of high-enriched uranium. India operates 20 power reactors, has 7 more under construction, along with a breeder reactor program. By 2013, India's nuclear power was 4780 MW_e, 3.6 % of its grid. The reactors are mostly heavy-water moderated with natural uranium fuel, to avoid importing enriched uranium. The Cirus and Dhruva plutonium-production reactors, use heavy water and natural uranium to produce India's weapons-grade plutonium. India tested on 18 May 1974 and it tested 5 times on May 11 and 13, 1998, followed by Pakistan on May 28 and 30. After these tests, India ruled out "first use of nuclear weapons" and to limit its nuclear force to a "credible minimum deterrence."

India, the NPT exception? President George W. Bush and Indian Prime Minister Manmohan Singh agreed on 18 July 2005 to a policy that lifts US and global restrictions on nuclear trade in return for some nonproliferation commitments by India. This policy allows India to import nuclear power plants, uranium feedstock and reactor fuels. India agreed to continue its nuclear test moratorium, supporting the US on a fissile material test ban. India is separating its military and civilian programs and placing some reactors under safeguards, but not its enrichment and reprocessing facilities. There are two main downsides to the agreement:

1. Other nations are asking, "why not us, as we have been well behaved?" India was intended to be an exception, but will India be a path-breaker?
2. India increased its ability to import uranium, freeing up its present uranium stocks for a growing triad-based nuclear structure.

The Nuclear Suppliers Group went along with this, as did the US Congress. Pakistan complained loudly that it should have gotten the same deal, but clearly the A.Q. Kahn exports put Pakistan in a different category. At the 2010 NPT review conference, India's special situation was not well received by the 118 members of the Non-Aligned Movement. The Nuclear Suppliers Group will have to adopt rules to soften the impacts.

Modernization plans. After their 1998 nuclear tests, India stated it would develop a traditional triad of nuclear weapons. This was followed in 2005 with the nuclear accommodation with President George W. Bush, which gave India an *almost* NWS status in the NPT. A long-range Agni missile is underway, as well as tactical-ranged missiles, Prithvi and Agni missiles. The first SSBN has been launched, out of a series of at least 3, with 12 SLBMs each. India plans to be able to confront any nation, including China; this is different from Pakistan, which directs its ire only at India.

10.3.2 Israel

Proliferation links. Israel has long been reported to own nuclear weapons, ever since the late 1960s. It is rumored that Israel covertly obtained natural uranium from Europe. In addition it is reported that Israel in the 1960s clandestinely obtained 100 kg high-enriched uranium from the Nuclear Materials and Equipment Corporation (NUMEC) facility in Apollo, Pennsylvania. In 2011, Prime Minister

Benjamin Netanyahu repeated the refrain that "we won't be the first to introduce nuclear weapons into the Middle East." This quote does not define when a nuclear weapon is assembled, or almost assembled. The strongest affirmation by Israeli leadership of their nuclear status was given by Prime Minister Ehud Olmert, when he criticized Iran for aspiring "to have nuclear weapons as America, France, Israel, Russia." Later, a spokesperson for Olmert said that he had been listing *responsible nations* and not *nuclear weapons states*.

Many have suggested that the 22 September 1979 signal from the Vela surveillance satellite is evidence of a nuclear explosion over the South Atlantic. To some, the optical bhangmeter meter data indicates a 3-kton explosion and to others it indicates a Zulu event, a one-hump signal caused by a tiny meteorite. If Israel did this test, it would be a violation of its membership in the Limited Test Ban Treaty. The strongest proof of Israeli ownership of nuclear weapons comes from Mordechai Vanunu when he published photos of the Israeli Dimone nuclear weapons facility in the *London Times* of 7 October 1986. Shortly after that he was drugged and taken to Israel for solitary confinement until 22 April 2004.

Weapons complex. Israel's nuclear weapons research is primarily done at the Negev Nuclear Research Center at Dimona. This un-safeguarded facility has a plutonium production research reactor, using a heavy water moderator with natural uranium fuel. The Dimona reactor of 40–150 MW produces some 11–19 kg of plutonium a year. The Dimona site also has plutonium PUREX extraction facilities. A second reactor of 5 MW is located at Soreq. Assuming 4 kg of plutonium for a warhead, Israel could have 400 warheads, but it is usually assumed Israel has about 80. Vanunu, Seymour Hersh and others believe that Israel has advanced weapons technologies, including boosting.

Kristiansen and Norris estimate that "Israel has a stockpile of approximately 80 nuclear weapons for delivery by 2 dozen missiles, a couple squadrons of aircraft and perhaps a small number of sea launched cruise missiles." Over the years, the US sold aircraft platforms to deliver these nuclear weapons beginning in 1968 with the sale and delivery of F-4 Phantom aircraft. This was followed with the sale of F-15 and F-16 aircraft in the 1980s and 1990s. The F-16 has a range of 1600 km, and the F-15 has a range of 3500 km.

Modernization plans. Israel is upgrading its land-based missile force from 25 to, perhaps, 100 Jericho-II with a range of 1500 km and Jericho-III with a range of 4000 km. In the future, Israel's air delivery might evolve to the stealth F-35 fighter-bomber. It is reported that Israel has cruise missiles with a range of 350 km on its new Dolphin class submarine, the first of, perhaps, six submarines.

10.3.3 North Korea

Proliferation links. The story of the *hermit nuclear-kingdom* is lengthy and interesting, we only summarize it. The US has 30,000 troops in South Korea, but it has no formal diplomatic relations with North Korea, mostly refusing to meet with

North Korea on a bilateral basis. The Democratic People's Republic of Korea (DPRK) has been led only by a single-family; starting with the grandfather, Kim il-Sung (1948–94); the father, Kim Jong-il (1994–2011); and now the son, Kim Jong-un (2011–). The DPRK nuclear program started towards plutonium weapons in the 1950s with help of Russian exports and expertise. North Korea acceded to the NPT on 18 April 1985 after strong coaxing by the Soviet Union, with the hope to sell North Korea power reactors. It took seven years for Pyongyang to approve the IAEA safeguards agreement. The inspections began in May 1992 to examine the 5-MW_e reactor and the 50-M_e plant under construction. The 200-M_e plant under construction was halted under the Agreed Framework in 1994. In addition, DPRK had an 8 MW_t research reactor, a critical assembly from the Soviets, and various fuel fabrication and reprocessing facilities. In October 2002, DPRK revealed it had a centrifuge plant of 2000 centrifuges, imported from A.Q. Kahn of Pakistan. In principle, this plant could be used to make both reactor fuel and uranium nuclear weapons. This was followed when North Korea expelled IAEA inspectors and removed IAEA monitoring equipment on 8000 fuel rods in December 2002. This escalated to North Korea's withdrawal from the NPT in January 2003, similar to its almost withdrawal in June 1993. By 2005 most analysts believe that North Korea could produce up to 9 nuclear weapons on the basis of its plutonium production rates. David Albright predicts DPRK could have up to 100 warheads by 2020. The three DPRK nuclear tests of 2006, 2009 and 2013 gave a modest, total yield of 10 kton.

Agreed framework. DPRK, US, South Korea and Japan signed the *Agreed Framework* of 21 October 1994. The agreement was a basic trade, with North Korea agreeing to freeze and eventually dismantle its nuclear facilities and eliminate its nuclear weapons capability, and remain a member of the NPT, with full compliance of IAEA safeguards, once a "significant portion of the light water reactor project is completed but before delivery of key nuclear components." In exchange for these actions, North Korea was to receive two modern power reactors, mainly from South Korea, and heavy-fuel oil from the US to replace the shutdown reactor power. The current negotiations have six parties: North Korea, South Korea, Japan, China, Russia and the US.

Modernization plans. The ultimate resolution of North Korea's weapons program will come down to what China will forcefully allow, but this diplomatic pressure has been a long time coming. North Korea's weapons program is still in its infancy. Its potential delivery systems include the SCUD-C and No Dong short-range missile and the Taepo Dong long-range missile. DPRK has sold missiles to Iran, Pakistan, Syria and Yemen. Thus far DPRK has not tested reentry vehicles. For a nation with few financial resources and fewer friends, it is most likely they would use an aircraft.

10.3.4 Pakistan

Proliferation links. Two major events in the 1970s and the Atoms for Peace Program gave a lasting impetus to Pakistan's nuclear weapons program that started in 1972:

- India's 1971 defeat of Pakistan allowed East Pakistan to become independent Bangladesh. This devastated the remaining West Pakistan, with considerable hatred for India.
- The May 1974 nuclear explosion by India compounded hate with fear.
- To add to this mix, the US Atoms for Peace program trained the first fifteen Pakistanis at Argonne National Laboratory in nuclear power in 1954. Seventy Pakistanis were trained by the US Atomic Energy Commission by 1971.

The two quotations on the first page of this chapter combine the fervor of Bhutto with A.Q. Khan's nuclear-Walmart store. A.Q. Khan returned to Pakistan a year after the 1974 Indian explosion with mechanical drawings of centrifuges in hand with the experience to manufacture and assemble. Pakistan's nuclear weapons program used a smuggling network to acquire clandestine technologies for the Kahuta centrifuge plant. By 1985 Pakistan was able to produce high-enriched uranium, in spite of pledges to the US that it would not do so. Pakistan was sanctioned for its nuclear activities in 1979 by the Glenn–Symington amendment. This denied military aid to nations with enrichment facilities outside of international arrangements. By 1986 Pakistan had enough HEU to make a uranium weapon. It now has 3000 kg of HEU and 150 kg of weapons grade plutonium. The 1985 Pressler amendment blocked military sales to Pakistan if the US president could not certify on an annual basis that Pakistan did "not possess a nuclear explosives device." Presidents Ronald Reagan and George H.W. Bush made these certifications between 1985 and 1990 until President George H.W. Bush was unable to make the certification (and the Soviets were leaving Afghanistan). Pakistan has repeatedly blocked the Fissile Material Cut-off Treaty in the UN Conference on Disarmament.

Nuclear tests. India had its initial nuclear test in May 1974, but did not test again until May 1998. Those of us that worked on sanction legislation feel the threat of sanctions refrained India and Pakistan from testing for 24 years until 1998. Pakistan tested some 2–6 weapons on May 28 and 30 with a cumulative yield of 8–20 ktons. The 1994 Glenn–Pell–Helms sanctions were applied after these nuclear tests, costing Indian and Pakistan billions of dollars. President George W. Bush waived the sanctions on 22 September 2001.

Kahn's Walmart. This is the house that A.Q. Khan built. Kahn's skills were in uranium enrichment, but he managed to take over the entire nuclear complex. Most observers believe that Kahn's trips to North Korea, Libya, Iraq and Iran were well known by the country's leadership. Nevertheless, when he was caught he was only put in house arrest, and he is now free. It took two decades for Pakistan to progress from centrifuge drawings to nuclear tests, but he had support from Pakistan's leadership. Not only did Kahn sell centrifuge parts, but Iranian scientists and others came to Pakistan to learn the technology. When Kahn confessed about his nuclear sales, it was in English, but on Pakistan state television. This maximized humbleness to Pakistan's allies but minimized what Pakistani citizens could understand.

The religious fervor in Pakistan and its coupling to their intelligence programs gives one pause. Serious observers feel that Pakistan's instability with 100 weapons is the world's greatest proliferation problem, greater than that of Iran or North Korea. Because of the unstable political climate in Pakistan, there is concern that nuclear weapons might be stolen and perhaps used. The US permissive action link (PAL) technologies can make warheads more robust to prevent theft, however the technology is classified and can't be shared. In other ways the US government spent $100 million to make Pakistan's weapons more secure.

Modernization plans. Pakistan is spending considerable funds to modernize its nuclear forces by developing Shaheen medium-range ballistic missiles, air-launched cruise missiles, ground-launched cruise missiles and short range rockets. Pakistan is adding its fourth plutonium production reactor and upgrading its enrichment and reprocessing facilities. Pakistan will have both uranium and plutonium warheads. Pakistan has 3 power reactors (725 MW, 5.3 % of grid), with two under construction, and plans for a breeder program.

10.4 Iran, a Work in Progress

10.4.1 Iran

Proliferation links. Iran does not possess nuclear weapons, but has been trying for four decades. In spite of this, Iran remains in the NPT and is negotiating with the *P5 + 1* (China, France, Russia, UK, US, plus Germany) on its current robust centrifuge program. In the 1970s under the Shah, Iran began to use laser isotope separation of uranium. In the 1970s, Iran planed to build 22 nuclear power plants with substantial enrichment capacity, but this was quickly reduced to 2 GW at Bushehr. The transfer of A.Q. Kahn's centrifuges made a great difference. A big impetus for Iran's nuclear program was Iraq's use of chemical weapons against Iran in the 1980s and the beginnings of the Iraqi nuclear program.

Hardball drama. The present environment is full of drama. Iran's centrifuge program was made public by Iranian citizens that blew the whistle on the government. This was followed by massive economic sanctions, that had a major affect. The Stuxnet attack on the centrifuges, reportedly by Israel and the US, was a further prompter. Stuxnet is a computer virus (worm) brought into the system on a memory stick. It attacked the Siemens control circuits of the IR-1 centrifuges. First, Stuxnet gave the control room false signals, hiding the attack from the owners. By shaking the centrifuges near their resonant frequencies of 1064 and 1410 Hz, about 25 % of the centrifuges were destroyed, 800 of the 4700 at Natanz. The actual situation is more complicated as the virus played hide and count games with the Siemens controllers. To this witch's brew, add in the murder of Iranian scientists. Sanctions have had their effect, the current government wants to partake of Western commerce and might be able to deliver an agreement, but nothing is assured on either side of the table.

Present negotiations. Recall that it takes 60 % of the total separative work to produce 4 % low-enriched LWR fuel (as compared to 90 % enriched), 25 % of the total to enrich the fuel from 4 to 20 %, and another 15 % of the total to raise 20–90 %. Thus, a stockpile of 20 % enriched material can significantly reduce the time a country requires to make a nuclear weapon. It is here that the P5 + 1 constrained Iran's 20 % stockpiles, as well as restrict the number and size of centrifuges that are operable in the near-term. This will not satisfy all parties since some believe Iran should not have *any capability to enrich*. On the other hand, without a war, it is not likely that all of Iran's centrifuges will be removed. They say that diplomacy is the art of the possible.

Iran's centrifuges produce 0.8–4 SWU/year depending on the model. Iran claims the right to pursue uranium enrichment in line with the NPT text. But its claim has been in dispute since Iran cheated by not reporting its previous clandestine enrichment activities to the IAEA. Iran must suffer some penalties to stay within the system, but what? These developments prompted UN Security Council resolutions, followed by economic sanctions. Other non-weapon states enrich uranium, such as Brazil, but they negotiated reporting and inspection agreements with the IAEA and are allowed to enrich. Figure 10.2 shows centrifuges Iran installed and operated at its large Natanz facility since 2007. Natanz and the Fordow facilities together house 19,500 centrifuges.

Iran/P5 + 1 agreement. A temporary agreement known as the *Joint Plan of Action* between Iran and the P5+ was signed on 14 July 2015. Iran allows constraints to be placed on its nuclear program in exchange for a reduction in billions of dollars in economic sanctions. The final agreement plans to increase the "breakout time" needed to obtain sufficient HEU to build a nuclear weapon from some two months to a year by the following actions:

The comprehensive nuclear plan with Iran contains these elements (K. Davenport, Arms Control Today, May 2015):

Fig. 10.2 P5 + 1 negotiations with Iran. Iranian President Mahmud Ahmadinezhad and centrifuges (*left*), US Secretary of State John Kerry, European Union Negotiator Catherine Ashton, Iranian Foreign Minister Mohammad Javad Zarif (*center*), Israeli Prime Minister Benjamin Netanyahu and President Obama debate the Iran agreement (*right*) (Pond 5, State Department)

- Reduce Iran's 19,466 operating centrifuges to 5060 to obtain uranium fuel at 3.67 % enriched, reducing the 10,000 kg stockpile to 300 kg for 15 years. This precludes 20 % enriched uranium. This increases the breakout time from 2 to 3 months to a year.
- 13,000 dismantled centrifuges are under IAEA seals. No uranium enrichment at Fordow, only research on four new types of 700 existing centrifuges and 348 on isotope work with Russia. Limit R&D on advanced centrifuges for 10 years.
- Destroy and replace the Arak heavy-water reactor so it no longer produces weapons-grade plutonium. Commit not to separate plutonium from spent fuel.
- Allow continuous monitoring of uranium mines for 25 years.
- Allow continuous monitoring of enrichment sites 20 years.
- Implement the IAEA additional safeguards protocol on weapons grade materials.
- Cooperate with IAEA to investigate allegations that Iran's nuclear program had military dimensions.
- $100 billion in frozen assets are released to Iran when criteria have been fulfilled. Conventional military embargo removed after 5 years and ballistic missile embargo removed after 8 years. Snap back sanctions if there are violations.

Monitoring centrifuges. An inspection regime would weigh and examine cylinders containing feed, product, and wastes to determine enrichment levels. Radio-frequency identification tags can track the cylinders. Similarly, tags, seals, and surveillance can be used to keep track of nuclear activity. Uranium stocks, records, and receipts can be audited. Environmental samples of soil and water can be obtained in and outside a plant. Unannounced inspections are useful to detect undeclared material and operations.

The IAEA, Russia, and the US have had considerable experience in monitoring centrifuges and enriched uranium. The technology used to monitor the *blend-down* of 500 tonnes of Russian HEU to reactor fuel is instructive. The density of ^{235}U in the uranium hexafluoride gas is determined by first measuring the emission of 186-keV gamma rays emitted by uranium-235. Secondly, the transmission of 122-keV gamma rays from a cobalt-57 source through the UF_6 gas determines the total uranium density. The ratio of the densities, obtained with the 186-kev and 122-keV intensities, gives the enrichment level. Twenty years of monitoring by Oak Ridge and Los Alamos National Laboratories successfully verified the dilution of the equivalent of 20,000 Russian HEU warheads.

Saudi Arabia. Saudi Arabia has long supported its Sunni Islam colleague, Pakistan, on its nuclear weapons program. The level of hatred between Sunni and Shiites is extremely tense and violent. The P5 + 1 have long hoped for an agreement to reduce the size of Iran's program and to strongly monitor it to calm the Middle East. It is to be expected that Israel would like to forbid all enrichment activities and possession in Iran. Blow back from Saudi Arabia is also to be expected, but can it be contained? Saudi officials on 13 May 2015 had this to say while attending a meeting with President Obama at Camp David: "We can't sit back and be nowhere as Iran is allowed to retrain much of its capability and amass its research." This was followed by Former Saudi Intelligence Chief, Prince Turki bin Faisal: "Whatever the Iranians have, we will have, too."

Iran politics: Congress passed a law that gives it an opportunity to review the verifiability of the Executive Agreement with Iran. This gives Congress the opportunity to block removal of US sanctions on Iran that were established by Congress, but allows those sanctions that were created by the President. It is true, that in ten years Iran would be able to return to revive its bomb program, particularly because it is unlikely that Russia and China would further support sanctions in ten years or now, if it fails. It is also true that 10 years of working with the West might have changed the political landscape. At this point six nations have agreed to constrain Iran for a reduction in sanctions, if it fails it is unlikely to be re-established. These are not easy choices.

10.5 Eight NNWS Successes (Argentina, Brazil, Iraq, Libya, South Africa, Belarus, Kazakhstan, Ukraine)

10.5.1 Argentina

Proliferation links. From the 1960s to the 1980s, Argentina and Brazil had programs to produce fissile materials for nuclear weapons. These were modest programs, to keep *options open* for building the bomb. But, in 1991, Argentina and Brazil agreed to establish a four-party safeguards organization, the first of its kind, which ended their weapons programs and set a hopeful standard for the world with the *Brazilian-Argentine Agency for Accounting and Control of Nuclear Materials* (ABACC). Argentina currently has no source of weapons-grade material, but it had a secret gaseous diffusion program at Pilcaniyeu from 1978 to 1993. This facility is now closed after producing small amounts of low enriched uranium. Argentina is a member of the NPT, the Treaty on Tlatelolco, the Missile Technology Control Regime (MTCR) and the Nuclear Suppliers Group. Argentina is self-sufficient in the nuclear area and is a second-tier supplier with reactor technology exports to Germany, Algeria, Egypt, Iran, Libya, South Korea and Australia. Argentina has ended its Condor missile program.

Brazilian-Argentine agency for accounting and control of nuclear materials. The driving force behind the ABACC success story was indigenous, it was not a result of Western diplomacy. If anything, overly strident voices against their existing nuclear commerce tended to anger participants and prevent diplomacy. On the other hand, if these voices had not been raised, there might have been an over abundance of weapons useable materials. Progress followed the leadership of the Argentine presidents, Raul Alfonsin and Carlos Menem, and Brazilian presidents Jose Sarney and Fernando Collor de Mello. In 1987, a Brazilian delegation visited the gaseous diffusion plant in Argentina, and in 1988 an Argentine delegation visited the centrifuge plant in Brazil. The bilateral agreement was signed 18 July

1991 and the quadripartite agreement was signed in Vienna on 13 December 1991 between Argentina, Brazil, ABACC and the IAEA. In 1994 the agreement entered into force as both countries signed the Treaty on Tlatelolco to ban nuclear weapon in Latin America.

Inspections. Both the ABACC and the IAEA draw on independent conclusions. These two bodies coordinate their activities to avoid unnecessary duplication. The verification process has three distinct stages: (1) Examine materials supplied by the countries, (2) collect information from ABACC and (3) assess the information. During this time, they audit documents, count and identify items, carry out non-destructive measurements, apply and verify surveillance equipment and obtain samples of nuclear materials for comprehensive analysis. In 1996, Argentine facilities were inspected 79 times and Brazilian facilities were inspected 81 times. During 2013, 118 inspections were carried out.

Why was ABACC successful? It would be useful if the ABACC process was applied to other difficult matched pairs; such as Israel/Iran, Pakistan/India, N.-Korea/S.-Korea, India/China and others. The US and the Soviet Union/Russia duo has been successful, but at a more difficult level. Brazil and Argentina had a testy relationship that was headed towards nuclear weapons. But, they spoke an almost, common language and they were mostly from the same religion. These countries had been sold a bill of goods by their military leaders, making it seem that nuclear weapons were easier to make and necessary to survive. But, the world now has many other constraints on these programs, including economic sanctions, which can dampen the fervor for nuclear weapons. And the CTBT makes it more difficult to test nuclear weapons. And the other, smaller Latin nations might field nuclear weapons if Argentina and Brazil had them. In 1978, I met with the Chilean Nuclear Regulatory Commission, which consisted of an Army general, an Air Force general and a Navy admiral, sitting at a villa overlooking the Pacific. It was my lot to remind them of the difficulties of going ahead with nuclear weapons. Another factor was that Germany adopted the full-scope safeguards requirement for exports in 1990. Since full-scope safeguards are needed German exports, it had a positive effect on Brazil when it was considering ratification of the safeguards agreement. Another factor was the public input to this issue. The American Physical Society honored two Brazilian physicists (Fernando Barros and Luis Rosa) and two Argentine physicists (Luis Masperi and Alberto Ridner) for working on *constitutional bans* on nuclear weapons in their two countries.

10.5.2 Brazil

Proliferation links. In the 1970s and 80s, Brazil created programs that showed its interest in procuring nuclear weapons. This was not a strict nuclear weapons program but the start of an option to the bomb. All the Brazilian Armed Forces played a role: The Brazilian Navy worked on enrichment technology at Resende and Aramar. The Air Force developed laser enrichment and built a nuclear test site with a 300 meter-deep shaft for underground tests at an Amazon military base. The Army worked on a graphite-moderated reactor, which could produce weapons-grade plutonium.

Brazil's first power reactor, the Angra-I, was ordered from Westinghouse in 1971. This was followed with an order for 10 reactors from West Germany, which included a German Becker nozzle enrichment facility. Angra-II joined the grid in 2000. The nozzle technology was not successful, so Brazil shifted to gaseous centrifuges. By 1987 Brazil had a module of 48 centrifuges and by 1991 it had 500 centrifuges. This project was NPT compliant since it was reported to the IAEA, and Brazil agreed to constrain enrichment to 20 %. (It is agreed that non-compliant Iran will not produce more than 5 % enriched.) Brazil terminated its short-range missile programs in the 1990s. Brazil ratified the NPT, the CTBT, the Nuclear Suppliers Group, the Treaty on Tlatelolco and the Missile Technology Control Regime. Brazil has a nuclear submarine program that caused some concern since it could use high-enriched uranium for its submarine reactors.

10.5.3 Iraq

1991 Iraq War. Iraq's nuclear weapons program started in the 1970s and ended in 1991. The 7 June 1981 Israeli destruction of the Osirak reactor showed that Iraq had weapons ambition. Two wars were held between Iraqi and the US and its allies. Saddam Hussein invaded Kuwait on 2 August 1990, under the illusion that the US would not respond. The first Gulf War lasted from 17 January 1991 to 27 February 1991. UN Security Council Resolution 687 of 3 April 1991 declared a formal cease-fire. Iraq was forbidden from acquiring or developing nuclear weapons, or nuclear weapon materials or components.

2003 Iraq War. All the available evidence indicates that Iraq did not restart or maintain a nuclear weapons program after the 1991 war. Charles Duelfer, director of the US government's Iraq Survey Group, concluded that "Iraq did not possess a nuclear device nor had it tried to reconstitute a capability to produce nuclear weapons after 1991." Duelfer's successor, David Kay, stated that Iraq's program "had been seriously degraded. The inspections in the early 1990s did a tremendous amount" to prevent re-starting Iraq's nuclear weapons program.

Saddam Hussein wanted to look strong after losing the 1991 Iraq War. It was his desire to bluff his way to apparent strength, but this had an unfortunate unintended consequence in that many people believed him. The US allowed its intelligence data to be misused by political leaders. The charge that Iraq attempted to purchase uranium from Niger was shown to be false. The aluminum tubes discovered in Iraq were ascribed to a centrifuge program, but experts in the US Department of Energy officials did not agree. This intelligence failure came about partially because agencies vote on the basis of one vote for each agency. The Department of Energy was not heard as numerical voting prevailed over science. The data had been stove-piped through the bureaucracy too quickly in its route to the top, to cause a false war.

IAEA inspection results. The IAEA conducted 218 inspections at 141 sites, including all those identified in overhead satellite imagery as having suspicious activity. Mohamed El Baradei reported to the United Nations Security Council on 7 March 2003 (12 days before the war) the following:

- There is "no indication of resumed nuclear activities… nor any indication of nuclear-related prohibited activities at any inspected sites."
- "There is no indication that Iraq has attempted to import uranium since 1990." The documents that indicated Iraqi attempted to purchase uranium from Niger were declared "in fact not authentic."
- "There is no indication that Iraq has attempted to import aluminum tubes for use in centrifuge enrichment." Even if it had, "it was highly unlikely that Iraq could have achieved the considerable redesign needed to use them in a revived centrifuge program."
- Though the question was still under review, there was "no indication to date that Iraq has imported magnets for use in a centrifuge program."
- "During the past 4 years, at the majority of Iraqi sites, industrial capacity has deteriorated substantially due to the departure of the foreign support that was often present in the late 1980s, the departure of large numbers of skilled Iraqi personnel in the past decade, and the lack of consistent maintenance by Iraqi of sophisticated equipment. At only a few inspected sites involved in industrial research, development and manufacturing have the facilities been improved and new personnel been taken on."

Iraq's nuclear weapons program. Iraq signed the NPT in 1969 but soon deviated from its obligations with a multibillion-dollar nuclear weapons program. Iraq's first goal was to produce plutonium in its Osiraq research reactor purchased from France in 1976. The Israeli attack of 7 June 1981 disabled the reactor, shifting Saddam Hussein from plutonium to uranium enrichment. Iraq pursued most of the viable uranium enrichment processes. As was revealed during the first Gulf War, Iraq's most successful approach was to use electromagnetic isotope separation (EMIS), which had been long considered too inefficient in the ionization stage. Iraq also had enrichment programs with gaseous centrifuges, chemical enrichment, gaseous diffusion and laser isotope separation. At the same time, Iraqi had a small reprocessing program, producing 6 g of plutonium. Iraq did not have long-range missiles after the 1991 war.

10.5.4 Libya

Proliferation links. Libya's mercurial, Colonel Mu'ammar Gadhafi had sought to purchase nuclear weapons and missiles for three decades. The US government was well aware of Libya's nuclear plans and its connection to the downing of the commercial airliner over Lockerbie, Scotland in 1988. Economic sanctions in the late 1990s were having an effect but not a reversal of Libya's policies. The decisive event that got Libya to abandon its nuclear program happened by accident on

19 December 2003. British and American officials boarded a German cargo ship heading to Libya from Dubai in October 2003. They discovered centrifuge parts based on Pakistani designs and manufactured in Malaysia. This discovery gave impetus for 11 inspections to undisclosed nuclear sites to remove and destroy key components of the nuclear weapons program. Ultimately Gadhafi chose national economics over nuclear weapons that had no realistic, specific purpose for Libya.

By 2002 Libya had a 9-centrifuge cascade completed, with 19-machine and 16-machine cascades close to completion. Libya also had a uranium conversion facility that had been cold tested in 2002. Libya had a limited SCUD-B missile (300 km) supply obtained from the Soviet Union in the 1970s, as well as a handful of North Korean SCUD-Cs (600 km). Libya made several unsuccessful attempts to purchase Soviet SS-23 and SS-21 missiles and Chinese DF-3A, M9, and M-11 missiles.

10.5.5 South Africa

Proliferation links. I attended the *coming-out party* on 23 July 1993 at the South-African Embassy in Washington given by its nuclear leader, Waldo Stumpf. He was Chief executive officer of the Atomic Energy Corporation of South Africa. South Africa proved that a small nation could produce HEU and make weapons for modest amounts of money, some $250 million over 10 years. Contrast that with Saddam Hussein's Iraq program of over $10 billion and little progress.

South Africa's first prototype warhead was ready in the 1980–82 timeframe, probably too late to explain the September 1979 event, observed by the Vela satellite near South Africa's Prince Edward Islands in the southern Indian Ocean. Many analysts believe interactions between Israeli and South African scientists explain the 1979 Vela event, that it was a 3-kton Israeli device and not a South-African device. It is conventional wisdom that the first South African bomb was not finished until the early 1980s, after the 1979 event. The South-African bomb design was a copy of the Hiroshima device, an inefficient HEU gun-design, to be delivered by airplane. Their stockpile peaked at six weapons by 26 February 1990 when President F.W. de Klerk issued internal orders to end the nuclear program and dismantle all nuclear weapons and the Pelindaba enrichment facility (formerly Valindaba). South Africa exceeded to the NPT on 10 July 1991, 2 years before revealing the program. South Africa concluded a fullscope safeguards agreement with the IAEA in September 1992. Nelson Mandela was elected president of South Africa in 1994.

South Africa claims it was driven to the bomb because of the 50,000 Cuban soldiers in Angola. They claimed a three-phased policy to deter potential adversaries and to compel Western nations to become involved. Phase one was deterrence without confirming or denying a program. Phase two was to reveal their program when under imminent attack, to coerce Western support. Phase three would involve overt nuclear testing to demonstrate South Africa's ability and willingness to use nuclear weapons. This logic could not deliver nuclear weapons on Cuba or Moscow, threatening only open African space, thus it was not convincing.

In mid-1977 a Soviet satellite revealed preparations for an underground nuclear test site at a Kalahari Desert site. This data was shared by the Soviet Union with the US, showing the two superpowers could work together on nonproliferation. Prodding by the US and the Soviet Union closed the two shafts in 1977 and no nuclear tests were conducted there.

South Africa adapted the German Becker nozzle process to enrich to HEU. The Y plant was commissioned in 1974, operating its first cascade in 1977. Stumpf claims that South Africa did not have quality HEU at the time of the 1979 event, as their uranium was only 80 % enriched at that time. The UF_6 gas and the hydrogen carrier gas can have catalytic reactions that froze the Y-plant for 18 months. The Y-plant is now closed, but South Africa still retains 500 kg HEU, which could be down-blended to fuel their two 900-MW_e Koeberg reactors.

10.5.6 Ukraine, Belarus, Kazakhstan

On 1 December 1991, Ukraine voted to leave the Soviet Union and become an independent nation. This was followed by similar acts of separation by Russia and the other 13 Soviet republics. On 23 May 1992, the US and the four nuclear capable Soviet republics (Belarusian, Kazakhstan, Russia, Ukraine) signed the Lisbon Protocol, which made all five nations State-Parties to the START Treaty. In addition, all Parties agreed to join the NPT. In Table 10.3, we list strategic weapons of the four nuclear former Soviet Republics at the time of the 1992 Senate ratification of START-I.

When Ukraine voted to leave the Soviet Union, it originally stated it would give up all their Soviet nuclear weapons located on its territory. However, it became apparent to all, when the US bought 500 tonnes of HEU from Russia, there was money to be gained from these stocks. On 20 December 1991, Ukraine changed its mind, shocking the meetings that I attended in Kiev. Ukraine stated it wanted compensation for having helped produce this windfall in uranium economics. When the Ukrainian Parliament, the Rada, ratified START I on 18 November 1993, it declared a number of stiff conditions, which Russia would not accept. After

Table 10.3 Location of nuclear weapons in four former Soviet republics

	Kazakhstan	Ukraine	Belarus	Russia	USSR
ICBM/WH	104/1040	176/1240	54/54	1067/4278	1401/6612
SLBM/WH	0/0	0/0	0/0	940/2804	940/2804
Bomber/WH	40/370	34/416	0/0	88/800	162/1600

The ratio is launchers divided by warheads. The Soviets had 62 SSBN submarines in 1992 (*The START Treaty*, Senate Foreign Relations Committee, 18 September 1992)

extensive negotiations, Ukrainian President Leonid Kravchuk, US President Bill Clinton and Russian President Boris Yeltsin signed the Trilateral Statement in which they agreed that all warheads would be withdrawn from Ukraine to Russia. As part of the situation, Ukraine was paid economic benefits and given security guarantees. This was accepted by the Ukrainian Rada on 16 November 1994. The Belarus parliament ratified START I on 4 February 1993 and Belarus acceded to the NPT on 22 July 1993. Kazakhstan ratified START I on July 1992 and formally acceded to the NPT on 14 February 1994. With these actions, three former (partial) nuclear weapon states returned to the status of nonnuclear weapon states. The US Operation Sapphire airlifted 581 kg of high enriched uranium in 1994 from western Kazakhstan. This was previously stored only behind a padlocked door.

Conventional forces 1991. Before nuclear weapons could be reduced in Europe, it was necessary to control conventional weapons to make the Warsaw Pact and NATO feel secure. To do this, the *Conventional Forces in Europe Treaty* (CFE) was ratified by the Senate on 27 November 1991. The CFE Treaty limited 5 types of treaty-limited equipment (TLE) made up of tanks, armored carrier vehicles (ACV), artillery pieces, combat aircraft and attack helicopters. Within the area from the Atlantic to the Urals, neither NATO nor the Warsaw Pact could field more than 20,000 tanks, 30,000 armed combat vehicles, 20,000 artillery pieces, 6800 combat aircraft and 2000 attack helicopters. Thus, NATO and the Warsaw Pact could each have 78,800 TLE. When the CFE was first considered in 1989, NATO had 76,000 TLE, the Warsaw Pact had 197,000 TLE and the Soviet Union had 151,576 of that. Thus, the Warsaw Pact had to destroy 120,000 TLE, while NATO had to destroy essentially nothing. The US transferred some of its older TLE to other NATO states. However these numbers are misleading since some of the Warsaw Pact equipment was older and should have been scrapped. The ratio of equipment held by the two blocks and the Soviet Union changed swiftly with these three major events, as shown in Table 10.4:

Table 10.4 Total TLE and ratios belonging to CFE Treaty parties

	January 1989	November 1990	EIF + 40 months
USSR	151,576	72,645	52,975
Be/CZ/H/P/R	35,000	35,684	25,825
GDR	11,000	10,674	0
Warsaw Pact	197,000	119,003	78,800
United States	15,000	15,246	13,172
NATO minus US	60,000	60,505	65,628
NATO	76,000	75,751	78,800
Ratios			
USSR/NATO	2.0	0.96	0.67
WTO/NATO	2.7	1.57	1.00
SU/21-CFE	1.2	0.59	0.51

CFE Treaty, Senate Foreign Relations Committee, 19 November 1991

- the CFE Treaty,
- the collapse of the Warsaw Pact,
- the collapse of the Soviet Union,

The ratio of TLE between the Warsaw Pact and NATO was initially 2.7. The final TLE ratio between the Russian Federation by itself and the other 21 CFE parties was 0.51. This ratio change from 2.7 to 0.51, a factor of 5.3, is a tsunami-sized change.

10.6 Eight More NNWS Successes (Germany, Japan, Sweden, Switzerland, South Korea, Taiwan, Algeria, Syria)

Germany (***World War II***). On 3 June 1940, Germany's Paul Hartek failed to produce neutron multiplication in the Hamburg reactor, made up of 185 kg of natural uranium and 15 tonnes of dry ice, CO_2. This was two and one-half years before Fermi went critical at Stagg Field, University of Chicago on 2 December 1942. Hartek also worked on building centrifuges, but they blew apart at high speeds. The Germans also built a reactor at Haigerloch and harvested heavy water in Norway. Werner Heisenberg was the leader of the Nazi project. It was debated whether Heisenberg purposely slowed progress for Hitler by not calculating a critical mass, or whether Heisenberg just failed to do the proper calculations. At a 4 June 1942 conference, Reich Minister for Munitions, Albert Speer, curtailed the project from nuclear weapons to nuclear reactors. After reading of the Farmhouse Tapes, I believe it was the latter situation for Heisenberg. Germany's ratification of the NPT on 2 May 1975 was encouraged because the US nuclear umbrella played the role of deterrence for Germany. Would Germany be an NPT member without this umbrella? Germany has significant reactor, reprocessing and enrichment facilities and skills.

Japan (***World War II***). Japan did gaseous diffusion and centrifuge studies during World War II. Japan ratified the NPT on 8 June 1976. Would Japan be an NPT member without the American nuclear umbrella? Japan has significant nuclear facilities.

Sweden (***1945–72, revealed in 1985***). A clandestine government corporation built three reactors, starting in 1951, and was planning a fourth pressurized-heavy-water reactor of 100 MW. This was to produce enough plutonium for 100 warheads, at the rate of 5–10/year. Sweden had a sophisticated program with uranium mining, plutonium castings and flash X-ray machines. Sweden at one point tried to buy nuclear weapons from the US, which would have been air-delivered, or in artillery shells. Sweden, responding to the NPT global environment by ratified the NPT on 9 January 1970.

Switzerland (***1950–60s***). The Swiss did studies on air delivered nuclear weapons. The Swiss Federal Council of Ministers decided, first in secret and then openly, to equip the Army with nuclear weapons in 1958. However, the process slowed when the Swiss realized that Swiss possession of nuclear weapons would encourage Germany to follow suit. Switzerland ratified the NPT on 9 March 1977.

South Korea (1970s). After the fall of South Vietnam in 1975, President Park Chung-hee of South Korea mentioned South Korea's nuclear weapons aspirations during a press conference. At that time, South Korea had a contract with France to import a reprocessing plant to obtain plutonium for reactor fuel. Since South Korea had a modest number of nuclear reactors at that time, this would not have been an economical decision. The French reluctantly canceled the contract. South Korea Ratified the NPT on 23 April 1975.

Taiwan (1970s). The development of nuclear weapons by the People's Republic of China has long been a contentious issue for Taiwan. During the 1970s, Taiwan had an active program to produce plutonium with heavy-water reactors. In September 1976, Taiwan agreed to dismantle its nuclear weapons program. Taiwan ratified the NPT on 27 January 1970.

Algeria (1990–95). The secret construction of the Es-Salam, 15 MW, heavy-water reactor from China in the 1990s caused concern in the US intelligence community. Algeria joined the NPT on 12 January 1995.

Syria (2000s). Israel bombed a Syrian facility it believed was a reactor under construction on 6 September 2007. Western press claimed a North Korean freighter delivered a reactor to Syria for its nuclear weapons program. The Institute of Science and International Security identified the site in eastern Syria. Syria ratified the NPT on 24 September 1969.

Problems

10.1 **Rank-ordered nuclear states**. Rank order 27 nations along various categories. This is a difficult, but do it anyway.

10.2 **US nuclear umbrella**. What are the upsides and downsides of the nuclear umbrella? Which states "need" the umbrella or they would go nuclear?

10.3 **Fractional arsenals**. What is the ratio of today's US and Russian nuclear arsenals, compared to their peak sizes.

10.4 **Tech transfer graphic**. Draw in some additional lines on Fig. 10.1.

10.5 **US force structure**. What is US strategic force structure at 1550, 1000 and 500 warheads?

10.6 **Triad to diad**. Which triad leg would you retire first? What are the pros and cons?

10.7 **PRC stockpile**. Plot PRC's warheads as a function of time. How is China shifting its arsenal for the future? Why did China enlarge so slowly?

10.8 **Heavy bombers**. Which NWSs intend to build new heavy bombers, which are not?

10.9 **India and Israel triad**. What is the status of the triad in these two nations?

10.10 **Dimona reactor**. What is the history of Dimona? Who supported Israel?

10.11 **Vanunu's London Times**. Look up the 7 October 1986 issue. What did you learn?

10.12 **NUMEX in the 1960s**. What does the internet say about this event?

10.13 **DPRK imports**. What did North Korea import from the Soviet Union and China?

10.14 **China and DPRK**. China fears a DPRK collapse if they lean on them too much. Is this viable? Why can't they keep their borders firm and avoid the deluge?

10.15 **Russian help to Iran**. What has Russia promised for Iran accepts reduced numbers of centrifuges in the P5 + 1 negotiation?

10.16 **ABACC details**. What does the internet tell us?

10.17 **Fullscope imports from FRG to Brazil**. Why would adoption of fullscope safeguards have an effect on Brazil? Why didn't the Becker nozzle project succeed?

10.18 **Other Countries**. We have only discussed 27 of the 188 NPT states. Which other states might consider having nuclear weapons, according to the internet?

10.19 **Cooperative measures**. How might India and Pakistan, on a unilateral-reciprocal basis, develop cooperative measures to observe aggressive motion towards nuclear weapons?

10.20 **Sweden's program**. What does the internet say about this serious program?

10.21 **Iran + P5 + 1 Status**. What is the status now? What failed and what was successful? Did Congressional pressure help or hurt?

10.22 **US enrichment**. What is the current US capacity in SWU/year? Have lasers produced SWUs yet? New contract with Russia for reactor fuel?

10.23 **Back-Sliding**. Which states might back-slide NPT commitments?

10.24 **Arguments for War**. What are some statements to coax us into war over proliferation?

Seminar: Proliferated States

ABACC, Atoms for Peace successes and failures, AQ Kahn status and history, P5 modernization, defacto 4 modernization, intelligence stove piping, intelligence voting on science issues, Iran negotiation status, Iran stocks versus centrifuges versus timely warning, Klaus Fuchs thefts, Libya weapons program, Manhattan Project spies, North Korean weapons status, nuclear umbrella for Japan and Germany, Pressler amendment, permissive action link help with Pakistan, SLBM fleets, South African HEU supplies, South African enrichment and weapons program, Sweden reactor and weapons program, Swiss weapons program.

Bibliography

Albright, D. (2014). Status of Iran's nuclear program and negotiations. In P. Corden, D. Hafemeister & P. Zimmerman (Eds.), *Nuclear Weapon Issues in 21st Century, AIP Conference of Proceedings* (Vol. 1596, pp. 171–179). AIP Publishing.

Albright, D. (2015a). *ISIS analysis of Iranian safeguards report*. Washington, DC: The Institute for Science and International Security.

Albright, D. (2015b). *Future directions in DPRK's nuclear weapons program*. Washington, DC: The Institute for Science and International Security.

APS/AAAS Nuclear Forensics Group. (2008). *Nuclear forensics*. Washington, DC: American Physical Society.

Ball, P. (2014). *Serving the Reich: The struggle for the soul of physics under Hitler*. IL: University of Chicago Press.

Burns, N. (2014). Passage to India. *Foreign Affairs, 93*(5), 132–141.

Cirincione, J. (2002). *Deadly arsenals: Tracking weapons of mass destruction*. Washington, DC: Carnegie Endowment for International Peace.

Cohen, A. (1998). *Israel and the bomb*. NY: Columbia University.

Commission Intelligence Capabilities of US Regarding Weapons of Mass Destruction. (2005). Washington, DC.

Cooke, S. (2009). *In mortal hands: A cautionary history of the nuclear age*. NY: Bloomsbury.

Davis, J., & Kay, D. (1992). Iraq's secret nuclear weapons program. *Physics Today 45*, 21–27.

Devolpi, A., Minkov, V., Simonenko, V., & Stanford, G. (2005). *Nuclear shadowboxing* (Vol. 2). Kalamazoo, MI: Fidlar Doubleday.

Doyle, J. (Ed.). (2008). *Nuclear safeguards, security, and nonproliferation*. NY: Elsevier.

ElBaradei, M. (2011). *Age of deception: Nuclear diplomacy in treacherous times*. NY: Metropolitan.

Ferguson, C., & Potter, W. (2004). *The four faces of nuclear terrorism*. Monterey, CA: Monterey Institute of Strategic Studies.

Gallucci, R. (2014). Negotiations with North Korea: Lessons learned. In P. Corden, D. Hafemeister & P. Zimmerman (Eds.), *Nuclear Weapon Issues in 21st Century, AIP Conference of Proceedings* (Vol. 1596, pp. 160–163).

Garwin, R., & Charpak, G. (2001). *Megatons to megawatts*. NY: Knopf.

Gilinksy, V., & Mattson, R. (2010). Revisiting the NUMEC affair. *Bulletin Atomic Sciences, 66*(2), 61–75.

Hecker, S. (2011). Adventures in scientific nuclear diplomacy. *Physics Today, 64*(7), 31–37.

Hersh, S. (1991). *The Samson option*. NY: Random House.

International Panel Fissile Materials. (2012). *Global fissile material report*. Princeton, NJ.

Jointer, T. (2010). The Swedish plans to acquire nuclear weapons. *Science Global Security, 18*(2), 61–86.

Jones, R., & McDonough, M. (1998). *Tracking nuclear proliferation*. Washington, DC: Carnegie EIP.

Joshi, S. (2015). India's nuclear anxieties. *Arms Control Today, 45*(4), 14–19.

Kristensen, H., & Norris, R. S. (2013). Chinese nuclear forces. *BAS, 69*(6), 79–85.

Kristensen, H., & Norris, R. S. (2014a). Israel nuclear weapons. *BAS, 70*(6), 97–115.

Kristensen, H., & Norris, R. S. (2014b). Worldwide deployments of nuclear weapons. *BAS, 70*(5), 96–108.

Kristensen, H., & Norris, R. S. (2014c). Russian Nuclear Forces. *BAS, 70*(2), 57–85.

Kristensen, H., & Norris, R. S. (2015). US nuclear forces. *Bulletin Atomic Scientists, 71*(2), 107–119.

Leslie, S. (2015). Pakistan's nuclear Taj Mahal. *Physics Today, 68*(2), 40–46.

McDonnell, T. (2013). Nuclear pursuits: non-P-5 nuclear-armed states. *Bulletin of the Atomic Scientists, 69*(1), 62–70.

Mearsheimer, J. (2014). Why the Ukraine crisis is the West's fault. *Foreign Affairs*, *93*(5), 77–89.

Mian, Z. (2014). Nuclear programs in Pakistan and India. In P. Corden, D. Hafemeister & P. Zimmerman (Eds.), *Nuclear Weapon Issues in the 21st Century*, *AIP Conference of Proceedings* (Vol. 1596, pp. 164–170).

Padvig, P. (2001). *Russian strategic nuclear forces*. Cambridge, MA: MIT Press.

Perkovich, G. (1999). *India's nuclear bomb*. Berkeley, CA: University of California.

Pregenzer, A. (2014). Evolution and resiliance of the nonproliferation regime. In P. Corden, D. Hafemeister & P. Zimmerman (Eds.), *Nuclear Weapon Issues in the 21st Century, AIP Conference of Proceedings* (Vol. 1596, pp. 152–159).

Reed, T., & Stillman, D. (2009). *The nuclear express: A political history of the bomb and its proliferation*. Minneapolis, MN: Zenith Press.

Richelson, J. (2009). *Inside NEST*. Norton, NY: American's Secret Nuclear Bomb Squad.

Sarotte, M. (2014). A broken promise? What the West really told Moscow about NATO expansion. *Foreign Affairs, 93*(5), 90–97.

Scheinman, L. (1987). *The international atomic energy agency*. Washington, DC: Resources for Future.

Shea, T. (2015). The verification challenge: Iran and the IAEA. *Arms Control Today*, *45*(5), 8–15.

Sigal, L. (1998). *Disarming strangers: Nuclear diplomacy with North Korea*. Princeton, NJ: Princeton University.

Spector, S. (1990). *Nuclear ambitions: The spread of nuclear weapons*. Boulder, CO: Westview Press.

Squassoni, S. (2010). The US–Indian deal and its impact. *Arms Control Today*, *40*(6), 48–52.

Toki, M., & Pomper, M. (2013). Time to stop reprocessing in Japan. *Arms Control Today*, *43*(1), 22–29.

Weiss, L. (2005). Turning a blind eye again? The Kahn network's history and lessons for US policy. *Arms Control Today, 35*(2), 12–18.

Weiss, L. (2011). Israel's 1979 nuclear test and the US cover-up. *Middle East Policy Journal, 18* (4), 83–95.

Weiss, L. (2012). The Vela event of 1979 (or, the Israeli nuclear test of 1979). In *Historical dimensions of South Africa's nuclear weapons program*. South Africa: Pretoria.

Chapter 11
Nuclear Testing and the NPT

> *.... to seek to achieve the discontinuance of all test explosions of nuclear weapons for all time and to continue negotiations to this end,*
>
> [NPT Preamble]
>
> *Each of the Parties to the Treaty undertakes to pursue negotiations in good faith on effective measures relating to cessation of the nuclear arms race at an early date and to nuclear disarmament....* [Article VI].
>
> [Treaty on the Non-Proliferation of Nuclear Weapons, 1970]

11.1 Comprehensive Nuclear-Test-Ban Treaty

The CTBT bans nuclear explosions of any yield, in all places, and for all time. CTBT is an arms control measure that constrains the five nuclear weapons states from developing new weapons. Much has been learned from testing nuclear weapons: According to the Comprehensive Test Ban Treaty Organization, the US conducted 1054 nuclear tests, followed by Russia (715), UK (45), France (210), China (45), India (3), Pakistan (2) and North Korea (3). Figure 11.1 displays these events.

CTBT signatories are constrained by the treaty even though it has not yet entered into force. The US failed to ratify the CTBT, but the US is bound not to test until it formally withdraws from Treaty. The CTBT is also a nonproliferation measure since the test ban raises a barrier to the development of first-time nuclear weapons. This is recognized in the quotes at the beginning of this chapter. The 1998 tests by India and Pakistan and the three North Korean tests highlight the need for a CTBT for all nations.

The CTBT has 183 signatories (July 2015), which amounts to practically all the nuclear capable nations, except for India, Iraq, North Korea, and Pakistan. Non-nuclear weapons states (NNWS) view the CTBT as a "quid pro quo" by nuclear weapons states (NWS) to the nonproliferation regime to gain their support.

As of July 2015, 164 states have ratified CTBT. Forty-four nuclear capable states must ratify CTBT before it can enter into force. Eight nations must ratify to do this;

D. Hafemeister, *Nuclear Proliferation and Terrorism in the Post-9/11 World*,
DOI 10.1007/978-3-319-25367-1_11

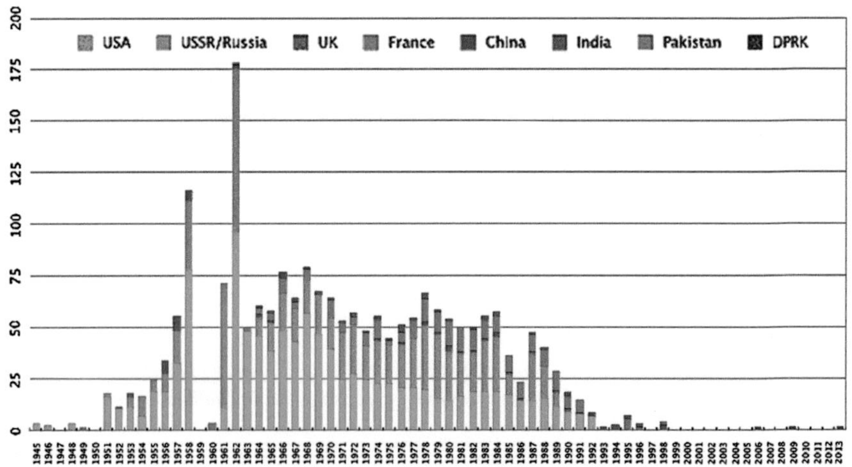

Fig. 11.1 Global nuclear tests (1945–2013). *Left* to *right* U.S., USSR, UK, France, China, India, Pakistan and N. Korea (Comprehensive Nuclear-Test-Ban Treaty Organization, 2013)

China, Egypt, India, Iran, Israel, N. Korea, Pakistan and US. In October 1999, the US Senate rejected the CTBT with a vote of 51–48. After the defeat, the National Academy of Sciences was asked by the Clinton and Obama Administrations to convene panels of experts to examine technical issues that may affect the viability of a test ban (National Academy of Sciences 2002, 2012). The 2012 panel was asked to assess:

- The plans to maintain the safety and reliability of the US nuclear stockpile without nuclear explosion testing;
- US capability to detect, locate, and identify nuclear explosions;
- Commitments necessary to sustain the stockpile and the U.S. and international monitoring systems;
- Potential technical-advances countries could achieve through evasive testing and unconstrained testing.

The panels stated that several factors were key in its analysis on the ability of the US stockpile stewardship program to maintain the safety and reliability of US nuclear weapons without testing. These were as follows:

- Confidence will require a high-quality workforce and adequate budgets.
- Stockpile stewardship and enhanced surveillance must examine nuclear components.
- Remanufacture to original specifications is the preferred remedy for age-related defects.
- Primary yield that falls below the minimum level needed to drive a secondary is the most likely potential source of nuclear-related degradation.

- Based on past experience, the *majority of aging problems* will be found in the *non-nuclear components,* which can be fully tested under a CTBT. (NNSA has stated that the *nuclear* plutonium pits have a *minimum* lifetime of 150 years (2012) with "no life-limiting factors.")
- A highly disciplined process is needed to make changes in nuclear designs.
- In the past, confidence tests were limited to one per year, as most tests were carried out to critique new designs. Other than a question of temperature ranges for ALCMs, no errors were discovered.

The US secretaries of Defense and Energy are required to make an annual certification on the status of the nuclear stockpile and whether the stockpile stewardship program is maintaining the warheads without testing. One approach is to obtain a fixed reliability level, but that requires performing a tremendous number of tests each year to determine the reliability. This has never been done since the US did only a few reliability tests per year with a stockpile of many types of weapons. The number of tests depends on the choice of reliability level. This was described three decades ago by Steve Fetter:

> Current levels of testing are insufficient to determine the actual reliability of the particular type of weapon with a high degree of certainty. Even if 10 confidence tests were performed and all were successful, there would be a 30 % chance that the weapon would be less than 90 % reliable and a 10 % chance that it would be less than 80 % reliable. To be 95 % sure that the weapon is at least 95 % reliable would require nearly 60 tests, all successful.

Another approach is to examine a tremendous amount of scientific data and determine the confidence in the reliability level by examining the various margins of safety. It is generally believed that the reliability of warheads is significantly higher than the reliability of missiles. One also must determine how much reliability is needed for various warhead missions. The main role of nuclear weapons is to deter attacks by others. Beyond that role, there might be the following unhappy roles:

1. Nuclear warheads could be used to respond to an attack by Russia or China.
2. They could be used to attack Russia or China first.
3. They could be used to respond to an attack by a smaller nation.
4. They could be used to threaten or attack a smaller nation first.

Option (1) has questionable need for very high reliability since most of the foreign silos would be empty and the cities on both sides are very vulnerable. A *launch on warning* strategy could be used, but that is highly dangerous. Option (2) needs very high reliability weapons to attack foreign strategic targets first, but this would not be wise or successful. A rag-tag foreign second strike would greatly endanger the US. Options (3) and (4) have less need for very high reliability since the US has a great number of weapons to over-target. Ultimately, the very main purpose for nuclear weapons is deterrence, which needs the perception of great reliability. Stockpile stewardship is discussed in Sect. 2.1 and mutually assured destruction is discussed is in Sect. 4.5.

11.2 NPT–CTBT Connection

Two tiers. The Nuclear Non-Proliferation Treaty regime has obvious weaknesses, as we look at India, Iran, Iraq, Israel, Libya, North Korea, Pakistan, South Africa, Syria, and more. Can the world forever be divided into two camps; one side with nuclear weapons and one side without? The present division is split into the Nuclear-Weapon States, the Big Five from World War II, and the Non-Nuclear Weapon States, the states that are not the *P5*. The list of new nuclear weapon states will grow unless we are diligent. The US should not flaunt NPT norms with acts that are legal but defy the spirit of the NPT treaty-law, or this will come home to roost. If we supply airplane and submarine nuclear weapon platforms to other nuclear states, that is not consistent with the intent of the NPT.

Proliferation issues. The NPT was extended by consensus in 1995 only because the five nuclear weapon states promised they would deliver a CTBT. Since the NPT was changed to last in perpetuity, forever, the CTBT ought to last the same time, in perpetuity, forever. Of course, most nations will not go to the bomb just because they know they don't want to live in a nuclear neighborhood. This is the case for Norway and Sweden. But this argument doesn't carry everywhere. With this in mind, I contacted Tom Graham, the long-time, top legal mind on arms control treaties and the author of *Disarmament Sketches*. Graham was the Chief Negotiator for the US for the extension of the NPT, a complex task. The outcome was an article in *Disarmament Diplomacy* called "*Nuclear testing and proliferation—an inextricable connection*."[1]

> In a letter dated 19 April 1995 from France, Russia, the United Kingdom and the United States (China agreed later) to the 1995 NPT Review and Extension Conference, the NWS coupled a determination to complete the CTBT with a request to the NNWS that the NPT provisions be made permanent (the quid pro quo): *'We reaffirm our determination to continue to negotiate intensively, as a high priority, a universal and multilaterally and effectively verifiable comprehensive nuclear–test–ban treaty, and we pledge our support for its conclusion without delay... We call upon all States parties to the [NPT] to make the treaty provisions permanent. This will be crucial for the full realization of the goals set out in Article VI.'*
>
> As in 1968, the non-nuclear weapon states in 1995 chose to back having a strong, durable NPT and thereby gave up the leverage of holding the treaty hostage over its extension. In the Statement of Principles and Objectives on Nuclear Non-Proliferation and Disarmament that accompanied the resolution indefinitely extending the NPT, all NPT parties agreed to conclude a CTBT in one year. The 1995 NPT Review and Extension Conference agreed on the following objective: *'The completion by the Conference on Disarmament of the negotiations on a universal and internationally and effectively verifiable Comprehensive Nuclear-Test–Ban Treaty no later than 1996. Pending the entry into force of a Comprehensive Nuclear–Test–Ban Treaty, the nuclear –weapon States should exercise utmost restraint.'*
>
> After this was agreed, the NNWS fulfilled their part of the bargain and renewed the NPT without a time limit. Without the CTBT promise, it might have been necessary to settle for a fixed renewal of the NPT, with proposals ranging from 10 to 25 years. A ten–year NPT

[1]Graham and Hafemeister (2009).

would have expired in 2005, when the Review Conference failed completely. If the NPT had been renewed for 25 years, the NPT would be nearing its expiration in 2020.

By way of comparison: if the US Constitution was nearing expiration and had to be renegotiated by the 50 states, there would likely be chaos and instability in the United States. Large states like California might insist on having more power in the Senate than smaller states like Wyoming or Rhode Island. Such arguments could foreseeably wreck the careful balances and, once undone, it would be difficult if not impossible to renegotiate and achieve a better US Constitution than the one we have.

11.3 Nuclear Tests in Atmosphere and Space

CTBT slows the development of proliferating states as they try to produce their first viable nuclear weapons, and particularly their development of boosted and hydrogen weapons. It also constricts existing nuclear weapons states by halting tests on new types of nuclear warheads. The International Monitoring System is able to detect tamped warheads over 0.1–0.2 kton in the Northern Hemisphere and it is able to detect warheads exploded in a cavity over 1–2 kton. CTBT could improve the political climate among nuclear weapon states, and its adoption is consistent with the Nuclear Non-Proliferation Treaty (NPT). In this section we examine detection of nuclear warhead tests in space and the atmosphere.

Nuclear tests in space. The *Limited Test Ban Treaty of 1963* forbids testing of nuclear weapons in space, in the atmosphere and underwater. Monitoring must be able to detect nuclear debris from an explosion above the Earth. The neutron, x-ray, and prompt γ-ray fluences from weapons exploded in space are measured at distances of 20,000 km, where GPS satellites reside. The results from the Los Alamos Vela satellite program of the 1960s showed that these signals are readily observable by satellites.

Nuclear tests in the atmosphere. Atmospheric nuclear explosions give unique, double-peaked optical signatures. The initial burst of x-rays lasts about 1 μs, after which they are absorbed in the first few meters of air, creating a million-Kelvin fireball of heated air. Shortly afterwards a shock wave carries energy beyond the fireball. Since the high-temperature shocked gas is ionized, it is opaque to the fireball optical radiation, reducing the luminosity of the fireball. After the shocked gas further expands and cools, it again becomes transparent to light from the fireball with its optical intensity rising again for a second pulse. Optical "bhangmeters" detect double pulses above the intense brightness background of earthshine. Empirical algorithms give the yield of weapon tests from the critical times of the blast. In Fig. 11.2, the first maximum is at 0.3 ms, the first minimum is at 12 ms and the second maximum is at 130 ms. The shocked region from a 20-kton explosion has a temperature of 100,000 K during the millisecond regime when absorption takes place in the first meters of air. This gives a shock velocity 20 times greater than the sound velocity at standard temperature and pressure at the breakaway minimum, when the shock wave passes the fireball. For a 19-kton weapon, the breakaway time of 12 ms is in good agreement with Fig. 11.2.

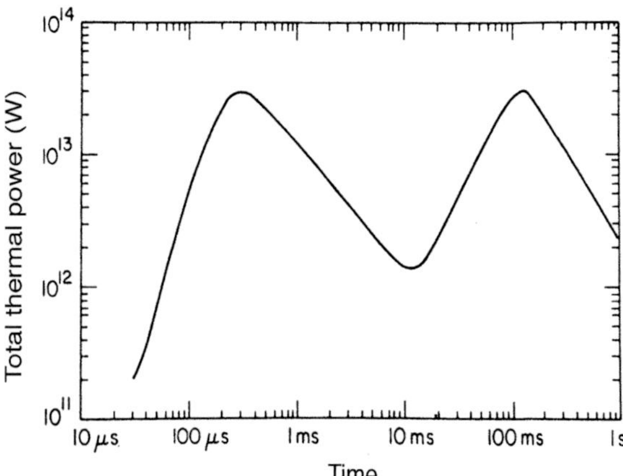

Fig. 11.2 Optical bhangmeter. The characteristic double-peak signal is from a 19-kton atmospheric nuclear test. The optical photometers record the luminosity of the fireball as a function of time (H. Argo in *Arms Control Verification*, 1986)

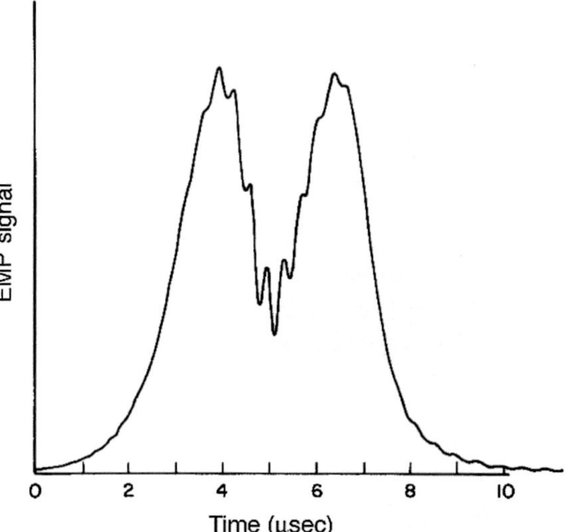

Fig. 11.3 EMP detection. A simulated EMP signal is seen by a satellite receiver tuned to 50 MHz with a 2-MHz bandwidth. The ionosphere disperses the signal as a function of the frequency, complicating the interpretation of the received signal. The second peak is a ground reflection (H. Argo in *Arms Control Verification,* 1986)

EMP signal. Additional confirmation of an atmospheric explosion is obtained if the bhangmeter signal arrives at about the same time as the signal from the electromagnetic pulse that accompanies a nuclear explosion (Fig. 11.3). This evidence is available from the GPS system that observes any spot on Earth with 4–8 of its 24 global positioning satellites. Accurate GPS clocks allow triangulation for accurately determining the position of an atmospheric nuclear explosion. Had these instruments functioned in 1979, they might have removed ambiguity from the possible nuclear explosion "event" over the South Atlantic. An independent panel of the Presidential Office of Science and Technology Policy reported in 1980 that the signals were probably not from a nuclear explosion, but other scientists disagree decades later (Fig. 11.3).

11.4 Underground Nuclear Tests

Nuclear tests have been confined to underground locations by the United States and Russia since 1963, as well as by UK (1958), France (1974), China (1980), India (1974, 1998), Pakistan (1998) and North Korea (2006, 2009, 2013). Seismographs remain the primary tool for monitoring underground tests, but other technologies supplement the seismic data. Seismic traces from nuclear explosions differ from earthquake traces in several ways. Nuclear explosion seismographic data display higher frequency components because the duration of explosions is much shorter compared to earthquakes. The ratio of the short-period, *pressure body wave magnitude* m_b to the long-period, *surface* wave magnitude M_S is larger for weapons than for earthquakes (Figs. 11.4 and 11.5).

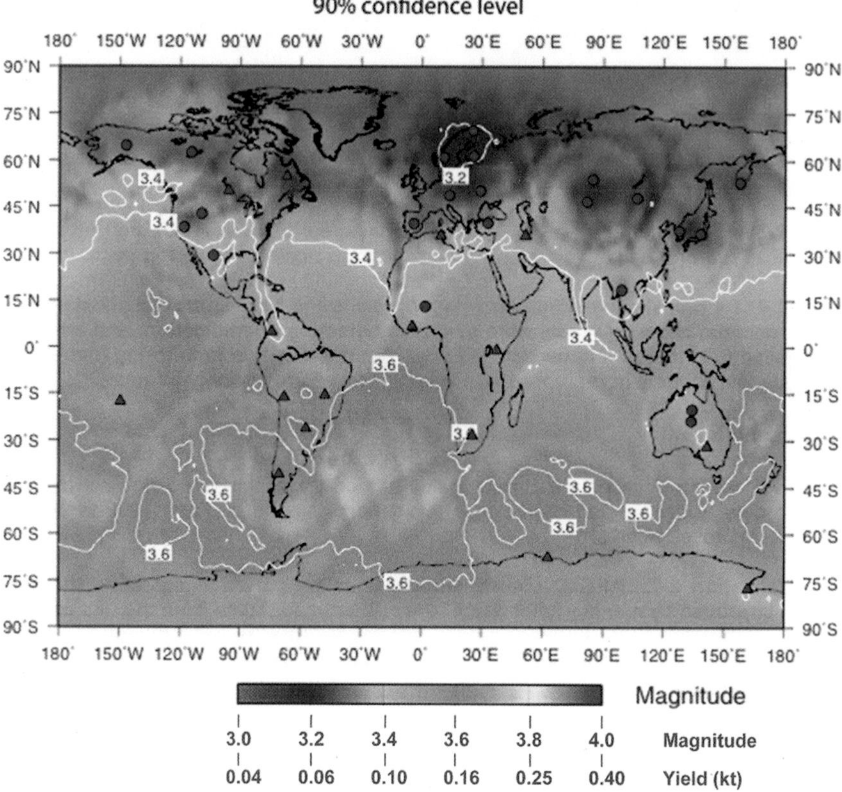

Fig. 11.4 IMS seismic monitoring limits at 90 % confidence. Detection capability of the IMS primary seismic network in late 2007 with 38 primary stations. For 90 % detection for the entire world, the detection level is m_b = 3.8, which corresponds to 0.22 kton in hard rock with better propagation and 0.56 kton for poorer propagation. For 90 % detection of Asia, Europe and N. Africa, the levels are 0.09 and 0.22 kton. Completion of this network to 50 stations would reduce these magnitudes by 0.1 or 0.2 units for Asia, much of Africa and the Indian Ocean. The IMS system with 33 stations detected 0.025-kton explosions at the semipalatinsk test site (2012 NRC/NAS study, p. 50)

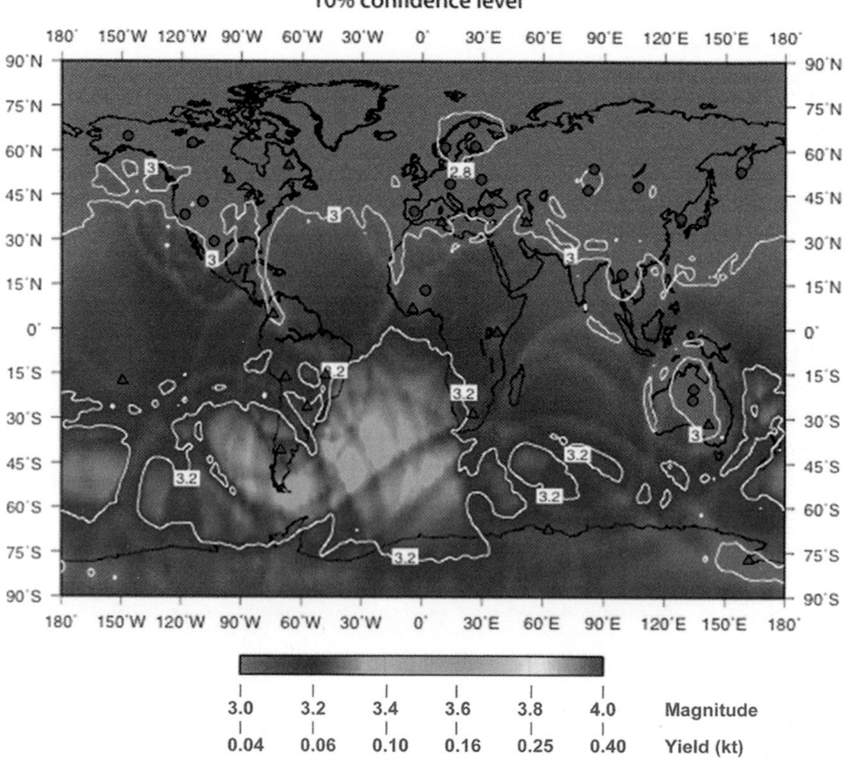

Fig. 11.5 IMS monitoring at 10 % confidence. Lowering the confidence in the detection capability of the primary seismic network to 10 % lowers the threshold magnitudes to 2.8, 3.0 and 3.2. This corresponds to fully coupled device yields of 0.22, 0.035 and 0.056 kton, respectively in regions of better propagation (2012 NRC/NAS study, p. 106)

11.5 National Academy Conclusions on CTBT Monitoring

The 2002 and 2012 NAS reports concluded that the IMS has the capability to monitor explosions with high confidence (90 % certainty) to a 3.5 m_b seismic level, which corresponds to a tamped explosion of about 0.1 kton in hard rock throughout Eurasia and North Africa. This is a factor of ten better than the 1 kton that was originally projected for the IMS, an assessment that was too cautious since it did not take into account the growing number of close-in, high-quality regional stations. The 2012 NAS report determines a threshold of 3.4 m_b (0.1-kton in hard rock, 0.2 kton poor propagation regions) for Asia, Europe and N. Africa, and a global threshold of 3.8 m_b (0.2–0.6 kton). In many instances, the regional threshold is 2.8 m_b and the test site threshold is 2.2 m_b.

Table 11.1 Detection versus evasion probabilities for fully coupled underground nuclear explosion tests

Yield in regions of hard rock (kton)	Yield in regions of poor propagation (kton)	Detect probability fully coupled (%)	Evasion probability fully coupled (%)
0.09	0.22	90	10
0.03	0.09	10	90

Three stations or more, average for Asia, Europe and North Africa—illustration based on capabilities using only IMS Primary Stations in 2007 (National Academy of Sciences Report, The Comprehensive Nuclear-Test-Ban Treaty, 2012, pg. 105)

Regional monitoring is at distances less than 1600 km with a threshold of 0.02 kton, as compared to global monitoring of 0.2 ktons recorded at distances typically greater than 3000 km. Monitoring at test sites, with additional transparency measures, can bring sensitivity down to about 5 tons, or 0.005 kton.

Success rate and threshold level. These threshold limits are based on a 90 % success rate to discover clandestine testing, which also means a 10 % failure rate. However, a lowered success rate requirement *comes with* a lowered threshold level. For a greatly lowered detection success rate of 10 %, the detection threshold is lowered by 0.5 m_b, a reduction of a factor of three in yield. In other words detection statistics do not give a sharp threshold value. For example, if the threshold is 0.1 kton, that does not mean that tests below 0.1 kton cannot be detected. The threshold level at a 10 % detection rate is about 0.03 kton, with 0.06 kton for poor propagation media (Table 11.1).

Other possible measures. A neighboring state can place regional seismographs close to suspected regions, to improve monitoring. Finally, chemical explosions are usually identifiable because they are ripple-fired along a line to reduce costs, these are not singular spherical explosions. CTBT voluntary notifications of chemical explosions larger than 0.3 kton reduce suspicions about chemical explosions.

Seismology has improved over the decades with these advanced technologies:

- analog to digital seismographs,
- narrow-band to broad-band seismographs,
- single axis to triple axis to array seismic stations,
- from magnitude picks to full seismic patterns to correlated template patterns,
- increased density and quality of seismic stations,
- from teleseismic data to close-in regional data,
- improved earth models for regional seismology with improved algorithms,
- spectra above 6 Hz discriminates source term,
- understanding geological bias factors with preferential absorption,
- improved ability for other technologies to assist seismology.

International monitoring system. The IMS has deployed monitoring stations without CTBT entered into force. IMS will consist of 50 primary and 120 auxiliary seismic stations, as well as 60 infrasound stations (1-kton global atmospheric threshold detection), 11 hydroacoustic stations (less than 1-kton global oceanic detection) and 80 radionuclide stations (less than 1-kton, global atmospheric detection when released). As of July 2015, there were 281 installed stations, 20 installed but not yet certified, 18 under construction and 18 in planning. In addition, the US uses satellite optical bhangmeters, particle detectors, EMP detectors and National Technical Means.

North Korean tests. The 2006 North-Korean test (0.6 kton, 4.1 m_b) was promptly detected by 22 IMS seismic stations, later confirmed by radioxenon detection. The 2009 North Korean test (3 kton, 4.5 m_b) was detected by 61 IMS seismic stations. The 2013 North Korean test (10 kton, 4.9 m_b) was detected by 94 IMS seismic stations, 2 infrasound stations and 2 radioxenon stations (^{131m}Xe and ^{133}Xe in Japan and Russia). The absolute value of the yields could be off by 50 % because of unknowns in local geological coupling, while the ratio of yields is much more accurate because they all have the same geological coupling. The yield of the 2013 test is about 3 times the 2009 yield, and it is 15 times the yield of the 2006 test. The 2013 P-wave amplitudes were 200 times the raw background noise. Radioxenon was detected from the 2006 and 2013 events, but not from the 2009 test. The 6 Janauary 2016 test was similiar to the 2013 test. The Korean peninsula is particularly good for seismic monitoring, down to the level of a few tons yield. Since the many aspects of CTBT monitoring are complex, we simplify matters with the regime in Table 11.2 (Fig. 11.6).

National technical means. NTM consists of satellite reconnaissance, human intelligence (humint), and other "ints" which are combined to make intelligence gathering greater than the sum of its parts. A nation's fear of being spotted by the IMS and NTM deters it from cheating, and these measures will be buttressed by on-site inspections. The 2012 NAS report commented on US NTM capabilities:

> NTM gives the United States significant additional information beyond what is available to other countries that do not have a robust NTM program. U.S. NTM can focus on monitoring countries of concern to the U.S. The United States global monitoring capabilities are generally better than those of the CTBTO because they can go beyond data available to the CTBTO with classified capabilities. However, the inclusion of classified means and data limits the extent to which analysis and even results may be shared and used openly. Drawing on all available assets is important because there are CTBTO installations in locations where the United States cannot readily deploy stations, as well as thousands of stations that operate independently of U.S. NTM and the IMS.

Global auxiliary network. The 2002 NAS study concluded that the monitoring confidence level could be lowered to 0.1 kton if the *IMS auxiliary network* of 120 stations was more fully utilized, working with the primary network of 50 stations. The NAS concluded that thresholds "would drop generally by about 0.25 magnitude units in Europe, Asia, and North Africa, and by about 0.5 magnitude units in some regions (such as Iran)." A magnitude drop of 0.25 corresponds to a factor-of-two reduction in the yield threshold. The Auxiliary network is now used in a spotlight mode (when needed) and not in a continuous mode. With more

Table 11.2 CTBT—monitoring capabilities

Method	Description	IMS assets (when complete)
Seismic	NAS concluded that explosions above 0.1 kton in hard rock can be detected in Asia, Europe, North America and North Africa. Tests in cavities can be detected above 1–2 kton for advanced nuclear weapon states, with risk of venting and excursion yields. This limit is perhaps 0.1 kton for new nuclear nations	IMS will use 50 primary and 120 auxiliary seismic stations. Arrays of seismographs and regional seismographs can obtain lower threshold yields. In addition, thousands of non-IMS stations have data that could trigger an on-site inspection
Hydroacoustic	NAS concluded that explosions above a few kg can be detected in Southern Hemisphere, and above 1 ton for all oceans	IMS will use six hydrophone arrays and five T-phase monitoring stations
Infrasound	NAS concluded that explosions above 1 kton in the atmosphere can be detected, and above 0.5 kton over continents	IMS will use 60 infrasound monitoring stations
Radionuclide	NAS concludes that explosions above 0.1–1 kton can be detected if released to identify the event as a nuclear explosion. The 0.6 kton North Korean test was detected at 7000 km distance	IMS will use 80 particulate monitoring stations, and 40 of these will also detect radioxenon. NTM sensors can be placed on airplanes for close approaches to suspected test sites
InSAR (Interferometric synthetic aperture radar)	InSAR can measure subsidence as low as 0.2–0.5 cm in many locations, with yields above 1 kton at 500 m depth. InSAR can determine locations to 100 m	United States has four classified SAR satellites. Europe, Canada and Japan sell unclassified SAR data for as low as $1000 each
On-site inspections (OSI)	Any CTBT party can request an OSI, which needs 30 of 51 votes in the executive council	Photos and radioactivity obtained by air and ground. Mini seismic arrays can observe aftershocks. Magnetic anomalies, SAR, soil data obtained with GPS locations
Confidence-building measures	After CTBT enters into force, states could locate more sensors at test sites or nearby to lower thresholds further	Close-in sensors could detect seismic, infrasound, electromagnetic pulse, radionuclide and other data indicative of a test
National technical means	US NTM technologies have considerable reach and precision	NTM sensors are located in space, in the atmosphere, on the ground, in the oceans and underground

By February 2015, the IMS system was 83 % operational (281 stations), 6 % installed (19), 6 % under-construction (19) and 5 % planned (18). The completed IMS will consist of 337 facilities (321 monitoring stations and 16 radionuclide laboratories): seismic (50), auxiliary seismic (120), infrasound (60), radionuclide (80, 40 with noble gas), hydroacoustic (11)

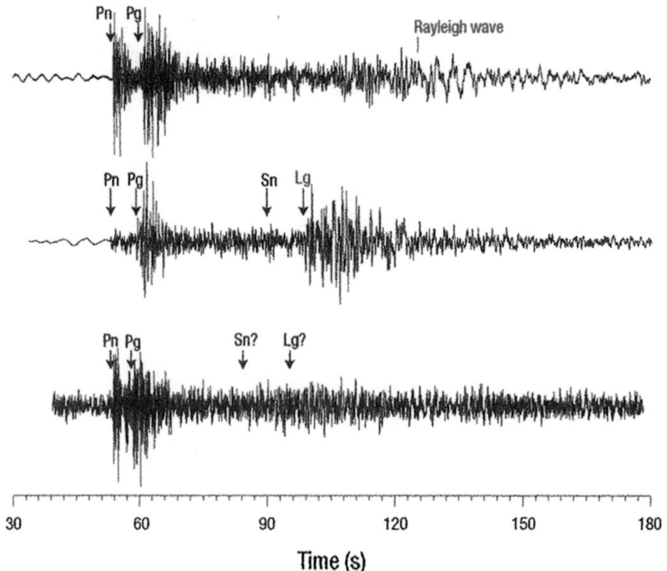

Fig. 11.6 North Korean 2006 test, earthquake, and chemical explosion. Seismograms with vertical ground velocity in microns/second recorded at Mudanjiang, China. (*top*) North Korean test (9 October 2006, magnitude 4.0, distance of 373 km, 0.7 kton), (*middle*) earthquake (16 December 2004, magnitude 4.0, 342 km), (*bottom*) small underground chemical explosion (19 August 1998, magnitude 1.9, 289 km, 0.002 kton). The strong P-wave amplitude in the 9 October 2006 trace shows the event was an underground explosion. The waveform for small chemical explosion has the same structure as the data from the underground nuclear test, begun with a strong P-wave signal (Richards and Kim 2007)

seismic stations, closer access to the seismic sources reduces geometric spreading and absorption. A higher density of seismic stations gives more data and better data to observe the more-complex regional waves that contain more information than long-distance teleseismic waves. The regional data is useful to rule out background earthquakes, which increase in number by a factor of ten as the threshold is reduced by one magnitude unit. Greater proximity reduces seismic attenuation, particularly at higher frequencies, and is useful in detecting explosions in cavities.

False positives from global network. Lowering the detection level for events increases the number of unidentified false positive events. IMS use of auxiliary stations on a continuous basis changes the obligation of CTBTO for their maintenance. What about the more than 3000 non-IMS seismic stations? The 2012 *Global Seismic Network* had 153 high-quality, seismic stations located in 70 host-counties. Can CTBTO use these assets to further improve their results? Yes, but with reservations. The data from the non-IMS stations can be submitted, but it carries less weight since these stations have not been certified by IMS, and quality varies from station to station.

CTBT monitoring limits. The CTBT treaty is effectively verifiable to about 0.1 kton for tamped explosions in hard rock in all of Eurasia. The National Academy of Sciences convened a bipartisan panel of experts to carry out studies during 2000–2002 and 2009–2012. The panels concluded that "the only evasion scenarios that need to be taken seriously at this time are cavity decoupling and mine masking." Successful covert testing involves at least seven issues, each with a differing probability of success:

1. Violators need excellent yield estimates to avoid yield excursions.
2. Violators need to hide removed materials from satellites.
3. Crater and surface changes from testing are observables.
4. Radioactive releases from tests often observed in former USSR.
5. Regional signals at 10 Hz improve detection.
6. A series of tests is needed to develop significant weapons.
7. Human and other intelligence can give information.

Because the net success probability for hiding a covert test in a cavity is the product of the individual success probabilities, the NAS panel did not use a decoupling factor of 70 times the 0.1-kton limit to obtain a maximum cheating limit of 7 kton. Rather, it concluded the following: "Taking all these factors into account and assuming a fully functional IMS, we judge that an underground nuclear explosion cannot be confidently hidden if the yield is larger than 1 or 2 kton."

Threshold test ban monitoring. The TTBT requires that yield be measured to determine if an explosion exceeded the treaty's 150-kton threshold. Because the plate below the Nevada site is relatively young, its seismic waves are absorbed more than waves at the Shagan River Test Site in Kazakhstan, which resides on a much older plate. Because of this difference, US explosions appear smaller compared to explosions of the same yield at the Soviet site. Ignoring this bias difference, the United States incorrectly charged the Soviets with a "likely" violation of the TTB treaty.

Yield estimates. The magnitude of an explosion is calculated from the body wave seismic magnitude (the pressure wave) from $m_b = a + b + c \log(Y)$, where $\mathrm{m_b}$ is the magnitude, a is the 4.1 magnitude of a 1-kton explosion, b is the bias correction for a test site, c is the slope of 0.74, and Y is the yield in kilotons. A 150-kton yield at the Nevada Test Site has an m_b value of 5.71, while a 150-kton explosion at the Soviet site with a bias of 0.4 is 6.11. The United States initially assumed there was no bias between the two sites ($b = 0$), which gave the United States a false impression that the Soviet's explosion at 6.11 $\mathrm{m_b}$ was a serious violation of 520 kton. These differences show that the choice of bias is exceedingly important in judging compliance. Security classification of seismic data prevented a thorough discussion by experts. Sectors of the US policy community purposely ignored geological bias to maintain the "likely violation" charge against the

Soviets.[2] The designation of "likely violation" on nuclear testing greatly hindered negotiations on the CTBT in particular and arms control in general.

11.6 Covert Cavity Tests

Explosion in a cavity. A nuclear test, exploded in a cavity, is muffled to appear as a lower-yield explosion, or not observed at all. There are little data on nuclear tests in cavities. A fully-decoupled test needs a sufficiently large cavity to minimize the observed yield by, at most, a factor of 70. The only fully decoupled test took place in 1966 when the 0.38-kton Sterling explosion was exploded in a Mississippi salt cavity with a 17-m radius. The cavity was created by the 5.3-kton, Salmon explosion. The Soviets carried out a 9-kton test in a former nuclear cavity at Azgir in 1976. This test was only partially decoupled, as the weapon was too large for the cavity's 36-m radius, itself created by a 64-kton previous test. If a nuclear weapon is placed in a cavity of sufficient size, the blast pressure on the cavity wall will fall below the material's elastic limit, reducing the seismic signal strength by a theoretical factor of 7 at 20 Hz and 70 at lower frequencies. For more details, see Appendix E of the 2012 NAS report.

Covert cavity testing is complicated by possible radioactive venting that could be detected. The Soviets had 30 % of its tests vent during its first decade of underground testing. The United States also had severe venting problems. Radioxenon was detected from the 2006 and 2013 North Korean tests, but not from the 2009 test. Perhaps, North Korea learned to use gasketed, hair-pin turns to contain radioxenon and other radionuclides? Smaller tests are often harder to contain than larger ones, as the last four US tests that vented had yields less than 20 ktons. It is hypothesized that smaller explosions do not sufficiently glassify cavities, and do not rebound sufficiently to seal fractures with a stress cage, thus making them easier to vent radioactivity. In addition, a nation new to testing is not likely to have adequate knowledge to predict the yield of the test, and thus they cannot rule out yield excursions that exceed the decoupling range.

Detection of cavity test. The probability of a successful clandestine effort is the product of at least seven detection probabilities for (1) venting, (2) yield excursion, (3) hiding removed materials, (4) subsidence of the surface, (5) regional seismic detection, (6) more than one test in a series of tests, (7) NTM detection. If two of the steps, such as yield prediction and venting, could be successful with medium

[2]*The author was technical lead on TTBT compliance in the State Department (1987), testified before the Senate Foreign Relations Committee on TTBT (October 1988), was lead technical SFRC staff (1990–1992) on TTBT ratification and Mitchell-Hatfield testing ban, and was the lead NAS technical staff on the CTBT study (2000–2002).*

confidence (50 % successful) and the other five were successful with high confidence (90 % successful), the total probability of successfully hiding a 1–2 kton test would be $(0.9)^5(0.5)^2 = 15$ %.

If blast pressure exceeds the elastic limit of the cavity's wall material, sufficient energy is absorbed to crack the wall, increasing coupling to the wall and increasing the seismic signal. Critical cavity size depends on explosion depth, but it is usually assumed to be about 0.5–1 km. The critical radius for decoupling a 30-kton explosion is 60 m (a 20-story building), an extraordinary engineering challenge when one considers the secrecy requirements. The critical radius at 1 kton is 20 m. It is easier to clandestinely mine a cavity in salt using water solvents than to mine granite cavities, but only a few nations have salt deposits thick enough to consider this kind of violation.

Problems

11.1 **NPT versus CTBT**. What are the differences between these two treaties?

11.2 **Monitor CTBT**. Describe the four IMS monitoring technologies.

11.3 **Enter into force**. What has to happen for EIF of the CTBT? Give a scenario for this path through the eight countries.

11.4 **Seismic advances**. What are the reasons that CTBT became easier to monitor?

11.5 **Stockpile stewardship**. Describe this program (Chap. 2). What are danger spots and requirements?

11.6 **Remanufacture**. What are the tradeoffs between the two extreme views?

11.7 **ALCM temperature**. Why did the ALCM warhead become vulnerable?

11.8 **Certification**. Describe the annual process to determine warhead operational status?

11.9 **100-year life**. How big a facility is needed for new pits if they last 100 years?

11.10 **Four missions**. Describe the 4 missions for warheads. How does reliability fit into this?

11.11 **NPT extension**. Could the NNWS have been mislead by the NWS?

11.12 **Spirit of the NPT**. What actions by NWS violate the spirit of the NPT?

11.13 **Space monitor**. What is monitored from tests in space?

11.14 **Atmosphere monitor**. What is monitored? What is a bhangmeter?

11.15 **Yield of tests**. How is yield determined in space, atmosphere and underground?

11.16 **NAS conclusions**. What were NAS conclusions on the CTBT?

11.17 **NTM**. What are some NTM techniques? How do they compare with the open literature?

11.18 **Subsidance**. How determined by InSAR (Chap. 6)? How could InSAR reduce location errors? What are the limits on InSAR for this?

11.19 **False positives**. Why are these dangerous? How overcome the downsides?

11.20 **TTBT versus CTBT**. What are the major differences?

11.21 **Confidence building measures**. What measures could help the CTBT without revealing national security?

11.22 **Cavity cheating**. What must the cheater avoid to be successful?

Seminar: NPT–CTBT Connection

Auxillary IMS network, bhangmeter, cavity test tasks for success, cavity data, CTBT aids NPT extension with risks, EMP signal with Bhangmeter, Global Seismic Array, IMS technologies, North Korea test data, NPT threatened, NPT discusses CTBT, NTM technologies, NTM versus IMS, NWS test causes what?, regional seismology, reliability tests gives confidence in reliability estimate, replace NPT with what, seismology improves 1–0.1 kton, weapon uses and reliability needed.

Bibliography

Corden, P., & Hafemeister, D. (2014 April). Nuclear proliferation and testing: The tale of two treaties, *Physics Today 67*(4), 41–46.

Dahlman, J., et al. (2011). *Detect and deter: Can countries verify the nuclear test ban*. NY: Springer.

Drell, S., & Purifoy, R. (1994). Technical issues of a nuclear test ban. *Annual Review of Nuclear and Particle Science, 44*, 285–327.

Drell, S., & Schultz, G. (2007). In *Implications of Reykjavik Summit on 20th Anniversery*. Stanford, CA: Hoover.

Dunn, L. (1980). *Arms control verification and on-site inspection*. MA: Lexington Press.

Fetter, S. (1988). *Towards a comprehensive test ban*. Cambridge, MA: Ballinger. 89.

Graham, T. (2002). *Disarmament sketches: Three decades of arms control international law*. Seattle, WA: University of Washington.

Graham, T., & Hafemeister, D. W. (2009). Nuclear testing and proliferation: An inextricable connection. *Disarmament Diplomacy, 91*, 15–26.

Hafemeister, D. (2007). Progress in CTBT monitoring. *Science Global Security, 15*(3), 151–183.

Jeanloz, R. (2000). Science-based stockpile stewardship. *Physics Today, 53*(12), 44–50.

Krepon, M., & Caldwell, D. (1991). *Politics of arms control treaty ratification*. NY: St. Martin's Press.

Krepon, M., & Umberger, M. (1988). *Verification and compliance*. Westport, CT: Ballinger.

Lynch, H., Meunier, R., & Ritson, D. (1989). Some technical issues in arms control. *Annual Review of Nuclear and Particle Science, 39*, 151–182.

Moynihan, M. (2000). The scientific community and intelligence collection. *Physics Today, 53* (12), 51–56.

National Academy Sciences. (2002). *Technical issues related to the comprehensive nuclear test ban treaty*. Washington, DC: NAS Press.

National Academy Sciences. (2012). *The comprehensive nuclear test ban treaty: Technical issues for the United States*. Washington, DC: National Academy Press.

Nelson, R. (2002). Low-yield earth-penetrating nuclear weapons. *Science Global Security, 10*, 1–20.

Office of Technology Assessment. (1991). *Verification technologies*. Washington, DC: OTA.

Pifer, S. (2010). *The next round: The U.S. and nuclear arms reductions after new START*. Washington DC: Brookings.

Richards, P. G., & Kim,W. Y. (2007 January). Seismic signature. *Nature Physics,* 3, 4–6.

Richelson, J. (1998). Scientists in black. *Scientific American, 278*(2), 48–55.

Schaff, D., Kim, W., & Richards, P. (2012). Seismological constraints on low-yield nuclear tests. *SAGS, 20*, 155–171.

Schultz, G., et al. (2008). *Reykjavik revisited: Steps toward a world free of nuclear weapons*. Palo Alto: Hoover Press.

Tsipis, K., Hafemeister, D., & Janeway, P. (Eds.). (1986). *Arms control verification.* Washington DC: Pergamon.

Chapter 12
Terrorism

I am the one in charge of the 19 brothers... I was responsible for entrusting the 19 brothers... with the raids.
[Tapes aired by Al Jazeera, contained Osama bin Laden speaking (2006)]

It is in the character of growth that we should learn from both pleasant and unpleasant experiences.
[Nelson Mandela, president of South Africa (1994–1999)]

12.1 Terrorism in 21st Century

Islamic war of religion is not new a phenomenon. We forget that 450 years ago, in the 1560s, North-West Europe faced bloody revolts by Calvinists, during the Wars of Religion that lasted 150 years. At one point a quarter of the population of Germany was killed. These wars determined which religion, Catholicism or Protestantism, would gain power and which would play a secondary role. Today's Islamic revolutions appeared in Egypt (1952), Iran (1979), Al Qaeda attacks (2001), the Arab Spring (2011), the Islamic State of Iraq and Syria (2014, ISIS) and Nigeria's Boko Haram. The Islamic structure is divided between its Sunni and Shiite sects, to complicate matters. When the US negotiates with Iran's Shiites on their nuclear program, the Saudi Sunnis state they will go to nuclear weapons if Iran does. It is difficult to find a win–win situation.

Terrorism is *ugly*. It is ugly because it involves sneak attacks on innocent strangers, without taking prisoners. Terrorism plays hardball for political gain. Terrorism does not selectively kill men in uniform, but is aimed at the general, anonymous public, without a filter. Terrorism is not deterred by the threat of mass retaliation because terrorists don't strongly defend territories, and they are ready to die. Terrorists, inflamed with religion or ideology, don't care a lot about logic, since they have issues that might not be addressed if they did not attack. The spread of technology favors both sides, but surprise attacks can be aid the offense more. The response to terrorism can be ugly, as some politicians, journalists, ideologues and others use fear as a tool to advance their issues. Terrorism can be defined in many ways:

D. Hafemeister, *Nuclear Proliferation and Terrorism in the Post-9/11 World*,
DOI 10.1007/978-3-319-25367-1_12

- Terrorists are frustrated political actors, wanting to correct *flaws* of the state.
- Terrorists use violence, not just to kill, but to get immediate public recognition.
- Terrorists are rational actors in that they realize to kill too many might make them suffer from responses to their attacks.
- Terrorists have limited budgets and means to make nuclear, weapons, but they could make crude, inefficient weapons with stolen materials or biological weapons.
- Terrorists are driven by religion, economics, nationalism and more.
- Terrorists may have valid issues, and they play hardball.

Two sides to a story. On the other hand, many of today's respected governments launched political quests for democratic rule by using terrorism. Some events in the past seem much less dangerous than those of today. John Adams was not a terrorist, but he encouraged the Boston Tea Party in 1765 with statements that the *Stamp Act* was illegal since there was "no representation" on the decision to tax the thirteen colonies. The British referred to this event as a rebellion, but we believe it was a quest for freedom. In the US, we savor dumped tea, while the New England Patriots won the Super Bowl. The Jewish patriot David Ben–Gurion blew up a British hotel in 1948, which led him to become Israel's first Prime Minister in 1948. But, many Palestinians do not respect him since there is their side of the story. And the list goes on. History is written by winners who carve in stone on Mount Rushmore. Good treaties salvage respect and land for the losers, or else the region will fester with newer conflicts. We hope global norms will respect human rights during these conflicts.

Some terrorist attacks of the past are listed below:

1960s: Hijacked planes (now magnetometers),
1988: PanAm 103 baggage bomb (270 dead, now baggage scans),
1993: World Trade Center truck-bomb (6 dead, 1042 casualties),
1995: Oklahoma City bombing (168 dead, 500 casualties),
1995: Tokyo subway nerve agent (12 dead, 5000 casualties),
1998: US Embassies in Kenya and Tanzania (212 dead, 5100 casualties),
1999: Columbine High School attack, CO (15 dead, 20 casualties),
2000: Algerian *Armed Islamic Group* and *Islamic Salvation Army* (>60,000 dead),
2001: World Trade Center planes (3062 dead, now scan baggage and persons),
2001: Richard Reid shoe bomb on plane (now remove shoes),
2002: Bali tourist bombing (190 dead),
2004: Madrid bombing (191 dead),
2006: Heathrow liquids plot (now liquids ban),
2009: Non-metallic bombs (now trace detection, whole body image, pat down),
2010: Printer cartridge bombs (now cargo trace detection),
2013: Sandy Hill School, Newton, CT, attack (28 dead, failed gun laws),
2013: Boston Marathon bombing (3 dead, 264 casualties),
2015: Emanuel African Methodist Episcopal Church, Charleston, SC (9 dead).

12.1.1 Terrorism Advantage to the Offense

Nuclear terrorism. Modest spending by terrorists can force the defense to spend much more in response and recovery. Terrorists have many means at their disposal. Nuclear weapons are much more difficult than biological weapons, but they might be used, using stolen weapons or stolen nuclear materials. Nuclear warheads can be small. The W48 artillery shell had a mass of 50 kg, a length of 1.4 m, and a diameter of 15 cm. Small designs fit in suitcases. Dirty bombs (Chap. 13), using stolen radioactivity, are more likely than actual nuclear weapons, but they have not been used over the past two decades. Dirty bombs would probably cause more disruption than destruction. Mass chaos in New York or Washington from modest radiation exposure would be crippling. Or one might choose to destroy facilities that have radioactivity, such as reactors.

Fear of terrorist nuclear weapons can be used to hype a political issue, such as an invasion of a country. Fear of nuclear weapons can also warn society to do work to lessen the problem. The second Iraq war of 2003 is an example of the former. On the other hand, there is modest evidence of terrorists thinking about using radioactivity from intelligence sources. Nuclear weapons of 10–20 kilotons made by terrorists seem unlikely, except if key materials, such as confiscated nuclear weapon parts or full weapons, are obtained. Insiders with knowledge of nuclear weapons can lower the threshold to do this. On the one hand, we have gone 15 years since 9/11 and 25 years since the collapse of the Soviet Union without a terrorist nuclear weapon. But, a small annual risk over a period of long time can be a certainty.

Non-nuclear terrorism. Many varieties of non-nuclear attacks could be devised with chemical or biological weapons spread on crop dusters, airplanes, drones, missiles, water supplies or with information technology cyber attacks on tapped phones, tapped email, hacked banking forgeries, and things that we do not yet know. It does not take a *rocket scientist* to envision crippling attacks to our society. Chats over coffee can devise crippling events. We have become more vulnerable as technology advances, making us more dependent on others. Technology centralizes transportation networks, electrical grids, internet communication and business transfers. As people, we have become narrower, many of us barely know how to pick apples from the trees.

Disruption. Instead of terming some options as *weapons of mass destruction*, we now call them *weapons of mass disruption*. Exposure to radioactivity may not be lethal, since you can remove yourself from Manhattan Island or the Washington Mall. Terrorists have gotten our attention in the past, and they could do that again. It has been over a decade since the 9/11 attack, and it is surprising there has not been an encore event. This result is partly due to the Department of Homeland Security (DHS) and other enforcers, but there may also be some good luck in this. The Oklahoma City bomb of 2000 kg was carried in a rental box truck, producing 80-psi pressures at 30 m. The largest truck bomb used against an American target was 6000 kg in Beirut in 1983, generating 250 psi at 30 m. It is difficult to monitor all the U-Haul trucks. It is also difficult to monitor all the sales of chemical fertilizers over tens kilograms, which can be combined into tons of explosives. It is difficult to monitor all trucks that could contain a short-range missile aimed at the Capitol. In the battle between the OFFENSE and DEFENSE, it is difficult not to conclude that the first-attacker, the OFFENSE, has the advantage. There are many ways to attack, and it is difficult to police all the pathways.

12.1.2 Future Perfect Storms

The dependence on more efficient, but more vulnerable, technologies enhances the opportunities for terrorists to disrupt these infrastructures:

- transportation networks,
- electrical grids,
- natural gas pipelines,
- water resources,
- health-care resources,
- global finance, and
- global shipping.

Combining the increased vulnerability of infrastructures with increased urban population of megacities and with increased availability of means of attack makes Earth *increasingly vulnerable*. To re-emphasize this, the three factors that make today's society more vulnerable are as follows:

1. Increased vulnerability of technical infrastructures,
2. Increased availability of weapons technology to smaller entities, and
3. Increased urbanization with higher population densities.

As we have become specialists in our own trades, we have learned that broader skills are needed to survive. By 2040, 80 % of *global energy* will be consumed in cities, compared to 66 % in 2010. This is a different world with 70 % of the population living in cities of 2050, up from 50 % in 2007, as seen in these *global urban trends* (Shell Oil in *NY Times,* 2014):

- 1970: 3 cities over 10 million (M), 1 city 10–20 M, 0 city over 30 M,
- 2010: 23 cities over 10 M, 1 city 10–20 M, 1 city over 30 M, and
- 2030: 41 cities over 10 M, 10 cities 10–20 M, 3 cities over 30 M.

These trends exacerbate concerns that citizens treasure most, such as freedom of speech, privacy and mobility, the very things that advancing technology has aided in the past.

12.2 Attack of 11 September 2001

The 9/11 event was historic because it changed the world we live in. Did we overreact? This is what Osama bin Laden wanted us to do, playing into his hands to do things that are too-far reaching, un-related and un-necessary. Was invasion of Iraq the worst outcome, greater than that of the 9/11 attack itself?

Al Qaeda. The terrorist attacks on the World Trade Center and the Pentagon killed 2977, causing a *paradigm shift from the fear of nuclear weapons to the fear of terrorists*. Sub-national terrorists are more likely to use conventional explosives and biochemical weapons as compared to nuclear weapons used by nations. It is far easier to use trucks full of fertilizer, divert airplanes, or drop anthrax aerosol bombs in subways than it is to obtain nuclear materials and assemble a viable nuclear weapon. There are reports of Al Qaeda attempting to obtain nuclear materials. Terrorists could steal plutonium or HEU to sell to proliferating nations. Terrorism involves many possible paths and many possible responses. How can we know the daily contents of 15 million shipping canisters on the high seas? Plutonium and HEU can be identified at airports, but only in larger amounts. We will not find *total* solutions to global terrorism, only progress in improved detection and interdiction. We should also learn what motivates the terrorists.

12.2.1 Four Airliner Missiles

Osama bin Laden, founder of the militant Al-Qaeda organization, was born in 1957 to the billionaire bin Laden family with close ties to Saudi Arabian royalty. Osama bin Laden was banished from Saudi Arabia in 1992 because of his Al Qaeda activities, moving to Sudan in 1992 and Afghanistan in 1996. Al Qaeda declared war on the US and others, initiating bombings and attacks. He was a fundamentalist, Sunni Moslem, opposing the Shiites. He believed US foreign policy harmed Muslims in the Middle East, saying "They hate us for what we do, not who we are." Osama bin Laden opposed music, fornication, homosexuality, intoxicants, gambling, and usury. Osama bin Laden's strategy was to use violence against large enemies, such as the US and Russia. He believed violence could lure the West into a long war of attrition, "bleeding America to the point of bankruptcy."

Osama bin Laden believed that Afghanistan under the Taliban was "the only Islamic country" in the Muslim world. The US provided early money and weapons in Afghanistan, while Pakistan's Intelligence Service trained Al Qaeda militants. After the fall of the Soviet Union, bin Laden worked against Saudi Arabia, Egypt and Iraq, which are all Moslem states. Osama bin Ladin funded the 1997 Luxor massacre, killing 62 Egyptians, the 1998 attacks on US embassies in Africa, killing several hundred, and the revolution in Algeria, killing over 100,000.

The Clinton administration tried to kill Osama bin Laden in 1998 with cruise missiles launched by the US Navy in the Arabian Sea, striking bin Laden's training camps in Afghanistan and narrowly missing bin Laden by a few hours. Osama bin Laden claimed responsibility for the 9/11 attacks on the US which hijacked four commercial passenger aircraft: American Airlines 11 and United Airlines 175 flew into the World Trade Center, American Airlines 77 flew into the Pentagon and United Airlines 93, aimed at Washington DC but it was crashed into a Pennsylvania field by brave passengers. Tapes aired by Al Jazeera in 2006 contained Osama bin Ladin saying, "I am the one in charge of the 19 brothers… I was responsible for entrusting the 19 brothers… with the raids." President George W. Bush demanded that the Taliban turn over Osama bin Ladin. The Taliban responded they would try him before an Islamic court if evidence of bin Laden's involvement was provided. The war began on 7 October 2001 with the aid of the UK. Ten years later, the Obama administration carried out operations from helicopters, killing bin Laden in his Abbottabad compound in Pakistan on 2 May 2011.

The attack. Khalid Sheikh Mohammed presented the idea of the 9/11 attack to bin Laden in 1996, who approved it in 1998. The 19 conspirators were upper–middle class Moslems, not from disenfranchised lower classes. They started assembling in 1999 in Hamburg, Germany, and reassembled in 2000 in San Diego, Arizona and South Florida where they took pilot training. In retrospect, it now seems obvious they were doing things out of the ordinary since they were not trained to take off or land. There was evidence at the time of their presence in the US, but not enough to flag the issue with gusto.

Conspiracy theories. The 110-story World Trade Center was designed to withstand a Boeing 707, but it could not withstand prolonged 1100 °C (2000 °F) temperature for several hours from copious amounts of burning jet fuel. Steel melts at 1500 °C, but it loses half its strength at 700 °C, allowing beams to sag and break. Building 7 of the World Trade Center with 47 stories also collapsed, although not hit by a plane. It experienced 2 s of almost free-fall acceleration of gravity. Some claim that building 7 was taken down with demolition explosions, but there is no evidence of that. It is hard to imagine that this could be kept secret, as it would take many people to do the demolition. When asked what happened to flight 77, the conspiracy buffs claim "the plane was destroyed and the passengers were murdered by Bush operatives." Theories abound with claims that the 9/11 attack was carried out by the CIA or claims that Saddam Hussein helped the 9/11 attack. The 9/11 Commission Report concluded that most of the terrorists entered the US illegally and that there was no evidence of a meeting between Mohamed Atta, the hijackers' leader, and an Iraqi intelligence officer in Prague in April 2001. Several conspiratorial books have been written on this subject. For a nice discussion of this 9/11 topic, see the Skeptic Article by Michael Shermer in the June 2005 *Scientific American.* A listing of past conspiracy issues is located at Venusproject.org.

Tipping point: The 9/11 attack on the WTC was the major tipping point, which drew the US into conflicts in Afghanistan and Iraq, increased the US role in the Middle East. Some of this might have taken place without the 9/11 attack, but the attack had much to do with our military commitments in this area. The record of successful democratic nation-building in the Middle East is not large. Can the creation of the Islamic State in Iraq and Syria (ISIS) be linked to the Iraq-II war?

Afghan war (2001–2016): The war began on 7 October 2001. The Taliban were displaced from cities across the country by December, when they escaped to Pakistan and the remote Afghan mountains. In August 2003 NATO took over command of the International Security Assistance Force (ISAF) of 43 countries in Afghanistan. In October 2004 Hamid Karzai was elected president, but he failed to unify his people. The Taliban resurgence began in March 2005. Obama added 30,000 surge troops in December 2009. ISAF forces peaked at 140,000 in July 2011. On 2 May 2011, US Navy Seals killed bin Laden in Pakistan. In May 2012, NATO stated its plan to withdraw by December 2014. The US followed with a similar plan in May 2014, with complete withdrawal by 2016. By March 2015, Afghan forces numbered 340,000, with 10,000 US troops for advisory and training purposes. The Taliban has been defeated politically, but not militarily. Only a broad US-based, domestic and military force could improve on that. It is politically unlikely that the US will send a new large force in the future.

Iraq war II (2003–2011): The October 2002 vote in the House of Representatives was 296 in favor (215 R, 81 D) and 133 opposed (6 R, 126 D). The invasion of Iraq was driven by two issues: (1) The fear that Iraq was moving towards nuclear weapons, and (2) Iraq was involved with bin Laden's attack of 9/11. Both of these motives were discredited. The US, UK and several allies started the *shock and awe* invasion on 16 March 2003, without declaring war. Iraqi forces soon surrendered. Saddam Hussein was captured in December 2003 and executed 3 years later. Bush increased US forces with a surge of 21,500 troops on 10 January 2007. US formally withdrew all combat troops from Iraq by December 2011.

ISIS. The 2011 civil war in Syria gave a platform for a broad civil war between Muslims. Because of the hard-line, pro-Shiite actions of Iraqi Prime Minister Nouri al-Maliki, the disgruntled Sunni Muslims joined the Islamic State of Iraq and Syria (ISIS). During the earlier surge in Iraq, the mainline Sunni's joined the fight against al Qaeda. But Maliki's actions caused them to join ISIS, which was famous for its inhumane treatment of civilians. President Obama stated that the US was to fight ISIS with air power and troop training, but not ground US troops. As of January of 2015 there were 3000 US troops in Iraq. ISIS had 31,000 fighters in August 2015, with half of them from 80 other states. It is generally agreed that ISIS is a very different kind of threat, but yet it is vulnerable to the loss of financial resources. ISIS controls some 60 % of Syria's oil production, gaining a billion dollars a year. Its reach in Iraq and Syria covered 10 million citizens. Iran has become more involved, hoping to make this a Shiite region from Iran to Iraq to Syria (and elsewhere). ISIS would like to draw the US into the conflict, hoping to gain in the process. As of March 2015, a very loose coalition of 60 countries has reduced the ISIS-controlled area by 25 % and ISIS has lost 75 % of its oil revenues. But yet, the recruits keep coming.

In August 2015, 3400 American troops assist Iraq's military, while 9800 support Afghanistan's military. At this time, the US is shifting some of assets from the fight against Al Qaeda to that opposing the Islamic State. These personnel come from American troops and from the intelligence community with some 100,000 employees.

This shift comes from the realization that during the 14 years after 9/11 there were no further successful attacks on the US. The increased use of social media by the Islamic State has resulted in increased recruiting of younger terrorists. Some 80 % of those arrested were younger than 30 and 40 % are younger than 21. IS encourages individual gory attacks on the West, which has happened.

12.2.2 Torture

"*Enhanced interrogation techniques*" are used on prisoners to force them to reveal their nation's secrets. These include water-boarding, in which water is poured over a cloth, covering the face and breathing passages of immobilized prisoners, causing them to experience sensations of drowning. Water-boarding causes extreme pain and panic, dry drowning, and damage to lungs and brain from oxygen deprivation. Other *enhanced interrogation techniques* include sleep deprivation, chaining in uncomfortable positions, replacement of body fluids, beating and threatening. CIA Executive Director John Brennan admitted that the CIA "did not always live up to the high standards we set for ourselves." Sen. John McCain, who was tortured while spending five years in North-Vietnamese prisoner-of-war camps, stated that torture "compromises that which most distinguishes us from our enemies… [the CIA's conduct] stained our national honor… [and has done] much harm and little practical good." These comments followed the December 2014 release of the Senate Intelligence Committee Majority Report (Democrats) on the use of torture on captives by the CIA. The minority (Republicans, except for McCain) did not agree with the conclusions by saying "the report does not adequately consider the whole context of the time."

The leadership of both political parties was aware of the facts, they were approved by Pres. George W. Bush. Previously, Obama banned the use of torture, while deciding in 2012 not to prosecute CIA agents who had participated in its use. The *Economist* (13 December 2014) concluded with the following:

> The people who were meant to scrutinize America's intelligence services were instead co-opted… they appear to be exercising 'hindsight, not oversight…. Mr. Obama, meanwhile, has refrained from criticizing the agency harshly. He has been in charge of it for 6 years. It and America's other intelligence agencies no longer torture people. Instead of capturing suspected terrorists, Mr. Obama prefers to have them blown to pieces by drone-fired missiles.

Complications. These matters are complicated by a number of issues. Americans involved in torture can be prosecuted abroad, in principle, by the *International Criminal Court.* In practice, this political hot potato is not likely because the US is large and has not signed the treaty. This was tried by Spain in 1988 to prosecute the former Chilean dictator, Augusto Pinochet, but it failed. Global support of US policies to fight terror will be muted by these results. The former Polish president and its prime minister both admitted they allowed the CIA to run *black sites* on Polish soil. They insist they did not know it would include torture methods. Less friendly nations like China and Iran are pointing out that America should not overly moralize when abroad. The forward of the majority report concluded the following:

> [we] could understand the CIA's impulse to consider the use of every possible tool to gather intelligence and remove terrorists from the battlefield, and CIA was encouraged by political leaders and the public to do whatever it could to prevent another attack. Nevertheless, such pressure, fear and expectation of further terrorist plots do not justify, temper or excuse improper actions taken by individuals or organizations in the name of national security. The major lesson of this report is that regardless of the pressures on the need to act, the intelligence community actions must always reflect who we are as a nation, and adhere to our laws and standards.

The chair of the Senate Intelligence Committee majority, Sen. Dianne Feinstein, introduced legislation on 6 January 2015 to provide in law, beyond Obama's Executive Order, the ban on CIA from holding detainees "beyond a short-term, transitory basis." The majority found that the CIA "impeded effective White House oversight" of the interrogation program by providing the Bush administration with "extensive amounts of inaccurate and incomplete information." They found that former Secretary of State Colin Powell was kept in the dark out of concerns that he would "blow his stack if he were briefed on its details." Feinstein hopes to reverse this by requiring the CIA to designate a specific agency officer responsible and accountable for the covert CIA operations, thus placing the responsibility on one individual for the actions taken.

12.2.3 Did Torture Prevent Attacks?

The minority disagreed with the conclusions of the majority conclusion that little useful information was obtained by using torture. The majority report give these examples:

Killing Osama bin Laden. "Within days of the raid on Osama bin Laden's compound, CIA officials represented that CIA detainees provided the 'tipoff' information on Ahmad al-Kuwati. A review of CIA records found that the initial intelligence obtained, as well as the information the CIA identified as the most crucial—or the most valuable—on Ahmad al-Kuwati was not related to the use of the CIA's enhanced interrogation techniques."

Thwarting dirty bombs and capture of José Padilla. "A review of CIA operational cables and other CIA records found that the use of CIA's enhanced interrogation techniques played no role in the identification of 'José Padilla' or the thwarting of the Dirty Bomb or Tall Buildings plotting."

12.3 Long-Term Response

Response to terrorism is complex because terrorism doesn't have the logic of war between nations, where the attacked can attack the attacker. Since a terrorist may act without support of the host nation, there is no easy place to respond. The specificity is not available to control terrorism. Terrorists, residing in faraway caves, feel less constrained to protect their brethren that reside in cities. Physicists and political scientists prefer linear theories to increase global stability, but terrorism is non-linear and illusive.

12.3.1 Department of Homeland Security

The Department of Homeland Security (DHS) was created in 2002 with five core missions:

- prevent terrorism,
- secure borders,
- enforce immigration laws,
- safeguard cyberspace, and
- increase resilience to natural disasters.

DHS was given considerable funding, but outside reviews discovered confusion in the ranks. Intelligence agencies were reorganized under the new *Director of National Intelligence* (DNI), who supervised the CIA, rather than reside there. The 2001 and 2008 Foreign Intelligence Surveillance Acts (FISA) allow phones and emails to be tapped abroad and at home, depending on the FISA court and other issues. Edward Snowden made this appearance in 2013. A lower court ruled in 2015 that the executive branch does not have the right to collect domestic information, Congress must be given that authority if it is to be done.

The desire to attack preemptively is strong, after the success of the Israeli attack on the Osirak reactor in 1981. Ballistic missile defense was funded more heavily as concerns increased over missile attacks. The courts tend to favor the Executive Branch over the Congress on foreign policy issues, giving way to the State Department and the president. The 2015 DHS budget authority was $65 billion. DHS was combined from 22 units of 12 executive branch departments in the following manner (2016 proposed budget):

- **Secret service** ($2.2 billion)
- **Border and transportation**:
  Immigration and Customs Enforcement Service ($6.3 B), Customs and Border Protection Service ($14 B), Citizen and Immigration Services ($4 B), Animal and Plant Health Inspection Service, Coast Guard ($10 B), Transportation Security Agency ($7.3 B).

- **Emergency preparedness and response**:
  Federal Emergency Management Agency ($13 B); Chemical, Biological, Radiological and Nuclear Response; Domestic Emergency Support Team; Nuclear Incident Response; Office of Domestic Preparedness; National Domestic Preparedness Office.
- **Science and technology**:
  Civilian biodefense research programs, National Biological Weapons Defense Analysis Center, Animal Disease Center, Domestic Nuclear Detection ($0.36 B).
- **Information analysis and infrastructure protection**:
  Critical Infrastructure Assurance Office, Federal Computer Incident Response Center, National Communications System, National Infrastructure Protection Center, National Infrastructure Simulation and Analysis Center.

12.3.2 Total Information Awareness

The Department of Defense created the Total Information Awareness (TIA) Program in 2003 to combine diverse data on individuals to predict what these persons might do in the future. The TIA program was led by Admiral John Poindexter until TIA was disbanded by Congress 10 months later after media reports stated that the "government was attempting to establish Total Information Awareness over all citizens." However, the basic thrust of TIA lived on to become reality. The goal of the TIA program was to greatly enhance the ability of the US to detect and identify foreign terrorists and to understand their plans in a timely manner, enabling the US to preempt and defeat terrorist acts. This was to be done by the following:

- Increased information coverage by more than an order of magnitude. This includes information technology data and more technology. The Boston Marathon bombers were quickly identified with camera surveillance. A work in progress is the Biometric Optical Surveillance System (BOSS) that mates computers and video cameras to identify persons. This DHS project is intended to help the police identify persons in crowds. At present, it can identify one individual in 30 s, but this is too slow for crowd detection. In addition, BOSS must have a low false positive rate or it will cause many false charges and extra problems. Researchers claim it will be ready in five years.
- Provide warning within an hour after a triggering event occurs or when an evidence threshold has been passed.
- Automatically queue pattern matches that cover 90 % of previously known foreign terrorist attacks.
- Supports collaboration, analytical reasoning and information sharing to mitigate future attacks.

TIA depended on the technical thrust from much cheaper computer capacity and much cheaper computer memory. These two aspects gave the capacity for huge statistical calculations using Big Data to determine the details in our lives. Moore's law states that the density of transistors on a surface doubles every two years, which has held up for almost fifty years.[1] These same trends have held for the cost of computer power and for computer memory.

A wide range of groups opposed the TIA program. *New York Times* reporter and author William Safire called it "the super-snopes dream: a Total Information Awareness about every citizen." The American Civil Liberties Union predicted TIA implementation would "kill privacy in America" because "every aspect of our lives would be catalogued." There are legal advantages for calling the US response to 9/11 a *war on terror*, as it gives the government more room to act. It is our obligation to ensure we maintain democratic processes as we sort through these issues. Chapter 14 returns to cyber terrorism issues.

12.3.3 Terrorism Pathways

There are many ways terrorists can attack society. DHS issued a list of possible terrorist attacks and natural events, and the likely fatalities from these attacks. These are scenarios and not rigorous calculations, but they could kill thousands and costs billions:

- 10 kton nuclear device in a central city (100,000 killed),
- aerosolized anthrax from a van in five cities (13,000 killed),
- influenza pandemic spreads from Asia (87,000 killed; 300,000 hospitalized),
- pneumonic plague released in 3 locations in a city (2500 killed),
- blister agent from a small plane over a stadium (150 killed; 70,000 hospitalized),
- attack an oil refinery with grenades and bombs (350 killed),
- sarin nerve gas in ventilators of 3 large buildings (6000 killed),
- destroy chlorine gas storage (17,500 killed; 100,000 hospitalized),
- 7.2 magnitude earthquake in city of 10 M (1400 killed; 100,000 hospitalized),
- category 5 hurricane (160 mph wind, 7 m surge in urban area, 1000 killed),
- dirty bombs with ^{137}Cs in 3 cities (180 killed),
- truck bomb and suicide belts at a stadium (100 killed),
- anthrax distributed in beef and orange juice (300 killed),
- hoof-and-mouth disease to cattle,
- cyber attacks on US financial infrastructure.

---

[1]Over 50 years, this gives 25 doublings, or $2^{25} = 2^{10} \times 2^{10} \times 2^{5} = (1024)^2 \times 32 = 33$ million.

12.3.4 Response to Terrorism

Complex planning is needed to defend against terrorism, to reduce vulnerabilities of varied pathways. Here we discuss technical issues, but we readily acknowledge that the non-technical issues, the more political ones, are more important. This is not something new, as history is full of such examples. Pragmatic leadership has tried to skate between two extreme endpoints. President Obama chose to avoid sanctioning Russia when Vladimir Putin took over the Crimea, since that issue was made murky by Nikita Khrushchev's single-handed transfer of Crimea from Russia to Ukraine, and with Crimea's large indigenous Russian population. However, when Putin sent troops into eastern Ukraine, Obama responded with financial sanctions. This caught Russia at a vulnerable time, making a bigger difference since petroleum prices were low. Is it ad hock, to respond to the fear of the day? On the one hand, to look the other way and do nothing encourages Russia to take more Ukrainian land, but it allows Russia to continue to participate in the family of nations. To some, this could sound like appeasement to Hitler on Czechoslovakia in 1938. On the other hand, to take extreme actions, beyond financial sanctions, pushes Russia to abandon arms limitation treaties and start a new arms race. In the Middle East, sacred principles may be at stake, pushing us to support those that are reviled by others. This is why we have elections to sort out these important, nonlinear policy issues. We get our feet wet when we consider the non-technical issues. Trade-offs are also involved with making policy on terrorist matters. On the one hand, Iran aided Hezbollah terrorists, but Iran may (or may not) be edging towards a nuclear accommodation with the *P5+1*. Iran is helpful in the battles with ISIS (Islamic State in Iraq and Syria), but yet this extends the reach of the Shiites. Too much accommodation with Iranian Shiites threatens the Sunni states of Saudi Arabia and Egypt, perhaps to build nuclear weapons.

Some issues. There has been progress on estimating the timing and severity of terrorist events in Iraq and Afghanistan, when the US was on the ground, but this is now more difficult. Political events, such as the ISIS arrival in 2014, degrade the analytical prowess of the past. A DHS study, "Using DHS risk assessments to inform resource decision making," might be helpful. Should DHS fund Midwestern towns, far from coastal vulnerabilities? The intelligence community wants more human intelligence and boots on the ground, not to rely too much on technology. Funding for bio-security increased as more was learned about the bio-threat. Funding for cyberspace programs increased with the success of Stuxnet attacks on Iranian centrifuges in 2010. Will these types of attacks generate copy-cat attacks in the US? The attacks on Saudi Arabian Armaco computers for oil production and attacks on banks and companies by the Chinese army unit show that nations have gained cyber-war prowess (Chap. 14). A former DHS Inspector General recommended these technical measures in 2007:

Aviation. Backscatter X-ray machines, multi-view baggage scanning, trace-gas explosive-detection, more liquid detection, inspect 100 % of cargo on passenger planes, only Americans work at airports, all employees screened at each checkpoint.

Seaports. Inspect for nuclear radiation on all cargo ships bound for the US.

Borders. Triple the number of Border Patrol Agents. Supplement with sensors, cameras and unattended aviation vehicles (UAV). End the visa-waver program from the EU and elsewhere. Add exiting information to the automated border entry system, to learn when terrorists left the US.

Mass transit. Provide funds for mass transit authorities to deploy police patrols, bomb-sniffing dogs and technology, surveillance cameras, and permanent random bag searches.

Intelligence. The intelligence community should better provide DHS with relevant intelligence. It has become apparent that too much reliance on National Technical Means (satellites, etc.) reduced the amount human intelligence to learn broader global trends. Shortly after 9/11, CIA placed advertisements in the *Economist* with photos of persons with widely varying nationalities:

> CLANDESTINE SERVICE; The challenging situations that CIA professionals face in the Clandestine Service demands superior intellect, ingenuity and courage. That makes this more than just a job. It's an adventurous opportunity to face complex, unstructured circumstances that might test your understanding of a foreign regions history, culture, politics and economy.

Preparedness. A clear chain of command must exist at the federal, state and local levels. Some elements of the response structure are as follows:

- control borders with low-tech and high-tech fences,
- *Brave-New-World* tapping of foreign phone lines and email,
- claimed address tapping only in US,
- national identification card,
- big-data compilation analysis,
- biometric identifiers (face, voice, iris, DNA, body odor),
- listening devices (directional mike, a tiny bug, laser beam on windows),
- small drone monitoring (radar, IR, sound, visible),
- surveillance cameras in UK, Boston and elsewhere,
- vehicles located with GPS and electronics,
- chemical markers,
- cell phone monitors,
- examine garbage,
- radio-frequency identification (RFID) tags, and
- encryption (secret key, public key, authentication).

12.4 Vulnerability to Terrorism

Estimated fatalities from biological or nuclear weapons attacks are given below. These results are strongly scenarios dependent. The estimated likelihoods in the press are only relatively ranked (Table 12.1).

Table 12.1 Fatalities and likelihood of attack

Type of attack	Possible fatalities	Estimated likelihood
Efficient bio attack: Ebola, smallpox, anthrax	1,000,000	Extremely low
Nuclear bomb in major city	100,000	Very low
Attack on nuclear or toxic waste plant	10,000	Very low
Inefficient BW/CW attack skyscraper/stadium	1000	Low
Conventional attack on a train or airplane	250–500	Low
Attack with explosives or firearms in mall or street	50–100	Modest

12.4.1 NRC Report, Making the Nation Safer

The *National Research Council* combined the assets of the *National Academy of Sciences*, the *National Academy of Engineering* and the *National Academy of Medicine* to determine the science community's best advice to protect the nation from potential terrorist attacks. The primary topics in the 2002 report, *Making the Nation Safer*, are listed below. Four of these topics will be deferred to Chaps. 13–15.

- Nuclear and Radiological Threats (Chap. 13),
- Information Technology, Cyberwar (Chap. 14),
- Human and Agricultural Health Systems (Chap. 15), and
- Toxic Chemicals and Explosive Materials (Chap. 15).

Other topics discussed are as follows:

- Energy Systems,
- Transportation Systems,
- Cities and Fixed Infrastructure,
- Response of People to Terrorism,
- Complex and Interdependent Systems,
- Significance of Crosscutting Challenges and Technologies,
- Equipping the Federal Government to Counter Terrorism, and
- Essential Partners in a National Strategy.

These issues are complex, a multi-dimensional approach is needed to try and put our arms around the larger pictures. Each policy decision has to be examined to estimate what it might accomplish or what it might hinder, versus the potential positive and negative results from no action being done. To do this, one must estimate the *risk*, which is the *product of the probability of the event times the severity of the event.* This is difficult:

$$\text{risk} = \text{probability of event} \times \text{severity of event}$$

Table 12.2 Responsive phases to terrorist attacks (National Academy of Sciences)

• ***Intelligence and surveillance***: Observe persons and groups, and their motives, as well as their means of destruction.
• ***Prevention***: Disrupt terror networks to keep the means of destruction from would-be terrorists.
• ***Protection***: If detection and prevention fail, then enhanced protection is needed to harden targets to minimize damage. This would be done with improved technical designs and procedures to make borders, buildings, airplanes and critical infrastructure more difficult to disrupt. Protection also includes enhancing public health by increasing resistance to diseases.
• ***Interdiction***: Disrupt impending catastrophic terrorism with preemptive attacks on terrorists and improved crisis management.
• ***Response and recovery***: Limit the level of damage and casualties with emergency responses that enhance public health measures and restore critical functions after the terrorist attack.
• ***Attribution***: Identify the perpetrators to allow retaliation and prosecution.

The NRC report indicated six response phases that should be considered in addressing the vulnerabilities, some of which are listed in Table 12.2. Each of these recommendations involves a trade-off between gained safety versus lost revenue and freedoms.

12.4.2 General Guidelines

Each vulnerability has its own characteristics. The NRC study described guidelines to make each system more secure:

- Identify and repair the ***weakest links*** in vulnerable systems and infrastructures,
- Use ***defense-in-depth*** (do not rely only on perimeter defenses or firewalls),
- Use "***circuit breakers***" to isolate and stabilize failing system elements,
- Build ***security*** into basic system designs where possible,
- Build ***flexibility*** into systems so they can be modified to address unforeseen threats. Pay attention to human factors in the design of all systems, particularly those used by first responders; and
- Take advantage of ***dual-use strategies*** to reduce vulnerabilities of private-sector targets while enhancing productivity or providing new commercial capabilities.

Energy systems include electrical generation and the electrical grid, as well as the petroleum, coal, natural gas and transportation systems. Physical barriers should be installed where they do not already exist. In the longer term, technology should be developed, tested and implemented to develop an intelligent, adaptive electrical grid. The smart grid should have the ability to fail gracefully, minimizing damage to the electrical components and enabling a rapid recovery of electrical power. This requires fast-acting sensors and controls to isolate attacked parts of the grid and generation facilities. These systems should be able to differentiate between the failure of a single component and a larger failure of many systems, caused by an attack.

Another issue is that of *grid replacement parts*, which might take months or years to replace damaged components. The Department of Energy and the Electric Power Research Institute (EPRI) should develop systems to provide temporary replacements when key components are destroyed. Chapter 14 discusses cyber terrorism, which includes protection of the *supervisory control and data acquisition* (SCADA) systems from cyber worms. Encryption, improved firewalls and cyber-intrusion detection technologies should be used to improve security and reduce the potential for hacking and disruption. This also applies to oil and gas SCADA systems that are vulnerable. It will be necessary to enhance surveillance technologies and their usefulness in defending energy systems against terrorist attacks. Simulation models of these attacks and of inter-dependent infrastructures may provide knowledge to strengthen the energy systems. Simulations should be carried out on the interdependence between the energy sector and the communication, transportation and water-supply systems. The smart grid is clearly meant to be flexible, but will it introduce more attack points for cyber intrusion?

Transportation systems should have a layered security-system with connected multiple security features that provide back-up for each other. Defeating a single layer should not shut down the transportation system. New approaches might be considered for marine shipping containers, with the possibility of moving inspections out to sea, away from US ports, to avoid explosive damage. The Transportation Security Agency (TSA) should create a multimodal research and planning office, working with the Federal Aviation Administration and the Federal Highway Administration. Recognition of human factors is important in designing the layered, multimodal security-system, which should leave the perception that terrorists will get caught.

TSA has tightened the search for guns at airports. In 2014, TSA discovered 2212 guns, as compared to 660 a decade earlier. This increase was mostly due to enhanced detection methods, but it is also partially due to changed habits. The usual explanation is “I forgot.”

Cities and their infrastructure have become more vulnerable with the increased urbanization. The buildings industry has long been decentralized around the country, without a unified focus to make buildings more blast-resistant and fire-resistant. The National Institute of Standards and Technology (NIST) should develop standards to incorporate strengthening features in the building codes. Large buildings are also vulnerable to infectious or toxic materials that circulate in their heating, ventilation and air-conditioning (HVAC) systems. Sensors should be installed in air ducts to determine whether the air is safe, or not, and how the HVAC system should respond in an emergency. A city’s response to a terrorist attack should be coordinated at its emergency operations center (EOC) and by first responders, dispatched to the scene before the EOC can determine causes. The Federal Emergency Management Agency (FEMA), in conjunction with state and local officials, should develop threat-based simulation models and training modules for training, for identifying weaknesses in systems and staff, and for testing and qualifying the EOC teams. Differing human factors and responses should be part of the design and simulation process. The models and designs should be challenged by red teams to prove their worthiness.

Partners in a national strategy includes the states and cities, industry, universities and professional organizations. The DHS undersecretary for technology should play a role in this coordination. A successful long-term effort against terrorism must be cost-effective and not place extra burdens on the basic conveniences of life or constraints on civil liberties. This is a tall order and cannot be totally successful, as there will be impacts on all the systems. The research universities should contribute to the process with newly trained graduates. One concern is that the number of American students interested in science and engineering has declined.

12.5 Insider Threats

Employees, behind security fences, can threaten the foundations of the institutions where they work. Certainly the National Security Agency was embarrassed by their consultant, Edward Snowden's revelations. Insiders can easily multiply the prowess of terrorists. The insider knows the locations of the dangerous weak points in physical systems and in computer software. The Nuclear Regulatory Commission's design-basis threat uses scenarios that incorporate a single insider, but not two insiders for good reason. It is difficult to cover all the possibilities with increased insider threats. The work of Matt Bunn and Scott Sagan, *A worst practices guide to insider threats: Lessons from past mistakes*, is intended for the owners of nuclear power plants, but their results can be generalized to apply to the private and public sectors. The managers of these large systems should learn from their own mistakes as well as the mistakes of others. The summarized findings, given below, are based on known mistakes and bad practices from the past:

1. Don't assume that serious insider problems are not in my organization.
2. Don't' assume that background checks will solve the insider problem.
3. Don't assume that red flags will be read properly.
4. Don't assume that insider conspiracies are impossible.
5. Don't rely on single protection measures.
6. Don't assume that organizational culture and employee disgruntlement don't matter.
7. Don't forget that insiders may know the security measures and how to work around them.
8. Don't assume that security rules are followed.
9. Don't assume that only consciously malicious insider actions matter.
10. Don't focus only on prevention and miss opportunities for mitigation.

Problems

12.1 **Examples of terrorism**. Describe additional terrorist acts, beyond those in the text.

12.2 **Treaty blowback**. Describe a treaty that was overly harsh on the losers, which encouraged angry blowback.

12.3 **Oklahoma City attack**. Why did Timothy McVeigh and Terry Nichols kill 168 and injure 680 others on 19 April 1995? How did they get explosive materials and how did they assemble it?

12.4 **San Jose grid attack**. Who carried out the 11 September 2014 attack on the PG&E substation on Metcalf Road? What were the ramifications of the attack? How could it have been prevented?

12.5 **Survivalist skills**. Compare and contrast your skills and those of your great-grand parents who farmed and stayed warm in the North Dakota winter.

12.6 **Global urbanization**. Analyze some urbanization data on the UN Population Department website.

12.7 **Future vulnerabilities**. Give examples that take into account advancing technologies and population trends.

12.8 **Impact of Osama bin Laden**. In what ways did bin Laden succeed and fail?

12.9 **Osama bin Laden details**. Give details on bin Laden's life in Saudi Arabia, Afghanistan, Pakistan.

12.10 **Afghan War**. Give some details on the Afghan war.

12.11 **Iraq-II War**. Give some details on the second Iraq war.

12.12 **NATO in Afghanistan**. How did NATO become involved? What was NATO's impact on the situation? What was the impact on NATO?

12.13 **Torture methods**. Describe some torture methods, beyond the text examples.

12.14 **Intelligence community gossip**. Go to the website, www.FAS.org/expert/steven-aftergood/.

12.15 **Additional terrorist pathways**. What are additional attack pathways for terrorists, beyond the text. What are ways to minimize these new vulnerabilities?

12.16 **Non-linear thinking**. What are some ways that some issues can expand and become more dangerous in the future?

12.17 **IC changes**. How has the intelligence community changed since 9/11? Are these useful?

12.18 **Chain of command**. What is the chain of command, after a nuclear or chem/bio attack?

12.19 **New radiation detectors**. What are some current ideas?

12.20 **New chem/bio detectors**. What are some current ideas?

Seminar: 9/11 Terrorism

Airline flights after 9/11, assassination legality of foreign leaders and of US citizens, big data in *Science Mag.* (Jan. 30, 2015, p. 536), deaths from terrorism (1960–2010), defense in depth examples, DHS budget data, DHS domestic nuclear detectors, FISA court criteria, intelligence community re-organization, Iraq-II war reasons, population data/UN, trends, religious terrorism, terrorism disincentives, terrorism incentives, terrorism motives, torture beyond text, torture legalities, Transportation Security Administration data and criteria, TSA detectors, World Trade Ctr. conspiracy data.

Bibliography

Allison, G. (2004). *Nuclear terrorism: The ultimate preventable catastrophe*. NY: Henry Holt.

APS/AAAS Nuclear Forensics Group. (2008). *Nuclear forensics*. Washington, DC: American Physical Society.

Bunn, M. (2006, September). A mathematical model of the risk of nuclear terrorism. *The Annals of the American Academy of Political and Social Science*, 103–120.

Bunn, M., & Sagan, S. (2014). *A worst practice guide to insider threats: Lessons from past mistakes*. Cambridge, MA: American Academy of Arts and Sciences.

Cronin, A. (2015, March). ISIS is not a terrorist group. *Foreign Affairs*, 87–97.

Davis, J. (2014). Post detonation nuclear forensics. In P. Corden, D. Hafemeister & P. Zimmerman (Eds.), *AIP Conference Proceedings on Nuclear Weapons Issues in the 21st Century* (Vol. 1596, pp. 206–209).

Ferguson, C., & Potter, W. (2004). *The four faces of nuclear terrorism*. Monterey, CA: Monterey Institute for Strategic Studies.

Gates, R. (2014). *Duty: Memoirs of a secretary at war*. NY: Knopf.

Harmon, J. (2015, March). Disrupting the intelligence community. *Foreign Affairs*, 99–107.

Institute of Medicine. (2014). *Nationwide response issues after an improvised nuclear device attack: Medical and public health considerations for neighboring jurisdictions*. Washington, DC: National Academy Press.

Katz, J. (2006). Detection of neutron sources in cargo containers. *Science Global Security, 14*, 145–149.

Katz, J., Blanpied, G., & Borozdin, K. (2007). X-radiography of cargo containers. *Science and Global Security, 15*, 49–56.

Kelly, H., & Levi, M. (2002). Dirty bombs: Response to a threat, Senate Foreign Relations Committee (6 March 2002) and Scientific American. www.fas.org/ssp/docs/021000-sciam.pdf

Kissinger, H. (2014). *World order*. NY: Penguin.

Levi, M. (2007). *On nuclear terrorism*. Cambridge, MA: Harvard University.

National Research Council. (2002). *Making the nation safer*. Washington, DC: National Academic Press.

National Research Council. (2012a). *Terrorism and the electric power delivery system*. Washington, DC: National Academic Press.

National Research Council. (2012b). *Export control challenges associated with homeland security*. Washington, DC: National Academic Press.

Owen, J. (2015, May). From Calvin to the Caliphate. *Foreign Affairs 94*(3), 77–89.

Panetta, L. (2014). *Worthy fights: A memoir of leadership in war and piece*. NY: Penguin.

Schneier, B. (2013). *Carry on: Sound advice from Schneier on security*. NY: Wiley.

Tenet, G. (2008). *At the center of the storm: My years at the CIA*. NY: Harper.

Chapter 13
Nuclear Terrorism

Could 3 or 4 people smuggle into New York City the parts necessary to build a nuclear weapon?
Of course, it could be done.
[J. Robert Oppenheimer, 1946 Senate Hearing]

We judge that there is a high probability that al Qaeda will attempt an attack using a CBRN weapon (chemical, biological, radiological or nuclear) within the next two years. There is little doubt that al Qaeda intends to and can detonate a weapon of mass destruction on US soil.
[John Negroponte, US Ambassador to UN, 2003].

No threat poses as grave a danger ... as the potential use of nuclear weapons or materials by irresponsible states or terrorists.
[US National Security Strategy, 2015]

13.1 Improvised Devices

These statements on nuclear terrorism by our national leaders, academics and the press are part of our everyday life since 9/11. They give momentum to policy decisions, such as the Iraq war of 2003. Are these statement on nuclear terrorism on-target and correct or overblown? This chapter contains both sides of this debate. Time will tell.

Dirty bombs of cobalt-60 or other isotopes are easier to make than a nuclear weapon device. Dirty bombs can be harder to detect if properly shielded and easy to deliver in the dark of the night. Investigative journalists have shown that pseudo-attacks using depleted uranium have not been detected, as they penetrated barriers intended to keep nuclear materials away. The perception is that these attacks are easy to do against a public that is very fearful of radioactivity. This makes dirty bombs a low-hanging fruit for terrorists.

However, fifteen years have passed since 9/11 without radiological attacks. Has the radioactive silence been caused by the good works of the Department of Homeland Security? Have protections against radioactivity greatly improved? Or

D. Hafemeister, *Nuclear Proliferation and Terrorism in the Post-9/11 World*,
DOI 10.1007/978-3-319-25367-1_13

have we been lucky? Or is this harder to do than we thought? Terrorists have the advantage, since they dictate the timing, location and method of attack. However, it is important to remember that an attack is made up of many actions in a chain of events. This allows the defense some latitude to address each link in the chain of events. Thus far we are unscathed from dirty bombs or terrorist nuclear bombs, but this doesn't mean that it will always be so. It also true that it is possible that this threat has been over-hyped.

13.1.1 Improvised Attacks

This chapter addresses improvised attacks (NAS 2002, 2014):

- ***Stolen nuclear weapons*** can be used to attack cities. This is not trivial to do, but it is easier if insiders are involved to get around the computer access links that control them.
- ***Improvised nuclear devices* (IND)** could be fabricated from stolen special nuclear material (SNM) or weapon components. Plutonium is more difficult to work with and easier to detect, while high-enriched uranium is easier to hide and easier to work with since it emits far fewer spontaneous neutrons.
- ***Radiological dispersal devices* (RDD)**, or dirty bombs, are constructed with stolen radioactivity that does not fission. An RDD is easier to construct than an IND, but they are easier to detect if sensors are available in the neighborhood.
- ***Improvised explosive devices* (IED)** use conventional-explosives to attack facilities, such as government buildings, housing complexes and hospitals. If nuclear power plants are attacked, this could unleash the internal radioactivity in reactors and storage sites, spreading it in air, water and ground.
- ***Unattended aerial vehicles* (UAV)**, or drones give terrorists a cheap way to launch attacks. This is easiest for dirty bombs, which could involve a few kg of material in a drone. It is harder for nuclear weapons of 50–100 kg or more, but they can be carried in suitcases, backpacks, larger drones, small airplanes or cars.

13.2 Improvised Nuclear Devices

13.2.1 Mathematical Model of Risk

Matthew Bunn devised a pedagogical model to estimate the likelihood of a nuclear terrorist attack with an IND on a major city in the next decade. Bunn points out that estimates of IND attacks appear in the literature with rates between 1 and 50 % per decade, but with no mathematics. Bunn *bravely* devises a logical mathematical model, not to get an accurate answer, since the numerical parameters are not known, but to understand the various probability paths to do the estimate. Bunn uses the following parameters:

- Number of plausible nuclear terrorist groups = 2,
- Yearly probability of an acquisition attempt by a particular group = 0.3,
- Probability of choosing an acquisition attempt based on outsider theft = 0.2,
- Probability of choosing an acquisition attempt based on insider theft = 0.3,
- Probability of choosing to attempt to purchase black-market material = 0.3,
- Probability of choosing to attempt to convince a state to provide material = 0.2,
- Probability that an outside theft attempt will succeed = 0.2,
- Probability that an insider theft attempt will succeed = 0.3,
- Probability that a black-market acquisition attempt will succeed = 0.2,
- Probability that an acquisition attempt from a state will succeed = 0.05,
- Probability of being able to convert acquired items to nuclear capability = 0.4,
- Probability of delivering and detonating bomb once acquired = 0.7,
- Consequence of terrorist nuclear attack = $4 trillion.

Bunn obtained a 29 % probability for a terrorist IND in the next decade (2006–2016). He estimated the event would cost $4 trillion, for an average of $1.2 trillion/decade (0.29/decade × $4 trillion) or over $100 billion per year. Bunn concluded the following: "The uncertainties in estimating the risk of nuclear terrorism are large. But the very uncertainty of the danger highlights what we do not know—including the possibility that a major nuclear theft could be in the planning stages at any time. There is, in short, no time to lose." As companion reading, I suggest the reference by Len Weiss, "On fear and nuclear terrorism." Weiss points out that decades have gone by and there has been little in the way of terrorist nuclear weapons or dirty bombs. Weiss disagrees with presidents and academics, by saying that this is not a major threat.

13.2.2 IND Details

PAL. Most US nuclear weapons are controlled with *permissive action links* (PAL) that prevent the weapons being used without knowledge of the security code. Not all US weapons have PALs, nuclear weapons based on submarines or tactical weapons do not possess PALs. In addition, weapon ballistic weapons and gravity bombs require passage on a flight path to arm them. Nine nations have nuclear weapons. Some of these weapons are weakly protected or not protected by a PAL. By law the US can't be helpful by giving PAL secrets to other nations, but the US can be supportive in indirect ways.

No silver bullets. There is no single silver bullet to prevent vulnerability of states to stolen nuclear weapons or INDs. Nonetheless science and technology can develop ways for a multi-layered defense system with the following steps (NAS 2002):

- Evidence of terrorist group members and intentions.
- Materials accounting and physical security for weapons and nuclear-useable materials in their home state.
- Detection and interdiction of weapons and nuclear materials with technology and intelligence, especially across Russian and US borders.
- Detection and movement of IND's inside the US.
- Effective response to nuclear and radiological attacks.
- Attribution characteristics to identify the origin of weapons and nuclear materials.
- Detection sensors placed at national chokepoints, such as border transit points, global cargo-container ports, major airports, major bridges and tunnels, and other significant locations.
- FEMA emergency response plans to respond to nuclear and radiological attacks. These would include medical resources, field hospitals, public information on protection against radiation fallout, technical procedures for decontaminating people, land and buildings.

13.2.3 Acute Radiation Syndrome

An improvised nuclear device is more likely to cause radiation problems than a dirty bomb, which is more psychological than dangerous. Acute radiation syndrome (ARS) is the most immediate health affects from radiation exposure from a nuclear device. The Institute of Medicine (2014) based their scenarios on the explosion of a 10-kton improvised nuclear device in an urban environment. ARS appears after whole-body irradiation of more than one gray of physical absorbed dose (1 Gy = 100 rad, similar to 1 Sievert = 100 rem of biological dose equivalent). IND destruction can extend for miles, while RDD destruction extends only over many city blocks. There are three main types of symptoms: gastrointestinal (nausea, vomiting), hematopoietic (drop in the number of blood cells) and neurovascular (dizziness, headache, decreased levels of consciousness). ARS is monitored with blood counts and absolute lymphocyte counts. If victims are given cytokines in a timely manner after a lethal exposure, lives might be saved, depending on dose and other circumstances.

13.2.4 Response to IND Explosion

FEMA advises persons within a 50-mile radius to take shelter in a nearby building and await instruction from authorities. Many fission fragments are short-lived, it may be best to wait a few days for these to decay. A three-zoned approach is advised by the federal government to save lives, while minimizing risk to emergency response workers.

Severe: Persons in the severe damage zone (SDZ) probably won't survive the threats from radiation exposure, trauma from collapsed buildings or thermal radiation. The SDZ extends out to 0.5 mile for a 10-kton explosion, with doses of 1–10 rem per hour.

Moderate: The moderate damage zone (MDZ) extends from 0.5 to 1 mile from the explosion, with doses up to 0.1 rem per hour.

Light: The light damage zone (LDZ) extends 1–3 miles from the detonation. Almost all windows are broken, making glass injuries likely, with doses up to 0.01 rem per hour.

Dangerous fallout: The dangerous fallout zone can extend up to 25 miles downwind from the explosion, with doses up to 10 rem per hour. A lethal dose for 50 % of individuals is 400 rem. There would not be a great number of deaths from radiation beyond a mile or two from a 10-kton explosion, but this depends on plume dynamics.

The zoned approach sends first responders to the middle zone, the MDZ, of the 3 zones. The Institute of Medicine estimated that a 10-kton detonation in Washington DC would likely kill 45,000 immediately and 100,000 would be at risk. This clearly depends on scenario-dependent parameters. In addition, 320,000 persons would have serious injuries and 175,000 would have minor injuries. The response to this hypothetical attack shows a lack of acute care hospitals and housing for those fleeing the city. Evacuates probably would not have funds to pay for food shelter and transportation, while theft and other crimes can confuse matters. Reunifying families would be a difficult task. The US has experienced only a few widespread displacements in the past, the closest being that of Hurricane Katrina in 2005. Cities are usually rebuilt and repopulated after earthquakes, hurricanes and other natural disasters. This would be more difficult after a nuclear explosion because of radioactivity, many would conclude the city is gone and no one is coming back.

FEMA. The Federal Energy Management Agency (FEMA) has special teams to work in heavily contaminated environments. DHS also coordinates the Interagency Modeling Atmospheric Assessment Center that estimates the geographic area affected by radiation and other hazards. The US military has a strong tradition of support in a crisis. The US Northern Command has more than 18,000 military responders that could be initially activated to the IND site. The response force is called into action by state governors through their state National Guard and the federal Northern Command. Local officials call into action fire departments that have had special training for IND incidents. A quick response by fire departments would help reduce the impulse for citizens to leave the radioactive zones too quickly, inhale too much and make roads impassable. The integration of command-and-control between local, state and federal agencies would be a challenge after an IND attack. If a local emergency operations center is destroyed, there could be complications. The three levels of government that train the responders will need protocols to implement their actions.

13.2.5 Nuclear Forensics

If a nuclear explosion took place in the US, questions would asked (American Physical Society study).

- Was it a nuclear explosion?
- Was it Russian or American, because they have the most, or somebody else?
- Was the device stolen, sold, smuggled, lost, built by a rogue state or built by knowledgeable terrorists?
- What was the yield of the explosion?
- Will there be more than one explosion?
- Can we reduce the chance of another explosion?
- How did the weapon get there?
- Where did it come from and who did it?

It would be helpful to have answers to these technical questions. Prompt answers might allow diplomacy or militancy to have an impact on the situation.

- Did the device use uranium and/or plutonium?
- What are the isotopic ratios of U and Pu?
- Did the device produce 14 MeV fusion neutrons?
- What was the efficiency of the device (yield per kilogram)?
- What was the fissile mass of the device (yield/efficiency)?
- What type of reactor was used to make plutonium?
- When was the plutonium reprocessed?
- What type of enrichment technology was used?
- What type of weapon design was used?
- How sophisticated was the design?
- What is the relevant intelligence data?

A variety of devices are used to determine these technical answers: single and coincident particle detectors, induced radiation after neutron and particle absorption, mass spectroscopy, radio-chemical approaches, scanning electron microscopy, transmission electron microscopy, X-ray spectrometry, X-ray diffraction, thermal ionization spectroscopy, X-ray fluorescence, gas chromatography, mass spectrometry, optical emission spectroscopy, Kr and Xe abundances, and chain of custody approaches (fingerprints, fibers, pollens, dust, DNA, explosives).

Gamma ray spectroscopy of explosion samples can determine the fuel type (U or Pu) and determine sophistication. If the device is not yet exploded, it can be examined with sensors or X-ray machines. The Nuclear Emergency Support Team (NEST) from the national laboratories is "prepared to respond immediately to any type of radiological accident or incident anywhere in the world." It could deploy up to as many as 600 people for a radiological incident, but deployments usually do not exceed 45. These are not full-time NEST employees but they are on leave from the National Nuclear Security Agency to NEST in an emergency. NEST has been warned of over 125 nuclear terror threats and has responded to 30. All have been false alarms.

It is possible to determine the type of reactor that produced plutonium from its isotope ratios. If many reactors are of the same type, it would not be clear which nation housed that type of reactor. It can reveal whether light-water reactors or heavy-water reactors were used. Uranium isotope ratios give information on the enrichment type used and which uranium-ore type was used. The Department of Homeland Security regularly carries out exercises with government leaders to determine its ability to respond and to recover from a large terrorist act.

13.3 Dirty Bombs

The 9/11 attack showed terrorists being capable of doing bad things in a big way, such as detonating radioactive sources with conventional explosives in large cities. Explosives would vaporize or aerosolize radioactive materials to disperse them. Since nuclear radiation is often feared beyond its lethality. A *radiological dispersal device* (RDD) explosion in a large city would disrupt activities a great deal. How much radioactivity is needed to convince a person to leave home? Fear of radioactivity can be illogical. Often the most educated are the most terrified. It is for this reason that RDDs have been called the "new WMD," *weapon of mass disruption*, more psychological than lethal, but still effective.

Sources of radioactivity. Rather than importing radioactivity, it probably would be stolen domestically from the extensive nuclear industry. Some common isotopes are cobalt-60, cesium-137, strontium-90, plutonium-238 and americium-241. Radioactive sources, listed in Curies (Ci), can be obtained from a variety of commercial sites (LANL):

- Sterilization: Co^{60}, Cs^{137} (100–10,000,000 Ci),
- Radioisotope thermoelectric generator (RTG): Pu^{238}, Sr^{90} (5–500,000 Ci),
- Teletherapy: Co^{60}, Cs^{137} (2000–20,000 Ci),
- Blood irradiators: Cs^{137} (1000–2000 Ci),
- Radiography: Co^{60}, Cs^{137} (1–100 Ci),
- Well logging: Cs^{137}, Pu^{238}/Be, Am^{241}/Be (0.01–80 Ci),
- Brachytherapy: Pd^{103}, I^{125}, Cs^{137} (0.001–0.1 Ci).

Fewer radioactive sources: Accelerators and X-ray machines can be used in place of radioactive sources. If solar flux is sufficient, solar panels can replace radioisotope thermoelectric generators. US has licensed approximately two million radioactive sources, many are small, in the range of millicuries. The National Nuclear Security Agency has worked to reduce the number of idle radioactive sources and tighten protection and control of those that remain in use. NNSA estimates that there are more than 2300 buildings that house sufficient radioactivity to make a dirty bomb. NNSA installed additional security equipment at 650 of these critical buildings, enhancing physical security with magnetic door locks, window bars, steel doors, biometric authentication, motion detectors, video cameras and radiation monitors. NASA collected more than 36,000 surplus or unwanted large radioactive sources since 2007. NNSA worked with the Russian Federation to help them dispose of 800 radioisotope thermoelectric generators of 10,000 curies that are usually located in the open. There are two classes of materials that are at high risk. Category one materials cause permanent injury to a person in contact with them for more than a few minutes. Of the 554 buildings with category 1 materials, 273 of these had their security upgraded. Category two materials are harmful to a person exposed for several minutes to hours. Sensors were placed in critical spots and traffic junctions to detect large sources of radioactivity. License-management and regulations have been tightened. It is preferable that radioactive sources not be soluble in water or not in powder form, such as cesium chloride and strontium chloride. Global sources are often left in the metallic state to minimize dispersion in dirty bombs. Isotopes with shorter half-lives are preferred to lessen long-term contamination.

13.3.1 Dirty Bomb Scenarios

Scenarios can describe possible dirty bomb attacks on cities. The number of fatalities depends widely on released radioactive strength, population density within the plume, bomb efficiency in spreading, wind velocity, actions taken by those in the affected areas (to shelter, or to leave the area), and the strength of government response to evacuate, shelter and cleanup. The two general types of radioactive doses people would receive are internal, from inhalation and swallowing, and from radiation external to the body.

Inhalation. It is important to avoid inhaling radioactivity because small particles and insoluble elements can remain in the lungs for a lifetime. If one were to receive a dose of 10 millirem/hour (0.01 rem/h) at a constant rate, one would absorb 88 rem over one year, and considerably more over the succeeding years. It is for this reason that individuals near the explosion should stay sheltered to avoid inhalation.

External radiation. The decision to shelter or to leave should be determined by experts drawn to the scene. These deciders would compare the inhalation dose, obtained from leaving the area, versus the external source dose from sheltering in place. The figures below from Henry Kelly and Michael Levi of the Federation of American Scientists give radiation-dose maps for external radiation and not radiation from inhalation, which is scenario dependent. These figures display mortality rates in terms of cancer deaths per person, assuming the *populace spends all their time* in the plume area. If the annual residence time is the 40-h working day, then the fatality rate should be reduced to one-fifth the projected level (2000 of 8766 annual hours).

Cancer rates. If the cancer rate exceeds one cancer death per 10,000 of those who remain full-time within a specified region, EPA would recommend decontamination or destruction of the area. The Nuclear Regulatory Commission has a less strict standard for decontamination or destruction, which is triggered when one cancer death in 500 *fulltime occupants* is exceeded. For comparison sake, the US background rate is 3.5 mSv/year (350 mrem/year), which gives a total dose of 0.28 Sv (28 rem) over 80 years. The *continual* 350-mrem/year dose-rate gives a 1 % cancer death rate from natural radioactivity. This is based on the National Academy of Science's BEIR standard of 20 Sv (2000 rem) spread over many persons to cause one excess cancer death. The standard for the general public is different since it is the decision to no longer remain in the radiation field. This decision is triggered when natural levels are exceeded by 100 mrem/year, one-third the natural level of 350 mrem/year.

Clean-up. The decision to determine what and how to clean would be a complex choice of many options. Individuals in the plume should be decontaminated quickly by washing their skin and hair thoroughly and by disposing of contaminated clothing. Potassium iodide pills would not help since excess iodine is useful only to diminish the effects of iodine fission-fragments, which are not in a dirty bomb. Some isotopes, such as Cs^{137}, chemically bind to concrete, asphalt and glass and remain in place for a long time. Cs^{137} is observed on sidewalks decades after the 1986 Chernobyl disaster. Mechanical techniques, such as pressure washing, sand-blasting and vacuuming, would be needed for serious situations (Figs. 13.1, 13.2 and 13.3).

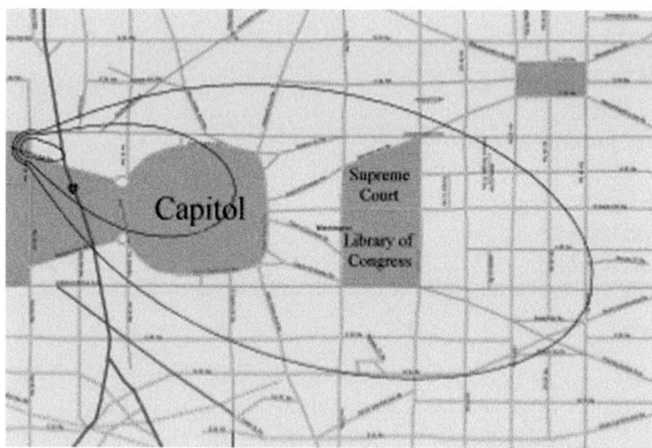

Fig. 13.1 Long-term contamination due to cesium bomb in Washington, DC. *Inner Ring* One cancer death per 100 people from radiation (5 % increase). *Middle Ring* One cancer death per 1000 people from radiation (0.5 % increase). *Outer Ring* One cancer death per 10,000 people from radiation (0.05 % increase); EPA recommends decontamination or distraction (Federation of American Scientists, H. Kelly and M. Levi)

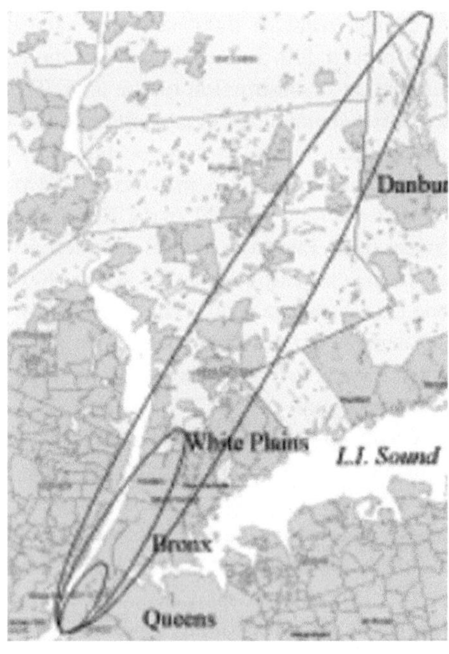

Fig. 13.2 Long-term contamination due to cobalt bomb in New York City; EPA standards. *Inner Ring* One cancer death per 100 persons from radiation (5 % increase). *Middle Ring* One cancer death per 1000 people from radiation (0.5 % increase). *Outer Ring* One cancer death per 10,000 people from radiation (0.05 % increase); EPA recommends decontamination or destruction (FAS, H. Kelly and M. Levi)

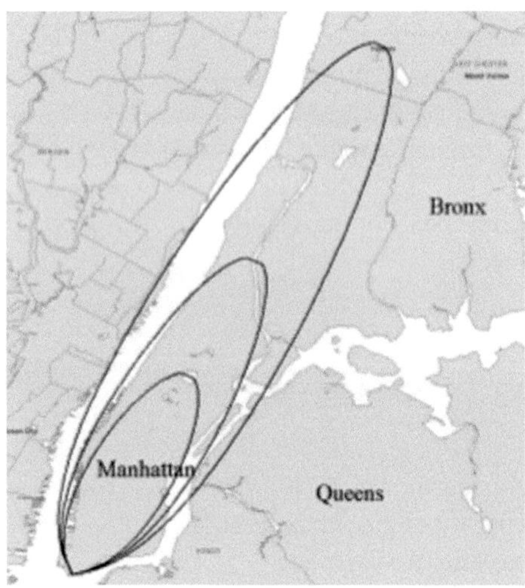

Fig. 13.3 Long-term contamination due to cobalt bomb in New York City; Chernobyl comparison. Inner *Ring* Same radiation level as permanently-closed zone around Chernobyl. *Middle Ring* Same radiation level as permanently-controlled zone around Chernobyl. *Outer Ring* Same radiation level as periodically-controlled zone around Chernobyl (FAS, H. Kelly and M. Levi)

13.3.2 Improvised Explosive Device

Improvised explosive devices (IED) do not contain radioactivity, but they can do tremendous damage with conventional explosives, TNT, fuel oil or fertilizers. The

9/11 attack on the World Trade Center and the Pentagon was done with an IED, using four airplanes. There is a long history of IED use in Vietnam, Northern Ireland, Afghanistan, Lebanon, Chechnya, Iraq, Israel, Palestine, UK, Spain, and the US (Oklahoma City bombing, 9/11, Boston Marathon). The explosive pentaerythritol tetranitrate (PETN) is used because it is very powerful, the most effective plastic explosive and it is hard to detect because of its strong molecular stability. Few PETN molecules escape into the air, so dogs and sensors find it difficult to detect. PETN has been incorporated into the underwear bomb, the print cartridge bomb and other devices. Since it only takes 300 g of the PETN to destroy an airliner, PETN can be a serious threat. There is some evidence that PETN might be surgically implanted into humans or pets, or placed in cameras and external hard drives.

IED device. A typical IED consists of five parts:

1. switch (activator),
2. initiator (fuse),
3. container (body),
4. charge (explosive) and
5. power source (battery).

If IEDs are intended to attack armored targets, they could penetrate with shaped charges or by propelling metal pieces. Robust armament can be overcome with depleted-uranium shells. If IEDs are intended to attack humans, it might contain ball bearings or nails. If the IED has not yet exploded, it would be useful to have a portable X-ray machine to examine suspicious objects in the field. IEDs might also carry chemical or biological weapon materials. IEDs might also contain incendiary materials to produce chemical reactions that spread fire rapidly. IEDs might be delivered by car, boat, airplane, drone, bicycle, cart, rocket, animal, suicide-bomber or surgical implant. IEDs might be triggered by a signal from a timer, wire, radio, cell-phone, infrared or by victim-operation. Because IEDs can be fabricated from so many types of components and directed by a variety of signals, they are difficult to defeat.

13.4 Nuclear Industry

Nuclear power produces 20 % of US electricity with 100 GW_e capacity since the 1990s, but growth has stopped and it is now slowly declining. Global nuclear power capacity has remained fairly constant over the past 5 years at 375 GW_e (2012). Seventeen nations have a higher percentage of nuclear power than the United States, although the United States has the largest capacity. French capacity is 63 GW_e, which is 78 % of its grid, while Japanese capacity was 44 GW_e for 35 % of its grid (uncertain after Fukishima). Nuclear power produces 16 % of global electricity. The growth of low-cost electricity from natural-gas, combined-cycle gas turbines removes nuclear power as a strong competitor. The continuing saga over spent fuel disposal forestalls building new nuclear plants.

Chapter 3 discussed reactor safety and nuclear accidents that could interest terrorists. When a nuclear power plant shuts down, it continues to produce heat (from beta decay) in the first minute at 7 % the heat rate it had when in operation (200 $MW_{thermal}$ for a $1GW_e$ plant). If there is no cooling water, a bare reactor core will start to melt destructively in a minute. The Nuclear Regulatory Commission requires nuclear power plants to be protected against a "design basis threat," that involves a ground attack by several external armed terrorists, aided by one inside collaborator. The 2002 National Academy of Sciences study, *Making the Nation Safer*, examined nuclear power vulnerabilities to potential terrorist attacks, which we summarize below.

Nuclear power plants:

Threat: Ground or air attacks.
Threat level: 100 US power reactors, 400 global power reactors.
Consequences: Reactor shutdown to core meltdown, with large release of radioactivity.
Probability of occurrence: High in the near-term.
Policy challenges: Stop airplane attacks that deliver energy on target.
Mitigation: Harden facilities, increase safety measures, and analyze vulnerabilities.

Research reactors:

Threat: Ground or air attack.
Threat level: High, 36 operating reactors.
Consequences: Little or no release of radioactivity likely.
Probability of occurrence: Unclear in the near-term.
Policy challenges: Provide security against all types of attacks.
Mitigation: Minimize fuel stored onsite.

Spent nuclear fuel [wet (78 %), dry (22 %)]:

Threat: Ground or air attack on ponds; dry cask attack less likely.
Threat level: High, at all commercial sides.
Consequences: Less danger with dry storage, pool fires much less likely.
Probability of occurrence: Rerack or remove spent fuel to reduce heat density in ponds.
Policy challenges: Stop airplane attacks that deliver energy onsite.
Mitigation: Finish re-racking to dry storage to lower heat density.

Radioactive waste:

Threat: Dirty bomb component.
Threat level: Very high waste is abundant worldwide and not well protected.
Consequences: Trivial, most available radioactive waste has low specific activity.
Probability of occurrence: High, readily available and few preventions in place.
Policy challenges: Train first responders.
Mitigation: Improve first responders and public education.

13.5 Drones

Drones are unmanned aerial vehicles (UAV) or remotely piloted aircraft (RPA). Drones are an extension of the war on terror to faraway places from a nation that is tired of war. Drones are a response to save both American lives and money. The close approach of drones can lead to more precise bombing, reducing collateral damage to innocent people and buildings. Drones can carry dirty bomb radioactivity or BW/CW materials. In 2015, a one-kg drone crashed into the White House, it could have carried dangerous materials. The most common role of drones is surveillance. Drones carry out surveillance in the visible, infrared and radar. Large drones attack lethally with hellfire missiles. Larger drones could carry an improvised nuclear device. Pilots sit at desks in Nevada or in Djibouti to operate the drones. The US withdrew its main contingent of 125 special operations advisors that work with drones from Yemen on 22 March 2015 because of the ISIS war. This crippled US intelligence obtained during drone attacks against ISIS.

Drone attacks have primarily been carried out by the US, UK and Israel. The number of drone attacks peaked at 120 in 2010, mostly in Pakistan, Somalia and Yemen. In Pakistan there were 400–500 drone strikes between 2004 and 2013. These killed more than 4000 people, including 400–950 civilians and 200 children. Drones have the advantage that they can hover more than a day, tracking individual persons and their cars with visible and infrared cameras, or other sensors. When the time is right, the Predator/Reaper launches Hellfire missiles to erase the target. By 2013 the US had 6000 drones, and is projected to spend $37 billion in the next decade. Some 50 nations seek drones, as well as local police and companies. The University of North Dakota has a degree program in unmanned aerial systems. "The sky is going to be dark with these things," says Chris Anderson, former editor of *Wired*. Sales have risen to 7500 per quarter, with prices as low as $200, compared to renting a helicopter at $1100/h. Current law appears to give little privacy protection from prying drones.

Management of military drones has shifted from the Pentagon to the CIA, as drones mostly do cloak and dagger surveillance, rather than traditional military missions. But in 2013 drones returned to the Pentagon as the role of Hellfire missile increased. The US took the lead on drones, but Russia and China and others are following. The US uses six criteria that must be fulfilled before an attack is made:

1. action must be authorized by US law,
2. threat must be serious and non-speculative,
3. threat must be urgent,
4. attack must be planned carefully to avoid civilian casualties,
5. US citizens under drone attacks have Constitutional protection and due process of law, and
6. there is no mention about US needing the consent of the nation under attack, such as Pakistan, Afghanistan and Yemen, but it seems they had quiet consent.

Drones have an advantage because they are difficult to shoot down by less sophisticated opponents in an asymmetric war. This advantage disappears as attacked nations learn to cope with drones. But, drones can make killing too easy and bloodless. This is a complex area, involving, as yet, unsettled issues of domestic and international law, as well as ethics and policy. The commander in chief is expected to use all options to defend the country, but the arguments are more complicated than this. The ease of drone attacks could change our values as a nation. DoD spent over $30 billion between 2011 and 2015 on drones. Manned aircraft and drones have few limits under arms control agreements. The number of allowed airplanes was capped in the Conventional Forces in Europe (CFE) Treaty, which regulated military aircraft between the Atlantic Ocean and the Ural Mountains. The New START treaty monitors the number of US and Russian strategic bombers. It appears it will be difficult to control drones with international treaties, because they are inexpensive at less than $1000. Some rules of the road could be introduced to reduce escalation to war.

13.5.1 Drone Technology

Reaper, formely predator, is built by General Atomic with a length of 8 meters and a wingspan of 15 m. It flies to an altitude of 8000 m, with a range of 730 km, staying aloft for 40 h. Its top speed is 220 km/hour and it costs, fully loaded, about $40 million. The Predator is a reconnaissance vehicle, using radar, visible, infrared and SIGINT electronic eavesdropping devices. It also is an attack vehicle, carrying laser-guided Hellfire missiles of 100 pounds that can penetrate armored tanks.

Golden Hawk, built by Northrop Grumman, is the most expensive and capable UAV. It is jet powered, staying aloft for 36 h. It obtains quality intelligence, surveillance and reconnaissance, with radar, infrared and visible, generating time-critical targets. Its synthetic aperture radar has a resolution of 1.8 m over 10 km^2. The Golden Hawk is expected to replace the classic U-2 reconnaissance aircraft and the Orion submarine hunting aircraft. It can survey 100,000 km^2 in a day.

ScanEagle, built by Boeing, carries visible and IR cameras on 3-m wings and stays aloft 24 h at 6-km altitudes. Its 1.5 kW engine gives a speed of 90 km/h at a cost of $100,000.

Hunter has 7-m wingspan and flies for 12 h. It is used to gather imagery at 5 km elevation in support of US Army ground troops or immigration officers along the border.

Shadow was built for the US Army. It has a wingspan of 3 m and flies for 5 h at 5 km altitude to obtain imagery.

Dragon eye fits into a backpack and is used in urban environments. It has a wingspan of 1 m and can stay airborne at 100 meters with a range of 5 km for up to an hour for surveillance.

Toruk AP10 can be purchased from *Home Depot* for $700. The 40-cm diameter drone is a quadcopter with four propellers. It flies at 300 m altitude at 25 m/s velocity for 25 min with a range of 15 km. The 16-Mpixel camera rotates and communicates with WiFi. It is rainproof and its autopilot can return it to home.

Taleuas develops 60 W of solar power. It flies at 64 km/h and comes in two sizes, 4 and 2.7 kg. The *Puma* solar drone flew for 9 h, compared to 3 h for the battery version.

Phantom 3, built by Da-Jiang Inovations of China, has 12 Mpixel photos, motorized mount, high-definition video, GPS, and no-fly zone hardware, all for $1000. It crashed into the White House by accident in 2015.

Dromida Kodo, quadcopter, 15 cm, gyro stabilized, 3-axis gyro 3 accelerometers, photos (1280 × 960 jpeg), movies (720 × 480 avi files), 2.4 GHz radio, battery and memory chip, 10 min, $60.

Micro satellites, as small as 8–15 cm, are being built. Experimental satellites called *sprites* have gotten smaller than this, since satellites do not need propulsion. The Space X rocketry firm launched the Cornell-University-built micro satellites, with a mass of 5 g. Sprites can carry solar and magnetic sensors, a gyroscope, and communications equipment in a 3 cm square flat object.

E-Fan, built by Airbus, is a two-seat aircraft, flown 100 % with lithium batteries. The two 32-kW electric motors fly for 90 min, enough to cross the English Channel.

13.5.2 Legality of Drone Attacks

Privacy. Because drones do not have a pilot, there is concern that their use might be more frivolous when compared to airplanes with a human onboard. Individuals in the US have few privacy protections from UAV surveillance. The US Supreme Court held that individuals do not have the right of privacy from police observation from public airspace. The American Civil Liberties Union warned of a nightmare scenario in which the police use mobile phone tracking data with video data from drones to make a database of peoples daily movements. Since drones are stealthier and can get closer than police cars, there could be more privacy restrictions placed on drones.

Authority to kill. The Justice Department stated that using drones to kill an American citizen can be justified if government officials determine the target poses "an imminent threat of violent attack." A Justice Department white paper expands upon this by stating that the condition that an individual presents an imminent threat of violent attack against the United States does not require the United States to have clear evidence that a specific attack on US persons and interests will take place in the immediate future… instead an informed high-level official of the US government may determine that the targeted American has been recently involved in activities posing a threat of a violent attack and there is no evidence suggesting that he renounced or abandoned such activities. Senator Rand Paul asked a formal question of the government; "Does the President have the authority to use a weaponized drone to kill an American not engaged in combat on American soil?"

Attorney General Eric Holder answered in the negative. Some think this answer was Obama's preference, and not the law. The Pakistani military allowed drone attacks in Pakistan, but the Pakistani public has become increasingly unhappy over this.

Constraints on drones. The Federal Aviation Agency (FAA) was requested to develop restrictions on drones. The FAA has allowed use of drones by police, universities and some commercial applications. Amazon would like to deliver their packages with drones. Others would like to deliver pizzas, while farmers use drones to investigate the status of their far-flung lands. The FAA and State governments have restricted drone use near airports, public stadiums and at higher altitudes. The FAA will try and follow a path that allows some commercial use, but with adequate safety and privacy protections. The FAA is building a test site for drones of less than 55 pounds (25 kg), operating at altitudes below 400 ft (120 m). The United Nations is considering adding drones to its conventional weapon registry. GPS chips in drones might prevent their operation in restricted zones. FAA released its draft regulations on drones of less than 25 kg on 18 February 2015. Agriculture and commercial interests will want looser regulations, compared to the FAA provisional regulations listed below.

- visual contact maintained with drones at all times, preventing package delivery,
- drone operators must be licensed,
- drones will not be certified by the US,
- drones will only fly below 150 m (500 ft) altitude,
- drones will fly only in the daytime,
- drones will not fly over crowds of people,
- looser regulations are being considered for smaller drones.

Control of armed drones. The Missile Technology Control Regime (MTCR) is a voluntary arrangement of 34 nations, intended to reduce the spread of ballistic missiles. It was amended in 1992 to cover unmanned aerial vehicles. Category-1 UAVs travel more than 300 km with a payload of more than 500 kg. The Reaper and the Golden Hawk are Category-1 UAVs. The US sold some Category-1 UAVs to UK, Italy and South Korea. There is diplomatic effort to reduce the 500 kg threshold, while at the same time designing new Category-1 drones so they cannot carry weapons.

Problems

13.1 **Oklahoma City bombing**. What actions did Timothy McVeigh and Terry Nichols do to gather explosive materials, deliver the weapon and escape?

13.2 **HEU easier than Pu**. Why is building plutonium weapons difficult for new nuclear nations and terrorists? Why is HEU easier to work with?

13.3 **IND, RDD, IED**. What are the differences between these 3 types of weapons?

13.4 **Estimating attacks**. Examine Bunn's model. Which 3 parameters are most uncertain?

13.5 **PALs for Pakistan**. Why should the US refrain from giving PALs to Pakistan? How might the US be helpful but not too helpful?
13.6 **Mton to MWatts**. Give details on down-blending and monitoring.
13.7 **Leave after IND/RDD attack?** What factors must be considered when deciding to leave the basement?
13.8 **US radioactive terror disasters**. What have been the US's biggest radioactive terror disasters?
13.9 **Nuclear forensics**. How would nuclear forensics be helpful?
13.10 **Radioisotope thermoelectric generator**. What technologies could replace nuclear RTGs? When are nuclear RTGs the only likely technology?
13.11 **Avoided chemistry**. What types of chemicals or materials should be avoided to prevent dirty bomb construction?
13.12 **Radioactive sensors**. Where would you deploy 500 radioactive sensors in the US?
13.13 **FAS dose maps**. Which general types of radioactive sources are considered in these maps and what did they not consider?
13.14 **EPA versus NRC standards**. Which would you favor and why? Give both sides of the argument.
13.15 **IED parts**. What are the 5 parts of an IED? What are variations for each part?
13.16 **Advantage offense**. Why is the defense at a disadvantage to the terrorist offense? How might this disadvantage be (partially) overcome?
13.17 **Design basis threat**. What is the design basis threat for reactors? If you wanted to shut down the industry, how could you modify the design basis threat? Is this fair or not?
13.18 **Spent fuel ponds**. The lesser radioactive rods are now placed in air-cooled storage. What are two reasons this is useful? Can this be sufficient to prevent a fire, explain.
13.19 **Regulated drones**. What are useful regulations for drone commercialization?
13.20 **Amazon delivery**. Should Amazon be allowed to deliver books by drones?
13.21 **Drone sensors**. What sensors could be added beyond visible, infrared and radar sensors?
13.22 **Drone privacy**. What rules should be applied? When must the police obtain a writ of habeas corpus from a court? When can the police or FBI do a general search over the entire population?
13.23 **Presidential authority**. Does the president have the authority to authorize the killing of a US citizen if there is evidence that this citizen is about to attack US national security?

Seminar Nuclear Terrorism

Acute versus low dose radiation, Bunn rate-calculation by other approach, clean up after RDD, detect IND in a package, drone attack advantages versus. disadvantages, drone privacy rules, evacuate versus shelter after nuclear explosion, FAA drone restrictions, FEMA plans, fuel-pond crowdedness, IED designs, inhalation vs. exterior radiation, Nuclear Emergency Search Team approaches, terror scenarios with IND/RDD/IED/drones, permissive action link, Transportation Security Administration rules to prevent IED, useful insoluble radioactive sources, US government right to kill US citizens with drones.

Bibliography

Allison, G. (2004). *Nuclear terrorism: The ultimate preventable catastrophe*. NY: Henry Holt.

APS/AAAS Nuclear Forensics Group. (2008). *Nuclear forensics*. Washington, DC: American Physical Society.

Bunn, M. (2006, September). A mathematical model of the risk of nuclear terrorism. *American Academy of Political and Social Science*, 103–120.

Davis, J. (2014). Post detonation nuclear forensics. In P. Corden, D. Hafemeister, & P. Zimmerman (Eds.), *AIP Conference Proceedings on Nuclear Weapons Issues in the 21st Century* (Vol. 1596, pp. 206–209).

Ferguson, C., & Potter, W. (2004). *The four faces of nuclear terrorism*. Monterey, CA: Monterey Institute for Strategic Studies.

Gubrud, M. (2014). Stopping killer robots. *Bulletin of Atomic Scientists, 70*(1), 32–42.

Institute of Medicine. (2014). *Nationwide response issues after an improvised nuclear device attack: medical and public health considerations for neighboring jurisdictions*. Washington, DC: National Academy Press.

Katz, J. (2006). Detection of neutron sources in cargo containers. *Science Global Security, 14*, 145–149.

Katz, J., Blanpied, G., & Borozdin, K. (2007). X-radiography of cargo containers. *Science Global Security, 15*, 49–56.

Kelly, H., & Levi, M. (2002). Dirty bombs: Response to a threat, senate foreign relations committee (6 March 2002) and Scientific American (www.fas.org/ssp/docs/021000-sciam.pdf)

Kramer, D. (2014). Are the makings of a dirty bomb in your neighborhood? *Physics Today*, August 22–23, 2014.

Levi, M. (2007). *On nuclear terrorism*. Cambridge, MA: Harvard University.

Lin, P., & Abney, K. (2011). *Robot ethics: The ethical and social implications of robots*. Cambridge, MA: MIT Press.

National Research Council. (2002). *Making the nation safer*. Washington, DC: National Academy Press.

Schlosser, E. (2015, 9 March). Break-in at Y-12, New Yorker, 46–69.

Sherman, J. (2005). The drone wars. *Bulletin of Atomic Scientists, 61*(5), 28–37.

Singer, P. (2009). *Wired for war: The robotics revolution and conflict in the 21st century*. NY: Penguin.

Weiss, L. (2015, March). On fear and nuclear terrorism. *Bulletin Atomic Scientists*, *71*(2), 75–87.

West, G. (2015, May). The sky's the limit—If the FAA will get out of the way. *Foreign Affairs*, *94* (3), 90–97.

Chapter 14
Cyber Terrorism

The US has crossed a Rubicon in cyber space. It launched a cyber attack that has caused another nation's sensitive equipment to be destroyed. It has legitimized that behavior. In the process it delayed the Iranian nuclear program by many months, but only by months. And, because Stuxnet escaped into the hands of hackers throughout the world, the US has also launched what is likely to be a cyber boomerang, a weapon that will someday be used to attack some of America's own defenseless networks.

[Richard A. Clarke, former National Coordinator for Security, Infrastructure Protection, and Counterterrorism under four presidents, *Cyber War*, 2010].

14.1 Cyber Introduction

Cyber and the movies. The Stuxnet attack on Iranian centrifuges is a well-known example of a cyber attack, not yet a cyber war. As we read below, Stuxnet was delivered to Iran's Natanz centrifuge factory on a memory stick, raising and lowering the rotation frequencies close to several resonant frequencies, destroying a thousand centrifuges.

The film industry is paying attention to these cyber events with three recent movies. The *Imitation Game*, a movie about Alan Turing's creation of the first computers to crack the German Enigma code. Seven decades later, The *Imitation Game* evolved to *Citizenfour*, a documentary about Edward Snowden's revealing the National Security Agency's (NSA) programs to read domestic emails and letters, and listen to phone calls. The government stated they examine the address list without FISA Court approval, but they must get FISA Court approval to examine the contents. To add to this excitement, recall Edward Snowden's revelations on NSA tapping German Chancellor Angela Merkel's phones and e-mails, making her furious. Later we find out that the German government has done the same. In November 2014, North Korea attacked the cyber networks of Sony Pictures

D. Hafemeister, *Nuclear Proliferation and Terrorism in the Post-9/11 World*,
DOI 10.1007/978-3-319-25367-1_14

Entertainment because its film, *The Interview*, gave a negative treatment of Kim Jong-un, North Korea's leader.

In parallel to those events, the Chinese People Liberation Army (unit 61398) hacked US commercial firms. From 2006 to 2012, there were at least 18 attacks on information technology firms, 16 on aerospace, 12 on satellites and telecoms, 10 on scientific research and 8 on energy. Today's paper told us that hackers, employed by Syria's Bashar al-Assad, sent siren-song emails with malware to lonely rebels. Posing with fake feminine pictures, the hackers drained the rebel computer memories, learning about "tactical battle plans and troves of information about friends and fellow fighters." Tomorrow's paper will tell us about new malware events.

Bulk collection of metadata of emails, and information from banks, credit-cards and phone calls and letters is legally supported under US law with the *Foreign Intelligence Surveillance Act* (FISA) of 1978 and the *Patriot Act* of 2001. The government stated they examine the messages of friends and allies only when there is a compelling national-security reason. The private sector has much more cyber prowess than the central government, and for this reason it stores much of the information. There are differences of opinion between the government and information technology companies, as the government does not want robust encryption to be used since this blocks government access to the information.

According to published sources, the US spent $5 billion in 2014 on cyber offense (to attack others) and cyber defense (to protect us), as part of the $80 billion intelligence budget. There have been diplomatic discussions to establish a code of conduct and rules of engagement when rival nations use cyber to attack, but with little progress so far. It is generally believed that theft of trade secrets is not considered an act of war. A major attack on military or government infrastructure would probably be considered an act of war. Using deterrence theory, countries are establishing attack modes as a response if cyber-units of other nations attack. The cyber version of MAD (Mutually Assured Destruction) is not nearly as dangerous as the nuclear version of MAD, but it could be part of a bigger attack that could lead into a wider war. We should be prepared to do

> what is reasonable and proportional to halt or retaliate against to constrain cyber attacks on the nation,

but international rules of the road would be far preferable. What is reasonable diplomacy has become difficult. The evolution of cyber-defense is similar to the evolution of defensive programs for nuclear, chemical and biological threats, namely keep research going in the name of strengthening the ability to retaliate. There is more to the argument than that, as each side wants to keep up in the various races. In the meantime, computer networks are being hardened with fewer entry points and nodes, more encryption of necessary subroutines, stronger barriers, and random-number approaches. Cyber-security is a broad subject, it includes Internet governance, Internet freedom, online privacy, cyber espionage, cyber crime, and cyber war.

14.1.1 Total Information Awareness

Shortly after the *9/11 attack*, Admiral John Poindexter and Brian Hicks encouraged the Pentagon to establish the *Total Information Awareness* program. Their idea was (1) to utilize the immense analytical power of computers which had become much cheaper and (2) to access the stored data, which also had become much cheaper. Using this capacity to examine the big data sites, they searched for correlations (not causality) in the data. This approach discovers past actions and personal contacts between a person's friends, citizens and foreigners, to determine what individuals might do in the future. This determines the probabilistic chance that citizen X might develop shoe bombs to affect our national security. Of course, it would have problems with false positives and the defaming of innocent citizens. The Congress, fearing the loss of privacy of its citizens, outlawed deployment of *TIA* on US citizens and reduced *TIA* funding. But tools are needed to track terrorists abroad. TIA research went forward. These ideas are not unique to the government, as the trove of data collected in the commercial sector by *Google, Facebook* and the rest of the *IT* industry is being examined for commercial advertising. What we do with our credit card and our social media like *Facebook* is used to determine which of us might need a new car or a new life.

It is well known that Netflix uses our past movie viewing habits to recommend other movies to us. Google uses GPS location data to give traffic advice in real time. The power of using correlations to make associations from mega amounts of metadata has been dramatically shown in a recent article in *Science* by de Montjoye et al. (2015). About 60 % of payments are made with credit cards. Among Americans, 87 % consider credit card information moderately private or extremely private, as compared to 68 % for health information and 62 % for location information. The MIT researchers obtained credit card information for 1.1 million individuals over three months. The information includes the location in the US, the date, the price, but not the user identification. Knowing only locations and date for four events allows 90 % identification. Knowing the price allows essentially all the individuals to be identified. They have shown that it is significantly easier to identify women (93 %) as compared to men (89 %). Wealthy people are easier to identify than poor: high (93 %), medium (91 %) and low (88 %). The appendix to the article gives technical details.

Edward Snowden. The applications and ideas of TIA were revealed on 6 June 2013, when alarming details about the security leaks of Edward Snowden, a National Security Agency (NSA) consultant, were published in the *Washington Post* and *Manchester Guardian*. These leaks described NSA's *PRISM Program*, which tracks global email, web searches and other Internet traffic and NSA's *Boundless Informant Program*, which tracks global phone calls. Other programs were discussed:

- *PrePol*: Predictive Policing Program.
- *Microtargeting*: Political targeting of centrist voters.
- BehaviorMatrix and *CrowdVerb*: Links on-line profiles with voters.

US response to Snowden. On the same day, President Obama acknowledged that phone-call and email lists of US-citizen messages are kept, but he stated, they are not accessed for US domestic citizens until evidence is presented to the FISA court (which mostly says yes). We will discuss these issues more at the end of this chapter. Snowden's identity was revealed four days later on June 10 by the press. The US Department of Justice charged Snowden with violating the Espionage Act with a maximum of 30 years in prison on June 14 and revoked his passport on June 22. Living abroad at the time, Snowden avoided extradition to the US from Hong Kong by flying to Moscow on June 23. Snowden spent 6 weeks in Moscow's Sheremetyevo Airport, applying for asylum in 21 countries. Russia first granted him a one-year temporary asylum on August 1, and a year later Russia issued a three-year residency permit, allowing him to travel freely within Russia, and to go abroad for no more than 3 months. The documentary *Citizenfour*, directed by Laura Poitras, gives you a ringside seat to this drama. Snowden is accompanied by *Manchester Guardian* journalist, Glenn Greenwald. Julian Assange, the Editor-in-Chief of WikiLeaks plays a small role. Assange and Chelsea Manning had previously released a half-million classified State Department and Pentagon cables and documents on the Afghanistan and Iraqi-II wars. The tapping of their e-mail and phone calls angered German leader Angela Merkel, the European Union and other world leaders. Policy in this area is in a state of flux, much needs to be negotiated within the US and within the EU and international communities. In May 2015, a lower court disapproved the collection of domestic communications, stating that the Congress must act to give this permission. The two polar positions are indefensible: On the one hand, the total loss of privacy, as dramatized in George Orwell's novel *1984*, is certainly not acceptable. On the other hand, establishing too-tight limitations on the nation's ability to defend itself from unpredictable terrorist attacks is not desirable. We will return to this conundrum at the end of this chapter.

Google glass privacy. *Google Glass* is a wearable head-mounted computer display in the form of a pair of glasses. It displays information in a smartphone-like, hands-free format that allows the owner to observe computer memory while out and about. Google Glass also has a camera and a microphone that allows it to film and transmit movies of conversations with your friends and coworkers. Certainly Google Glass can be used surreptitiously to obtain relevant data for court trials, and it could be misused. Various applications are being developed for facial recognition, exercise, photo manipulation, translation, and sharing on social networks. It received one of *Time Magazine's Best Inventions of 2012 Awards*. This very clever device is selling for $1500 and became available to the public in May 2014. In January 2015 Google stopped selling it because of privacy concerns, the public didn't want to be subject to being photographed and recorded at all times. Google Glass is now sold only to certified partners for commercial trials in places like hospitals and factories. This is not the last chapter in this story.

Identifying a face is more difficult than detecting the face. When comparing two photos, humans can identify the correct pairings 98 % of the time. The DeepFace computer program does about the same at 97.35 %. However, it is one thing to identify a known person in a crowd, and it is more difficult to identify an unknown person in a crowd.

Similarly, voice imprints can be used for identification. Your voice can be your password, if you are expecting the call, but it is much more difficult to identify the voice from someone you are not expecting. *Nuance Communications* has collected 35 million voiceprints in the past two years. I have found the latest version of *Dragon Dictate* marvelous as it correctly spells cities, correctly uses the word "too" and uses appropriate commas as in "I, too, am going to Wauwatosa and Vladivostok."

The world we live in. Micro sensors are becoming cheaper and smaller. In the health area, the *FitBit* costs $95 to measure energy, distance, steps taken, elevation, and sleep quality. Spend $199 for *Basis* and it also measures skin temperature and perspiration. Spend more money and you can measure breathing, heart activity (EKG), blood glucose, blood pressure and posture. It is well known that the GPS on your Smart Phone gives your location. It is also well known that the webcam on your computer can be hacked so others can watch and listen to you, so put tape over your camera and turn off your computer microphone if you are bashful.

If you don't want to have your data analyzed by market forces, you can throw away your *Smartphone* and use only dumb phones. You should also use ghost emails, by having several email accounts, some for friends, some for work and some for your various interests. Your computer is available for hacking and examination if you allow cookies to enter your computer, but then most of us do allow this. What have we lost? You can reject cookie provisions in each of your dealings. As a remedy, you might use private browsing modes that prevent browsers from storing information from web sites during browsing sessions. Free browser extensions are available, like *Adblock Plus,* which blocks advertisements and disables online tracking. You might use multiple browsers, *Goggle Chrome*, *Apple Safari* and *Mozilla Firefox*, each for a different task. Lastly, if you pay by cash, there will be no record of your doings, but then you would lose *frequent flyer miles*. At the end of this chapter, we describe more defensive practices than these. The 20 June 2013 *New York Times* reminds us that Hester Prynne in the *Scarlet Letter* warned us in 1850 that "We must not always talk in the market place of what happens to us in the forest."

Internet sectors. The Department of Homeland Security developed a list of 18 critical infrastructure sectors for the US. All these sectors are heavily dependent on the Internet and are vulnerable to cyber war attacks:

Agriculture and food
Banking and finance
Chemical
Commercial facilities
Communications
Critical manufacture
Dams
Defense industrial base

Emergency services
Energy
Government facilities
Healthcare and public health
Information technology
National monuments and icons
Nuclear reactors, materials and waste
Postal and shipping
Transportation systems
Water

Byte units. The amount of information in a file can be listed in kilo bytes, and national totals can be in tera and peta bytes. One byte is 8 bits (0 or 1), which is a letter or number. A kilobyte is 10^3 byte is the information on a book page. A gigabyte is 10^9 byte which can be a compressed 2-h film. A terabyte of 10^{12} byte contains all the books in the US Library of Congress. A petabyte of 10^{15} bytes is 20 % of annual US letters. In 2015, the fastest US computer, Oak Ridge's *Titan*, obtained 27 petaflops/second, which is exceeded by China's *Milky Way* at 55 petaflops/s.

A zettabyte of 10^{21} byte was the electronic information available in 2010. Newspapers don't like zettabyte, so they use the term trillion gigabytes. In 2013 there was 4.2 zettabytes in memory, and by 2020 this is projected to rise to 40 zettabytes. The *New York Times* of 20 June 2013, projects that the 2020 data will be divided geographically in this way: China (8.6 zettabyte), US (6.6), Western Europe (5.0), India (2.9) and rest of world (16.9).

14.2 Stuxnet

Malware is malicious computer software that can destroy data, corrupt data, or steal data. By data, we mean zeros and ones, so malware is electrons killing electrons. But more than that, malware can destroy physical objects that have mass, conduct electricity, or affect sensors that allow people to live. A malware application (*app*) usually has the capacity to transmit and multiply worms, which are stand-alone mini-programs that replicate. Worms conceal themselves among other programs and are hard to locate. We live in a vulnerable society of conventional and nuclear power plants, electric grids, transportation networks, communication networks and financial marketplaces, which are made up of physical objects that are more complex than zeros and ones. We are aware of hacking financial information, but we are much less aware of malware's potential to kill people and destroy property.

The Stuxnet computer worm was detected for the first time in 2007. In Chap. 10, we discussed the 10,000 operating centrifuges in Iran and the attempt to prevent them from making nuclear weapons. In 2010, Stuxnet became part of this policy discussion by changing the operating rotational-frequency of these centrifuges, to be near the resonant frequencies, destroying 1000 centrifuges at Iran's Natanz facility. Not only did Stuxnet destroy centrifuges, but this fact was hidden by sending false signals to the owners, telling them that their centrifuges were running normally. This watershed event proved malware can destroy physical property in a big way. This kind of attack was expected by experts, as many realized it was possible, but they preferred not to think about it. Stuxnet was not perfect since it did not remain in the shadows, as the link between Stuxnet and Iran's centrifuges was established in 2010.

How did Stuxnet operate? (The U.S. government has never confirmed the existence of Stuxnet or its participation to sabotage Iranian centrifuges. The following account is based on public sources. The malware attacked a window-based Microsoft operating system that used Siemens controlling systems, which are called *supervisory control and data acquisition systems* (SCADA). Much has been written on Stuxnet beyond the *New York Times*. Two excellent general references on cyber war are those by Paul Rosenzweig, *Thinking about Cyber Security* (*The Great Courses*) and Richard Clarke, *Cyber War* (*HarperCollins*). Both authors had high-level positions in the government, Rosenzweig in the Department of Homeland Security and Clarke in the White House under four presidents. They know about defending the country against a cyber war.

There are at least two phases to an attack, first it must be delivered and second it must operate to destroy and hide. Computers that handle sensitive data or control sensitive equipment are usually not directly connected electronically with the Internet. This connection is avoided by having an *air gap* between the Internet or research computers and the computers that control the SCADA. To circumvent the air gap, one has choices: An insider spy can insert the malware on the sensitive side of the air gap. This was probably done with a memory thumb drive, entering the Siemens SCADA. The developers of Stuxnet had considerable information about Siemens SCADAs and the situation at the Natanz plant. Stuxnet was designed to enter the Natanz plant by using the specifics of the site. It has been claimed that Stuxnet infected 100,000 computer servers with the Stuxnet virus, 60 % of them in Iran. Stuxnet was designed to refrain from attacking the non-centrifuge SCADAs, because its presence would have blown its cover. Stuxnet knew where to attack the holes, nodes and backdoors in the Siemens systems. Stuxnet was capable of counting the number of centrifuges in cascades and much more. Stuxnet tested to learn if it was at the proper Natanz facility and not elsewhere.

Stuxnet had several ways it could spread throughout the system. Stuxnet was capable of sending out signals saying everything is okay, at the same time it was changing the operating frequency to destruction. Stuxnet spun the centrifuges near 1000 Hz to destroy the rotors and near 2 Hz to destroy the vertical enrichment tubes. It has been reported that at least 1000 IR-1 centrifuges were destroyed. The *New York Times* reported that Israel and the US cooperated to produce Stuxnet. According to Rosenzweig, Stuxnet contained footprints that are traceable to Israel. One of these was the number 19790509, which represents 9 May 1979, a date when

Iran executed a prominent Iranian Jew. Rosenzweig and the *Washington Post* pointed to the US/Israeli *Flame* program, which collected intelligence in preparation for the Stuxnet attacks. According to Rosenzweig, the US Department of Homeland Security practiced Stuxnet by destroying a diesel generator in Idaho in 2007.

Why was Stuxnet done? It was done because of dangers posed by the Iranian nuclear weapons program. Was it successful in softening Iran? Many in Iran would like to see an accommodation with the West to return their country to normal commerce. We have seen pleasant pictures of Secretary of State John Kerry sitting next to Iranian Foreign Minister Mohammad Javad Zarif and Baroness Catherine Ashton of the European Union, as they negotiated the numbers of centrifuges, sizes of centrifuges, stockpile sizes of 5 and 20 % enriched uranium, the lifetime of the treaty, reactor modifications, as well as complex monitoring issues. Stuxnet can be thought of as a shot across the bow, letting Iran know that strong forces are aligned against them. Was Stuxnet successful in coaxing Iran to accept negotiations with the P5+1, or was Iran just grasping for time, to later refuse treaty language at the last critical moment? This is a major power ball game between political parties. Israeli Prime Minister Benjamin Netanyahu speaks before the House of Representatives in a break with protocol, followed by the Saudi King snubbing a meeting at Camp David. The Congress is ready to vote on sanctions to kill the negotiations. This raises political pressure on all parties, which can nudge Iran into compliance or something bigger in response.

Is Stuxnet a big deal? This chapter begins with a quote from Richard Clarke, former National Coordinator for Security, Infrastructure Protection and Counterterrorism in the White House under presidents Ronald Reagan, George H. W. Bush, Bill Clinton and George W. Bush. Clarke tells us that the United States crossed the Rubicon of cyberspace with its Stuxnet attack on Iran. Opponents of the US would argue that this legitimizes cyber war attacks on similar US infrastructures. Perhaps this behavior was to happen anyway, we will never know. In the meantime we have to see if we can make cyberspace more robust against attack and establish rules of the road so nations can take steps to coexist peacefully. This is a tall order.

The very fact that the Hiroshima and Nagasaki explosions took place gave the Soviets confidence they could make nuclear weapons. Will the existence proof of a successful Stuxnet give others confidence to establish a cyber war program? Will cyber-war attacks be a major weakness in our Internet-based society? Clearly we have to design stronger barriers around our SCADA controllers. Each industry will have its own vulnerable places that could be enflamed quickly and irreversibly. Clearly we all want to take advantage of the Internet and the strengths of cybernetics, but we have to protect our assets and make concessions on some issues, which won't be pleasant or easy. To establish a global norm on cyber behavior is more difficult than to establish a global norm on nuclear weapons. There is much ground to cover if Earth is to achieve cyber stability.

14.3 Cyber Details

WWW size. As of 2012 there were more than 2.5 billion Internet customers, which is 35 % of the world population. Each minute of the day, there are 30 million Google searches. About 200 million e-mail messages are sent each minute, or 80 quadrillion messages a year. The attachments to these e-mails could contain a worm virus. The Library of Congress digitizes 75 terabytes of information each minute. YouTube adds 15 years of video each day. The CEO of Google stated "every 2 days, we now create as much information as we did from the dawn of civilization up until 2003." (Is some of this information useless?) Perhaps, hundreds of websites are maliciously hacked in the same minute. This is a revolutionary change, make no mistake about it. More and more, the workplace is decentralized to homes, hotel rooms and automobiles.

Hacking. Hacking into another's website and inserting a worm is repugnant to most of us. Why mess around with the personal lives of people you know or don't know? At first, worm programs were considered an oddity and a potential problem. A friend of mine was one of the first to publish on worms (*Scientific American,* March 1985, page 20) with his program called *Apple Worm*. Now we see that worms have evolved to Stuxnet and much more. How should society respond? The stakeholders are studying worms at universities with their computer-science hacker clubs and in industries to protect their customer data and industrial secrets and in governments that worry about smooth-running commerce or national security. This chapter tries to address some of the weaknesses of the web and some partial fixes. At the end of this chapter we describe the modest tries at diplomacy to establish global norms to minimize cyber problems. The Internet is vulnerable.

Telephone networks versus cyber networks. The phone system and the Internet are very different: The phone network is made up of hub and spoke situations, where the phone connection is controlled by the central switching point hub. When a phone call goes from Uncle Gus to Aunt Helen, they are connected at the central hub by connecting a wire from Uncle Gus's wire to Aunt Helen's wire; this can be done with two alligator clips or by plugging a wire into the two receptacles. The central authority can deny a phone call or allow it to pass, initially with human intervention and later automatically. The central authority can decide whether to add extra functionality to the message, such as call waiting, caller ID or a conference call. The central authority has lots of power to make the system stable or recessive. The central authority can work with the federal government to decide which phone messages are to be tapped and listened. At some point the phone system is involved with the FISA Court (Foreign Intelligence Surveillance Act) or with local police and judges. Some have called for a better FISA secret court, by authorizing the judges to appoint lawyers to the public interest when difficult issues come before it.

Phone messages are transmitted in whole, not divided, going in one continuous batch, catching our words as well as our hiccups. There is direct data transfer in these messages, not just zeros and ones that have to be decoded. The telephone

Table 14.1 Phone system versus internet, simplified (D. Hafemeister and M. Cahill)

	Phone system	Internet
Geometry	Hub and spokes	Random spider web
Digital or analog	Analog	Digital
National borders	Yes	No
Control on apps	Strong	Weak
Attribution	More	Very weak
Data transfer	Slow	Deluge
Dynamic–powered apps	Small	Great
Governance	Strong	Weak

system has a great deal of network control, which is reasonably robust to remain stable under attack. Of course our phone calls can be tapped without too much effort. On a group basis where we encounter *big data* with masses of computers, the phone system is more robust than the Internet (Table 14.1).

Cyber networks. Rather than a central hub that connects wires, cyberspace deals only with the transmission of zeros and ones. A message is divided into smaller packages and then assembled into one package at the final receiving point. The cyber network is a crazy spider web, the paths vary each time a message goes between Gus and Helen. There is no central command authority that says you can or cannot send a message, as long as you stick with the rules and protocols that the web requires. You must use a formatted message address, such as babe. ruth@newyorkyankees.com. There is usually no authority that determines whether you can or cannot send a certain piece of code, a picture, or a glorified app that can shut down the financial system. But the web is far broader than web pages where you can buy a car. The Internet can carry all kinds of functionality, far beyond what we might expect. Stuxnet is a good example. Who would have thought that it could spin up centrifuges and hide the results to the owners, after determining that the worm was correctly placed at Natanz? Cyber space has a rigid set of protocols to make sure the system works, mostly without supervision. No one usually decides whether a message can go forward or not, as all the mini-packages go forward. That is the reason the cyberspace or the Internet has been so successful. It doesn't involve human intervention on whether a message can be sent, or to determine what functionality goes forward. We don't know where cyberspace is taking us. Certainly the people who invented cyber space hadn't thought of spinning centrifuges or interfering with the financial systems of countries. This is a new world.

Search engines. The other day I put in my father's name, Lester Hafemeister, into a Google query. Dad has been gone 30 years, but you never know what you might find. I got only one hit but it was treasured, as he attended a school board meeting in Milwaukee around 1970. Search engines can do more than find old memos. They can answer questions. I will now put in Google, *1957 World Series pitcher winner of 3 games.* The answer came back in a second, stating that Lew Burdette had died at the age of 80 in Florida in 2007. Lew Burdette was a spunky guy. It was rumored he threw spitters. I saw him in the 1957 World Series in the pre-game outfield, goofing around catching long fly balls behind his back, endangering his body at a critical moment. And, Lew Burdett won three full-games (4–2, 1–0, 5–0). Google does this by picking the most likely choices. But this was an easy one since the World Series 1957 narrative narrowed the choices. Let's try harder one, *Milwaukee Braves right fielder in 1957*. Quickly we are told it is MVP winner Hank Aaron. The search engine bases its choices on using pure probabilistic logic. It could've interpreted Milwaukee Braves as a discussion on bravery in Milwaukee, or Indian braves who lived near a mill. But Google wisely understood that the following phrase *right fielder is* a baseball term and Google decided that it should be the Braves that play on diamonds. The Google search *app* uses probability to determine which choice to emphasize at the top of the selection list and which to place further down. The search apps on iPhones have raised the level of debate in coffee-houses with requests to Siri. An open and free network allows the growth of new ideas, no one wants to vote against that. But, a very free unsupervised Internet also enhances its vulnerability, creating future societal chaos from bad guys doing bad things. As usual it's a contest between near-term goals and long-term goals. It is easy to appear strong and moral to pursue one of these goals, but not both of them.

Nodal points. Each cell phone or computer is a nodal point in the very complex spider web. Each nodal point accepts or sends messages. But each of these points gives hackers a place to attack. For example, malware can be installed on our cell phones, such that an outsider can turn the phone on and into a microphone, without making a call to the phone. An employee can inadvertently be used to obtain secret business or government information with this open microphone, listening to others. Since IPhones have GPS abilities, it can turn the phone off when outside of work and turn it on when at work. In order to invent *apps* for this, you have to know about the many layers in cyber domain.

Vulnerability gateway. The web structure is both good and bad, both a virtue and a vice. We like the web's 100 % accurate, almost-instantaneous, communication across vast distances. And it is flexible, readily distributing computer codes to do many things. However, the billions of nodal points are available gateways for deviant hackers. Your iPhone can be turned on while driving, so your business partners can check up on your behavior when you are near certain GPS locations. Here are five gateways to consider:

1. ***Time and space***. The moon is only a second away for electromagnetic waves, traveling at the speed of light, making communication essentially instantaneous. Distances between conventional armies matters, but now the whole planet is instantaneously available for attack.
2. ***Asymmetric medium***. Smaller countries can attack the cyber structure of larger countries and get away with it. The advantage is clearly with the offense, it is much more effective to attack many nodes than it is to defend them. A small group of dissident actors can make a difference. The group *anonymous* has done that.

3. ***Anonymous actors***. The patterns of zeros and ones in computer messages make it very difficult to determine the authors. It is easy and cheap to send out 1 million scam e-mails, even if they fool only a few people. *Anonymous* and smaller nations like North Korea can challenge the bigger states. They can use *botnets*, which are *robotic network* commands that direct computers to do things, taking advantage of address lists to send out robotic messages that infect other computers to infect others. The botnet *Conficker* infected over 10 million computers, all under the control of that group. It is unclear who runs *Conficker*. It is part of the culture of the Internet to believe in anonymity, but attribution should be folded in at some point. It shouldn't take a year to determine who runs the *Conficker* group and what to do about it. The fact of the matter is that bad actors are much harder to locate then good actors, who leave return addresses.
4. ***No internet borders***. For hundreds of years, it has been recognized that sovereign nations control all forms of traffic over their borders. This is not true for the Internet since a message may be written in France, sent in pieces to Italy, Germany and Canada, to be repackaged in the US. No border authority acts as a checkpoint on the Internet. Recently, the European Union stated it would love to break up Google, but legal people don't believe the EU can do this. Countries with fewer access points, such as China, Australia and New Zealand have limited connections across their borders, and can exert some muscle on Internet traffic. The US has so many connections that this is virtually impossible. Traditional laws and regulations and treaties are more difficult to apply to cyberspace. At the end of this chapter, we explore treaty possibilities to establish a global norm, a very difficult task.
5. ***Distinction***. When you look only at zeros and ones, it's not clear where they originated. It also is not clear whether the zeros and ones are commercial information, recreation information, academic information or malpractice. The executable portion of the computer code is not apparent. It is not apparent as to how it would interact with a Siemens controller, since it is not clear it is headed for a Siemens controller. However, if this program has been seen before, we will know what it does from past experience. In this sense, those that have been stung will know what to look for. Malware that has been directed towards commercial airlines or electrical grids can be shared among airlines and electrical grids, so they all can learn. But malware can be rewritten to do other things the next day. And the initial attack usually gets through your defenses.

14.3.1 Cyber Attacks

Denial-of-service attack. Anyone can do this. You obtain a program in which you enter the web address of the server you want to attack, you hook up to the Internet and begin to flood the target with requests to connect. A major website might be shut down if it got 100,000 requests per second. If your computer sends 1000 requests per second, then you will have to recruit 100 friends with computers to make this happen. This converts a *denial of service attack* into a *distributed denial of service attack*.

Not all attacking computers do this voluntarily. Bad actors use lists of networks of computers to send robotic commands, without the knowledge of their owners. A robot network is inserted into innocent computers with a botnet (robot network) virus that forces the innocent computer (the zombie) to make the requests. They can be dormant and leap into action at a particular time, defined by the botnet, sometimes called the *botnet herder of zombie computers*. It is possible to rent computer networks and e-mail address lists from herders. E-mail address lists are harvested by WebCrawler, or spider botnets. Conficker.net has 10 million zombie computers. Spam requests can trick you to go to websites that have embedded malware. This can turn your computer into a zombie. A malicious version of this attack uses a *Trojan horse* in which the malware is hidden inside the program and looks like innocent information, but it can be turned into action by an outside signal.

Phishing and spear-phishing. A random approach that obtains data maliciously is called *phishing*. If a fishing expedition is directed towards a certain class of people, companies or skills, it is *spear-phishing*, a directed attack on a particular fish type, not the random ocean. A phishing expedition could gather computer passwords, financial information, or turn on cameras and microphones. For example, China used various cyber techniques to keep an eye on the Dalai Lama.

Zero-day vulnerability. Malware can be more effective if it uses newly created malware. Then there will be no time to observe its operation and engineer against it. The new malware would be turned on upon command or with a time certain. It is possible to modify the new malware with a random number generator, so it would be very new. Such malware might be invulnerable on the zero day, unless the defensive computer is hyperactive.

How large the vulnerability? Norton estimated that losses from cyber attacks in the 24 largest countries was \$114 billion, with total losses at \$388 billion. Since Norton is in the anti-virus business, this might be a high-ball estimate. The US Government Accountability Office estimated in 2005 that annual computer crime cost \$67 billion to the US. These are difficult estimates to make and verify.

Attribution. Attribution is hard to determine if the sender hides his identity. It may take 6 months to determine the owner of a botnet, which is of little use for most attacks. The US government has made significant investments to address the problem of attribution. One approach looks at the beginning of the process, examines the decisions made to join a conversation, rather than analyze the situation after the attack. Another approach raises the threshold, making it more difficult to attach to a conversation. Another defensive approach uses *trusted identities*. This limits the participants to those who volunteer to display an authentic identity symbol, narrowing the conversation only to guaranteed volunteers. Available software programs allow authors to make their messages anonymous; this is used by journalists, actors, cops and robbers. One of these is *TOR*, which stands for *the onion router*. Another approach is for websites to display attribution symbols and processes that would show that the website is authentic. When you go to united.com, you would be able to test the web page to learn if it really is United Airlines. A robust attribution system would go through a chain of events, but each chain is only as strong as its weakest link. Another approach is for the Internet to issue death certificates to domain addresses that have shown malevolent behavior.

Hacktivist insurgency. An insurgency is more similar to a war than a crime; it intends to force change on others, and not to make money. The hacktivist (hacker activist) insurgency looks like war since its weapons corrupt a website, or deny access. *WikiLeaks* released 250,000 classified State Department cables to undercut US military actions in Iraq and Afghanistan. This seems reminiscent of the release of the Pentagon Papers to undercut the Vietnam War. The main response came from the commercial sector when *MasterCard, PayPal* and *Amazon* stopped selling Internet services to WikiLeaks. *Anonymous* responded to this by laying down copious *distributed denial of service* attacks. Interpol responded by arresting 25 members of Anonymous in 2012. When cyber security firms began investigating the identity of Anonymous members, Anonymous retaliated with cyber attacks. It gets more complicated if the activists reside in Iran or North Korea.

Cyber war. There was a prelude to a cyber war between the US and Iran, with both sides acting on the offense. A more full-blown event was the August 2008 war between Russian troops and Georgian troops, contesting their borderlands. An aggressive cyber attack that could not be contained could be followed by conventional war. A coupling of cyber war with conventional war concerns many. The Russians accompanied their troop movements into Georgia with cyber attacks on Georgian websites, defacing them with messages that glorified Russia and diminished Georgia. These political attacks were done by Russian civilians, which were employed by the military. But the attack was more than political rhetoric, as Georgian leadership lost communication with its military units.

Who will be cyber warriors? Most observers believe the most likely candidates are the US and China. A Department of Defense report had this to say in 2010:

> Numerous computer systems around the world, including those owned by the US government, continued to be the target of intrusions that appear to have originated within the [People's Republic of China]. These intrusions focused on exfiltrating information, some of which could be of strategic or military utility. The …skills required for these intrusions are similar to those necessary to conduct computer network attacks.

Reportedly, China stole terabytes of data, including designs for the F-35 stealth jet fighter from defense contractors. In 2010, Chinese Internet-providers broadcast erroneous network routes, with the result that the Internet traffic from the US and others were routed through Chinese servers, amounting to 15 % of the total traffic.

Vulnerability to massive cyber wars. Stuxnet was developed and delivered by wealthy countries and not by terrorists. How vulnerable is the US to attacks by China? Certainly the US is vulnerable if China launched a cyber war against the US. But China is also vulnerable if the US launched a cyber war against China. Will this happen, hopefully not. It begins to look like *mutually assured disruptive deterrence* (MADD), backed by the nuclear version of *mutually assured destruction* (MAD). Digital weapons are much less lethal than nuclear weapons, thus more disruptive then destructive. A massive digital cyber war would put lives at risk, accompanied with great financial losses, but not leveled cities. But, a massive cyber war could escalate to military responses.

14.3.2 Deterrence and Defensive Strategies

Volatile human minds have few limits in concocting cyber war offense and defense. Deterrence can work to some extent, but not as completely as in the nuclear case since a smart group of persons with computers can cause problems while not being vulnerable to a responsive attack. The 2011 White House stated its policy on appropriate retaliatory responses and threats to attacks on the Internet:

> When warranted, the United States will respond to hostile acts in cyberspace as we would to any other threat to our country. All states possess an inherent right to self-defense, and we recognize that certain hostile acts conducted through cyberspace could compel actions under the commitments we have with our military treaty partners. We reserve the right to use all necessary means—diplomatic, informational, military, and economic—as appropriate and consistent with applicable international law, in order to defend our nation, our allies, our partners, and our interests. In so doing, we will exhaust all options before military force whenever we can; will carefully weigh the costs and risks of action against the costs of inaction; and will act in a way that reflects our values and strengthens our legitimacy, seeking broad international support whenever possible.

There is no one silver bullet to strengthen the Internet, many approaches will be needed in parallel. Here is a partial list of actions that could be used to protect the Internet.

- Firewalls to close off routes of access,
- Encrypted files to hide some of the code,
- Protect systems with air gaps,
- Block USB flash drive ports,
- Educate computer personnel to avoid scams,
- Reward staff for reporting suspicious activity,
- Remunerate internet staff sufficiently,
- Attempt to develop international norms to constrain cyber misbehavior,
- Develop monitoring and verification methods to warn of attacks and strengths and agreements,
- Develop a national cyber policy that strives to be consistent.

14.4 Cyber Governance

The National Security Agency and the US Cyber Command are located at Fort Meade, Maryland, which would be vulnerable in a massive attack. The US might try to establish a regulatory regime for the Internet, but who would be in charge?

To make the Internet more robust could make it less open, which would be resisted. What standards would be used to regulate the web? Presumably, one would establish performance standards, in which the upside and downside risks and benefits are examined, based on experience and theory. The regulatory process is slow, to try to be thoughtful and avoid politics.

Some suggest keeping the Internet as self-governing, *but* requiring insurance to cover liabilities, resulting from bad decisions. In the present system, government warns of dangers, gives advice, but does not give directions through regulations. When it is crisis time, the Federal Government bends wrists to get it what it wants. This is what Stuart Baker, former Assistant Secretary for Policy at the Department of Homeland Security said in 2012.

> This broad structure is meant to solve the problem of how to regulate a fast-moving and complex technology. It does so by leaving as much direction as possible in the hands of the private sector… [This] doesn't call for government simply to tell industry what security technologies to adopt. The point of the process is to identify the risks, warn industry of those risks, and challenge industry to develop standards and adopt measures that industry finds best adapted to the risks.

International governance. When the Internet was born in the 1970s, it was considered a novelty and mostly ignored by sovereign nations. It is tempting to call for a broad, new agency in the United Nations, similar to the International Atomic Energy Agency, or the Comprehensive Test Ban Treaty Organization, but this has risks. Cyberspace is huge, with two and one-half billion users and 90 quadrillion annual e-mails. Not only is cyberspace huge, it does not have distinct national borders. There are great differences between nations participating in cyber governance.

Electrical grid governance. The grid is vulnerable to cyber attacks, but difficult to regulate in shared regions such as Eastern Canada and Eastern US, who are in the same grid region, but in different countries. If the US regulated its electronic control computers, it would not be regulating the Canadian computers, unless this was pre-negotiated.

Cyber space governance. Issues become more complicated with 190 nations in cyberspace. The international organization that sets rules for domain names is called the *Internet Corporation for Assigned Names and Numbers*. ICANN is a nongovernmental organization (NGO), which is truly international, but some think it has too much American control and don't trust it. Another NGO is the *Internet Engineering Task Force*. The IETF is an "open international community of network designers, operators, vendors and researchers," dedicated to making the Internet work better. This group primarily consists of engineers who define better as "making the Internet technically more effective."

FCC decisions. In February 2015, the Federal Communications Commission voted to regulate the internet as a utility, but claims this will be lightly done. The internet will be open, meaning no constraints (except for national security or fighting crime). The FCC says it is banning "blocking, throttling and paid prioritization" which means no blocked sites (except for national security or crime fighting), no slow lanes and fast lanes, and no pricing that depends on amount consumed by large companies. Google and Facebook applauded the decision while cable television and the telecommunications industry did not like the decision.

14.4.1 Issues for International Governance

Too much control in cyberspace? If you google *Tiananmen Square* when you are in China, you get a null result. This kind of control is also used to prevent child pornography, as was done in Australia, but recall the difficulty in defining pornography. Greater control is exerted in Belarus, which requires all individuals and companies be registered in Belarus to use domestic domains for services or e-mail.

Data sovereignty. There are uncertainties on assigning nationality to data storage facilities. It can be defined by the location of the storage servers. This can be confusing in cloud storage, which might be defined by the location of the corporate headquarters. It is difficult when data packages are split between nations, making it difficult to assign sovereignty to data.

Russian versus American views on governance. Russia would outlaw cyber war attacks on civilian targets, like the grid or Wall Street. This seems reasonable, but the US only enters into treaties that are verifiable by effective monitoring, to make sure the terms of treaties have not been violated. The 1989 Hague convention dealt with protection of civilians, civilian property, deserters, prisoners of war, hostages, pillaging, spies, truces and prisoner exchange, but it contained no monitoring provisions. It is difficult to monitor and verify terms of cyber treaties. The US compromised and began talks with Russia on cyberspace negotiations in 2009, but with little success. There is a conflict between an anonymous private Internet and a robust, stable Internet.

Direction in negotiations? It is unlikely that the technical leadership of the *Internet Engineering Task Force* (IETF) would promulgate standards that made it easier to censor the Internet. Freer nations like this position, but more regressive nations like more control. A middle ground is difficult to obtain. Will the two international NGOs (ICANN and IETF) lose some regulatory power with the FCC decisions of 2015?

Government proactive-defensive efforts. The US Constitution directs the government to "provide for the common defense," which includes the common defense of cyberspace. Let's examine how this is being done. The government created the Einstein 2.0 system to detect intrusive code, which was deployed by 2008 to protect the federal cyber network. Einstein 2.0 uses a lookup system to spot malicious code by comparing it to previous malicious code. This is like looking for particular kinds of DNA by comparing to existing DNA. Einstein 2.0 sends an alert to the receiving network, letting it know they are under attack. It does not alleviate the situation, other than to give a warning.

Einstein 3.0. The next generation of Einstein goes beyond alerting of impending malpractice, as it acts proactively to slow or stop malpractice. Einstein 3.0 uses a probabilistic method that deals with variations in code in the lookup system. Einstein 3.0 intercepts all Internet traffic bound for federal computers before it is delivered. It delays temporarily for screening, and then passes it along or quarantines it, as appropriate, thus preventing dangerous deliveries. Most companies and persons wouldn't want this degree of intrusiveness, which could examine content of messages. Einstein 3.0 was being deployed only on federal systems. Einstein 3.0 is being expanded to the large defense contractor networks, such as Raytheon and Boeing, if they give permission. These defense companies deal with secret government programs, which require conditions to obtain government contracts. It is unlikely that Amazon or Google would accept these conditions. Einstein 4.0 is intended to be operated by contractors, to remove the objection to government reading content, with contractors being able to read content. Legal issues complicate this adoption by industry.

Address and content of messages. Addresses on mailed letters or internet e-mails do not have constitutional privacy protection, since the confidentiality of the address was removed once it was mailed or sent. The government retains its right to keep these lists, which is not appreciated but certainly legal. The content of these messages is protected by constitutional privacy restraints. The secrecy protection of the contents is not valid for federal and corporate networks, which gave up their right of privacy to obtain use of those networks. This exception is given with signed releases as a condition of employment. Sometimes the line between content and non-content can be blurred. For example, a message sent to aa.org, the address for Alcoholics Anonymous, may give away the secrecy of your drinking habits. One concern for the future: To gain access to the Internet, one could be asked for permission to have the contents of your messages searched for malware, which local providers already do. This would expand the role of the government. Section 14.1 discussed the reality that citizens' messages are subject to examination pursuant to FISA court rulings. This should be scrutinized to make sure it functions as intended.

Big data. Google searches our obscure facts, both accurately and quickly. It is partially true, *privacy is dead, get used to it*, but we hope considerable personal privacy will be maintained. Defining the gray areas will be difficult. An incredible amount is known about our many small decisions, such as where we buy coffee and which political candidates we support with donations. In the past this open information was somewhat available but difficult to access and combine. Now, the mega information of stored big data is available with considerable analytic prowess. The result is that others can learn our sociological personalities and quirks. As you move about town with your cell phone, big data is collecting where you've been and for how long. Has your time been spent in church or the nearest bar? In addition, your credit card, telephone calls, travel and vacations are already scrutinized and correlated, with probabilistic analysis. One can imagine this prowess could be used to search the innards of judicial appointments and more. Political, legal battles will be fought in considering whether to constrain this analytical powerhouse, intended by the Department of Homeland Security to constrain terrorists. It is a competition between the public good and the private good.

The big data transition has come because of the impressive growth of economical computational power, while at the same time there has been an incredible drop in price of digital storage. *Moore's law* predicts the future prowess of electronic chips. Moore's law successfully predicted an increase by a factor of two in the density of transistors per unit area every 1.5–2 years. This gives an increase by a factor of million in the density of electronic switches since 1970. Couple this advance with a huge drop in price of electronic storage. In 1984 it cost $200 to store a megabyte, while today, 40 years later, you can store 100 megabytes for a penny. This is a drop in cost for a single transistor by a factor of a million, which is faster than Moore's law.[1]

In the past, only the federal government, such as the National Crime Information Center, had resources to gather obscure information and perform serious analysis. Now big companies, like Google, do that and the size of the companies able to do this is much smaller. Much of big data is not considered private since it has already been released to others. Is government metadata available to others outside of government? No. A Freedom of Information Request to the Department of Justice for this government information was denied. A bipartisan vote of the Supreme Court upheld the denial of this request, concluding the following:

> Plainly there is a vast difference between the public records that might be found after a diligent search of courthouse files, county archives, and local police stations throughout the country located in a single clearinghouse of information and a computerized summary… [The court concluded that] the privacy interests in maintaining the practical obscurity of rap-sheet information will always be high.

This decision strengthened privacy, but not much as companies buy this type of information from others who obtain it from open sources. One company, *Acxiom,* estimated it holds an average of 1500 data items on each adult American. These private sources of information could become more contentious if the federal government subpoenaed these records held by corporations. At present, Supreme Court decisions state that information held by third parties, such as Facebook or your phone provider, are not protected by the Fourth Amendment (need probable cause for searches). This means you have no privacy interest in the data since you voluntarily gave it to the phone company and others in the past. This issue is not totally settled.

New rules on privacy? The1970 rules on privacy need strengthening because of the ever increasing power of computers and digital storage. We begin by discussing privacy aspects of government cyberspace, followed with privacy aspects of commercial cyberspace. In 2012 the Supreme Court examined a part of this issue in

---

[1]The price of a electronic storage dropped by two million: $200/Mbyte to $0.01/100-Mbyte. Moore's law has existed for 50 years (2015–1965), which is 25 doubling periods of 2 years. This gives an increase in electronic surface density of transistors by a factor of $2^{25} = 2^{10} \times 10^{10} \times 2^5 = 1024 \times 1024 \times 32 = 34$ million over a longer period of time.

US versus Jones. In this case, US agents attached a GPS tracker to Jones's car because they thought he was a drug dealer, which he was. However, the agency did not get a proper warrant before it placed the GPS tracker on the car. Jones appealed that his constitutional privacy rights had been violated. The Supreme Court agreed unanimously, nine to zero, requiring a valid warrant for the GPS-tracking. This was a narrow ruling on privacy, but it shows the Supreme Court is examining this area.

Fair information practice principles. *FIPP*, the basis of current law, dating from the 1970s, concludes that governments should limit their collection of personal information to what is necessary. Electronic collection should be used only for specific and limited purposes and it should be transparent and open to the public as to how the information is collected and used. The individual under surveillance should be allowed to see the collected data and correct it, if it needs to be corrected. But, present, high-tech examination of big data, obtained from credit cards and phone calls, does not follow this 1970s standard. We need to think more about the meaning of privacy. When we vote, our privacy is guaranteed with sealing the envelope or closing the election booth entrance. We expect protection of privacy when we pray and when we speak behind closed doors. A pure form of privacy is secrecy, but a lesser form of secrecy is anonymity when we use our credit cards. A definition of privacy will have to consider what are the unjustified consequences if our privacy is violated, such as with the release of phone records.

Gray areas. Who will determine gray areas in the attempt to balance national security and the loss of individuals rights? The FISA (Foreign Intelligence Surveillance Act) Court is the present approach, but it has weaknesses since persons under surveillance are not represented at the court; only one side presents data on consequences. Should there be an ombudsman to defend the accused? Privacy decisions intersect the Department of Homeland Security's *Privacy Officer* and its *Office for Civil Rights and Civil Liberties* and the Intelligence Community's *Civil Liberties Protection Officer*. When the nation is under attack, the gray area is moved towards national security protection and away from privacy protection. Even Abraham Lincoln curtailed habeas corpus protections during the Civil War, something he had to do for the nation to survive. Ultimately we will need watchers to watch the watchers. In the future we might expect *no-tracking rules* with the phone company, similar to *no-call rules* with telemarketers. It is not clear how effective this will be. The standards within FISA court proceedings might change to define when a party is guilty of violating privacy by cyber malfeasance or mistake, and the consequences.

Secret agents. The intelligence community is concerned that big data analysis will reveal the identities of its cloak-and-dagger agents. Some technical operations can be revealed by use of Google Earth and big data. This analysis can be done by commercial entities, smaller nations, or someday, wealthy terrorists. These issues will not go away.

Going dark: encryption and packaged messages. Law enforcement and the Intelligence Community are concerned that two things can undercut attempts to combat counter-terrorism and anti-crime operations: (1) Robust encryption has become easier to use and cheaper to buy. (2) Internet tapping has become less available, as other pathways evolved. This loss of espionage and crime information to the government is referred to as *going dark*.

Encryption. With brilliance, Alan Turing broke German's Enigma code at Bletchley Park in World War II. He used statistical methods on intercepted messages to and from German submarines. He used this information on a purloined Enigma machine, obtained by Polish rebels. The earliest codes used a cipher letter for each letter. One approach replaces message letters with the letter, two letters after the letter in the alphabet, in the code-message. The word *it* would be replaced with the new word *kv*. This code can be broken easily. An improved code might use cipher letters from paired set of letters in a table of matched pairs (message letters and replacement letters). This can be broken by observing the frequency of letters in the message. The letter *e* is the most common letter, and so forth. It is easy to imagine more complicated algorithms, making it more difficult to decode, even as computers use more sophisticated statistical methods. However, public-key encrypted codes are essentially unbreakable, unless the code breaker has the private key to combine with the public key to break the code. This approach multiplies two extremely large prime numbers, which are combined with one-way mathematical functions. To multiply by two and then divide by two returns you to the beginning value, this is not true with one-way functions. You can buy these codes on the Internet for $100.

Clipper chip back door. In the 1990s the FBI attempted to require manufacturers to include back door entrances for their clipper chip. The government believed this was necessary to overcome the new prowess of encryption. There are downside risks from this policy: If it was known that US equipment had back doors for the US government, it would destroy their global business as no one would buy it. Backdoors might also be misused by the federal government. Conversely, the government itself is more vulnerable since its information would be more easily hijacked. Finally, privacy advocates, opposed to the clipper chip, were successful in removing this policy. Both the US and China have been accused of inserting less-effective encryption code in hardware.

Less wiretapping. Using new technologies, it is easier to prevent your message from being intercepted. After all, Internet messages are divided into packages, which can be arranged to make interception more difficult. There are other ways to protect messages, such as using voice communication through Skype. This web-based video-conferencing system uses old-fashioned analog communication, without zeros and ones.

Going dark for how long? If your message is intercepted but its encryption is not broken, your secrets will remain private. If your message can be decoded, but it is not intercepted, your message will remain private. It is only when your message is both intercepted and decoded that your secrets are revealed. The government usually gets its way if there is a large national security concern, and the loss of privacy is not great. The government may not get its way if the national security concerns are smaller, or the loss of privacy is larger. It is in the gray areas that we face difficult policy choices. Some are proposing a FISA court with an ombudsman, representing the absent defendant, and watchers watching watchers. At this time, the law does not support widespread, non-filtered message reading. US policy is inconsistent in that *internationally* we favor *internet freedom* to encourage democracy movements around the globe, while domestically the government wants more information from computer-code backdoors.

Imported chips with malware. This is already taking place, by other countries and by the US. The back doors in the malware chips are permanent and cannot be disabled. In 2007, the Defense Science Board made these dire comments:

> The current systems-designs, assurance-methodologies, acquisition-procedures, and knowledge of adversarial-capabilities and intentions are inadequate to the magnitude of the threat.

The Department of Homeland Security testified before the House of Representatives in 2011 that it was aware that imported electronic equipment contained preloaded malware. Conversely, Rosenzweig stated that the US exported gas turbo-generators to the Soviet Union in 1982 that were flawed. The conclusion is that severe threats to the US exist from cyber attacks from hardware malware, and they are far greater than attacks from software malware. When US equipment breaks down abroad, usually it is repaired by foreign nationals whom we do not know. These concerns apply to both government and private systems, as revealed by Android Mobile Phones, Barnes & Noble and AT&T. This problem is exacerbated because knowledge from the intelligence community cannot be shared with government and private purchasers. Here are some possible responses.

- ***Create a blacklist***. A seemingly good response by the government is to blacklist unacceptable products. This, unfortunately, has downside risks, as it would tend to reveal our sources and methods, used to compile a list of corrupted technologies. It could lead to lawsuits by those who felt the blacklist discouraged sales. To defend itself, the government would have to reveal something about the products and why they were a threat, or seek judicial approval to present evidence under seal.
- ***Bottom-up chip inspections***. More than 97 % of silicon chips are produced outside the US. These 10 quintillion transistors (10^{19}) are too numerous to inspect, it just can't be done. The US could reduce the number of inspections by combining deep disclosures and reporting, with spot checks. This approach is what gives the IAEA the capacity to track nuclear fuel. This is probably the best we can do.
- ***Top-down counterintelligence***. The US Foreign Investment and Interagency Committee and the Federal Communications Commission (FCC) are authorized to examine transactions with foreign-controlled chip-manufacturers and other technologies, to determine if these transactions harm national security.
- ***COTS hardware and software***. Unfortunately there are no easy methods to make sure that *commercial off-the-shelf technologies* (COTS) do not contain malware. At one point the Defense Department thought it could control this by using a DoD computer language, ADA, making sure that equipment was less likely to contain malware. Unfortunately, this was too expensive and

inconvenient. The engineers and the economists prefer to work with *open architecture designs*, which implies the hardware is compatible with equipment and computer codes of many different manufacturers. We like the ability to ward off the latest malware with updated operating systems, or at least to call a hotline for help.

- ***Select missions to be protected***. The Defense Science Board recommended that the government determine which missions and systems should be protected because they are the most critical to the nation. This leaves the less vital missions unprotected, which seems to accept defeat.
- ***Better intelligence and better investigations***. The US should do what it does, but do it better. Expand review authority for the government. Diversify the types of hardware and software used by the government, but this may raise costs and reduce flexibility. One could compel US industry to *diversify* its software and hardware, but this would be uncharacteristic and promote litigation. Lastly, evaluate the security of products provided by suppliers. Who owns the company? Who manages the company? What is the track record of the company? But then, we know that the products can be laundered to other companies. The bottom line is that these are small tools to confront a big problem.

Your protection in cyberspace. The Constitution isn't going to protect you from malware attacks. But many small actions can reduce your chance of being hacked. Several weak security layers on your computer system can dissuade attackers, as the hackers search for less-robust protected systems. When living in Washington DC, I bought a club for the steering wheel, to prevent our car from being stolen. Knowledgeable crooks can disrupt the club, but the crooks don't want to be bothered with a club if most cars don't have one. A number of smaller steps can reduce your vulnerability. Here are a few suggestions:

- ***Don't clicks***. Don't click on suspicious websites or URLs unless you are quite sure they are not suspicious. If things look too good to be true, then they must be too good to be true. Official messages from your bank, your credit card or the government usually are not what they appear. They could be phishing for your financial and privacy fish. The usual approach to enter a web site is to use HTTP, hypertext transfer protocol, language. A safer approach is to use the HTTPS, where the S means secure. By using the program called HTTPS, you can browse the Internet while your system automatically encrypts the two-way transmissions. At some sites, the owner deposits cookies in your computer to ease the hookup process. That may seem too aggressive to you, and if so, use a program called *Cookie Cleaner* to wipe out the cookies. Lastly, enable the password requirement to enter your home Wi-Fi, to prevent strangers from entering your network.
- ***Passwords***. Password-cracking programs can contain 1 million passwords, so avoid simple things like *Fido*. Passwords can be made very strong by using a phrase you like. I am fond of Vince Lombardi's, "when we have the ball, they don't have the ball." Vince's first letters are *wwhtbtdhtb*. To strengthen the password you could capitalize every other letter starting with the first letter, and

then add in the playing number of Aaron Rodgers, which is 12. This gives you a very strong password, *WwHtBtDhTb12*. Then you can insert new letters, such as, the letters replaced by the letter, two down the alphabet. Don't use the same password repetitively for other accounts. For enhanced security, store all your passwords in a vault in your computer, with a strong master password that is used only for this one task. Use a different password for your computers at work and at home. Change important passwords and credit card numbers on an annual basis. Consider not using electronic banking or Face Book. This is getting onerous.

- ***System concerns***. Have an effective firewall and intrusion detective system. The rumor that Apples are more robust to attack than Microsoft is only partially true, since Microsoft, being larger, has had more malware attacks. This is changing as more people use Apple products. Your cell phones and computers can be turned on without your knowing it, letting your platform be used as a camera or microphone. To avoid this, take the battery out, or leave the devices out of the meeting room.
- ***Encryption***. If you do secretive work, make sure you encrypt your results. Using 256-bit encryption is essentially uncrackable. When placing files in your computer's trash, they are not erased, only the address was removed. Programs, such as *eraser*, delete all the data.
- ***System robustness***. Governments and the Internet have to work together to make a more robust Internet. Rosenzweig laid out an approach in an analogy to the medical community. One expects some diseases will break out, but one hopes that the Center for Disease Control can prevent an epidemic.
- ***Redundancy***. Have available extra copies of servers, essential data and programs, with excess bandwidth capacity.
- ***Quarantine***. Diseased computers should be taken off-line. This is harder than it sounds.
- ***Antivirus programs***. Require servers to certify adequate antivirus programs to participate in the Internet.
- ***Isolation***. Key parts of the cyber world should be isolated from each other.
- ***Diversity***. Consider using different computer languages for different systems.
- ***Observe outbound traffic***. Flag infected systems.
- ***Keep changing what you do***. If your systems are changed occasionally, this can fool the attacker.
- ***Deterrence***. Nations that show the determination to respond in kind if attacked might be attacked less often. This sounds like nuclear deterrence, which has some problems. Active defenses are likely to be against the law because the defenders didn't have permission to enter the servers of others.
- ***Active defenses***. Honey pot traps catch botnets. False data can be made available, with beacons that lead to the attacker.
- ***Cloud computing***. Clouds are more resistant, but more authority is given to those in charge of the clouds.

14.5 Cyber Diplomacy

International law addresses conventional wars and is fairly well followed. How would provisions apply to cyber war? International law requires the following:

- Allows targeting of combatants who are participants in the war,
- Must carry their military weapons openly,
- Must be readily identified as combatants by their uniforms, and
- Must respect the rights of neutrals.

These four issues need significant discussion to translate them from conventional war to cyber war. Cyber war would probably be directed at the territory of the attacker. There is some available time for discussion on the attackers side, with lawyers playing a role. It is difficult to guarantee the rights of neutrals, as cyber signals pass through neutral nations. Transmission of data packets through neutral nations cannot be prevented. The First Amendment to the Constitution protects the right of anonymous political speech. This is confused for the Internet because transmission of messages travels through many nations.

Convention on cyber crime. This convention was opened for signature in 2001 and entered into force on 1 July 2004 with 45 member states (2015). Support for the CCC is very strong in Europe with 39 members, plus 6 other states including the US and Japan, but not China and Russia. The convention is the first international treaty on crimes committed via the Internet and other computer networks, dealing particularly with infringements of copyright, computer-related fraud, child pornography and violations of network security. Its main objective, set out in the preamble, is to pursue a common criminal policy aimed at the protection of society against cyber crime, especially by adopting appropriate legislation and fostering international cooperation. The treaty is a framework for future negotiations, as the articles on substantive matters all begin with the phrase,

> each party shall adopt such legislative and other measures as may be necessary to ensure….

The substantive topics in the Convention are as follows: illegal access, illegal interception, data interference, system interference, misuse of devices, computer-related forgery, computer-related fraud, child pornography and copyrights. This is a difficult area in which to legislate, as the international borders between nations are less relevant, attribution is difficult and it is difficult to establish intent. Cyber crimes are hard to prosecute and punish, and there is now little deterrence available from the law.

White house executive order of 12 February 2013. This Executive Order directs the Departments of Justice and Homeland Security and the Director of National Intelligence to enhance the flow of unclassified cyber threat information to private sector entities. It also tasks these organizations to establish a process to distribute classified information to certain entities. The Executive Order calls for expansion of programs that bring private sector subject matter experts into the federal service on a temporary basis. Unfortunately, the executive branch trains experts in this area, only to loose them for higher salaries in the private sector. The Executive Order states that offensive operations in cyberspace, intended to produce significant consequences outside the US, must have presidential approval, except in emergency situations. Since damage from a cyber attack is much less severe than that of a nuclear war, it is difficult to see that a proportional attack response in cyber space would trigger the use of nuclear weapons, but that might be wishful.

Cyber negotiations. China and Russia have proposed treaty language to outlaw attacks on public services, such as electricity, gas and water, but no verification provisions. Discussions have taken place, but with little progress. China and Russia are not members of the Convention on Cyber Crime. The 2013 UN report on cyber space recommended the following confidence building measures on exchanges of views and best practices, on creating consultative frameworks, on sharing of information, on increasing cooperation to address incidents that affect critical cyber systems, and on exchanging mechanisms for law enforcement cooperation. This rather light list of topics and Putin's support for Edward Snowden let's us know that true cooperation is far away. The UN expert's group is writing a new report.

Problems

14.1 **Snowden**. What are further details on Edward Snowden.
14.2 **Stuxnet**. Give us details on Stuxnet.
14.3 **Espionage standard**. What are the criteria for a guilty finding in this law?
14.4 **Merkel/EU**. What were they angry about? Justified? They also do it? Big picture?
14.5 **Internet capacity**. How much cyber power would it take under the following circumstances: A movie of 1 Gbyte taking two hours. Twenty-five percent of the 120 million US houses stream a movie over 6 h.
14.6 **SCADA**. What are some of uses for *supervisory control and data acquisition systems*?
14.7 **China**. How might China control thought on its Internet.
14.8 **Worm**. Describe a simple worm, eating away in your computer.
14.9 **Zero-day vulnerability**. What is this? Why is it useful for an attacker?
14.10 **Phishing**. Describe a phishing event you observed. What makes it spear-phishing?
14.11 **Trusted ID**. How could this make Internet conversation more reliable?
14.12 **Convention on cyber crime**. What topics does CCC cover? Details? Deterrence works?
14.13 **Public-code encryption**. How does this work?
14.14 **Cyber crime deterrence**. What are examples? Deters individuals and nations?
14.15 **Rules of war and cyber war**. What are the former? How apply to cyber war?
14.16 **Governance**. Details on IETF and ICANN?
14.17 **FISA court**. How does FISA work? Weaknesses and strengths? How improve?

14.18 **Push back on *Anonymous***. How did the Internet push back against *Anonymous*?
14.19 **Einstein 2-4**. How do these government programs remove cyber malware?
14.20 **US policy on cyber**. How is US policy inconsistent?
14.21 **Going dark**. How is the intelligence community going dark with technological progress?
14.22 **Commercial off-the-shelf**. What are COTS examples? What is being done to control it?
14.23 **Cloud stability**. How could cloud computing bring both stability and chaos?
14.24 **Active defense**. How find and destroy botnets, using honey pots, searchlights and more?

Seminar on cyber terror

Alan Turing and Enigma, Anonymous, attribution, big data, Clipper Chip hardware, commercial off-the-shelf, Convention on Cyber Crime, cyber negotiations with China and Russia, cloud stability, denial of service attack, deterrence to restrain cyber terror, Edward Snowden, Espionage Act, Einstein 2-4, FISA court and its rules, Fair Information Practices Principles, going dark in the intelligence community, Google Glass, honey pot, imported chips with malware, international internet governance, internet sectors, nodal points, privacy of address and address lists, NSA Prism program, public and private key encryption, safe passwords, steaming 100 M at a time, phising, SCADA, Stuxnet, telephone versus internet modalities, Total Information Awareness, WikiLeaks, zero-day vulnerability,

Bibliography

Clarke, R., & Knake, R. (2010). *Cyber war: The next threat to national security and what to do about it*. NY: HyperCollins.

de Montjoye, Y., Radaelli, L., Singh, V., & Pentland, A. (2015). Unique in the shopping mall: On the reidentifiability of credit card metadata, *Science 347*, 536–539.

Elazari, K. (2015). How to survive cyberwar. *Scientific American*, 66–69, April 2015.

Elazari, K. (2015). *Who are the hackers?* Nine Ted Talks. www.ted.com/playlists/10/

Glick, J. (2012). *The information: a history, a theory, a flood*. NY: Pantheon.

Healey, J. (Ed.). (2013). *A fierce domain: Conflict in cyberspace, 1986 to 2012*. NY: Atlantic Council.

Henschke, A., & Lin, H. (2014). Cyberwarfare ethics, or how Facebook could accidentally make its engineers into targets. *Bulletin Atomic Scientists*, 25 August 2014.

Lin, H. (2012). A virtual necessity: some modest steps towards greater cyber security. *Bulletin Atomic Scientists, 68*(5), 75–87.

Nye, J. (2010). *Cyber power*. Cambridge, MA: Harvard Belfour Center.

Nye, J. (2013). From bombs to bytes: Can our nuclear history inform our cyber future? *Bulletin Atomic Scientists, 69*(9), 8–14.

Owens, W., Dam, K., & Lin, H. (Eds.). (2009). *Technology, policy, law and ethics regarding US acquisition and use of cyberattack capabilities*. Washington, DC: National Academy Press.

Rid, T. (2013). *Cyberwar will not take place*. London, UK: Oxford University Press.

Rosenzweig, P. (2012). *Cyber warfare: How conflicts in cyberspace are challenging America in changing the world*. Santa Barbara, CA: Praeger.

Rosenzweig, P. (2013). *Thinking about cyber security: from cyber crime to cyber warfare*. Chantilly, VA: The Great Courses.

Schneier, B. (2003). *Beyond fear*. NY: Copernicus.

Schneier, B. (2013). *Carry on: Sound advice from Schneier on security*. NY: Wiley.

Science Authors. (2015). The end of privacy. *Science*, *347*, 479–540.

Singer, P., & Friedman, A. (2014). *Cybersecurity and cyberwar*. London, UK: Oxford University Press.

Wolter, D. (2013). UN takes a big step forward on cybersecurity. *Arms Control Today*, September 25–29, 2013.

Zetter, K. (2014). *Countdown to zero days: Stuxnet and the launch of the world's first digital weapon*. New York, NY: Crown.

Chapter 15
Biological and Chemical Weapons

I will try and inoculate the bastards with [small-pox on] blankets that may fall into their hands, and try and take care not to get the disease myself.
[British General Jeffrey Amherst at Fort Pitt, 13 July 1763]

In five years we will know less about biological weapons than we do now.

15.1 BW History

The five anthrax letters put in the US mail in 2001 greatly raised concerns about BW and CW. In general, chemical weapons are considered about as lethal per unit mass as conventional explosives and much less lethal than either nuclear or biological weapons (dirty bombs are much less lethal, Chap. 13). Biological weapons can be as lethal as nuclear weapons if dispersed effectively, but that is not easy to do. For BW to be effective, the bio-materials must be chemically robust, remaining stable, not decaying in sunlight, moisture and rain. This can be accomplished with spores and aerosols. BW material must be dispersed widely, not clumping in few locations. BW can be delivered by cluster bomblets from small missiles. BW can be delivered by helicopter or low-flying crop-duster, or BW can be released in many locations in many small spray cans. Bio-disease is much more devastating if it transfers from person to person through air contact, rather than by touch or with liquids. The bio-weapon developers determine the best effective drop size to enhance killing. It is with good reason that the global community strongly supports the Biological Weapons Convention.

Comparisons between the relative effectiveness of nuclear, chemical and biological weapons are needed to understand this topic. Comparisons are scenario dependent and change with advances in bio-weapons. Future advances will make it easier to start pandemics. Infectious diseases, such as Ebola, smallpox and the plague, are particularly dangerous because of infectious spreading, which is a multiplicative effect. Research enhances choices of pathogens, increasing transmission in air (Table 15.1).

D. Hafemeister, *Nuclear Proliferation and Terrorism in the Post-9/11 World*,
DOI 10.1007/978-3-319-25367-1_15

Table 15.1 Effects of various weapons

Agents	Nuclear	Chemical	Biological
Area affected	About 75 km^2	Up to 60 km^2	Up to 100,000 km^2
Time to affect	Seconds	Minutes	Days
Structural damage	Widespread	None	None
Other effects	Fallout on 2000 km^2	Contamination; days or weeks	Possible epidemic; new disease foci
Time to normal use	3–6 months	Limited for days or weeks	Variable
Effect on humans	90 % deaths	50 % deaths	50 % disease

Comparison is between (1) 1-Mton hydrogen bomb, (2) 15 tonnes of nerve agent and (3) 10 tonnes of biological agent. Each weapon can be dropped by a bomber on an unprepared population. (Messelson, 1970)

President Obama's National Security Council issued a policy statement on biological weapons in November. 2009. The NSC statement reveals BW dangers but not BW solutions.

> The effective dissemination of a lethal biological agent within an unprotected population could place at risk the lives of hundreds of thousands of people. The unmitigated consequences of such an event could overwhelm our public health capabilities, potentially causing an untold number of deaths. The economic cost could exceed $1 trillion for each such incident. In addition, there could be significant societal and political consequences that would derive from the incident's direct impact on our way of life and on the public's trust in government.....We must reduce the risk that misuse of the life sciences could result in the deliberate or inadvertent release of biological material in a manner that sickens or kills people, animals, or plants or renders unusable critical resources.

Black death, or the *great plague*, originated in China in 1334, spreading to Europe in 1347 when trading ships landed in Messina, Sicily. Ultimately, 60 % of Europe's population died from plague between 1347 and 1350. Plague was spread through the air as well as through bites from fleas and rats. In some cases there was not sufficient population alive to bury the dead. Some give credit for economic, technical and social progress to the reduced population levels. The *black death* was followed by successive plagues but they were less lethal.

Bio-weapons are called *mass casualty weapons* since they kill people without destroying buildings and cities. Mass casualty weapons also spread fear, which is what terrorism is all about. Bio-weapons use microorganisms (bacteria and virus, or derivative spores and toxins), which cause disease in man, animals and plants. Biological agents can be categorized into 4 groups: bacterial agents, viral agents, rickettsial agents, and toxins.

- ***Bacterial agents*** that cause anthrax and tularemia are single cell organisms that invade host tissues and multiply, or they produce nonliving toxins (poisons), or both of these options. Some bacteria become spores that are more resistance to temperature and humidity than the original bacteria, and can be spread over greater distances and longer times. Because of their toughness, spores are more effective as a BW agent.

- ***Rickettsial*** agents cause typhus and Q fever, they are parasitic microorganisms that live and reproduce inside cells. Antibiotics can be successful in stopping them.
- ***Viral agents*** include the smallpox virus. The viruses are sub-cellular organisms that depend on host-cells for survival. Host cells can lead to death of host cells. It has been reported that there are only two smallpox samples in existence, one in Russia and the other in the US.
- ***Toxins***, such as ricin and botulism, are poisons generated by bacteria, fungi, algae, and plants. Toxins are not alive and do not reproduce and spread and they are less dangerous than living pathogens. However many toxins are much more toxic than chemical nerve agents.

15.1.1 Historical Examples of Biological Warfare

- 200 BC: Hannibal of Carthage flings poisonous snakes on enemy ships.
- 1155: Emperor Barbarossa poisons water wells with human bodies, Tartona, Italy.
- 1346: Mongols catapulted bodies of plague victims over walls into Caffa, Crimea.
- 1495: Spanish mix leprosy-patient blood with wine to sell to French foes, Naples.
- 1650: Polish fire saliva from rabid dogs towards their enemies.
- 1675: German and French forces agree not to use *poison bullets*.
- 1763: English distribute blankets from smallpox patients to Native Americans at Ft. Pitt.
- 1797: Napoleon floods plains near Mantua, Italy to enhance malaria.
- 1940: Japan drops bubonic-plague-infected fleas and rice on Chuhsien, China.
- 2001: Letters containing anthrax spores mailed to two US Senators and to journalists.

15.1.2 Pandemic Threats

A pandemic is an epidemic of infectious disease that has spread through human populations across a large region or multiple continents. A pandemic can be caused by a natural epidemic or by malicious use of biological weapon materials. There have been pandemics caused by cholera, influenza, typhus, smallpox, measles, tuberculosis, leprosy, malaria and yellow fever. Pandemics might be caused by smallpox, viral hemorrhagic fevers, antibiotic resistance, SARS, influenza, Ebola and biological weapon materials.

Influenza. *The Spanish Flu* location was misnamed to calm US soldiers in the First World War. It was first reported in January 1918 in Kansas, while Spain was not in the war. Spanish Flu infected 500 million persons between 1918 and 1920, killing 50–100 million (3–5 % global population). Influenza virus kills by an over-reaction of the body's immune system (a cytokine storm). It is most deadly for those under 5-years and those over-65. The 1918 Spanish Flu and the 1976 Swine Flu were both H1N1, with the same H and N proteins, which help the virus invade host cells. The Hong Kong Flu killed one million in 1968. The 2002 SARS Bird Flu, H5N1, spread to 37 nations because of a lack of transparency. The 2009 H1N1 virus killed 18,000 world-wide. The 2013 H7N9 was less virulent, as nations worked together with more transparency, sharing data, samples and vaccines.

More than 500 million people died of infectious diseases during the past century. The Ebola pandemic in Western Africa heightened concerns in 2014–15. Persons infected with Ebola die with a probability of 25–75 %. Ebola is easily transmitted from person-to-person. A few days after contracting the Ebola virus, the patient typically has a fever, sore throat, muscle pain and headaches, which is followed with vomiting, diarrhea and rashes. This is followed with decreased functioning of the liver and kidneys. At this point the patient begins to bleed internally and externally. The 2014 Ebola pandemic in Western Africa reported 21,724 cases with 8641 deaths as of 18 January 2015.

Anthrax spores were included in letters sent to two US Senators and journalists a week after the 9/11 attack. The letters contained this message:

> You can not stop us. We have this anthrax. You die now. Are you afraid? Death to America. Death to Israel. Allah is great.

The anthrax letters caused the death of five persons; seventeen others contracted anthrax but survived. White House staff and others took the powerful antibiotic Cipro as a precaution. The anthrax attack generated a huge response by costing more than $200 million to clean two large post-office process centers by quarantining 1.8 million letters, magazines and packages and by interviewing 10,000 witnesses, with subpoenas for 5750 grand-jury witnesses. The FBI determined the sole perpetrator was Bruce Irvins, a researcher at the US Army Fort Dietrich BW Laboratory.

Anthrax is not usually directly spread to persons from infected persons or animals, but rather it is spread by spores, which is a dormant form of the bacteria. Spores are robust, they can survive harsh conditions for decades and centuries. Spores, lying on the ground, can be transferred to humans. An animal dying from anthrax contains naturally-produced anthrax spores. When inhaled, ingested, or in contact with a skin lesion, spores may become active and multiply rapidly. Effective vaccines against anthrax are available and anthrax can be controlled with antibiotics. The US and other nations researched anthrax bacteria that are immune to antibiotics. A dose of 10,000 anthrax spores can be lethal, while plague needs only 500 organisms. Implementing safe handling of Anthrax has had logistic problems.

Mistaken shipment of live anthrax bacteria took place 74 times to dozens of US labs and five foreign labs from Dugway Proving Ground in Utah.

Smallpox. Smallpox is considered eradicated, according to the World Health Organization in 1979. However, samples are stored in two laboratories in the US and Russia. Smallpox is much feared as it is transmitted by airborne virus from one person to another, regenerating it self in the new victim. Smallpox needs only 10–100 organisms for a fatality. Preventative vaccinations are highly effective. Drugs are of limited value for those infected with smallpox.

15.2 BW Control with BWC

Short history. The first modern efforts to control biological and chemical weapons took place with the 1874 International Declaration on the Laws and Customs of War, which banned the use of poisons on the battlefield. The 1899 Hague Convention stated that, in any war between signatories, the parties will abstain from using projectiles "the sole object of which is the diffusion of asphyxiating or deleterious gases." This was ratified by all major powers, except the United States. These declarations were severely violated by Germany in World War I, to which the Allies responded. A tremendous amount of CW was used, some 125,000 tonnes of chlorine gas (1915) and mustard agent (1917). This negative experience encouraged adoption of a new, stronger regime, producing the 1925 Geneva Protocol, a no-first-use doctrine. The Protocol was ratified by many countries, except the US and Japan. The Senate Foreign Relations Committee favorably reported the Protocol, but strong lobbying by the chemical industry and the American Chemical Society prevented a Senate vote.

The Germans prepared substantial chemical weapons but were deterred from their use by President Franklin Roosevelt, who declared US intentions in 1943:

> Use of such weapons has been outlawed by the general opinion of civilized mankind. This country has not used them, and I hope we never will be compelled to use them. I state categorically that we shall under no circumstances resort to the use of such weapons unless they are first used by our enemies…. Any use of gas by any Axis power, therefore will immediately be followed by the fullest possible retaliation [in kind] upon munitions centers, seaports, and other military objectives throughout the whole extent of the territory of such Axis country.

Gerhardt Schrader of I.G. Farben was working on a new class of insecticides, called organophosphates, when he discovered a compound too deadly to be used as an insecticide. Because of the close cooperation of German industry, academia and the military, this new agent, called Tabun, was soon undergoing full-scale production. The Germans also discovered other nerve agents, like Sarin. The Allies had nothing to compare to Germany's Tabun stockpile and were unaware of it, but nevertheless they sent word to the Nazis that poison gas use would result in overwhelming retaliation. It is generally assumed that the fear of strong retaliation by the Allies deterred Hitler from using them. Hitler was gassed in World War I, having first-hand experience of poison gas suffering. Also, Hitler wrongly assumed the US had developed nerve agents since all publication of US research related to pesticides stopped prior to the war. However, this was due to the discovery of DDT by the US and had nothing to do with nerve agents.

President Richard Nixon renounced the use of CW and BW in 1969, stating that the US should comply with the 1925 Geneva Protocol. President Nixon ordered all US BW stockpiles to be destroyed. At that time the E133 cluster bomb was the leading US bio-weapon. It contained 536 biological bomblets, each holding 35 ml of anthrax spores. This was a significant step. President Nixon submitted the Geneva Protocol to the Senate, which ratified it on 19 January 1975 under President Gerald Ford. The Geneva Protocol was limited by the ratification debate in the Senate Foreign Relations Committee. This was not a legal change to the Protocol but it modified it, in practice, by merely answering a question:

> ***SFRC***. Assuming the Senate were to give its advice and consent to ratification on the grounds proposed by the administration, what legal impediment would there be to subsequent Presidential decisions broadening the permissible use of herbicides and riot-control agents?
>
> ***Executive Branch***. There would be no formal legal impediment to such a decision.

In the 1990s, at the end of the Cold War, it was determined that the Soviets had a significant biological weapons program in violation of the BWC. By 2000 it was generally believed that there were a dozen countries that had biological warfare programs. The *dirty dozen* was thought to include Iraq, Iran, Israel, North Korea, Syria, Libya, Russia, and possibly India, Pakistan, China, Egypt, and Sudan. Most of these programs were research programs but 3 nations are believed to have produced and stockpiled agents (Iraq, Iran and Russia) and three other nations may also have done so (Israel, North Korea, and China). Previously South Africa and Taiwan had been on the BW-stockpile list but they are no longer so considered.

15.2.1 The BWC

On 10 April 1972, the US signed the *Biological Weapons Convention* (*BWC*). Pound for pound, biological weapons could be as effective as nuclear weapons in destroying life.

The key provisions of the BWC are as follows:

Article I: Never develop, produce, stockpile or acquire BW or BW-delivery systems.
Article II: Destroy BW and equipment not later than 9 months after 16 March 1975.
Article III: Never transfer BW, BW-equipment, or its means of delivery to any recipient.
Article VI: Lodge complaint with UN Security Council on violations of others.
Article X: All have the right to participate in full exchange of commercial equipment, materials and information *for peaceful purposes*.
Article XIII: BWC is for unlimited duration.

BWC does not have a Verification Protocol, which was rejected by US in 2001.

As of December 2014, 171 nations ratified or acceded to the BWC. Egypt, Syria and seven African nations signed but did not ratify. Israel and fifteen African nations have not signed the BWC.

15.2.2 BWC Monitoring and Verification

It is difficult to detect small, but significant, quantities of bio-weapon materials. This realization shifted the response from BW *detection and mitigation* to BW *prevention and protection*. Both the Chemical Weapons Convention and Biological Weapons Convention ban the production, acquisition, stockpiling, and transfer of these materials. The CWC has a verification regime, while the BWC does not.

Industrial role in treaty negotiations. It is well known that the chemical industry, led by the Monsanto Corporation, contributed to the defeat of the Geneva protocol in the 1925 Senate. To Monsanto's credit, they took a very aggressive public role in support of CWC negotiations and its ratification, which I observed as a Senate Foreign Relations Committee staff person on the issue in 1990–92. The CWC verification regime is discussed in Sect. 15.4, but for now we note the fact that the chemical industry is less vulnerable to loss of industrial secrets in CWC as compared to the pharmaceutical and biomedical industries in the BWC. Biotechnology is driven strongly by high technology research, as compared to the mass-production of chemicals. The pharmaceutical and biomedical industries did not support the draft BWC verification protocol for this reason. It will be difficult to devise an effective BWC monitoring regime without the loss of some bio-industrial trade-secrets. But, the international community should keep trying to develop a BWC monitoring regime. In the meantime, partial measures can be adopted to strengthen BW declarations by enhancing transparency between nations with confidence-building measures. This can improve implementation of the BWC.

BW detection. Sophisticated devices, such as the Handheld Advanced Nucleic Acid Analyzer, detect pathogens in the field by examining DNA of samples and comparing to known-DNA pathogen sequences. Gas chromatographs, mass spectrographs, and other sensors can identify many of the bio-materials in the laboratory. The US National Laboratories are working on these issues.

BW research. Unfortunately, current biological research for peaceful purposes can lead to findings that could be misused for hostile purposes. In response, the National Academy of Sciences recommended criteria for identifying particularly dangerous experiments. It is interesting to point out that the NPT, the BWC, the CWC and the Missile Technology Control Regime (MTCR) all allow exceptions for *peaceful research*.

Article X of the BWC gives the States Parties:

> the right to participate in, the fullest possible exchange of equipment, materials and scientific and technological information for use of bacteriological agents and toxins for peaceful purposes….This convention shall be implemented in a manner designed to avoid hampering the economic or technological development of States Parties to the Convention or international cooperation in the field of peaceful bacteriological activities….

Interaction of offense and defense. The five anthrax letters, mailed a week after 9/11, greatly raised US national security concerns. This happened, not so much by the five deaths and 17 casualties, but more by the perception of vulnerability to something large and uncertain in the future, something that we can't control. The public literature suggests that Al Qaeda and other terrorists are moving towards biological weapons. Their agents, anthrax and ricin. At the time of the anthrax attack, the civilian bio-defense 2001 budget was $414 million, but it rose by a factor of 18 to $7.6 billion in 2005. The National Institutes of Health asked to construct 6–10 new biosafety level-3 and level-4 facilities to augment the existing 7 level-4 facilities, those that work on the most dangerous pathogens for which there are no known cures. The argument goes that we must be up-to-date on what can be done to the US, so we can respond with antibiotics or other treatments to nullify newer types of poisons.

Anthrax research. If we develop a strain of anthrax that is resistant to antibiotics, we could try to develop a new antibiotic that would stop the antibiotic-resistant anthrax. This also means there is a possibility that the results of this research will not remain secret and others will learn how to make anti-biotic-resistant anthrax. US research on anthrax spores with antibiotics resistance was revealed in 2002, in response to investigations on the anthrax letters. The US was committed to declare these results, pursuant to the 1986 agreement on BWC confidence building measures, but it did not do so.

There also is a gamble that we might succeed to develop new antibiotics to curtail anthrax, but the enemy, in the meantime, has worked on the antibiotic-resistant strain of the plague. To our regret we could bet on the wrong horse if these bio-secrets leaked out. It would be useful to have an internationally recognized code of conduct that increased awareness and accountability of scientists to reduce risks from biological research and development. The National Academy of Sciences in 2003 recommended the National Institute of Health create a review system for seven classes of experiments that raise concerns. The experiments of concern are those that do the following:

- demonstrate how to render a vaccine ineffective,
- confer resistance to therapeutically useful antibiotics or antiviral agents,
- enhance virulence of a pathogen or render a non-pathogen virulent,
- increase transmissibility of a pathogen,
- alter host range of a pathogen,
- enable evasion of diagnostic/detection modalities and
- enable weaponization of a biological agent or toxin.

15.3 CW History

Biological poisons were first used in warfare, as described in Sect. 15.1. Chemical weapons were used in the distant past, as ancient Greeks mixed sulfur and pitch-resin to engulf enemies in poisonous fumes during the Trojan War. The dramatic shift to CW took place when Germany released chlorine gas on Allied troops at Ypres, Belgian in 1915. The German CW program was fathered by Fritz Haber, who later was awarded the Nobel Prize for inventing the Haber process, which fixed nitrogen by synthesizing ammonia from nitrogen and hydrogen gases. Haber and the Germans followed with mustard gas in 1917, and the Allies responded. A total of 125,000 tonnes of CW was used in World War I, killing 90,000, with 1.3 million casualties. This carnage led to the Geneva Protocol of 1925, a no-first-use doctrine.

World war II. The Italians used mustard gas in Ethiopia in 1935, killing 15,000. This was followed by Japanese attacks on China and Manchuria, killing thousands during World War II. CW was not used in Europe in World War, primarily because President Roosevelt stated the US would not use CW, but if Germany used CW, the US would respond in kind. Germany produced 11,000 tonnes of Tabun between 1942 and 1945; they did not know that the Allies did not have such a stockpile. Chemical weapons got a deservedly bad reputation "as the weapon that even Hitler would not use." President Roosevelt was asked to certify CW use on Iwo Jima during World War II. This would have been a particularly good military case for CW because the Japanese were holed in caves and vulnerable. However, Roosevelt was consistent with his previous statements of non-use, he did not certify the attack.

The US used 11,000 tonnes of herbicides and riot-control tear gas in Vietnam between 1962 and 1967. These gases were considered somewhat benign, but soldiers who suffered from Agent Orange, which contains dioxin, would disagree. The US was accused by many nations of chemical warfare in Southeast Asia by its massive use of defoliants and tear gas. From 1980 to 1988, Saddam Hussein's Iraq used CW against Iran and his Iraqi-Kurd citizens, killing 50,000. Iraq developed its CW capability with help of Western companies, including those in Germany and the US, often in violation of law.

Aum Shinrikyo, the Japanese Cult, became the first non-state actor to use potent CW. They attacked the Tokyo Metro with Sarin and cyanide, killing 13, seriously injuring 54 and wounding 1000 on March 20 and May 5, 1995. This showed that non-state actors can disseminate CW (with limited success in this case), becoming equalizers for marginal groups. Following this event, sensors to detect CW were deployed in the richer cities of the world. The cult was not as effective as they might have been but this could change. In response to these events, the CWC entered into force in 1997, banning use, storage, production and sale of CW. Very good progress has been made, reducing CW stockpiles of five nations.

Syrian CW. The Syrian Assad regime attacked its citizens in a civil war in 2013, killing 1000. This prompted Russia to forcefully coerce Syria to join the CWC. A dozen nations, including US and Russia, worked together to successfully destroy 1000 tonnes of Sarin and other materials in Syria within a year.

15.3.1 Types of CW

There are about 70 viable chemical CW agents. These can be difficult to handle because of reactivity and toxicity. Most CW agents are liquids that must be able to withstand long-term storage without deterioration. Old CW canisters have become dangerous, as was discovered in the area near American University in Washington, DC, where CW was researched in World War I. The chemicals must be robust, to retain effectiveness when delivered through atmospheric water vapor and free-oxygen. CW agents must be able to withstand heat generated by explosive detonation. Liquid CW agents should evaporate naturally for best dispersal. The weapons designers are concerned with droplet size to enhance their effectiveness.

Chemical weapons are categorized by their effects, rather than their chemical similarities. There are five general categories:

- ***Blood gases***, such as hydrogen cyanide, poison cells by blocking oxygen transport in blood.
- ***Blistering agents***, such as lewisite and mustard gas, penetrate mucous membranes and body tissues to react with proteins, enzymes and DNA to destroy cells. The eyes, skin and air passageways are very vulnerable.
- ***Choking agents***, such as phosgene and chlorine, damage the lungs and cause suffocation. Choking agents must be inhaled to harm the person.
- ***Nerve agents***, such as Sarin, Tabun and VX, diminish transmission of nerve pulses in the nervous system, causing death. Nerve agents spread quickly after skin contact or inhalation. The cost to produce VX in the 1960s was about $5 a kilogram, where 1 kg contain 2 million lethal doses of about 0.5 mg each.
- ***So-called incapacitants***, such as fentanyl, LSD or Agent BZ, are intended to disable enemies temporarily. The Russians used fentanyl in 2002 to attack Chechen hostage-takers in Moscow, killing 130 hostages in the process.

Delivery methods. Chemical weapons are disseminated primarily as liquid droplets or aerosols. The dispersion of a chemical cloud strongly depends on environmental conditions, such as wind velocity, turbulence and humidity. CW effectiveness is diminished if temperature is too low or there is rain. On the other hand, warm temperatures and high humidity enhance toxicity of the chemical cloud.

There are many military devices to spread CW: artillery shells, aerial bombs, cluster bomblets, big and small spray tanks, missiles, rockets, helicopters, crop-dusters, mines and grenades. A B-52 bomber can carry 20 tonnes of VX nerve gas, fatal over 200 km^2. The most likely delivery approaches are low-flying helicopters or airplanes and by artillery shells, containing CW bomblets. To this list, we now must include the use of drones (Sect. 13.5). On 26 January 2015, a $1000, 70 cm × 70 cm, quadcopter successfully penetrated the White House grounds. Its arrival was detected visually. The drone could have carried BW or CW.

Binary chemical weapons. Binary weapons combine two chemicals into a CW material after the shell is fired. Binaries were devised to make CW weapons safer for manufacture, transport and storage. The two originating chemicals are not nearly as toxic as the final combined CW agent. Since the chemicals are combined in-flight, the toxicity of the CW is not an issue until it is on target. The cylindrical projectile is divided in half, with a membrane, or rupture disc, between the two compartments. The firing of the artillery shell breaks the rupture disk to start the mixing process. The rotating artillery shell completes the mixing process. Both the US and the Soviet Union developed binary CW weapons, but they were never used. It was reported that the US Bigeye bomb had significant problems, which increased the political pressure for its demise. President George H.W. Bush withdrew the Bigeye bomb from development, as he promoted the Chemical Weapons Convention during 1990–1992. Both Iraq and Syria developed a variant in which one chemical was stored in the weapons, while the other was added and mixed with the first component just before the bomb or missile was deployed.

CW defense. Both the Soviet Union and the United States spent considerable funds preparing their troops against CW attacks. A tight-fitting gas mask reduces CW concentrations in air breathed in by a factor of 100,000. Exotic, protective clothing (MOPP suits) can be very helpful, but they are very cumbersome in the battle zone, particularly in hot climates. This issue played a role in the decision to start the second Iraqi War in March 2003. President George W. Bush did not want to delay the decision to avoid Iraq's CW weapons in the hot summer, which would hamper US troops. Ironically, most MOPP suits provided to US forces in Iraq were found to be defective, due to improper sealing of seams. More time was needed to look for Iraq's weapons of mass destruction and for diplomacy to have an effect.

15.4 CW Control with CWC

US interest in controlling chemical weapons was stirred in the 1980s by Iraq's massive use of CW against its Kurd citizens and those of Iran, killing 50,000. A draft Chemical Weapons Convention was submitted to the UN Conference on Disarmament in 1984. The verification measures included *managed access monitoring* and *mandatory challenge inspections*. Negotiations were accelerated in 1990 when the US and the Soviet Union signed an agreement on chemical weapons. The two countries agreed not to produce chemical weapons, to reduce stocks of CW to 20 % of their current levels, and to begin CW destruction in 1992. They also agreed to have less than 5000 tonnes of CW by 2002. The bilateral chemical weapons agreement never entered into force but it showed a strong willingness by the two largest CW nations to work together to eliminate CW.

As political differences between the superpowers diminished, other issues became larger. Several Arab nations linked CW disarmament to progress on nuclear disarmament, an approach aimed in part at Israel's nuclear program. The Australia Group, a group of nations opposed to BW and CW weapons, wanted to strengthen export controls, while maintaining free trade. Lastly, the US withdrew its condition of maintaining a small CW stockpile to respond to violations.

CWC entered into force on 29 April 1997 with 87 states parties. By mid-2015 CWC had 191 members, representing 98 % of the global chemical industry. Israel signed but did not ratify the CWC, while Angola, Egypt, North Korea and South Sudan did not sign. CWC has established a strong global consensus to ban CWC. Under the CWC, a number of nations declared possession of chemical weapons; India, South Korea, Soviet Union, United States, Albania, Iraq, Libya and Syria. By 2015, 90 % of the global CW stockpile of 71,200 tonnes had been destroyed, which is slated to be completed by 2023.

15.4.1 Terms of the CWC

Prohibitions:

- develop, produce, acquire, stockpile or retain CW,
- direct or indirect transfer of CW,
- use of CW or military preparation for use,
- assist, encourage or induce other states to engage in CWC-prohibited activity,
- use of riot control agents as a method of warfare.

Required written declarations:

- CW stockpiles,
- CW production facilities,
- relevant chemical industry facilities and
- other weapons-related information.

CW Stockpile Definitions:

- Schedule-1: High-risk military agents with no- or low-commercial use.
- Schedule-2: Significant-risk precursors and toxic chemicals that are not produced in large quantities for commercial use.
- Schedule-3: Some-risk duel-use chemicals.

On-Site Activity:

- Routine Inspections,
- Challenge inspections, and
- Investigations of alleged use of CW.

Trade:

- restrict trade with non-state parties on Schedule-1 and Schedule-2 chemicals,
- measures to ensure Schedule-3 transfers to non-states parties are not CW,
- CWC encourages trade among states-parties.

15.4.2 Russia and US Help Syria Join CWC

The Middle East had three primary CWC holdouts, Egypt, Israel and Syria in 2012. In December 2012, reports alleged that Syria was involved in a major Civil War, using chemical weapons on its citizens. This was greatly amplified on 21 August 2013 when Syrian civilians were attacked with Sarin near the rebel-controlled, city of Ghouta, east of Damascus. It was reported that 1400 people died, including several hundred children. Photos of victims, frothing at the mouth and suffering suffocation, were displayed.

A UN team was dispatched to the area and confirmed the validity of the attack:

> Chemical weapons have been used in the ongoing conflict between the parties in the Syrian Arab Republic, also against civilians, including children, on a relatively large-scale.

Following a US-Russian agreement in early September 2013, Syria formally acceded to the CWC. This fast turnaround resulted primarily from strong Russian political pressure on its Syrian ally. The accession of Syria to CWC put in motion steps to destroy the Syrian CW stockpile. Under the ambitious plan, the US and Russia agreed to destroy the stockpile of 1000 tones in a year. This was unprecedented for its swiftness. To implement the plan, a joint operation was established between the CWC's implementing organization—the OPCW (Organization for the Prohibition of Chemical Weapons)—and the UN.

Early on, it was decided that the chaos and threats from the Syrian Civil War would prevent destruction of chemical agents in Syria and that the chemicals in the Syrian stockpile would need to be destroyed outside Syria. It would be difficult enough to transport CW to its seaport, Latakia. Realizing the difficulties in finding a land site for a destruction facility, the US modified an older freighter, the *MV Cape Ray*, to neutralize CW at sea. In November 2013 it was modified to carry the Field Deployable Hydrolysis Systems, to neutralize CW. A similar system had neutralized 1500 tonnes of mustard agent from the US stockpile at Aberdeen between 2003 and 2005. The Syrian stockpile contained sulfur mustard agent, Sarin nerve agent and precursor chemicals. Industrial chemicals in the Syrian stockpile were sent to commercial facilities for destruction.

The first CW batch left Latakia on 7 January 2014 and the last shipment left Latakia on June 23. The Cape Ray was not allowed into Syrian waters. Other ships transported the CW to southern Italy, over the protests of the local citizens. Sympathetic demonstrations of 10,000 took place in Athens, Istanbul and Cyprus. The Cape Ray began neutralizing CW on July 7. There were no major incidents onboard the Cape Ray. It delivered neutralized sulfur–mustard effluent to Bremerhaven, Germany and neutralized nerve agent precursor to Finland. As of October 20, 2014, 100 % of Category-1 chemicals (1047 tonnes) and 89 % of Category-2 chemicals (232 tonnes) were destroyed. The remaining 29 tonnes are to be destroyed by late 2015.

Unfilled missiles and bombs were destroyed in Syria. Chemical weapons production facilities were also destroyed in place. Overall, the elimination of Syria's chemical weapons program was an extremely complex operation, involving financial and in-kind contributions from a large number of countries. The degree of international cooperation was unprecedented and a key factor in the success of the operation

Problems

15.1 **BW effectiveness**. How could BW be made to be more lethal?
15.2 **Cluster bomblets**. Why are bomblets more lethal?
15.3 **Black Death**. Describe European plague epidemics. How did this help Newton?
15.4 **Anthrax spores**. What are they, how produced, why effective?
15.5 **Antibiotic resistant?** What research was done? Where? How? What are the trade-offs?
15.6 **Spores versus toxins**. What are the differences in production and use?
15.7 **BW versus TNT?** Differences in use, results and modalities?
15.8 **General antibiotic usage**. How can antibiotics be compromised? Data?
15.9 **Anthrax and small pox attacks?** How should the government respond?
15.10 **BW and CW budgets**. How much did the US and USSR spend on BW/CW?
15.11 **Herbicide and riot-control agents**. How much US usage?
15.12 **Yellow Rain**. What was this? Was it caused by humans or bees?
15.13 **Dual use of technologies**. What does Article X of the CWC do? Peaceful use exemptions in BWC/NPT/MTCR?
15.14 **Unratified and unsigned**. Which nations have not signed, or signed but not ratifed the BWC, CWC, and NPT?
15.15 **Deterence.** What are successful examples of deterrence with strength in BW and CW. What are negative examples?
15.16 **Negatives from research**. What are examples of R&D that exacerbated CW and BW?
15.17 **Code of ethics for scientists**. What CW or BW research should be avoided?
15.18 **CW stockpiles**. Status of US/RF/Korea/India/Syria CW stockpiles?
15.19 **MV Cape Ray**. Describe this ship and how it uses hydrolysis neutralization?
15.20 **Managed access**. Describe how managed access works in CWC inspections.
15.21 **Mandatory challenge inspections**. Describe how this works and its record.

Seminar on BW and CW

Anti-bacterial resistant BW, anthrax, bacterial agent, binary Buckeye bomb, bio-safety level 3 and 4, blistering agents, blood gases, botulism, BW budgets, BWC, BWC verification protocol and the tradeoffs for acceptance, BW monitoring technology, challenge inspection, chlorine, chocking agent, CWC, drones, delivery of BW bomblets on missiles, Fritz Haber, Geneva Protocol, Hague Convention, incapacitants riot control, influenza, managed access, Monsanto on Geneva Protocol and on CWC, nerve agent, non-members to BWC and CWC, poor man's nuclear weapon, Sarin, Schedules 1–3, small pox, Spanish flu, Syrian CW decommissioning, toxins, yellow rain, viral agent

Bibliography

Casagrande, R. (2002, October). Technology against terror. *Scientific American, 287*(4), 82–87.

Choffnes, E. (2002, September). Bio weapons: New labs, more terror? *Bulletin of the Atomic Scientists*, 29–32.

Cirincioni, J., Wolfstahl, J., & Rajkumar, M. (2002). *Deadly Arsenals, Chapter 4*. Washington, DC: Carnegie Endowment for International Peace.

Diamond, K., & Kadlec, R. (2012, December). Accelerating threat, stalled strategy: A call to action on biological weapons. *Arms Control Today*, 15–21.

Leitenberg, M., & Zilinskas, R. (2012). *The Soviet biological weapons program: A history*. Cambridge, MA: Harvard U.

Messelson, M. (1970). Chemical and biological weapons. *Scientific American, 222*(5), 15–25.

Miller, J., Engelberg, S., & Broad, W. (2002). *Germs: Biological weapons and American's secret war*. New York, NY: Simon and Schuster.

National Research Council. (2009). *Counting biological threats*. Washington, DC: National Academy Press.

National Research Council. (2011a). *Prudent practices in the laboratory*. Washington, DC: National Academy Press.

National Research Council. (2011b). *Trends relevant to the biological weapons convention*. Washington, DC: National Academy Press.

National Research Council. (2012a). *Biosecurity challenges of the global expansion of high-containment biological laboratories*. Washington, DC: National Academy Press.

National Research Council. (2012b). *Determining core capabilities in chemical and biological defense science and technology*. Washington, DC: National Academy Press.

National Research Council. (2014). *Assessing the biological weapons and bioterrorism threat*. Washington, DC: National Academy Press.

Roffey, R., Hart, J., & Kuhlau, F. (2006, September). A code of conduct for bio-defense scientists. *Arms Control Today*, 17–20.

Tucker, J. (2004, October). Biological threat assessment. *Arms Control Today*, 13–19.

Tucker, J. (2007). *War of nerves: Chemical warfare from WW I to Al Qaeda*. New York, NY: Random House.

Walker, P. (2014, December). Syrian chemical weapons destruction: Taking stock and looking ahead. *Arms Control Today*, 8–17.

Wheelis, M., & Dando, M. (2002, January). Back to bioweapons? *Bulletin of the Atomic Scientists*, 41–46.

Chapter 16
Conclusions

16.1 Three Main Issues

This chapter summarizes the text, covering enough topics to spin your heads. Many technical aspects of proliferation and terrorism will remain relevant into the future, while often political aspects are less permanent, shifted by events. Our goal for this course is to equip an upper-division history major to enter the State Department's Foreign Service to work on these issues.

Let's first summarize where the world is heading in 2015 in the three major areas (strategic nuclear weapons, proliferation of nuclear weapons, and terrorism). It is hard to be too cheerful in these summary remarks, but that has always been the situation. I recall looking at the ceiling above the Senate floor, realizing that I was sitting in the cross-hairs of Soviet ICBMs. There is much that can go wrong. Only by diligent work, with some good luck, can we improve the condition of the globe. Collectively, we have it in our power to improve these basic areas:

Strategic Weapons: President Obama was rebuffed by Russian President Vladimir Putin on additional controls on strategic nuclear weapons. Putin states and believes that the splintering of Ukraine from Russian markets and influence was Western aggression. This conclusion is not believed in the West, but it is true that the extension of NATO into Eastern Europe reduces Russia's comfort zone. The West believes Russia broke an agreement made at the end of the Cold War to maintain established national boundaries. This dispute will continue for some time, at least a decade. Putin's replacements will continue this path. Russia can't compete in global manufacturing markets, except on nuclear power. Russia's financial plight is exacerbated by reduced petroleum prices. The new world order, that began in 1991, ended by 2010. It is time to re-read George Kennan's 1947 State Department papers on containment.

China is building new islands in the South China Sea. This land-grab polarizes South and East Asia. Hopefully the desire for global markets, favorable to both sides, will dampen the political trajectories. China will deploy additional strategic

D. Hafemeister, *Nuclear Proliferation and Terrorism in the Post-9/11 World*,
DOI 10.1007/978-3-319-25367-1_16

nuclear weapons, to catch up with the US and Russia. China's military leaders wish that the US would resume nuclear testing, so they can catch up. There are many reasons why the three super-powers, the US, Russia and China, should work together on trade, while reducing nuclear deployments. It is difficult to see how this will happen. It behooves us to continue to try for restraint, as a robust arms race would not be helpful.

Proliferation of Nuclear Weapons: The NPT regime codifies the two classes of states, those with nuclear weapons and those without nuclear weapons. This separation of states is not cast in concrete as the legal nuclear weapon status of India survived the US Congress and the Nuclear Supplier Nations. There are many reasons to be nervous about the long-term effectiveness of the NPT regime. Rather than make a list of problems, let us begin with a big success story. Over 20 non-nuclear weapon states started nuclear weapon programs. Only four of these 20 NNWS dalliances developed into viable nuclear weapon programs (North Korea, India, Israel, Pakistan). Two of these are somewhat modified, with India now considered an almost NWS, and North Korea awaiting Chinese pressure to cease and desist on nuclear weapons.

Much depends on the success of the deal arranged between the P5 + 1 and Iran. The Joint Comprehensive Plan of Action (JCPOA) shuts down paths to nuclear weapons at the centrifuge plants at Natanz and Fordow and at nuclear reactors. It would be foolhardy for Iran to cheat against these strong verification provisions. In 15 years Iran will not have these strong constraints on its enrichment program, but working on nuclear weapons will continue to be against the NPT-law. The economic sanctions can be "snapped back" and other stronger measures can be applied. The hope is that a maturing Iran will enjoy western capitalism, its citizens will want that, but the leadership can be another matter. It is not clear that a regime of only economic sanctions would have brought Iran to its knees. We had observed a growing number of centrifuges while the sanctions were in place.

Watching in the wings are the Sunni Arab states, such as Saudi Arabia, claiming they will go to the bomb if Iran does. There are other states that will consider nuclear weapons, it will take diplomatic diligence to keep the NPT robust. Since the regime is asymmetric, the US should keep this in mind and do what it can to reduce the differences. Ratifying the CTBT would be a good start. A good second step would be to remove "spirit of the law" violations used by the US, such as the US providing weapon platforms for UK and Israeli nuclear weapons.

Terrorism: The cruel murder of fourteen bystanders in San Bernardino by terrorists on 2 December 2015 displayed the political leverage of terrorist events. At this point it is not known if the attack was locally managed or by ISIS. A large majority of Americans want a strong response to this attack, but few are calling for "US boots on the ground" to fight ISIS. There is a call to loosen the "rules of engagement" to strengthen the US response, and to send more equipment to the Kurds to do the fighting. Two or three individuals can cause major shifts in US foreign policy.

Since the 1 August 1966 University of Texas Tower killing of 16, there have been 124 mass shootings in the US, killing a total of 814, at an average rate of

17 per year. These attacks were carried out mostly by deranged US citizens, without international terrorist connections. The background data caused by mental health and guns does not diminish the vulnerability of the US to ISIS-lead terrorist attacks using nuclear, chemical, biological materials, or by cyber attacks on the institutional framework of the country with its electrical power grid, financial markets and banking systems. The attack on *supervisory control and data acquisition systems* (SCADA) that control technical systems has been begun. The world modernized, globalized, miniaturized and computerized to become more vulnerable to attacks by religious terrorists. The populace is more fearful of ISIS terrorism and less concerned about its privacy rights.

16.2 Chapter Conclusions

1. History of the Atomic Age

1.1 Knowledge in the nuclear sciences exploded from **Rutherford's 1909 discovery** that the nucleus is very small, to nuclear weapons in the 1940 s, to US and Russian overkill arsenals, to the world of 2015 with centrifuge enrichment, GPS locations, drones, nuclear forensics and space platforms.

1.2 Hitler's government demanded that ***Aryan Physics*** be taught in place of Einstein's ***Jewish Physics***. Anti-Semitism transferred significant physics brainpower from Germany to the US.

1.3 Leo Szilard was the first to realize and ***patent the idea of a run-away chain reaction***, producing a nuclear explosion. Szilard attempted to keep scientists from publishing facts that would lead others to the bomb. Szilard authored two letters to Roosevelt, with Albert Einstein's signature, that launched the US Manhattan Project. At the close of the war Szilard authored a third Einstein letter that suggested nuclear restraint against Japan, as a way to begin international control.

1.4 Enrico Fermi produced considerable artificial radioactivity when ***bombarding natural uranium with neutrons***. He speculated that the level of radioactivity would be reduced if uranium and the neutron source were placed underwater. He believed that water would reduce the flow of neutrons to the uranium target. Fermi was surprised when he obtained much *more* radioactivity with the water-bath experiment, rather than *less* radioactivity. What Fermi did not realize was that slow neutrons (thermal neutrons) interact 1000 times more strongly with uranium than fast neutrons from fission.

1.5 The released energy from a fission weapon is a ***million times larger than the chemical energy*** released from reaction of the same mass. This reduction in mass of by a factor of a million allows bombers and missiles to deliver fission weapons.

1.6 Experiments in 1939 showed that ***more than two neutrons are released*** when a uranium nucleus fissions (splits) with a neutron, fulfilling Szilard's requirement for an atomic bomb.
1.7 A German attempt at criticality failed in January 1941 at the Berlin "virus-house" reactor, using 300 kg of uranium with a graphite moderator. This reactor could have gone critical, but too many neutrons were absorbed in ***impure graphite, which contained boron***, a strong neutron absorber. These failed results convinced the Germans to base their reactor program on heavy water moderators.
1.8 START I was followed with the 1993 signing of the START II Treaty, lowering the limit to 3500 strategic warheads and ***banning land-based MIRVed missiles***, such as the Peacekeeper and the SS-18. In 1997 this was followed by START III pre-negotiations between Presidents Bill Clinton and Boris Yeltsin to reduce to 2500 strategic warheads. The START II and III agreements did not enter into force because of disagreements over defensive weapons, covered by the ABM Treaty. The 2002 Strategic Offensive Reductions Treaty (SORT) and the 2010 New START Treaty lowered the limit on deployed warheads to 2200 and 1550, respectively.
1.9 At the end of the Cold War, the ***Nunn-Lugar legislation*** helped control and maintain the accounting of 1200 tonnes of HEU and 130 tonnes of plutonium in the former Soviet Union. Without this financial and technical assistance, it is doubtful that these materials would have stayed out of the hands of terrorists or other nations, as the Soviet's million-person nuclear payroll could not be met.
1.10 ***Sanction legislation*** was tightened by Senators Glenn, Pell and Helms in 1994. Sanctions can only be truly effective when the full family of nations abides by them in concert. If solidarity is weakened, control by sanctions is weakened.
1.11 To be consistent with US restrictions on plutonium use abroad, the Carter administration ***mothballed the Allied General Nuclear Services reprocessing plant*** at Barnwell, South Carolina and it halted plans for the ***Clinch River Liquid Metal Fast Breeder Reactor*** (LMFBR) at Oak Ridge, Tennessee.
1.12 Former Secretaries of State Henry Kissinger and George Schultz, former Secretary of Defense Bill Perry and former Chair of the Senate Armed Services Committee San Nunn are committed to work towards the goal of ***removing all nuclear weapons from the Earth***.

2. Nuclear Weapons

2.1 ***Critical mass*** is the minimum amount of fissile material needed to sustain exponential growth of the neutron population, which depends on the isotope composition, its density and geometrical shape and situation.
2.2 ***Reflectors made of beryllium*** (***giving two neutrons***) ***or uranium*** reduce the critical mass from the "bare-sphere" values. Implosion increases the density of ^{239}Pu or ^{235}U, reducing the critical mass.

2.3 The ***Hiroshima uranium gun–weapon*** released 13 kton of energy, consuming 0.8 kg of ^{235}U. *Little Boy* consumed only 1.3 % of its 60 kg of ^{235}U. The 22-kton Nagasaki implosion weapon obtained a much higher efficiency of 20 %, consuming 1.3 kg of 6 kg of ^{239}Pu. The higher efficiency of *fat man* was obtained because it was imploded, increasing the density of ^{239}Pu.

2.4 Plutonium is favored over *highly enriched uranium* (HEU with 90 % ^{235}U) ***for military weapons*** since it emits more neutrons per fast fission ($\nu = 2.94$ vs. 2.53), more neutrons per neutron capture ($\eta = 2.35$ vs. 1.93), has a higher fast fission cross-section (1.7 barns vs. 1.2 barns) and releases slightly more energy (214 MeV versus 207 MeV). For these reasons, plutonium can make smaller primaries, which are essential for multiple-independently-targetable reentry-vehicles (MIRV) on intercontinental ballistic-missiles (ICBM) and submarine-launched ballistic-missiles (SLBM).

2.5 ***Plutonium is more difficult*** to make into nuclear explosives because of the high rate of spontaneously-emitted neutrons that are emitted by ^{240}Pu. These precursor neutrons can begin a chain reaction before plutonium reaches its most compact form.

2.6 ***Uranium used to be more difficult*** to enrich in the past, as compared to the production of Pu in reactors. The advent of better centrifuges makes this no longer true. U production is harder to locate since U emits much less radiation and U does not need reactors for production. There is a shift in preference for wannabe nuclear states to use U for weapons as compared to Pu.

2.7 A ***boosted primary*** contains $d + t$ gas (heavy hydrogen and heavy-heavy hydrogen) to magnify its yield. The extra energy from *dt* "burning or fusing" is small, some 1/3 of a kton, but the extra neutrons, released early in the cycle, allow the fission cycle to skip many generations, increasing the fraction of nuclei that fission, lowering the amount of fissile material needed. Without tritium, modern nuclear weapons would not function since *dt* reactions are needed to raise the yield to ignite the secondary stage.

2.8 The Department of Energy and the JASON group determined in 2012 that ***pit lifetime*** is in excess of 150 years. The Los Alamos Pit Production Facility has a capability to make 6–10 pits per year.

2.9 Stanislaus Ulam and Edward Teller discovered the ***concept of the H-bomb*** when they deduced that X-rays from the primary would reflect (re-emit) from the casing on to the secondary. The absorbed X-rays heat the casing and other materials to such high temperatures that they produce black-body X-rays, as well as reflected X-rays. Since X-rays travel with the speed of light, they are absorbed by the secondary before the mechanical shock arrives. The high temperature rise at the surface of the secondary creates radiation pressure as the hot surface of the secondary evaporates.

2.10 The first hydrogen device, **Big Mike, was a large thermos bottle** containing liquid deuterium, giving a yield of 10 Mton. Mike could not be delivered with bombers or ICBMs because of the extreme size of its cryogenics.

A deliverable H bomb was soon developed with deuterium in the LiD salt in the form of $^6Li^2H$. Since salt is a solid, there is no need for a cryostat. A neutron interacting with ^6Li gives a tritium to interact with deuterium ($^6Li + {}^1n \Rightarrow {}^3H + {}^4He$).

3. Nuclear Reactors and Radiation

3.1 A typical ***light water reactor*** (LWR) uses low-enriched uranium of 4.4 % ^{235}U for fuel and normal water as a coolant and moderator. The LWR produces 220 kg of plutonium per year (1 GW = 1 GW_e = 1000 MW_e). Since fuel rods remain in the LWR for 4 years, there is plenty of opportunity for the newly-produced ^{239}Pu to capture a second neutron to become ^{240}Pu. LWR reactors produce reactor-grade plutonium with over 20 % ^{240}Pu.
3.2 In a ***loss-of-coolant accident***, fission stops because the moderating water has run out, but what remains is beta-decay heating. During the first minute after fission has stopped with the loss of water, beta decay heat continues at 7 % of the 3 GW of thermal power. The 200 MW of beta-decay heat is considerable and rapidly raises the temperature of the core in one minute to the molten state at 2200 °C.
3.3 The 2011 triple reactor melt-down accident at ***Fukushima*** was not caused by the 9.0 earthquake, but it was caused by the 14 m tsunami, which removed electricity to pump water to remove heat from the reactors and spent-fuel ponds.
3.4 US commercial nuclear reactors have, thus far, ***not had a serious life-destroying accident***. They have caused a total radiation exposure of 5 person-Sv/year, causing 0.3 of a death per year.
3.5 Safety could be ***enhanced with smaller cores***, lower power densities and greater heat capacity to prevent loss-of-coolant accidents. The cost of these modular reactors could be high.
3.6 In 1977 the US government offered to ***accept utility spent fuel*** for a one-time repository fee of 0.1 ¢/kWh, amounting to \$0.75 billion per year. Since the underground repository has not been available over that time, the electrical utilities charged the government with not fulfilling its agreement. The courts agreed, requiring DOE to pay utilities to keep spent fuel on-site until such time as a repository can accept it, ending the payments to DOE.
3.7 ***Radiation from natural and human-made sources*** gave an average dose of 3.6 mSv/year (360 mrem/year) in the United States. This gives an absolute death rate of about 1 % for 60 years natural exposure to this level. The 1 % death rate is 5 % of the normal 20 % cancer death rate.
3.8 The 1979 ***Three Mile Island*** accident released 20 curies of the core's 64 million curies of ^{131}I. This is not surprising since containment was not breached, but it has been conjectured that the relatively small release of iodine was a result of iodine bonding to Cs, making nonvolatile CsI. At the other extreme, the 1986 Chernobyl accident released 150 MCi, which was widely dispersed by the burning of the carbon moderator.

3.9 The principal health ***risk from radon*** arises not from ^{222}Rn, which, as a rare gas, does not adhere to the lungs, but rather its four radioactive daughters that chemically attach to aerosols and are trapped in the lungs.

4. Missiles and War Games

4.1 ***Survival of missile silos*** depends on at least five parameters:

1. ***Accuracy of missiles*** in circular error probable (*CEP*),
2. ***Maximum overpressure*** a silo survives, called hardness (*H*),
3. ***Yield*** (*Y*) of weapons,
4. ***Reliability*** (*R*) of a weapon between 0 and 1,
5. ***Number*** (*n*) *of warheads targeted at a silo.*

4.2 ***Accuracy*** is very important, a *CEP* reduction by a factor of 2 allows a yield of 1 Mton to be reduced by a factor of 8–125 kton for the same destruction of silos.

4.3 When ***crater radius is similar to CEP***, the kill mechanism becomes cratering and not overpressure. US hard-target warheads produce craters with radii approaching their accuracy.

4.4 Two warheads can avoid ***fratricide*** by using one surface warhead and one high-altitude warhead, but the high altitude blast is less effective on the silo. *Fratricide*, of the second warhead is caused by blast waves and dust; electromagnetic pulse; and neutrons.

4.5 ***Launch-on-warning errors***. US and Russia each have about 1000 long-range strategic, nuclear weapons on alert, as a significant deterrent, letting others know they should not attack us. One problem with this scheme is that this allows the possibility of a miss-informed, accidental, launch-on-warning. Over the past decades there have been at least four cases in which human and technical errors significantly increased the possibility of accidental nuclear war.

4.6 ***De-alerting***. Bombers and submarines are already on de-alert. It was proposed to the Air Force that additional, verifiable physical barriers be inserted to increase the minimum response time of missile silos to a day, or so. It would be best if these actions could be verified to the other side.

4.7 In the 1960s Secretary of Defense Robert McNamara quantified ***mutually assured destruction***. MAD has been referred to as placing two scorpions in the same bottle. In the nuclear case, neither side would survive as a modern society. The first criteria of MAD is that a survivable force must be large enough to annihilate 25 % of the other side's population to deter that nation from attacking first. Calculations in 1980 showed it would take a survivable 50 one-Mton weapons to destroy 25 % of the Soviet population. A calculation showed that the US could lose 25 % of its population from 130 one-Mton weapons. The second criteria of MAD was to have a survivable force to destroy 50 % of the national industrial base.

4.8 The GAO in 1993 stated that "***SSBNs*** are in essentially constant communication with national command authorities and, depending on the scenario, SLBMs from submarine platforms would be almost as prompt as ICBMs in hitting enemy targets. Other test data show that the accuracy and reliability of the Navy's D-5 SLBM are about equal to DOD's best estimates for the Peacekeeper. Further, its warhead has a higher yield than the Peacekeeper's. In short we estimate that the D-5 has a hard target kill capability about equal to the Peacekeeper's, while its platforms remain virtually undetectable, unlike easily located silos."

4.9 The Navy has been encouraged to use ***20 % enriched reactor fuel*** rather than the 90 % fuel it uses now. This would reduce the demand for weapon grade materials, as called for in the Fissile Material Cut-off Treaty.

5. Ballistic Missile Defense

5.1 ***SDI*** was intended to attack ballistic missiles in their *boost phase*, before they released their reentry vehicles. *Boost phase attacks* on SS-18s could reduce the number of targets by a factor of 10 since ten RVs remain on the boost-phase SS-18 bus. A boost phase attack is potentially much better than an attack on the midcourse and reentry phases since the 10 RVs would have become separate targets. Such a strategy *requires space-based weapons* to be close to the SS-18 launch positions, some 10,000 km away.

5.2 The impact of ***SDI on Gorbachev*** was much less than that of economics and nuclear instability. SDI could have shortened or extended the Cold War by, at most, *6 months*. Six months after Reykjavik, the US abandoned beam weapons in favor of kinetic-kill vehicles. At that point SDI became irrelevant and did not prevent the Soviets from accepting the INF, TTTBT, CFE, START and CTBT treaties.

5.3 The ***offense can spoof the defense*** with decoys or with cooled RVs to reduce infrared signatures. The offense can also spoof with RVs in large balloons or by releasing small bio-weapon bomblets in the boost phase.

5.4 The Earth's magnetic field deflects charged-particle beams, while ***neutral-charged beams*** have straight-line trajectories. Neutral beams are formed by passing an energetic hydrogen ion beam (H^-) through a stripper gas to remove one of its two electrons, forming neutral hydrogen.

5.5 The American Physical Society concluded that the ***HF laser*** must have "output powers at acceptable beam quality need to be increased by at least two orders of magnitude for hydrogen-fluoride and deuterium-fluoride lasers for use as an effective kill weapon in the boost phase."

5.6 The ***X-ray laser weapon*** needed *pop-up missiles* based on submarines in the Arctic Ocean, quickly leaping into space to attack Soviet boost-phase missiles. Edward Teller and Lowell Wood contemplated powerful X-ray lasers, pumped with energy from a nuclear weapon. The limited tests of the X-ray laser were not successful, a fact that was kept from many policy makers. The system has serious timing problems.

5.7 The first ***kinetic kill vehicle*** was the space-based interceptor (SBI), which was to be based in groups of 10 in satellite garages. The KKVs were expected to attack the boost phase at a distance of 300 km from higher orbits in 50 s with a relative velocity of 6 km/s.
5.8 The ***airborne laser*** was developed to destroy the theater-based boost phase. The ABL planned to use 3-MW, oxygen-iodine lasers, located on a fleet of seven Boeing 747 s by 2011. Ultimately only one ABL was built.
5.9 In the 1980s, the United States deployed an ***ASAT missile*** on F15 fighters, lifting KKVs into orbit, while the Soviets deployed ASAT missiles with fragmentation explosives. Ground-based lasers can destroy satellites that are soft objects in predicable orbits.

6. Verification and Arms Control Treaties

6.1 The verification process has been occasionally politicized, but it has mostly been accurate. The Soviet ***Krasnoyarsk radar*** was an *unambiguous* violation of the ABM Treaty. An example that cuts the other way was the falseness of the US charge of a *likely* violation by the Soviets of the Threshold Test Ban Treaty (TTBT) in the 1980s, an issue with which the author was intimately involved as the State Department technical lead on testing issues.
6.2 Ambassador Paul Nitze defined ***effective verification***: ".... if the other side moves beyond the limits of the treaty in any militarily significant way, we would be able to detect such violation in time to respond effectively and thereby deny the other side the benefit of the violation."
6.3 An excellent measure from START is the ***reentry-vehicle on-site inspection*** (RVOSI) that counts physical RVs (warheads) on a missile. RVOSIs are an extremely important tool if an agreement limiting numbers of actual warheads is achieved.
6.4 Former Secretaries of State Henry Kissinger and George Schultz, former Secretary of Defense Bill Perry and former Chair of the Senate Armed Services Committee Sam Nunn are committed to work towards the goal of ***removing all nuclear weapons*** from the Earth. They readily admit they do not have all the answers in terms of verification and nuclear stability, but they also realize that the recommended steps are useful even if we do not get to zero. They concluded: "But the risks from continuing to go down the mountain or standing pat are too real to ignore. We must chart a course to higher ground where the mountaintop becomes more visible."
6.5 Reconnaissance ***optical systems*** improved in many ways:

- Aerial survey films are available with high contrast of 1 μ between lines.
- Charge-coupled devices allow real-time photography at excellent resolution.
- Chromatic aberration is reduced with computer-aided lens design and machining.
- Blurring by satellite motion is removed by moving camera with precision.

- Similar-appearing objects reflect and emit differently in the visible and IR.
- Two views of the same terrain at different angles give stereoscopic images.

6.6 ***Digital image processing***. The development of electro-optic CCDs and very large integrated circuits enhances the ability to monitor military activities and arms control treaties. Digital image processing restores and enhances photographs that are blurred by many causes.

6.7 ***Bolometric infrared detectors*** have solved the cooling problem. A weak IR signal marginally warms a small pixel, raising the temperature by milli-Kelvins. This temperature rise increases the electrical resistance, which can be measured. The small electrical current raises the pixel's temperature. Background noise can be eliminated by statistics. The bolometric sensors are used on Predator drones.

6.8 ***Subsidence*** of the Earth's surface is measured with an accuracy of 2 mm, using inter-ferometric synthetic aperture radar (InSAR) that combines two SAR images.

6.9 ***Reentry vehicle size*** is measured with ground-based radar using inverse synthetic aperture radar (ISAR), in which the moving and rotating RV supplies antenna motion.

7. Winding Down the Cold War

7.1 As part of the US Government's assessment of the verifiability of a warhead disarmament proposal, ***Project Cloud Gap*** carried out experiments at the Pantex facility for close monitoring of weapon dismantlement, at Rocky Flats they monitored the disassembly of warhead pits and separation of materials into plutonium, uranium and other residue, at Paducah they monitored the separation of materials into salvageable categories and the disposal of classified residue and At Oak Ridge, they monitored the disassembly of HEU parts and casting it into ingots.

7.2 ***Slava experiment***. On 5 July 1989 a team of Soviet and US scientists measured gamma-ray spectra from a Soviet warhead mounted on a SS-N-12 cruise missile on the *Slava* cruiser with a high purity germanium detector. The most detectable gamma transitions showed the presence of uranium-235, plutonium-239 and uranium-232. The presence of uranium-232 indicated that the uranium in the Soviet warhead had resided in a nuclear reactor before being used as feedstock for an enrichment plant.

7.3 Warheads were dealt with mainly through ***counting rules*** that attributed a certain number of deployed warheads to a particular delivery vehicle:

1. Ballistic missiles are the major delivery vehicle for nuclear warheads.
2. Ballistic missiles, silos, submarines and bombers are much larger and easier to count than nuclear warheads. They are much more difficult to hide than warheads or their fissile material components.

3. The number and characteristics of the Soviet and US deployed strategic delivery vehicles and launchers provide better measures of strategic significance of nuclear arsenals than the size of warhead or fissile-material stockpiles.
4. Modern strategic delivery vehicles cost ten times more to develop, produce and maintain than the nuclear warheads they carry.

7.4 The collapse of the Soviet Union raised fears for the security of thousands of Soviet nuclear warheads, their proliferation as ***loose nukes*** to other countries and as tonnes of weapons-usable Pu and HEU. In responding to these risks and opportunities, Russian and US leaders undertook remarkable initiatives that provided basic foundations for much more comprehensive arrangements to control nuclear warheads and materials.

7.5 ***Reciprocal-unilateral initiatives on nuclear weapons***. In 1991 President George H.W. Bush announced the withdrawal of all US ground—and sea-launched tactical nuclear weapons to the US. All of the ground-launched and about half of the sea-launched weapons would be destroyed. Soviet President Mikhail Gorbachev responded with the announcement that all Soviet tactical nuclear weapons would be withdrawn to the Russian Federation, and that nuclear artillery, ground-launched missile warheads and nuclear mines would be destroyed.

7.6 ***Cooperative Threat Reduction***. In December 1991, a bipartisan effort led by Senators Sam Nunn and Richard Lugar addressed the dangers of loose nuclear materials and nuclear weapons. The legislation authorized the president to transfer up to $400 million from the 1992 defense budget, making DOD the first major agency engaged in what became known as the Cooperative Threat Reduction (CTR) program.

7.7 ***Fissile Material Cut-off Treaty***. Four of the five nuclear weapon states officially declared they stopped production of HEU and plutonium for nuclear weapon purposes. In a major initiative, the 1992 Russian–US informal agreement to ban the production of fissile materials was expanded to create the concept of a multilateral Fissile Material Cut-off Treaty (FMCT). On 27 September 1993, President Clinton proposed at the United Nations a multilateral agreement to halt the production of HEU and plutonium for nuclear explosives.

7.8 ***Mayak Storage Facility Agreement***. In 1991 Minatom Minister Viktor Mikhailov requested financial help to build a large facility near Tomsk to store excess weapon-usable materials under secure conditions. In January 1996 US Secretary of Defense William Perry and Mikhailov agreed on the construction of a storage facility for excess weapon-usable fissile materials at Mayak.

7.9 ***NAS on Monitoring Warheads***. The 2005 National Academy of Sciences study, *Monitoring Nuclear Weapons and Nuclear Explosive Materials* concluded that “a system could be designed such that inspection of a relatively small sample could be sufficient to prove a high degree of confidence in the authenticity of all weapons and components.”

8. Nuclear Proliferation

8.1 ***Nuclear Weapon Efforts***:

- 5 NWS: US, FSU/Russia, UK, France, China;
- 4 defacto NWS: India, Pakistan, Israel, North Korea.
- 4 former defacto NWS: South Africa, Belarus, Kazakhstan, Ukraine, Iran.
- 14 former nuclear weapon programs: Algeria, Argentina, Brazil, Germany, Iran, Iraq, Japan, Libya, South Korea, Sweden, Switzerland, Syria, Taiwan.

8.2 ***Proliferation policy is more complicated*** than superpower strategic nuclear weapons policy. This is because nuclear power supplies commercial energy to 30 nations, which could grow to 50. On the other hand, the SALT/START/SORT/New-START treaties involve only two nations (United States and Russia) and the treaties do not impact commercial energy supplies and commerce. In 2012, 30 nations produced 16 % of the world's electricity with nuclear power for a total capacity of 374 GW_e.

8.3 ***Reactor-Grade Pu***. Reactor-grade plutonium (RgPu) with 20 % ^{240}Pu from civil power plants could be used to make lesser-quality nuclear weapons. The US and Russia could make significant 20-kton weapons from RgPu, but that takes sophistication and experience. Still, reactor-grade Pu can be made into smaller weapons of, perhaps, 0.5 kton, and reprocessing makes the process available.

8.4 ***Atoms for Peace***. President Dwight Eisenhower gave his "Atoms for Peace" speech to the United Nations (8 December 1953) in which he proposed that the IAEA should impound, store, and protect fissile materials. Eisenhower claimed that nuclear fuel could be proliferation-proof: "The ingenuity of our scientists will provide special safe conditions under which such a bank of fissionable material can be made essentially immune to surprise seizure."

8.5 The Iraqi, Iran, Libya and North Korea violations involved ***undeclared locations***. After the Gulf War, the IAEA changed its procedures to

- increase use of intelligence information from large nations;
- take environmental samples to search for clandestine enrichment and reprocessing plants; and
- establish special inspection procedures for undeclared sites.

8.6 ***Plutonium Economy***. The US abandoned the plutonium economy with President Carter's 1978 cancellations of the

- Clinch River Liquid Metal Fast Breeder Reactor (LMFBR) in Oak Ridge,
- Barnwell Nuclear Fuel Plant reprocessing plant in Barnwell, South Carolina.

8.7 ***Fuel Bank***. The fuel bank was first proposed in the Eisenhower regime and it continues to resurface as the world searches for proliferation solutions. Some $150 million has been committed by the European Union, Russia, US and Warren Buffett ($50 million) to help nations obtain reactor fuel if their commercial contracts are threatened for other reasons.

8.8 ***Export Criteria***. The Indian bomb of 1974 created the climate for the passage of the Nuclear Nonproliferation Act (NNPA) of 1978, requiring safeguards on all nuclear facilities before US exports are allowed. NNPA also established criteria for allowing reprocessing of US-origin spent fuel with a demand for enough timely warning to respond to acts of proliferation before they became militarily significant.

8.9 ***Spent Fuel Storage***. The 35,000 tons of US-origin spent fuel at foreign storage sites do not have a final destination. The glut of spent fuel has overcrowded spent fuel ponds, shifting older spent fuel to air-cooled surface-storage. It is important not to overcrowd the ponds because of potential thermal problems if cooling water is lost. The US does not want to accept US-origin foreign spent fuel rods, nor does it want them reprocessed since this creates more separated plutonium.

9. Proliferation Technologies

9.1 ***AQ Kahn's*** defection in 1978 from URENCO in the Netherlands to Pakistan showed the weakness of the international system. Kahn took centrifuge designs and the associated experience from URENCO, paving the proliferation path for HEU nuclear weapons. Not satisfied with helping Pakistan, Kahn expanded nuclear sales to Iran, Libya, and North Korea. Kahn also tried to sell centrifuges to Iraq's Saddam Hussain, but Saddam didn't trust Kahn. The shift to HEU allows easier weapons designs compared to plutonium-implosion weapons.

9.2 ***Laser Isotope Separation***. The slightly smaller volume of ^{235}U nuclei provides enough difference in electrostatic energy to produce slightly different energy levels in ^{235}U as compared to ^{238}U. Tunable dye lasers and other lasers can selectively ionize ^{235}U in uranium vapor or UF_6 molecules. The ionized ^{235}U is separated from neutral ^{238}U with electric fields. Laser isotope separation (LIS) can make weapons-usable HEU but with varying degrees of difficulty.

9.3 ***Uranium countries***: Argentina, Brazil, China, France, Germany, India, Iran, Japan, Netherlands, North Korea, Pakistan, Russia, UK and US currently enrich uranium. Commercially enriched stock is primarily supplied by Europe's URENCO (10 million SWU/year), Russia's Rosatom (26 million), and France's AREVA (10 million). Iran's centrifuges produce 0.8–4 SWU/year. Iran claims its right to pursue uranium enrichment in line with Article IV of the NPT. But its claim is subject to dispute, considering that Iran did not report past enrichment activities to the IAEA.

9.4 ***Megatons to Megawatts***. As the cold-war world subsided from 70,000 total warheads to 3000 operational warheads under New START (plus others). There was a concern about Russia's ability to manage its 130 tonnes of plutonium and 700 tonnes of HEU (after the sale to the US). As an encouragement to Russia, the US placed 12 tonnes of plutonium and HEU under IAEA safeguards in hopes that Russia would follow suit. The major barrier to weapons production is the availability of weapons-grade fissile materials, rather than the design and fabrication of weapons. To reduce dangerous proliferation, the US agreed to pay $12 billion over 20 years to purchase 500 tonnes of HEU, and blend it to low-enriched reactor fuel.

9.5 All nine nations with plutonium weapons ***chose WgPu and rejected RgPu*** for their nuclear weapons. Nevertheless, poor weapons of one-kiloton or good sophisticated weapons can be made from RgPu. For this reason, the United States constrains reprocessing of US-origin spent fuel and the IAEA maintains safeguards over both types of plutonium with equal authority.

9.6 ***Rokkasho Reprocessing Plant***. The Rokkasho reprocessing plant (RRP) was intended to produce plutonium for Japan's breeder program, but now it is more likely to produce plutonium for mixed oxide fuels for light water reactors. Perhaps the Pu will not used at all as some of Japan's nuclear power plants are being closed after Fukushima. RRP was designed to reprocess 800 tonnes/year of spent reactor fuel from 30 GW_e of LWR. RRP serves as a test-bed to examine advanced safeguard technologies, integrated through a collaborative process between IAEA, Japan and US. IAEA will have inspectors present at all times.

9.7 ***Plutonium produced in HEU research reactors***. Research reactors provide two paths to nuclear weapons. The first path is the diversion of HEU reactor fuel. The US worked to close this path by reducing ^{235}U content in the research reactor fuel from 90 to 20 %. The same reactor power can be maintained in most cases by increasing uranium density while reducing the ^{235}U-enrichment level, which has the effect of maintaining the same volume density of ^{235}U, but reducing the enrichment level considerably below 90 %.

9.8 ***Natural uranium research reactors***. At the other extreme, Israel's Dimona and India's Cirus research reactors use natural uranium fuel and a heavy water moderator to produce plutonium at a rate of 0.3 kg/MW_t-year. It takes about six months to obtain 5 kg for a Pu warhead from a 40-MW_t reactor, or about two warheads a year. The US urged conversion of natural uranium research reactors to a *higher* enrichment level of 20 % to substantially reduce plutonium production.

9.9 The issue of missile proliferation encouraged establishment of the ***Missile Technology Control Regime*** (MTCR) in 1987. MTCR constrains exports of missiles and subsystems to deliver *no more than* 500 kg, the size of a crude nuclear warhead at a Scud range of 300 km. In 1993 the throw-weight criteria was removed since biological weapons are much lighter and can be as dangerous.

10. Proliferated States

10.1 ***Proliferation began with the United States*** (after Germany). Figure 10.1 displays the United States as the center of the technology transfer pattern. This is to be expected because the US developed nuclear weapon first. The first spoke from the US hub extends to USSR/Russia, driven by competitive fear, as well as the spying of Klaus Fuchs and others. The threat of US dominance made Stalin uneasy. The second spoke is drawn towards the UK because UK and Canadian scientists were intimately involved in the Manhattan project. Klaus Fuchs (Germany, UK) returned to Britain as a professor, then landed in jail and, after release, moved to East Germany. The current British submarines and their SLBMs were produced in the US. The Canadian connection stimulated their CANDU reactors and heavy-water technology. Another spoke is drawn to India, which participated in Eisenhower's Atoms for Peace Research at Argonne and Oak Ridge, learning reactors, reprocessing and plutonium-practices.

10.2 ***Russian theft*** of US nuclear weapon plans by Klaus Fuchs and Ted Hall was the first covert act in the nuclear proliferation drama, shortening research time to four years for the Joe-I test explosion in 1949. The Soviets initially transferred weapons designs and enrichment technology to China, but then stopped since it was helping a rival in Asia. Since the signing of the NPT in 1968, the US and USSR cooperated many times over NPT issues, with the Soviets being firmer with its Warsaw-Pact client states, as compared to the US with its NATO and other friends.

10.3 ***China***: Beginning in 1955, the Soviet Union gave considerable support to China's nuclear weapons program by giving them designs of older weapons, uranium enrichment, nuclear reactors, and other items. By 1960 the Soviets ended their nuclear support, as China became a rival. China's first nuclear test took place in 1964, followed with its first hydrogen test in 1967. China's views on nonproliferation have changed *mightily* over the decades.

10.4 ***United Kingdom***: In January 1941, the Germans were doing reactor experiments with impure graphite in Berlin at the *Virus House*, while the *British Maud Committee* determined in July 1941 that a weapon could be made with 10 kg of ^{235}U. The US National Academy of Sciences endorsed the British results. British, Canadian and other scientists worked on the Manhattan project. On 27 January 1950, Klaus Fuchs, then a British citizen, confessed that he transmitted atomic secrets to the Soviets.

10.5 ***India*** has not signed either the NPT or the CTBT. India has long been concerned about its nuclear neighbors, China and Pakistan. The world was shocked by India's 1974 nuclear test, as this was the first nuclear test beyond the P5. Many hoped only rich nations could test a nuclear device, but that was clearly wrong. The US State Department responded with the creation of the Nuclear Suppliers Group in 1975. The Congress responded with the Nuclear Non-Proliferation Act of 1978 and Glenn-Symington sanctions.

10.6 ***Israel*** has long been reported to own nuclear weapons, ever since the late 1960s. It is rumored that Israel covertly obtained natural uranium from Europe. In addition it is reported that Israel in the 1960s clandestinely obtained 100 kg high-enriched uranium from the Nuclear Materials and Equipment Corporation (NUMEC) facility in Apollo, Pennsylvania. In 2011, Prime Minister Benjamin Netanyahu repeated the refrain that "we won't be the first to introduce nuclear weapons into the Middle East." This quote does not define when a nuclear weapon is assembled, or almost assembled.

10.7 ***DPRK's*** nuclear program started towards plutonium weapons in the 1950s with help of Russian exports and expertise. North Korea acceded to the NPT on 18 April 1985 after strong coaxing by the Soviet Union, with the hope to sell North Korea power reactors. It took seven years for Pyongyang to approve the IAEA safeguards agreement. Inspections began in May 1992 to examine the 5-MW_e reactor and 50-MW_e plant under construction. The 200-MW_e plant under construction was halted under the Agreed Framework in 1994. In addition, DPRK had an 8 MW_t research reactor, a critical assembly from the Soviets, and various fuel fabrication and reprocessing facilities. In October 2002, DPRK revealed it had a centrifuge plant of 2000 centrifuges, imported from A.Q. Kahn of Pakistan.

10.8 ***Pakistan***: Most observers believe that Kahn's trips to North Korea, Libya, Iraq and Iran were well known by the country's leadership. Nevertheless, when he was caught he was only put in house arrest, and he is now free. It took two decades for Pakistan to progress from centrifuge drawings to nuclear tests, but Kahn had support from Pakistan's leadership. Not only did Kahn sell centrifuge parts, but Iranian scientists and others came to Pakistan to learn the technology. When Kahn confessed about his nuclear sales, it was in English, but on Pakistan state television.

10.9 **Brazilian-Argentine Agency**. The driving force behind the ABACC success story was indigenous, it was not a result of Western diplomacy. If anything, overly strident voices against their existing nuclear plans tended to anger participants and prevent diplomacy. On the other hand, if these voices had not been raised, there might have been an over abundance of weapons useable materials.

11. Nuclear Testing and the NPT

11.1 As of 2015, 164 states ratified CTBT. ***Forty-four nuclear-capable states must ratify CTBT*** before it can enter into force. Eight nations must still ratify to do this; China, Egypt, India, Iran, Israel, N. Korea, Pakistan and US. In October 1999, the US Senate rejected the CTBT with a vote of 51–48. After the defeat, the National Academy of Sciences (NAS) was asked by the Clinton and Obama Administrations to convene panels of experts to examine technical issues that may affect the viability of a test ban:

- Safety and reliability of the US nuclear stockpile without nuclear testing;
- US capability to detect, locate, and identify nuclear explosions;
- Commitments necessary to sustain the stockpile and monitoring systems;
- Potential technical-advances countries could achieve through evasive testing.

11.2 The NAS concluded that the ***International Monitoring System*** is able to detect tamped warheads over 0.1–0.2 kton in the Northern Hemisphere and it is able to detect warheads exploded in a cavity over 1–2 kton. CTBT could improve the political climate among nuclear weapon states, and its adoption is consistent with the Nuclear Non-Proliferation Treaty (NPT).

11.3 The ***Limited Test Ban Treaty of 1963*** forbids testing nuclear weapons in space, in the atmosphere and underwater. Monitoring must be able to detect nuclear debris from an explosion above the Earth. The neutron, X-ray, and prompt γ-ray fluences from weapons exploded in space are measured at distances of 20,000 km, where GPS satellites reside. The results from the Los Alamos Vela satellite program of the 1960s showed that these signals are readily observable by satellites.

11.4 ***IMS monitoring at 10 % Confidence***. Lowering the confidence in the detection capability of the Primary Seismic Network to 10 % lowers the threshold magnitudes to 2.8, 3.0 and 3.2. This corresponds to fully coupled device yields of 0.22, 0.035, and 0.056 ktons, respectively in regions of better propagation.

11.5 ***North Korean Tests***. The 2006 North-Korean test (0.6 kton, 4.1 m_b) was promptly detected by 22 IMS seismic stations, later confirmed by radioxenon detection. The 2009 North Korean test (3 kton, 4.5 m_b) was detected by 61 IMS seismic stations but no radioxenon stations. The 2013 North Korean test (10 kton, 4.9 m_b) was detected by 94 IMS seismic stations, 2 infrasound stations and 2 radioxenon stations (^{131m}Xe and ^{133}Xe in Japan and Russia).

11.6 ***National Technical Means (NAS)***: "NTM gives the United States significant additional information beyond what is available to other countries that do not have a robust NTM program. US NTM can focus on monitoring countries of concern to the US. The United States global monitoring capabilities are generally better than those of the CTBTO because they can go beyond data available to the CTBTO with classified capabilities. However, the inclusion of classified means and data limits the extent to which analysis and even results may be shared and used openly."

11.7 ***Detection of cavity test***. The probability of a successful clandestine effort is the product of at least seven detection probabilities for (1) venting, (2) yield excursion, (3) hiding removed materials, (4) subsidence of the surface, (5) regional seismic detection, (6) more than one test in a series of tests, (7) NTM detection.

12. Terrorism

12.1 ***Terrorism can be defined*** in many ways:

- Terrorists are frustrated political actors, wanting to correct flaws of the state.
- Terrorists use violence, not just to kill, but to get immediate public recognition.
- Terrorists are rational actors who realize to kill too many might make them suffer from responses to their attacks.
- Terrorists have limited means to make nuclear, weapons, but they could make crude, inefficient weapons with stolen materials or biological weapons.

- Terrorists are driven by religion, economics, nationalism and more.
- Terrorists may have valid issues, and they play hardball.

12.2 ***Infrastructures***. The dependence on more efficient, but more vulnerable, technologies enhances the opportunities for terrorists to disrupt these infrastructures: transportation networks, electrical grids, natural gas pipelines, water resources, health-care resources, global finance and global shipping.

12.3 ***Urban Densities***: By 2040, 80 % of global energy will be consumed in cities, compared to 66 % in 2010. This is a different world with 70 % of the population living in cities of 2050, up from 50 % in 2007, as seen in global urban trends.

12.4 ***Osama bin Laden*** opposed music, fornication, homosexuality, intoxicants, gambling, and usury. Osama bin Laden's strategy was to use violence against large enemies, such as the US and Russia. He believed violence could lure the West into a long war of attrition, "bleeding America to the point of bankruptcy."

12.5 ***9/11 Attack***. Khalid Sheikh Mohammed presented the idea of the attack to bin Laden in 1996, who approved it in 1998. The 19 conspirators were upper-middle class Moslems, not from disenfranchised lower classes. They started assembling in 1999 in Hamburg, Germany, and reassembled in 2000 in San Diego, Arizona and South Florida where they took pilot training. In retrospect, it is now obvious they were doing things out of the ordinary since they were not trained to take off or land planes.

12.6 ***Enhanced interrogation techniques*** are used on prisoners to force them to reveal their nation's secrets. These include water-boarding, in which water is poured over a cloth, covering the face and breathing passages of immobilized prisoners, causing them to experience sensations of drowning. Water-boarding causes extreme pain and panic, dry drowning, and damage to lungs and brain from oxygen deprivation. Other enhanced interrogation techniques include sleep deprivation, chaining in uncomfortable positions, replacement of body fluids, beating and threatening. CIA Executive Director John Brennan admitted that the CIA "did not always live up to the high standards we set for ourselves."

12.7 The ***Department of Homeland Security*** (DHS) was created in 2002 with five core missions: prevent terrorism, secure borders, enforce immigration laws, safeguard cyberspace, and increase resilience to natural disasters. Intelligence agencies were reorganized under the new Director of National Intelligence (DNI).

12.8 The Department of Defense created the ***Total Information Awareness*** (TIA) Program in 2003 to combine diverse data on individuals to predict what these persons might do in the future. The TIA program was led by Admiral John Poindexter until TIA was disbanded by Congress 10 months later after media reports stated that the "government was attempting to establish Total Information Awareness over all citizens." However, the basic thrust of TIA lived on to become reality.

12.9 ***Risk*** is the product of the probability of the event times the severity of the event.

13. Nuclear Terrorism

13.1 Fifteen years have passed since 9/11 without dirty bomb ***radiological attacks***. Has the radioactive silence been caused by the good works of the Department of Homeland Security? Have protections against radioactivity greatly improved? Or have we been lucky? Or is this harder to do than we thought? Terrorists have the advantage, since they dictate the timing, location and method of attack. However, it is important to remember that an attack is made up of many actions in a chain of events. This allows the defense some latitude to address each link in the chain of events.

13.2 ***Types of terrorist weapons***: Stolen nuclear weapons, improvised nuclear devices, radiological dispersal devices (dirty bombs), improvised explosive devices.

13.3 Bunn obtained a ***29 % probability for a terrorist IND in the next decade*** (2006–2016). He estimated the event would cost US society $4 trillion, for an average over $100 billion per year. Bunn concluded the following: "The uncertainties in estimating the risk of nuclear terrorism are large. But the very uncertainty of the danger highlights what we do not know—including the possibility that a major nuclear theft could be in the planning stages at any time. There is, in short, no time to lose." As companion reading, I suggest the reference by Len Weiss, "On fear and nuclear terrorism." Weiss points out that decades have gone by and there has been little in the way of terrorist nuclear weapons or dirty bombs.

13.4 ***PAL***. Most US nuclear weapons are controlled with *permissive action links* (PAL) that prevent the weapons being used without knowledge of the security code. Not all US weapons have PALs, nuclear weapons based on submarines or tactical weapons do not possess PALs. In addition, ballistic weapons and gravity bombs require passage on a known flight path to arm them.

13.5 ***Nuclear forensics***. Was it a nuclear explosion? Was it Russian or American, or somebody else? Was the device stolen, sold, smuggled, lost, built by a rogue state or built by knowledgeable terrorists? What was the yield of the explosion? Will there be more than one explosion? Can we reduce the chance of another explosion? How did the weapon get there? Where did it come from and who did it?

13.6 ***Clean-up***. The decision to determine what and how to clean radioactivity would be a complex choice of many options. Individuals in the plume should be decontaminated quickly by washing their skin and hair thoroughly and by disposing of contaminated clothing. Potassium iodide pills would not help since excess iodine is useful only to diminish the effects of iodine fission-fragments, which are not in a dirty bomb.

13.7 ***Drones*** are a response to save both American lives and money. The close approach of drones can lead to more precise bombing, reducing collateral damage to innocent people and buildings. Drones can carry dirty bomb radioactivity or BW/CW materials. In 2015, a one-kg drone crashed into the White House, it could have carried dangerous materials.

14. Cyber Terrorism

14.1 The ***film industry*** is paying attention to ***cyber events*** with three recent movies. *The Imitation Game* is a movie about Alan Turing's creation of the first computers to crack the German Enigma code. Seven decades later, The Imitation Game evolved to *Citizenfour*, a documentary about Edward Snowden's revealing the National Security Agency's (NSA) programs to read domestic emails and letters, and listen to phone calls. The government states they examine the address list without FISA Court approval, but they must get FISA Court approval to examine the contents. To add to this excitement, recall Edward Snowden's revelations on NSA tapping German Chancellor Angela Merkel's phones and e-mails, making her furious. Later we find out that the German government has done the same. In November 2014, North Korea attacked the cyber networks of Sony Pictures Entertainment because its film, *The Interview*, gave a negative treatment of Kim Jong-un, North Korea's leader.

14.2 Shortly after the 9/11 attack, Admiral John Poindexter and Brian Hicks encouraged the Pentagon to establish the Total ***Information Awareness program***. Their idea was (1) to utilize the immense analytical power of computers that had become much cheaper and (2) to access the stored data, which also had become much cheaper. Using this capacity to examine the big data sites, they searched for correlations (not causality) in the data. This approach discovers past actions and personal contacts between a person's friends, citizens and foreigners, to determine what individuals might do in the future. This determines the probabilistic chance that citizen X might develop shoe bombs to affect our national security. Of course, it could have problems with false positives and the defaming of innocent citizens.

14.3 **The applications of TIA were revealed on 6 June 2013, with details about the security leaks of *Edward Snowden***, a National Security Agency (NSA) consultant, were published in the Washington Post and Manchester Guardian. These leaks described NSA's PRISM Program, which tracks global email, web searches and other Internet traffic and NSA's Boundless Informant Program, which tracks global phone calls.

14.4 ***Google Glass*** has a camera and a microphone that allows it to film and transmit movies of conversations with your friends and coworkers. Certainly Google Glass can be used surreptitiously to obtain relevant data for court trials, and it could be misused. Various applications are being developed for facial recognition, photo manipulation, translation, and sharing on social networks. It received one of *Time Magazine's Best Inventions of 2012 Awards*. This clever device is selling for $1500 and became available to the public in May 2014. In January 2015 Google stopped selling it because of privacy concerns, the public didn't want to be subject to being photographed and recorded at all times. Google Glass is now sold only to certified partners for commercial trials in places like hospitals and factories.

14.5 DHS developed a list of 18 ***critical infrastructure sectors*** for the US. All these sectors are heavily dependent on the Internet and are vulnerable to cyber war attacks: agriculture and food, banking and finance, chemical, commercial facilities, communications, critical manufacture, dams, defense industrial base, emergency services, energy, government facilities, healthcare and public health, information technology, national monuments and icons, nuclear reactors, materials and waste, postal and shipping, transportation systems, and water.

14.6 In 2010, ***Stuxnet*** became part of the policy discussion by changing the operating rotational-frequency of centrifuges, to be near the resonant frequencies, to destroy 1000 centrifuges at Iran's Natanz facility. Not only did Stuxnet destroy centrifuges, but this fact was hidden by sending false signals to the owners, telling them that their centrifuges were running normally. This watershed event proved malware can destroy physical property in a big way. This kind of attack was expected by experts, as many realized it was possible, but they preferred not to think about it. Stuxnet was not perfect since it did not remain in the shadows, as the link between Stuxnet and Iran's centrifuges was established in 2010.

14.7 ***Denial-of-service attack***. Anyone can do this. You obtain a program in which you enter the web address of the server you want to attack, you hook up to the Internet and begin to flood the target with requests to connect. A major website might be shut down if it got 100,000 requests per second. If your computer sends 1000 requests per second, then you will have to recruit 100 friends with computers to make this happen. This converts a *denial of service attack* into a *distributed denial of service attack*.

14.8 ***Attribution*** is hard to determine if the sender hides his identity. It may take 6 months to determine the owner of a botnet, which is of little use for most attacks. The US government has made significant investments to address the problem of attribution. One approach looks at the beginning of the process, examines the decisions made to join a conversation, rather than analyze the situation after the attack. Another approach raises the threshold, making it more difficult to attach to a conversation.

14.9 In the past, only the federal government, such as the National Crime Information Center, had ***resources to gather obscure information and perform serious analysis***. Now big companies, like Google, do that and the size of the companies able to do this is much smaller. Much of big data is not considered private since it has already been released to others.

14.10 ***Imported Chips with Malware***. This is already taking place, by other countries and by the US. The back doors in the malware chips are permanent and cannot be disabled.

15. Biological and Chemical Weapons

15.1 ***Anthrax*** is not usually directly spread to persons from infected persons or animals, but rather it is spread by spores, which is a dormant form of the bacteria. Spores are robust, they can survive harsh conditions for decades and centuries. Spores, lying on the ground, can be transferred to humans. An animal dying from anthrax contains naturally-produced anthrax spores. When inhaled, ingested, or in contact with a skin lesion, spores may become active and multiply rapidly. Effective vaccines against anthrax are available and anthrax can be controlled with antibiotics. The US and other nations researched anthrax bacteria that are immune to antibiotics.

15.2 ***Smallpox*** may exist in US and Russia laboratories. Smallpox is much feared as it is transmitted by airborne virus from one person to another, regenerating it self in the new victim. Smallpox needs only 10–100 organisms for a fatality. Preventative vaccinations can help. Antibiotics are of limited value for those infected with smallpox. Progress in BW research and new deployments are greatly feared.

15.3 As of December 2014, 171 nations ***ratified or acceded to the BWC***. Egypt, Syria and seven African nations signed but did not ratify. Israel and fifteen African nations have not signed the BWC. It is difficult to detect small, but significant, quantities of bio-weapon materials, this is easy to do for radioactive nuclear weapon components. This realization shifted the response from BW *detection and mitigation* to BW *prevention and protection*. The BWC does not have a verification protocol. BWC technology is driven strongly by high technology research, as compared to the mass-production of chemicals. It will be difficult to devise an effective BWC monitoring regime without the loss of some bio-industrial trade-secrets.

15.4 ***Aum Shinrikyo***, the Japanese Cult, became the first non-state actor to use potent CW. They attacked the Tokyo Metro with Sarin and cyanide, killing 13, seriously injuring 54 and wounding 1000 on March 20 and May 5, 1995. This showed that non-state actors can disseminate CW (with limited success in this case).

15.5 ***Syrian CW***. The Syrian Assad regime attacked its citizens in a civil war in 2013, killing 1000. This prompted Russia to forcefully coerce Syria to join the CWC. A dozen nations, including US and Russia, worked together to successfully destroy 1000 tonnes of Sarin and other materials in Syria within a year.

15.6 Chemical weapons are categorized by their effects, rather than chemical similarities:

- ***blood gases*** (hydrogen cyanide) block oxygen transport in blood;
- ***blistering agents*** (lewisite, mustard gas) penetrate membranes to destroy lungs, eyes, skin;
- ***choking agents*** (phosgene and chlorine) damage lungs;
- ***nerve agents*** (Sarin, Tabun and VX,
- ***incapacitants*** (LSD or Agent BZ); and ***defoliants*** (2,4-D) cause leaves to drop.

15.7 ***Binary chemical weapons***. Binary weapons combine two chemicals into a CW material after the shell is fired. Binaries were devised to make CW weapons safer for manufacture, transport and storage. The two originating chemicals are not nearly as toxic as the final combined CW agent. Since the chemicals are combined in-flight, the toxicity of the CW is not an issue until it is on target.

15.8 ***CWC has 190 members*** by 2015, representing 98 % of the global chemical industry. Israel signed but did not ratify, while Angola, Egypt, North Korea and South Sudan did not sign the CWC (Angola and Myanmar plan to join shortly).

Appendix A
Reflections on Nuclear Arms Control

Modus Opperendi. Appendices A and B on Nuclear Arms Control and Non-Proliferation are labeled *Reflections*. This is to differentiate the style of the numbered chapters, as fact-filled textual writing, versus the alphabetic appendices, with a more humane description. It turns out that the back-channel negotiations of Frank von Hippel and Tom Cochran made a difference, as they made contact with the Soviets on the road to arms control. This back channel obtained better measurements of Soviet warheads and Soviet test sites than the US Executive Branch had obtained at that time. I believe the life and death issues for planet Earth need more than a recitation of logic and facts. I hope A and B will do that for you. As we set out in life, it is never clear where we will wash to shore.

As a young post-doc, two events caught my attention. On 16 October 1964, China exploded its 1st nuclear bomb. It took China only 3 years to move from fission weapons to fusion weapons. Suddenly, the US had to confront proliferation beyond the Soviets. I wondered *if anyone was in control.*

The second event was the 17 January 1966 crash of a B-52 bomber near Palmares, Spain. Plutonium was spread across their tomato fields from the two US H-bombs. The other 2 H-bombs landed in the ocean. One bomb was found quickly, but the second one was very difficult to find. Finally, a Spanish fishermen told a US official that he thought he saw where the bomb entered the ocean. He was correct but the bomb was very close to falling into a deep ocean canyon. Finally, the submersible Alvin collected it. Again, I wondered *if anyone was in control.*

A.1 Soviet-American Back-Channel

On 23 March 1983, President Reagan gave his famous speech on the Strategic Defense Initiative, or Starwars. In this speech, Reagan called "to make nuclear weapons impotent and obsolete." One of these ABM weapons was the *Excalibur* X-ray laser, as pumped with a nuclear explosion. To base this device close to the Soviet boost-phase, it necessitated the placement of the X-ray weapon in orbit or to have a pop-up missile close to the Soviet Arctic coast. I was asked by the Federation of American Scientists to join an American-Soviet study group on these

D. Hafemeister, *Nuclear Proliferation and Terrorism in the Post-9/11 World*,
DOI 10.1007/978-3-319-25367-1

issues. In 1987 I traveled to Washington to meet my Soviet counterparts. I had never worked with a real Russian communist.

Our subgroup was tasked to examine nuclear power in Earth orbit, to make large amounts of electrical energy for US ABM systems. The SDI program had been heralded as a non-nuclear program (except for Excalibur), but the SDI statement did not consider nuclear power needed to energize the ABM devices. It was clear you would need nuclear power in orbit because of the large energy requirements for neutral particle beam weapons and rail guns. Our subgroup consisted of three American and three Soviet scientists. The Soviets were led by Roald Sagdeev, the Director of the Institute of Space Research of the Soviet Academy of Sciences. In addition, Sagdeev was the Chairman of the Committee of Soviet Scientists for Peace and Against the Nuclear Threat. Sagdeev was accompanied by physicists Oleg Prilutsky and Stanislav Radionov. The three Americans were Joel Primack, Physics Professor, UC Santa Cruz; Steven Aftergood, Director of the Committee to Bridge the Gap and myself. Our research resulted in a paper, "*Nuclear Power in Space*" in the June-1991, *Scientific American*, and 5 papers in *Science and Global Security*. These papers showed it was easy to monitor a ban on nuclear power reactors in space, since reactors emit copious amounts of infrared radiation, which is readily detectable. Data was obtained at the NASA Ames Laboratory, Moffett Field, California by participating in an overnight flight at 45,000 ft in the Kuiper Airborne Observatory. If the ban on nuclear power in orbit had been accepted it would have constrained SDI and saved considerable funding.

A.1.1 The ABM Treaty

Reagan's 1983 speech called for the Strategic Defense Initiative "to make nuclear weapons impotent and obsolete." The new SDI technologies depended on directed energy weapons systems, such as neutral particle beams, chemical lasers, X-ray lasers, free-electron lasers, rail guns and so forth. In my view, Agreed Statement D of the ABM treaty bans testing these new types of weapons in space. Judge Abraham Sofaer wrote the US legal finding on this for Secretary of State George Schultz, so that he could follow the broad interpretation of the ABM treaty, to allow tests in space "based on other physical principals". It was my task to develop a scheme to put monitoring limits on the ABM technologies, an approach suggested by Ambassador Paul Nitze.

APS Study. The American Physical Society was asked in 1985 by the *Presidential Office of Science and Technology Policy* to investigate SDI technologies. The APS report was declassified in 1987. The APS report on SDI technologies was the doing of science and public policy at its best. The panel of 17 scientists was extremely qualified, with excellent professional reputations. The panel was equally populated by those who were actively-involved in SDI research at DOE/DoD laboratories, and those from academia. This was an excellent combination because the practitioners had seen the data, written the papers and reports, and were prepared to determine new results, and the academic scientists were imminently well-qualified scientists. Let me quote the summary conclusion (*Reviews of Modern Physics*, July 1987):

> Although substantial progress has been made in many technologies of Directed Energy Weapon Systems over the past two decades, the study group finds significant gaps in the scientific and engineering understanding of many issues associated with the development of these technologies. Successful resolution of these issues is critical for the extrapolation to performance levels that would be required in an effective ballistic missile defense system. At present, there is insufficient information to decide whether the required extrapolation can or cannot be achieved. Most critical elements required for a DEW system need improvements of several orders of magnitude. Because the elements are inter-related, the improvements must be achieved in a mutually consistent manner. We estimate that even in the best of circumstances, a decade or more of intensive research would be required to proved the technical knowledge needed for an informed decision about the potential effectiveness and survivability of directed energy weapon systems. In addition, the important issues of overall system integration and effectiveness depend critically upon information that, to our knowledge, does not yet exist.

This paragraph does not say that that SDI will fail. However, the more you dig into this paragraph and the report, you find that it is unlikely the technical barriers will be overcome on such issues as the amount of fuel on board, accuracy in timing and location, countermeasures by the Soviets, reliability over decades, and so forth. Recall that the SDI system could never be tested in its entirety, but only on a piece-meal basis before it's called to defend the US. The APS report did not examine the more-complex, system questions. The APS mainly examined if there could be enough beam energy on target with the proper resolution to destroy the incoming missile within the time allotted and at ranges of the thousands of miles. The APS report was not strongly-worded, but stuck to numerical estimates, assuming best-case analysis for SDI. Let's briefly quote some APS conclusions:

> We estimate that chemical laser output powers at acceptable beam quality needs to be increased by at least one order of magnitude for HF/DF lasers for use as an effective kill weapon in the boost phase.

This might not sound insurmountable, but it really is when you consider volume and costs. And it ignores other problems with chemical lasers in orbit, such as its duty factor, burning a hole in the defense, traveling under the laser barrier and a variety of counter measures.

> … excimer lasers… need improvements by at least four orders of magnitude…. Free electron lasers operating near 1 μm will require validation of several physical concepts… Neutral particle beam accelerators must be scaled up by two orders of magnitude in voltage and duty cycle…. Ground-based laser systems for BMD applications need geographical multiplicity to deal with adverse weather conditions… Ground-based laser systems require techniques for correcting atmospheric propagation aberrations. We estimate that these techniques must be extended by at least two orders of magnitude in resolution than presently demonstrated. Phase correction techniques must be demonstrated at high powers… Nonlinear scattering processes in the atmosphere impose a lower limit on the altitude at which targets can be attacked with a laser beam from space… Detection and acquisition of ICBM launchers will pose stringent requirements for high detection probability and low

> false alarm rates.... For boost phase, infrared tracking of missile plumes will have to be supplemented by other means to support sub-microradian aiming requirements of DEWS.... For post-boost and mid-course, precision tracking will require active sensor systems... For midcourse, when the RVs are interspersed with penetration aids, interactive discrimination may be required... Housekeeping power requirements for operational maintenance of many space platforms for strategic defense applications necessitate nuclear reactor driven power plants on each of these platforms... During engagements, prime power requirements for electrically driven space-based DEW present significant technical obstacles.... Survivability is an essential requirement of any BMD system employing space-based assets; such survivability is highly questionable at present.... Survivability of ground-based facilities also raises serious issues... Directed energy weapons with capabilities below those needed for many ballistic missile defense applications can threaten space-based assets of a defensive system... X-ray lasers driven by nuclear explosions would constitute a special threat to space-based sensors, electronics and optics... Nuclear-explosion-pumped X-ray lasers require validation of many of the physical concepts before their application in strategic defense can be evaluated... Since a long time will be required to develop and deploy an effective ballistic missiles defense, it follows that a considerable time will be available for responses by the offense. Any defense will have to be designed to handle a variety of responses since a specific threat cannot be predicted accurately in advance of deployment.

The initial tests of Excalibur, the X-ray laser weapon, showed only very slight multiplication, and that was controversial. Where would these X-ray lasers be based? SDI was intending to place them on submarines, based near the Arctic coast of the Soviet Union. Excalibur was to be boosted into space quickly to make X-rays to destroy the Soviet missiles while still rising in the boost phase. The Soviets would have plenty of time to respond to Excalibur with a better attack plan. The easiest thing for them to do would be to base their ICBMs further south towards Kazakhstan. Since the earth is round, this means that the Soviet ICBMs would have more time to release warheads. They also could build fast-burn missiles, rising in 1 min, compared to 5 min for the SS-18. Or, use counter measures.

In 1987, DOD shifted the main focus of SDI from directed energy weapon systems (DEWS) to hit-to-kill, kinetic-kill vehicle (KKV) weapons. The APS report helped pave the transition from to KKV by showing the weakness of DEWS. Objects in low-Earth orbit have a kinetic energy density about the same as that of high explosives. Moving mass in orbit can destroy missiles and RV's. At that time, the KKV technologies were the ground based interceptor (GBI) and *Smart Pebbles*, which would remain in orbit for ten years, waiting for the call to attack.

A.2 Strategic Arms Reduction Treaty

From January to December of 1987, I was a visiting science fellow in the Office of Strategic Nuclear Policy in State's Bureau of Politico Military Affairs, working on START, the Intermediate Nuclear Forces Treaty (INF), the Anti-Ballistic Missile Treaty (ABMT) and the Threshold Test Ban Treaty (TTBT). Here are a few dates to put things in perspective:

December 8, 1983: Soviets suspend START negotiations because of US Intermediate Nuclear Force deployments.
October 14, 1985: Secretary of State George Schultz accepts the broad interpretation of the ABM Treaty as "fully justified" to allow SDI testing of *other physical principles* in space.
October 11–12, 1986: President Reagan and Soviet leader Mikhail Gorbachev agree to cut strategic weapons by 50 %. They almost agreed to ban all nuclear weapons, but the issue of SDI-ABM tests and limitations scotched the deal at Reykjavik, Iceland. When I arrived in January of 1987, the State Department was confused by the attempt to abolish all nuclear weapons. Twenty-five years later the spirit of Reykjavik blooms with the plans of senior diplomats, Kissinger, Schultz, Perry and Nunn, to abolish all nuclear weapons, but without specificity.
November 28, 1986: The 131st B-52 bomber is deployed with air-launched cruise missiles, ending SALT II. The US is the 1st country to publicly renounce a nuclear arms-control agreement.
December 1986: Ten Peacekeeper missiles become operational in Minuteman III silos.

The Office of Strategic Nuclear Policy (SNP) was the State Department's lead location for negotiations on arms control treaties. The negotiations on START had been going on for a half-dozen years, but the discussions at Reykjavik gave it a political push. I was asked to co-chair, with the Defense Department, a subcommittee for the START working group. The subcommittee was to examine the ratio of throw weight to launch weight for all missiles. Each of the super-powers had its own advantages in the missile race.

US had advantages in aiming accuracy, reliability and the invulnerable submarine fleet. Soviets had an advantage in the size of their missiles. The SS-18 was incredibly large and could have carried 20 warheads, twice its listed 10 warheads. And Soviet warheads had slightly more yield, at 500 kton, compared to US ICBMs of 350 kton and SLBMs at 450 kton. This difference is irrelevant when you consider that US missiles are much more accurate. The disparity in physical size can be worrisome, but this fact has been overused by those who believed the Soviets could carry out a successful first strike on the United States. Senator Scoop Jackson held a model of an SS-18 next to a model of a Minuteman to traumatize the audience. This was fraudulent, because accuracy is far more important than yield. If you reduce the accuracy from 200 to 100 m you can reduce the yield of warheads by a factor of eight (8!) for the same lethality. The US is stronger in this important respect since US accuracy is 100 m and Soviet accuracy is about 200 m.

To worry about Minuteman vulnerability by itself is not being complete. The Soviets would have to be suicidal to attack a Minuteman base. If they could get 100 % of the Minutemen, which they couldn't, they would still have to contend with 14 submarines with 24 missiles. Each SLBM now has 4 warheads but formerly they had 8 warheads, for a total of 2500 warheads. And the US has other warheads on bombers and other systems, which are better than the Soviets. The Soviet air defenses were shown to have holes on several occasions. The issue of throw-weight to launch-weight is a surrogate argument for cheating potential. A missile with a throw-to-launch ratio of 5 % is good, if it's 3 % it's not very good. A sensitivity analysis is needed to determine what survives of the total US nuclear force. My boss in the State Department co-chaired the interagency group and was interested in my spread-sheet analysis of Soviet Testing, but he was timid to present them in the interagency forum. I was told to pass it along to people in various agencies but they were not to be presented under the signature of the assistant secretary. These results were classified because the whole START process is classified, but they were of the type that had already been published. The only reason to classify them is so that the Soviets would not know what the US was thinking, but there was nothing that would endanger national security by putting them in the newspaper.

A.3 Conventional Forces in Europe Treaty

One of the perks of Senate ratification of the Conventional Forces in Europe Treaty (CFE) was an Army helicopter ride from Fort McNair over Washington and Baltimore to Aberdeen Proving Ground. CFE was an obvious triumph for NATO, even Senator Helms was supportive. These were fast moving events, but it was clear to me that CFE ratification was very positive, but it was the precursor to the ratification of START and the end of the Cold War.

I will quote liberally from the SFRC report on CFE. The treaty was signed on 19 November 1990 by 16 NATO states and 7 Warsaw Pact states. The reduction in conventional arms eliminated a fundamental cause of tension in Europe since the end of World War II. The huge numerical advantage of Soviet conventional forces threatened the security and prosperity of the West and it was integral to Soviet domination of Eastern Europe. Of course, it would have been suicidal (and stupid) for the Soviets to attack NATO, but the threat was always there. This was a false-numerical superiority, driven by the fact that the Soviets retained old equipment. But these conventional arms numbers fueled the nuclear arms race. In every strategic doctrine adopted by NATO, from massive retaliation to flexible response, nuclear weapons were intended to compensate for the Soviet edge in conventional arms. Indeed, tens of thousands of nuclear weapons were deployed to support these strategies. All this had a principal purpose, to deter the Soviet Union from threatening Western Europe.

The key effect of the CFE treaty was to force the Soviet Union to withdraw forces from Eastern Europe, to reduce the major armament holdings in Eastern European by more than half, to destroy a large quantity of these weapons, to place 20 % of the remaining weapons in storage, and to distribute the rest geographically to comply with all the ceilings and sub-ceilings. The CFE treaty provisions did not require NATO members to reduce armament holdings in a meaningful way. CFE set ceilings for each group of states within the area from the Atlantic to the Urals (ATTU). Neither group may field more Treaty Limited Equipment (TLE) than 20,000 tanks; 30,000 armored combat vehicles (ACV); 20,000 artillery pieces;

6800 combat aircraft and 2000 attack helicopters. This gives a total of 78,800 TLE for NATO and 78,800 for the former Warsaw Pact. No single country may exceed the following levels from the "sufficiency rule:" 13,300 tanks, 13,700 artillery, 20,000 ACV's, 5150 aircraft and 1500 helicopters. Senator Pell placed the final TLE numbers into the *Congressional Record* on the final day of ratification.

The Soviets unilaterally reduced their TLE in the ATTU from 152,000 to 73,000 between 1988 and 1990, for an early reduction of 50 %. It is true that the Soviets didn't throw away older equipment for bureaucratic reasons but these reductions are very significant. Once these old tanks were moved beyond the Urals, it was generally believed they would quickly rust into obsolescence, which was readily monitored by satellites and Open Skies Treaty aircraft.

The total Soviet TLE moved from the ATTU, destroyed or converted was 103,000, plus 21,000 from the 6 non-Soviet Warsaw Pact states, for a total of 124,000 TLE moved, destroyed, or converted. The total amount of Soviet TLE to be destroyed or converted was 57,700, plus 21,000 TLE destroyed by non-Soviet Warsaw Pact nations, giving 79,000 TLE destroyed or converted. This number included 10,700 East German TLE now owned by the Federal Republic of Germany. At the beginning of CFE negotiations, the Warsaw Pact had an advantage of a factor of 2.7/1 in TLE over NATO. The CFE treaty reduced this to a 1/1 ratio. The demise of the Warsaw Pact made a ratio of Soviet/NATO to 0.67/1. The exit of Ukraine, Belarus and Kazakhstan gives a ratio of Russia/NATO of 0.5/1. Thus, the ratio changed from an advantage of 2.7 to a disadvantage of 0.5, a reduction in ratios by a factor of 5. This major change happened in an open and peaceful manner, *a highpoint* for mankind.

CFE Chief Negotiator Jim Woolsey worked on these additional issues:

1. Individual TLE would not be tagged with bar-codes or other labels, but rather displayed in groupings of TLE, accessible to group counting from satellites or helicopters. This approach could easily detect a violation of a thousand TLE, which was much less than the NATO final advantage of 25,000 TLE.
2. The Soviet SS-23 missile had a range of 500 km, which barely qualified it for the INF regime, which had a lower limit of 500 km. The discovery of 72 SS-23's in East Germany, Bulgaria and Czechoslovakia were violations of INF. My speculation is that this over-loaded, politically confusing time caused this accidental violation, which was more political than significant. It is hard to debate that the Soviet's could expect to get away with this. I view it as a stupid bureaucratic Soviet error with little to gain by cheating. This issue wasn't solved until START was ratified. The Soviet Krasnoyarsk phased-array radar was a violation, but I also view that as a stupid bureaucratic Soviet error. Gorbachev admitted this, as he asked to turn the radar site into a furniture factory. The US held firm and he dismantled it. The Resolution of Ratification contained the phrase "The Senate declares that it will take into account, as part of its consideration of the START, [these two violations]".
3. The Soviets removed 80,000 TLE to the East of the Urals. This massive force was a concern to the US, even though it was older equipment, which was soon

to decline. A resolution of ratification condition required the President to inform the Senate if this force became militarily significant.

A.4 Cold War Ends

It would have been suicidal for the Warsaw Pact to invade Western Europe through the Fulda Gap to go to Frankfurt. But this was a powerful symbol of the Cold War. Such a hypothetical tank invasion could have been met with US tactical nuclear weapons, that was the goal of the neutron bomb. Supplying Soviet tanks far from their bases would have been difficult, it probably wasn't going to happen. But, in military international affairs, perceptions often drive policy. Once the disparity in conventional forces disappeared (which was exaggerated because of poorer quality Soviet equipment), progress could be made on reducing strategic nuclear weapons that threatened Moscow and Washington. All was going well until the Belarusian parliament closed, without indication that it would return. The CFE Treaty could not *enter into force* until all the parties ratified CFE. The long-time lead U.S. lawyer on these matters, Thomas Graham of the Arms Control and Disarmament Agency, conceived a solution to this problem. I quote from Graham's book, *Disarmament Sketches: Three Decades of Arms Control and International Law* (University of Washington Press, 2002, pp. 208–209):

> Why not amend the text of the provisional application protocol simply to provide that on a specific date in July the entire Treaty would actually be brought into force for the period of baseline inspections (4 months) on a provisional basis and as a minor administrative change?" After many phone calls the following was decided: "It was decided that Jim Timbie and I would meet with Senators Lugar and Pell (who was in Rhode Island but would be represented at the meeting by his aide David Hafemeister). I discussed the concept with ACDA Director Ron Lehman on Monday and he approved it. That afternoon Timbie and I went off to meet with Lugar and Hafemeister…. It was agreed that there would be an exchange of letters to this effect between Director Lehman on the one hand and Senators Pell and Lugar on the other. Ron would write asking for permission, setting forth the circumstances that necessitated this action and Pell and Lugar would reply giving their assent along with the aforementioned caveat that this was an extraordinary step required by events. This was accomplished the next day and our delegation at the Joint Compliance Group was instructed immediately to propose this and lobby for support. Not surprisingly there was resistance in Vienna because, to put it mildly, this was an unusual move. Nevertheless, Helsinki now was just a few days away and delegations increasingly came to support this procedure. Not without some anguish, however. The Italian ambassador told the US representative that his government's international lawyers in Rome 'had to swallow beach balls over this.' … The CFE treaty was formally announced to have entered into force on July 16. Baseline inspections began one week later, Armenia ratified later in the summer, and Belarus ratified on November 9, one week before the expiration of the four-month period. As I said, Czechoslovakia did split in two in January but now the Treaty was fully in force and the Czech Republic and Slovakia became successors states, bringing the number of CFE parties to thirty, which remains the number today.

A.4.1 Nunn-Lugar Cooperative Threat Reduction

It was clear that the demise of the Soviet Union could lead to chaos since Soviet plutonium and high-enriched uranium were vulnerable. As the KGB collapsed, these sensitive materials became loosely guarded. Some have said that potatoes were guarded more carefully than Soviet plutonium. In addition, the Soviets had not done proper materials accounting on its 130 tonnes of plutonium and 1200 tonnes of high-enriched uranium. And indeed, our government was asleep at the switch on this issue. It is easy to see why this happened as nonscientists are in control of science issues, and they often feel uncomfortable with technical matters. The Soviet's lead nuclear administrator, Victor Mikhailov, came to Washington in November 1991 after the ratification of CFE. When he met with Undersecretary of State Reggie Bartholomew, they did not make progress on the Russian dilemma of poorly-protected special nuclear material. There are reasons for this situation. There is not a science cone in the State Department, as promotions are more likely to come through the Bureau of Politico-Military Affairs and Economic Affairs, as well as the country desks.

Soviet Union in DC. Frank von Hippel, Princeton professor and Chair of the Federation of Atomic Scientists, and Tom Cochran, Head of the Nuclear Policy Project at the Natural Resources Defense Council, arranged for Mikhailov to meet with Congressional staff and various NGOs at Thanksgiving time in 1991. Mikhailov was clear that he was uncertain on where to store his nuclear materials, and this would take considerable funding. It was clear he was going to have to shut down his nuclear industry of 1 million workers. Presumably Mikhailov met with Senators Nunn and Lugar, giving impetus to their legislation, which passed on 27 November 1991. Five days later on December 1, Ukraine voted to become independent of the USSR, followed by 13 other Soviet Republics and Russia. The Nunn-Lugar legislation authorized $500 million to assist the Soviets to dismantle nuclear weapons and missiles, to enhance physical security, to protect these systems, to improve accountancy of these items, to establish negotiations, to buy reactor fuel obtained from Soviet HEU, and to retrain and encourage Soviet scientists to enter the private sector and not participate in a brain drain. This was a tall order.

Megatons for food. Tom Neff wrote an op-ed piece for the *New York Times* that suggested that the US send food to the former Soviet states in trade for surplus, lightly guarded plutonium and HEU. This op-ed stimulated much talk on the issue. Senators Alan Cranston of California and SFRC-Chair Claiborne Pell submitted a bill to try and address this issue, entitled the *Nuclear Warhead Security and Plowshares Act of 1991*. It was clear that HEU was far more dangerous than plutonium because it is much easier to make weapons from uranium than from plutonium, which must be imploded in a fat-man design. And uranium is much less radioactive than plutonium, therefore easier to work with and harder to detect. And HEU can be converted into 4 %-enriched reactor fuel that has economic value. On the other hand, plutonium has a negative economic value because of its radioactivity and it is less useful as reactor fuel. For these reasons, Senator Nunn didn't accept plutonium as a swap for food or MOX fuel.

Megatons for MegaWatts. Within a year, the Executive Branch began negotiations on denaturing HEU to reactor fuel, led by ACDA-Director General Bill Burns, with whom I worked closely on the National Academy study on *Beyond START*. The US and Russia formally agreed on 18 February 1993 to the sale of 500 tonnes of HEU turned into 4 % enriched reactor fuel. This sale covered 50 % of US reactor fuel needs over the past two decades. The issue of plutonium-for-food swaps was not addressed, as it was agreed to convert plutonium to mixed-oxide reactor fuel or bury it, this continues to flounder.

A.5 Soviet Union's Final Days

American delegation. Ukraine voted to leave the Soviet Union on 1 December 1991. Russia and the remaining 13 soviet republics followed Ukraine's exit shortly after. Arrangements were made for the new Russian Federation to take power on 1 January 1992. To assist in this transition, Frank Von Hippel and Tom Cochran arranged a unique international workshop on *Verified Storage and Destruction of Nuclear Warheads,* held in Moscow and Kiev on December 16–20, 1991. I stayed on until December 23 in preparation for the upcoming SFRC hearings on START ratification. This was a very unusual trip, indeed. The American delegation was a mixture of nongovernmental organization (NGO) scientists and Executive Branch scientists, along with former lab directors, the CIA, the Senate, and the press. These were tense times with a lack of communication between the two sides. Presidential National Security Advisor, Brent Schoolcraft condoned government participation in the trip. The following individuals participated: former director of Los Alamos National Laboratory Harold Agnew, former director of the Los Alamos Theoretical Division Carson Mark, the former vice-president of Sandia National Laboratory Jack Howard, three nuclear weapons designers and a CIA scientist. The NGO and academic scientists were Frank von Hippel, Tom Cochran, Tom Neff, Chris Paine, Steve Fetter and Alex DeVolpi (DOE Argonne) and myself from the Senate. This unique delegation included three reporters: William Broad of the *New York Times*, Jeffrey Smith of the *Washington Post* and Jonathan Schell of *Newsday* and author of *Fate of the Earth*.

Soviet delegation. The Soviet Union (soon Russian Federation) was represented by 31 delegates, including the following: Victor Mikhailov, Deputy Minister of the Ministry of Atomic Power and Industry (MAPI) in charge of the Soviet nuclear complex; Victor Karpov, Deputy Foreign Minister; General Vitali Yakovlev; General-Lieutenant Sergei Zelentsov; General Gelili Batenin, senior arms-control adviser to RF President Boris Yeltsin; Major General Alexi Leonov, CHETEK, first Soviet to walk in space; Victor Ivanoff, lead negotiator on nuclear testing and Deputy Director, MAPI Nuclear Weapons Development and Testing; Academician Yuri Trutnev, Deputy Scientific Director, Arzamas-16; Vadim Simonenko,

Director, Theoretical Department, Chelyabinsk-70; Victor Slipchenko, Ministry of Foreign Affairs; Anatoli Diakov, Director, Center for Arms Control, Moscow Institute of Physics and Technology (a Russian academic, who did serious arms control). The Ukrainian delegation had 15 members, led by General-Major Vadim Grechaninov.

A.5.1 CHETEK Offer to Destroy Chemical Weapons and Radioactivity

The Ministry of Atomic Power and its commercial firm CHETEK made a proposal to end two of Earth's problems in one stroke by incinerating large stocks of chemical weapons with surplus Soviet peaceful nuclear explosions (PNE). I quote from the trip report:

> At the arrival dinner, Trutnev presented a variation of the CHETEK-PNE waste disposal concept, purposing that one such explosion could also be used to destroy 5,000 plutonium pits from dismantled nuclear warheads. In response to a question regarding the disposition of the commercially valuable high-enriched uranium components from these destroyed warheads, one of the Russian scientists present stated that would be a 'simple matter' to 'unscrew' the uranium-laden secondary stage of the weapon from the plutonium-bearing primary stage. Air Force Major General Alexei Leonov served as toastmaster for the occasion. Leonov, twice named a hero of Socialist Labor, gained fame in the USSR as the first cosmonaut to walk in space and as Soviet commander of the joint Apollo-Soyuz Mission in the mid-1970's. The exact nature of Leonov's affiliation with CHETEK was also unclear, but it seems probable that Mikhailov, Trutnev, Tchernyshev and Leonov may be shareholders and/or directors of the company. An October 1991 report in the industry journal *Nucleonics Week* quotes CHETEK Vice-President Valery Siderov as saying that 'about ten experienced scientists' from Arzamus-16 are now CHETEK shareholders.

Many were hesitant that this would make peaceful nuclear testing viable and radiation would leak. On the other hand, after 25 years these two issues remain: very slow chemical weapon destruction and we are still faced with 100 tonnes of surplus plutonium with an uncertain future. Russia believes in plutonium more than the US does, it is unlikely that Russia would place 100 tonnes of plutonium into explosive cavities. More government-to-government discussions on plutonium disposition might have accelerated the slow progress on the plutonium disposition. On the other hand, meetings with Deputy Foreign Minister Victor Karpov and with Alexander Penyaguin, Chair of the Supreme Soviet Subcommittee on Nuclear Security showed Russian distrust of CHETEK. The CHETEK press release of December 11, 1991 stated the following:

> Scientists of this institute are conducting scientific research on nuclear explosion technology, the goal of using this technology for the destruction of chemical weapons, as well as highly toxic and radioactive waste.... A general agreement was finalized between the Ministry of Atomic Power and Industry and the international joint stock company CHETEK in May 1991... This project and its financial-commercial, component part is an

example of the high moral standards in approach by a commercial organization and cooperation with a government organization which, during a time of difficult economic situation of the country in regard to attracting funding from outside the budget (including from abroad), is solving a most important ecological and disarmament problem for mankind.... Signed by V. Mikhailov and V.B. Dmitriev, CHETEK

A.5.2 Where to Store Excess Soviet Plutonium and HEU?

The meeting with Victor Mikhailov at the Ministry of Atomic Power and Industry revealed Soviet thinking on Nunn-Lugar funding:

> The nuclear weapons of the FSU will be dismantled in the plants of Mr. Gorobrets... near Chelyabinsk or Tomsk. At one point, Mikhailov mentioned Sverdlovsk as a site, but he quickly corrected himself. He asked the delegation not to publicly discuss the potential sites for the new plutonium storage facility, because MAPI's plans had not yet been discussed with local officials in the affected region.

A year later the citizens from Tomsk strongly opposed building the plutonium storage facility near their city, closing-down our meeting for an hour. It was refreshing to see the public speak up in this formally closed society. Quoting from the December 1992 trip report:

> Mikhailov said that the entire $400 million should be spent on a single new MAPI facility, rather than spread around among several different projects... which would diffuse the impact of the funds. In this way the United States could assure itself that the money would not be spent on 'duel-purpose' equipment that might later be used to produce weapons as well as dissemble them.... The large uncertainty in the estimate 'from 10,000 to 20,000 warheads to be destroyed' stems from the vague language used by Gorbachev when he announced the reciprocal Soviet reduction initiative. In a side conversation, Mikhailov said that the weapons storage depots in Russia were filled to capacity and that trains carrying weapons to the depots had been stopped en route.

A.5.3 Plutonium

The 1991 trip report discusses interactions between MAPI and US scientists on plutonium issues:

> Jack Howard, a former vice-present of the Sandia National Laboratory gave two short presentations, one on permissive action links and another on methods of securing nuclear weapons in transport and storage against unauthorized access. Howard suggested going beyond reversible steps for disabling weapons by introducing a sticky material into the interior of the hollow plutonium core of the primary (trigger) stage through the channel reserved for tritium gas injected into the core prior to detonation. He later joined former Los

> Alamos Director Harold Agnew in private discussions with the Soviet MoD and MAPI officials, in which they suggested using a borated epoxy that would harden inside the sphere, eliminating the risks of a nuclear yield in the event of an accidental detonation of the high explosive surrounding the core, and enabling the plutonium cores to be stored closer together.

Mikhailov commented: "We have spent too much to just throw this plutonium away…. We need plutonium for small reactors in the North, for district heating." We returned to the plutonium issue in December 1992, but now the delegation consisted only of NGOs and myself, as the new Russian Federation preferred to deal with formal government-to government procedures. We discussed with the Russians plutonium storage, plutonium production and reprocessing, MOX fuel in light water reactors and breeder reactors. MAPI scientists did not waiver in their support for a plutonium economy. You might think, as Russia went from state-controlled markets to free markets, this would change, but they had invested their lives in plutonium. Additional commentary from the 1992 trip is as follows:

> A year ago a trip with 6 scientists and 6 DOE employees was ahead of the Executive Branch since Nunn-Lugar policies and funding were not yet in place. One year later, as one would expect, the Executive Branch has been negotiating on many of the dismantlement issues, but there are still great uncertainties, especially on the issue of the final disposition of plutonium. Russian Minatom officials stated that they plan to continue to reprocess and burn their plutonium in the new generation of breeders, an unlikely scenario. At this point, the oversight by the Congress has been limited to a one-hour hearing in SFRC on July 27, which took place before the decisions on LEU/HEU purchases and plutonium storage.

A.6 Monitoring Warhead Destruction

Discussions on monitoring warhead destruction, and extending arms control to count warheads took place at MAPI, the Soviet Foreign Ministry, the Russian Duma in their White House, the Ukrainian Rada Parliament and the Ukrainian Foreign Ministry. This was a divisive issue in the US delegation because some were interested in prompt Russian ratification of CFE and START and others had their eye on a future world without nuclear weapons. I was aware of the difficulty of the goal of warhead monitoring because of the strong legal requirement to protect warhead secrets and the fact that warhead monitoring technologies needed to be developed. I had observed that Frank Hippel was buffeted by Senator Malcolm Wallop before the Senate Armed Services Committee on the detection and identification of warheads in the field. Acting with the near term in mind, I passed the Senate Foreign Relations Committee CFE ratification report and START hearings to the Russian and Ukrainian Supreme Soviet staff members for their ratification of CFE and START, a first for the new Russia. The 1991 trip report states the following on warhead dismantlement:

> In what amounted to a preemptive strike against the principal objective of the sponsors of the US delegation—verification of the elimination of nuclear warheads removed from Ukraine—General Zelentsov lead off the workshop with the presentation of the Ministry of

Defense position in opposition to any verification measures that had not been formally agreed-upon 'by the two Presidents. Since Gorbachev and Bush had not raised the need for additional control measures when they undertook their mutual declarations, Zelentsov maintained that there was no political basis for the Ministry of Defense to pursue such measures. 'No inspection of nuclear weapons by strangers are allowed. Only trained Army-staff are allowed to get near the weapons. We cannot expand access to nuclear weapons' because strict procedures had been designed to exclude access by 'journalists, terrorists and other outsiders.' If tags were to be applied to U.S. as well as Soviet weapons, then this should be done by the Ministry of Defense personnel or a special inspectorate created for this purpose, Zelentsov noted. But he indicated that tags are unnecessary for tracking warheads through the dismantlement process because all Soviet warheads and their principal complements....are stamped with serial numbers, and the Soviet Army has quality-control inspectors who follow warheads from production to dismantlement.

In rebuttal, members of the US delegation (Cochran and DeVolpi) explained that (1) the purpose of *sealing* was not to prevent access to the warhead but to *reveal* unauthorized access; (2) fiber-optic seals could be applied *after* the weapons had been disabled and made safe prior to shipment to storage facilities in Russia, and (3) that the Soviet and Ukrainian military personal accustomed to handling nuclear weapons could themselves apply the tags and sealed with a minimum of direct involvement by the inspecting party. Cochran demonstrated 10 types of tagging and sealing technologies for warheads and warhead containers.

Monitoring warhead destruction became a flash point in Kiev when the Deputy Chair of the Ukrainian Parliament, Volodymyr Gryniov, made an offhand remark that perhaps the Soviet weapons in Ukraine might be dismantled in a third country, other than Ukraine or Russia. This drew a strong response from General Zelentsov of the MOD and Victor Ivanov of MAPI. General Zelentsov not only vigorously quashed this notion, arguing that the appropriate personnel and facilities for this task were in Russia, but he used the opening provided to renew his argument that only qualified MOD personnel could be allowed near the weapons, and that international inspectors would compromise their safe removal from the Ukraine. The latter comments sparked a strong response from Cochran, who warned Gryniov, 'don't be fooled by the General's argument' concerning the alleged dangers of verifying warhead elimination. Cochran repeated his suggestion that it would be a short and relatively simple task to tag and seal weapons before they left Ukraine. This comment angered Ivanov, who later accused Cochran of attempting to drive a wedge between Ukraine and the FSU ministries.

The US delegation indicated it would examine the science of warhead monitoring and hopefully the new Russian government would join in. This later happened under President Bill Clinton, but it took five years to obtain DOE agreement on how to retain warhead secrets. Warhead monitoring re-enters below in START ratification and the National Academy study on warhead monitoring.

On a tank with Yeltsin. Five months later, Bill Ashworth of SFRC and I had a memorable lunch with Nikolay Vorontsov, Soviet Minister for the Environment, who was famous for standing on the tank with Boris Yeltsin at the Russian White House as they defied the Soviet military at the time of the military coup and Gorbachev's capture. Quoting from Murray Feshbach's *Ecocide in the Soviet Union* (Basic Books, 1992, pp. 251):

> At noon on the first day of the putsch, Boris Yeltsin, the most popular politician in the Soviet Union, clambered to the top of an Army tank in front of the white skyscraper that housed the Russian Republic Parliament. Yeltsin used that unlikely podium—a symbol of what proved to be a decisive split with the Soviet military—to rally the crowd in front of him, the city around him and the world beyond to defy what he had sneeringly termed the 'junta' in the Kremlin. Vorontsov was the next to speak from the tank's deck to denounce the takeover attempt as 'illegal.' Having declared his defiance in public, he repeated it in private at an evening meeting of the Soviet cabinet. Even behind closed doors he was the only minister to voice uncompromising opposition to the men he later called 'incorrigible conservatives.' Two of his Kremlin colleagues—a deputy prime minister and the cultural affairs chief—declined to endorse the coup and its aims, but only Vorontsov truly resisted the conspiracy while the outcome was still in doubt.

A.6.1 A Bribe to Escape Moscow

The US delegation departed after the meetings in Moscow and Kiev, but I stayed for further discussions at the Soviet Ministry of Defense and at the Soviet On-Site Inspection Agency (OSIA). I needed specific information for the upcoming SFRC ratification hearings on START and on the breakup of the Soviet Union. I was accompanied by the US Military Attaché to the Soviet Union. From my trip report:

> The US military chief of the US OSIA in Moscow arranged for a briefing at the headquarters of the Soviet OSIA, which is located in Soviet Ministry of Defense. The communication facilities were examined. One thousand messages annually pass through the Nuclear Risk Reduction Center, which is co-located with the Soviet OSI. The Soviet leader, General Vladimer Medvedev pointed out that the Soviet bases in Eastern Europe had been closed and these facilities should be removed from the inspection list at the Standing Consultative Commission meeting. Gen. Medvedev stated that 'we have had just a few problems that would raise discussions, to the surprise of some people.

General Medvedev was a gracious host, giving me a piece of the last SS-20, which were destroyed pursuant to the INF treaty. Further details on INF implementation were given to us, which were transmitted to Washington by the attaché. I followed this with a cable from the US embassy, describing details of our US-Soviet meetings.

This extra work forced me to stay two days beyond my visa dates. I naively hoped that the Soviet airport authorities would honor my government passport and ignore this detail. Moscow's Sheremetyevo Airport was mobbed as Russians and foreigners were fleeing Moscow to avoid the chaos. The panic level was compounded by a lack of jet fuel as the Soviet system started to collapse. As I waited in line, a petite Russian woman came and beat on me with her fist as she thought (incorrectly) that I had a better place in line. Her husband apologized to me as his wife had panicked about obtaining a flight to Soviet Georgia for their family of four.

Finally the Air France line began to move. As I reached the head of the line, I noticed three ominous-looking Soviet soldiers, packing AK-47's. I held my breath and handed them my passport with its outdated visa. My heart sank as they quickly spotted the discrepancy. They pointed towards Moscow and said that I must get a new visa. But how could that really happen as the Soviet Union was crashing, while the new Russian government barely existed? It could take months. Where was I to stay? There were the forthcoming hearings on START, and my family in DC was awaiting my Christmas arrival.

The Bribe. At last I thought of something I had never done before (or since). Why not try good old-fashioned bribery? I stuck three ten-dollar bills into my passport and said something in English, which I knew they could not understand. At this time a $10 bill was worth about $1000 as the ruble crashed. The senior officer and the two young recruits huddled in the corner. I broke into a sweat. Will they send me to jail? Will they send me back to an uncertain Moscow? Happily, they motioned to me to enter the plane. Here we were, former enemies but yet four humans reaching out to survive. After all, these were decent soldiers put into an impossible situation. Their lives were going to get much worse before they got better. I view the $30 visa expense as a necessary evil for me and as a bonus for the three soldiers and their families. I hoped things worked well for them as I returned to the comforts of the U.S.

A decade later in 2001, I returned to Moscow to present a paper representing a National Academy of Sciences study on *Beyond START*, primarily on the issue of monitoring warhead dismantlement under a new treaty. I knew some of the Soviet scientists and government officials would know English but most would not. These presentations can be deadly dull if you use a translator who speaks with no eye contact, losing the audience. A friend mine at the National Academy translated my view graphs into Russian. At the Russian Academy I ran into Stanislas Rodionov from the previous FAS-Soviet back channel. I asked Stanislaus if he would read these sentences in Russian, which he did. I was later told that this was the most humane event of the two-day conference on monitoring warhead dismantlement.

A.7 Ratification of START

My previous involvement with START was primarily academic papers, the MIT verification book and as State co-chair with the Pentagon on the launch-weight to throw-weight committee. Starting in January 1992, my full-time job at the Senate Foreign Relations Committee (SFRC) was ratification of START. The issue was not in doubt, as the anti-START senators only got six votes. But it was an issue of substance and future directions. Could the anti-START forces attach an unsavory amendment to kill the treaty, or would hearings on a particular issue become overly clouded. It was nice to have the Senate-majority Democrats join with President George H.W. Bush on this mission, as we agreed on most START issues.

Minuteman vulnerability was more complicated than the political debates, described in the press. Even if the Soviets had been able to attack and destroy 900 of 1000 Minuteman missiles, there was still an over abundance to respond.

My various calculations used the parameters of hardness, yield, reliability, range, accuracy, number of warheads on a target and number of warheads of each type. This can be complex for hearings, but it does allow one to ask questions based on these concepts to prove a point.

SFRC had 20 hearings on START between January 1992 and ratification in September 1992, out of a total of 45 SFRC hearings on arms control with fifty physicists as witnesses over the two-year period. We will discuss the highlights from the 900-page START hearings.

A.7.1 START History with Scientist Hans Bethe and Negotiator Paul Nitze

Chairman Pell wanted Bill Ashworth and myself to get into the important details. It all begins with history. Who better than a pair of 85-year-olds, Hans Bethe on the history of nuclear weapons science and Paul Nitze on the history of nuclear weapon policy. Each had detractors, as Edward Teller argued with Hans Bethe, and arms-controllers argued with Paul Nitze. SFRC enjoyed the hearing as it put START into a broader perspective. Let's start with Bethe: I phoned Bethe at his house in Ithaca, New York and his wife responded he was at UC San Diego. I phoned San Diego and they said he was at Stanford, who said he was at Livermore, who said he was at the University of Washington, where we finally chatted. Eighty-five-year-old Bethe got around nicely. At the hearings, Bethe said the following:

> The Cold War is ended.... On the basis of the situation, President Bush, in September, took a most important unilateral step: he ordered withdrawal of all tactical nuclear weapons from Europe, with the prospect that many of them would be destroyed. He also ordered elimination of tactical nuclear missiles from naval ships, including sea-launched cruise missiles, which were a great obstacle to any arms control agreement. He stopped the alert of nuclear-armed intercontinental bombers and ordered several further measures contributing to relaxation. Gorbachev responded promptly and positively in the same spirit. These unilateral measures with responses are much to be welcomed and are much quicker than negotiations of a treaty. Some people believe that in this way, START has become irrelevant. In my opinion, START is as important as ever. It will provide a firm basis for the future.

At this point, Nitze was evolving from the author of stern memos on the Soviet threat and as a leader of the Committee on the Present Danger. He was evolving to the *walk-in-the-woods INF negotiator* in Geneva and soon to become a nuclear-abolitionist. I thought some of his arguments on such things as *Minuteman vulnerability* were not scientifically accurate, but he was kindly to me and I liked him. At the hearings, Nitze said the following:

> As nuclear weapons systems become increasingly survivable on each side, the prospects for counter nuclear targeting, or counterforce nuclear targeting, will continue to diminish. I would continue to target nuclear installations that could be attacked efficiently, such as

> bomber bases and SSBN ports. But I would not reserve the large numbers of warheads necessary to target mobile ICBMs, nor what I plan to target ICBM silos. The latter should be mostly empty by the time the United States retaliated, as I would expect fixed ICBMs to be the weapons of choice in a first strike, while the reserve force will rely mostly on mobile ICBMs and SLBM's.... I would also retain the option to target leadership, but with much more flexibility than we have had in the past in order to adjust to the broader range of possible conflict scenarios.... With all this in mind, I believe we can safely reduce our inventory of deliverable strategic warheads to no more than 3,000 to 5,000 in the near-term, provided that the states of the former Soviet Union agree to reduce at least that far.... In the long run, further cuts in the U.S. strategic inventory should be possible as conventional weapons gain additional effectiveness or if other nuclear powers shift from their current reluctance to participate in arms reductions.... Taken as a whole, U.S. strategic forces are sufficiently survivable to give us a strong deterrent.

A.7.2 Why Did the US Fail to Predict the End of the Cold War?

This question was interjected into the Bethe-Nitze hearing by Senator Daniel Patrick Moynihan:

> In 1979, *Newsweek* had a forum on…what will happen in the 1980's. I wrote a short piece which said that in the 1980's it should be obvious that the Soviet Union will break up along ethnic, national lines; and then you will have a problem of who gets control of the warheads if they begin to divide among ethnic, national lines. I wrote this and rewrote it, and argued it in the intelligence community. It comes out of an analysis of Marxist propositions about class solidarity. If you look at the history of the century and the behavior of the Marxists, they always stressed nationalism themes; even that nationalism would eventually lead to proletarian internationalism. But it never did. Nobody could hear you in Washington, but you could make the case that the Soviet Union was about to break up. And one did. And now it has.
>
> Nitze responded: "My view is that the situation is highly uncertain and all you can deal with today is probabilities, various probabilities. I've been asked that question recently: what probability would you put upon this and that and the other thing happening. I would put a probability of 1 in 10 on their being able to solve economic problems within the next 5 years.... I would put the probability as 1 in 2 that they will have a takeover with the military important in the takeover within the next 2 years."

A.7.3 Q and A on Minuteman Vulnerability

One thousand questions were asked of witnesses and others. The responses fill 150 pages of very small type. It was our mission to learn all aspects of START and its implementation. This is what Senators Lugar and Pell had to say:

> *Senator Lugar:* "I just say once again, for the record, that the Administration has now written answers to more than 1,100 questions that have been offered by Senators. That is a substantial amount of interest. The body has been characterized as apathetic. I just want to set the record straight-1100 inquiries, all answered in writing. Let me say simply there were 30 hearings held by 3 committees, and I say that there has been substantial understanding of the elements of this Treaty for years-not simply in the very intense period of the hearings and the questioning."

> *Senator Pell:* "If the Senator will yield for a moment, I would like to demonstrate the number of pages, over 1,000. They weigh several pounds, and one can see the amount of wordage in these responses by the Administration to the Committee's questions."

Some of these issues involved classified information, but if you wrote the question in general terms you could sufficiently learn the essence. For example, what is the effectiveness of a possible Soviet breakout from START by attacking Minuteman? The SFRC START report contains the following commentary:

> The Committee asked the Administration to quantify the marginal utility of additional hard-target killing warheads on silos for 2 cases: (1) Without the destruction of one warhead by another warhead, or without fratricide, and (2) with fratricide that was only effective for warheads beyond the first two warheads on a target. Under the worst case assumption that the Russian weapons are very reliable and very accurate, it was assumed that they could destroy 81% of the targets in a one-on-one attack (90% reliability and 90% single shot kill probability), the Administration was asked to determine the marginal destruction rates of the last Soviet warhead for various attacks on U.S. silos. The Administration response is given below in tabular form" for one to four warheads on targets of Minuteman II with one RV, Minuteman III with 3 RVs and Peacekeeper with 10 RV's. The Committee defined an effective attack as one in which the last warhead used on a silo destroyed one or more warheads on average. The Administration responded: There is no *effective attack* against Minuteman II (or any single warhead system). A one-warhead attack on a Minuteman III [with 3 RVs] and a one- or two-warhead attack against a Peacekeeper silo would be effective attacks.

This response shows that attacks of more than 2 warheads, even on the target-rich Peacekeeper, disarms the attacker since such attacks use more warheads than they destroy. Attacks with more than two warheads had been of concern since it is possible, but not likely, that the SS-18 could contain as many as 20 warheads. But the US Trident missiles could, in principle, contain 14 warheads, 6 more than the declared value of 8. These over-loaded missiles would be detected by the RV on-site inspections, which counted the RV's on each missile. Quoting from the SFRC START Report:

> Because the marginal return from the use of a third (or fourth) warhead is very small, with or without fratricide, the ratio of hard target (counterforce) warheads to hard targets greater than two is not militarily significant. Since counterforce weapons can only be effectively applied to silos at the rate of two per silo, and because counterforce weapons cannot successfully attack submarines at sea and bombers aloft and away from airfields, the effective use of counterforce weapons is definitely limited for those above twice the number of silos and launch control centers.

A.8 Verification Standard for START

The standard for *effective verification* of arms control treaties was established by Ambassador Paul Nitze when testifying on the INF Treaty in 1988:

> What do we mean by effective verification? We mean that we want to be sure that, if the other side moves beyond the limits of the treaty in any militarily significant way, we would be able to detect such a violation in time to respond effectively and thereby deny the other side the benefit of the violation.

The START report language gave the results of an exchange model that started with 10,000 US warheads but was reduced by a Soviet first-strike-attack to 3600 survivable warheads. This was a worst-case analysis, one that I didn't believe. It's interesting to point out that Senator Jesse Helms' staff organized the ratification of START II, using the same approach, but they probably believed in the worst-case analysis. Even with worst-case analysis, the US had 3600 warheads remaining, when ten warheads aimed at the 10 largest Soviet cities would in effect destroy the Soviet Union.

A.8.1 Presidential Report on Treaty Compliance Arms Control Treaties

The resolution of ratification required the president to determine whether arms control had more *costs from noncompliance* versus *benefits from reducing Soviet nuclear arms*. Condition number 7 required the President to do the following:

> (A) a listing and discussion of actions which are violations or probable violations of the obligations of the SALT I, SALT II, ABM, INF and START Treaties and the ultimate resolution of these issues;
> (B) a listing and discussion of the actions which are in compliance with the SALT I, SALT II, ABM, INF and START Treaties; and
> (C) a comparison of the military significance of those actions listed in subparagraphs (A) and (B).

Ultimately the bureaucracy produced the report, which I analyzed in *Presidential Report to the Congress: Net Benefit Analysis of US/Soviet Arms Control* (*Science and Global Security* 2005). This effort was necessary to understand Soviet violations, to learn that arms control (B) was far more important then Soviet violations that lacked military significance (A).

A.8.2 Biden Condition on Nuclear Weapons Monitoring

The NGO's movement to establish warhead monitoring at the end of the Cold War is discussed in Sect. 7.6. The lead Soviet General Zelentsov wanted no part of this, nor did U.S. Secretary of Defense Richard Cheney, who stated the following: "We wanted to make certain that what we were agreeing to were agreements that could be verified. And getting into the area, verifying the destruction of warheads is a much more complicated subject." Declaration Three, adopted by SFRC, instructed the President to move ahead with warhead destruction, but it did not require verification by the two countries. The Resolution of Ratification was *marked up* in the small SFRC hearing room in the Capital Building, which held 30 visitors. Things were going well as the Executive Branch and a few reporters were on one side of the table and the senators and staff were on the other side. SFRC was about to adjourn when Senator Joe Biden made a motion to add an Eighth Condition:

> .... the president shall seek an appropriate arrangement, including the use of reciprocal inspections, data exchanges, and other cooperative measures, to monitor (A) the number of nuclear stockpile weapons on the territory of the parties to this Treaty; and (B) the location and inventory of facilities on the territory of the parties to this Treaty capable of producing or processing significant quantities of fissile materials.

I knew the Biden amendment would create a problem because the next treaty, START II, did not have these monitoring provisions. But I also knew the amendment was vaguely worded since it wasn't clear to which treaty it was to be applied. Upon hearing the amendment, the Executive Branch's Tom Graham and Jim Timbie became ashen-faced. Ultimately, Senator Sam Nunn became involved, forcing Biden to agree that his amendment would not complicate the de-MIRVing of missiles in START II, but that it would apply to the future START III. Monitoring warheads gained stature when it was agreed *to be considered* in START III by Presidents Yeltsin and Clinton on March 21, 1997 in Helsinki. A joint Soviet-American work program was established by the weapons laboratories of the two nations during 1997–2001. New approaches were tried for monitoring warheads, using techniques called attributes and templates. This approach was dropped when President George W. Bush became president. Ultimately, I became involved with warhead monitoring, as the Study Director of the National Academy of Sciences study, *Monitoring Nuclear Weapons and Nuclear-Explosive Materials* (2005).

A.8.3 Team B and the Strength of the Triad

At the end of the Ford administration, the *Committee on the Present Danger* obtained access to CIA data on Soviet forces to write a *Team B* report on Soviet military prowess. At the end of the Cold War, it seemed useful to examine what

Team B said and how their results compared with what the General Accountability Office (GAO) determined about the relative strengths of the Soviet and American triads. Senator Glenn agreed to chair the hearing in the Government Affairs Committee with DoD Deputy Secretary Bill Perry and GAO to be the key witnesses. The un-classified GAO cover volume was backed with 8 classified volumes. The hearings became tense when the Air Force was caught in a bold lie to Congress about the magnitude of the B-1 radar cross-section. Glenn was particularly incensed because he had been a B-1 supporter. The hearing showed that the submarine-based weapons were as accurate, lethal and timely as those from ICBM's. The hearings further showed that on a survivability basis, the SLBM's were the best buy. The 1992 GAO report concluded the following:

> The Soviet threat to the US triad was overstated, that the performance of existing US systems was understated, and that the performance of new US systems was overstated….. In the case of the land leg…the claimed *window of vulnerability* caused by improved Soviet missile capability against [US] silo-based ICBMs was overstated on three counts. First, it did not recognize existence of sea and air leg deterrents, that is the likelihood that the Soviets would hesitate to launch an all-out attack on the ICBM silos, given their inability to target submerged US submarines or on-alert bombers and their thousands of warheads that could be expected to retaliate. Second, the logic behind the claim assumed the highest estimates for Soviet missile performance dimensions as accuracy, yield and reliability, while at the same time discounting very substantive substantial uncertainties about performance that could not have been resolved short of nuclear conflict. Third, it ignored the ability of US early warning systems to detect a Soviet ICBM attack, and thereby allow a reasonably rapid response. [*Science and Global Security*, 1997]

A.8.4 Four Days of Killer Amendments

Senators Claiborne Pell and Dick Lugar and their three staff sat in the *Well of the Senate* for four days, from 28 September to 1 October 1992, fighting killer amendments. The purpose of Senators Wallop, Smith, Symms and Helms was to pass at least one amendment that would change the bargain with the Soviets, in effect blocking the treaty from ever entering into force. These four senators did not have the votes to block START, as it ultimately passed 93 to 6. But they could cause mischief along the way. These four senators were too busy to write the 61 amendments, which were written by my counterpart David Sullivan, Senator Jesse Helms' key staffer on arms control treaties. As an example, we list Amendment Number 3260:

> The START Treaty, including the May 23, 1992 Protocol, the two Annexes, six Protocols, Memorandum of Understanding and Corrigenda, shall not enter into force until the President certifies that all MIRVed ICBM's, and all launchers for MIRVed ICBM's, shall be eliminated in accordance with the agreement in the Joint Understanding on deep cuts on June 17, 1992 signed by the President of the United States of America and the President of the Russian Federation.

This amendment may sound like apple pie, but it is mixing 2 different treaties to make it a killer amendment. President George H.W. Bush signed an agreement with President Yeltsin to ratify START II at a later date, which would ban land-based MIRV missiles. START I did not ban MIRV, which was going to happen in START II. This amendment was unacceptable to both treaty parties since it changes the basic bargains of START I. David Sullivan was too cute by one-half, but the amendment was handily defeated. There were 60 other amendments like this, but we won't discuss them.

A.8.5 Evolution of MIRV

MIRV stability. Multiple-Independently-Targetable-Reentry Vehicles (MIRV) were invented to save money. If the nation wants to attack with 1000 warheads, it is cheaper to place them on 100 missiles with ten warheads on each missile, as compared to 1000 missiles with one warhead per missile. But, the approach with 10 RV's per missile is unstable. The nation with 1000 warheads on 100 missiles could be attacked with 200 hard-target warheads, instead of 2000 warheads (2 on 1 targeting) if based without MIRV. One might look at the cost-exchange loss of five-to-one if attacked, or a gain of 5/1 if you launch an early pre-emptive attack. Going from 5/1 loss to 5/1 gain is a factor of 25. This ignores the fact that other warhead systems would level the attacker.

Senator Brook. My first visit to the Senate was in June 1969. The Senate chamber was essentially empty except for one lonely senator eloquently presenting his case on Senate Resolution 211 for a ban of the development of MIRV systems. The senator was from Massachusetts by the name of Edward Brooke, a rare individual in that he was a black Republican, the 1st black senator since the Reconstruction Era. This is what SR 211 said in part:

> Whereas development of multiple independently targetable reentry vehicles by both the United States and the Soviet Union represents a fundamental and radical challenge to such stability;
>
> Whereas the possibility of agreed controls over strategic forces appears likely to diminish greatly if testing and deployment of multiple independently targetable reentry vehicles proceed;
>
> Resolved further, that the president should propose to the government of the Union of Soviet Socialist Republic an immediate suspension…of the further development of all offensive and defensive nuclear strategic weapon systems subject to national verification or such measures of observation and inspection as may be appropriate.

Senator Brooke's actions were significant, in spite of the fact that the Senate was almost empty and it seemed unlikely to happen, even if it had no teeth. But ultimately SFRC reported favorably on SR 211 on 24 March 1970, and it passed the Senate on 9 April 1970 on a vote of 72 to 6. I didn't forget that moment, and I felt fortunate to be able to recall that moment some 23 years later, while working on the

SFRC report on START-1. At that point the proposed START-II treaty was to ban land-based MIRV missiles. For the United States, this meant the removal of the Peacekeeper with 10 warheads each on 50 missiles, and the Minuteman III with 3 warheads each on 450 missiles. For the Soviets this meant the removal of the SS 18 with 10 warheads on 310 missiles (cut in half by START-1), the SS 19 with six warheads, and other systems. Thus, de-MIRVing would reduce thousands of warheads. A month after the ratification of START-1, I decided to write a note of thanks to Sen. Edward Brooke:

> In order to place Declaration Number 1 for START-I in context we chose to point to Senate Resolution 211 which you introduced in 1968, because, after all, that was the point in time when the Senate correctly understood the danger of MIRV. Two decades have passed and at least a 3rd decade will be needed to fulfill SR 211, billions of dollars were wasted, and great dangers were taken with the instabilities of MIRV. If the US government had listened to you we would have been much better off….I felt fortunate to be able to recall SR 211 as we worked on the SFRC report on START-1.

> Sen. Brooke responded as follows: "You may consider yourself fortunate to have been present in the Senate at the introduction of SR 211, but I feel even more fortunate in that you remembered."

A.9 Nuclear Test Ban Treaties

A.9.1 Threshold Test Ban Treaty

My first involvement with nuclear testing was in the 1987 State Department, where I was the lead technical analyst on Soviet compliance to the Threshold Test Ban Treaty (TTBT). The issue is complex because Western US geography is younger, absorbing seismic signals more, while the Soviet Asian plate is older, absorbing seismic signals less. The Reagan and Bush-I administrations concluded that Russia was in "likely violation" to the TTBT. The day I arrived at the TTBT interagency meeting, the newspaper stated that the East-West bias-value leaked to the press with a magnitude of 0.2 m_b units. [The bias factor is now believed to be 0.4, reducing the apparent yield of US tests by a factor of 2–3 compared to their Soviet counterparts.] This was a tense time on these issues: In 1985 Gorbachev stopped Soviet testing, calling the United States to join the Soviets. The US House of Representatives blocked US nuclear testing. The Soviets ended their unilateral nuclear testing moratorium after 18 months in February 1987. Negotiations were not going well in Geneva.

My main effort was to examine spreadsheets of data by varying the parameters, including the bias factor. I wrote a lengthy memo, which showed that by using the 0.2 bias factor, the data would not give the conclusion of a *likely violation*. We have to remember that monitoring amplitudes of explosions is a statistical event. By assuming reasonable distributions of explosions, I could see that the few explosions at higher yield could not be explained as a likely violation. I suggested to my boss that we submit it to the interagency process. He did not want to ask the assistant secretary for a signature. But he told me I could spread my results to other interagency members, which I did.

My work moved the State vote from "*likely noncompliance*" to "*compliance with the possibility of a few tests with small excessive yield*." The final interagency result was a vote of 3–3, with DOD, JCS and ACDA in favor of noncompliance and State, DOE and CIA in favor of compliance. President Reagan broke the tie, making the Soviets in "likely noncompliance." I believe this was bad science and done to avoid dealing with Gorbachev on the Comprehensive Test Ban Treaty. There was nothing I could do, but I made sure my calculations and spreadsheets were deposited in a State Department vault.

Six months later I was at Stanford's Center for International Security and Cooperation. Fred Lamb, a physicist from the University of Illinois, told me that the House of Representatives was about to have a hearing on *TTBT compliance*. I sent a one-page commentary to chair of the House Committee on Foreign Affairs for a hearing on June 28, 1988. The three government witnesses were General Bill Burns, ACDA Director; Bob Barker, Secretary of Defense Richard Cheney's lead nuclear assistant; and Paul Robinson, the TTBT Chief Negotiator, all of whom I knew. At the end of the hearing, the Committee Chair, Dante Fascell of Florida, lectured the government witnesses:

> Without objection, I would like to insert in the record a letter from Professor David Hafemeister. He writes in his individual capacity about the question of Soviet compliance with the Threshold Test Ban Treaty. The whole letter will be put in the record for examination, but basically he says: 'Taking into account the uncertainties, I conclude that one cannot state that the Soviets have or have not been violating TTBT. Or to put it another way, the data is not inconsistent with Soviet compliance with the TTBT.' And then he goes on to support the OTA report on seismic verification of nuclear testing on this issue, and he refers to pages 17 and 19. I thought the Administration witnesses might want to take a look at the letter. We will give you a copy of the letter, of course.

SFRC on TTBT. The lead Administration witness before the SFRC on 6 October 1988 was TTBT Chief Negotiator, Paul Robinson. He was followed by three technical witnesses who disagreed in their testimony with the Executive Branch compliance decision; Mylo Nordyke of Lawrence Livermore National Laboratory, Greg Vander Vink of the Office of Technology Assessment and myself (below):

> In particular, I carried out a thorough statistical analysis of the seismic data taking into account the random and systematic errors and using the accepted U.S. government value for the bias factor.... If one applies the same standard of evidence to the testing programs in both the US and USSR, one can state that both sides are complying to the TTBT within the accuracy of the measurements.... Therefore I agree with the OTA report on this issue which states that: 'All of the estimates of Soviet and U.S. tests are within the 90% confidence level that one would expect if the yields were 150 kton or less.'

A.9.2 Nuclear Test Ban Passes

The US had been a holdout on the Comprehensive Nuclear Test Ban Treaty. There were questions on monitoring, on reliability of older weapons and on safety of nuclear weapons. There were also questions about the future testing by increasing numbers of nations. SFRC held the only hearing on nuclear testing, in conjunction with START, with witnesses Frank von Hippel, Ray Kidder and the Executive Branch. I worked with the Armed Services Committee, drafting an amendment for Senators Mitchell, Hatfield and Exxon.

Testing safety. Practically all the released radioactivity from nuclear weapons resulted from bomber crashes. A key paper, written by Steve Fetter and Frank von Hippel, showed that nuclear weapon accidents without criticality didn't cause as many deaths as suspected. One can estimate the probability of such accidents from the historical record. The thought occurred to me to require the Executive Branch to carry out a cost/benefit calculation on new, safer weapons and from this determine the cost to save a life by improving the warheads. This amendment did not state a numerical safety standard for the warheads. We let the Executive Branch decide *how much is safe*, but the Executive Branch should at least do the calculation. To my surprise this law stopped all US nuclear tests. The last US nuclear test took place on 23 September 1992. The final result on cost-benefit standards appears in the 1992 law:

> Subsection F: A plan for installing one or more modern safety features in each warhead identified in the assessment referred to in subparagraph (C), as determined after an analysis of the costs and benefits of installing such feature or features in the warhead, should have one or more of such features.

A.9.3 Two National Academy CTBT Studies

The first NAS study, *Technical Issues Related to the Comprehensive Nuclear Test Ban Treaty* (2002) was chaired by John Holden. The committee's charge was to review the state of knowledge about the three main technical concerns, raised during the Senate CTBT debate in October 1999: (1) the capacity of the US to maintain confidence in the safety and reliability of its nuclear stockpile; (2) the capabilities of the international nuclear-test monitoring system (with and without the national technical means); (3) the addition to their nuclear-weapon capabilities that other countries could achieve through nuclear testing at yield levels that might escape detection. Jo Husbands, the study director, and I, the technical staff, worked with the panel of experts in Washington and at Lawrence-Livermore National Laboratory to conclude the following:

> (1) We judge the United States has the technical capabilities to maintain confidence in the safety and reliability of its existing nuclear weapons stockpile under the CTBT, provided

> that adequate resources are made available to the Department of Energy's nuclear-weapon complex and are properly focused on this task.

The NAS Panel examined each of the potential problems for each type of warhead. The question was always asked, "Would nuclear testing help solve this problem?" The answer always came back, "No, we can do it without nuclear tests."

> (2) Underground explosions can be reliably detected and can be identified as explosions, using IMS data, down to a yield of 0.1 kton in hard rock if conducted anywhere in Europe, Asia, North Africa, and North America. In some locations of interest, such as Novaya Zemlya, this capability extends down to 0.01 kton or less. [If the explosion took place in a large underground cavity at great depths, the limit of cheating is about 1–2 kton for sophisticated nations.]

> (3) China and Russia might use the option of testing to make certain refinements in their nuclear arsenals. In the case of Russia, it is difficult to envision how such refinements could significantly increased the threats to the US security interests that Russia can pose with the previously tested nuclear-weapon types it already possesses. In the case of China, further nuclear testing might enable reductions in the size and weight of its nuclear warheads as well as improve yield-to-weight ratios. Such improvements would make it easier for China to expand and add multiple-independently targetable re-entry vehicles to its strategic arsenal if it wanted to do so.

The Academy was asked in 2009 to update the 2002 report. I testified in September 2009 on my studies at Stanford (SAGS 2007) and CTBTO presentations in Vienna [2007, Article XII; 2009 Science Study Program]. The panels of experts were made up of national lab directors, chairmen of the joint chiefs, and weapon experts. The second NAS panel gave these updated conclusions in 2012:

> [The stockpile stewardship program] has been more successful than was anticipated in 1999 [when the Senate last considered the CTBT]. The status of US national monitoring and the International Monitoring System has improved to levels better than predicted in 1999.... Constraints placed on nuclear-explosion testing by the monitoring capabilities of the IMS and... US NTM will reduce the likelihood of successful clandestine nuclear-explosion testing, and inhibit the development of new types of strategic weapons.... Other states intent on acquiring and deploying modern, 2-stage thermonuclear weapons would not be able to have confidence in their performance without multi-kiloton testing. Such tests would likely be detectable [even with evasion measures].

A.9.4 Connection Between CTBT and NPT

The Nuclear Non-Proliferation Treaty regime has many trouble spots in India, Iran, Iraq, Israel, Libya, North Korea, Pakistan, Syria, Saudi Arabia, and more. Can the world forever be divided into two camps; one side with nuclear weapons and one side without? The split the world into five Nuclear-Weapon States, four or more defacto nuclear weapon states and 185 Non-Nuclear Weapon States. The number of nuclear weapon states will grow unless we are diligent. If the US flaunts NPT norms with acts that are legal but defy the spirit of the law, this will come home to roost. The NPT was extended by consensus in 1995 only because the five nuclear weapon states promised they would deliver a CTBT. Since the NPT was decided to last in perpetuity, forever, then the CTBT ought to last the same time, in perpetuity, forever. Of course many nations will not go to the bomb just because they know they don't want to live in a nuclear neighbourhood, such as Norway and Sweden. But this argument doesn't carry everywhere. With this in mind, I contacted Tom Graham, the long-time, top legal mind on arms control treaties and the author of *Disarmament Sketches*. The outcome was an article that appeared in the summer 2009 issue of *Disarmament Diplomacy*, called "*Nuclear testing and proliferation: An inextricable connection*." (Sect. 11.2).

Appendix B
Reflections on Nuclear Proliferation

Growing up after World War II, we learned to live with the bomb. Our wartime leaders, Roosevelt and Churchill, didn't focus on nuclear proliferation, first introduced to them by Niels Bohr in 1944. They were too busy fighting for survival. The Cuban missile crisis of 1962 raised concerns on the stability of the system based on strategic nuclear weapons. The fifth member of the Big Five, China, proliferated as it exploded its first fission weapon in 1964, followed by its first hydrogen weapon in 1967. There were times when nuclear weapons were lost, such as the bomber crash near Palomares, Spain in 1966.

B.1 Selected Proliferation History

Indian nuclear test. On 18 May 1974 India set off a ten-kiloton nuclear explosion under the Rajasthan desert. This was remarkable because it was the first nuclear explosion beyond those produced by the P5; US, USSR, UK, France and China. The P5 had their heads in the sand, thinking that proliferation would not take place in their time frame. The 1974 Indian explosion focused opinion on the dangers of nuclear proliferation and the plutonium economy. In those days, Congress led the Executive Branch to develop a response on proliferation.

South African 1978 test site and 1991 abolition. The US exported high-enriched uranium to South African for its research reactor. South Africa was building two nuclear power plants with US help and developing gaseous nozzle enrichment with German help. On 6 August 1978 the Soviet Acting Ambassador Vladillen Vasev went to the White House with a message from Chairman Leonid Brezhnev. Soviet intelligence showed that South Africa was preparing to detonate a nuclear device in the Kalahari Desert. Brezhnev asked Carter to help stop it. Brezhnev also sent appeals to the British, French and West-Germany leaders. These 5 nations worked together to convince South Africa to promise "no nuclear explosive tests will be taken now or in the future."

On 10 July 1991, South Africa signed the NPT and destroyed its 6 nuclear weapons. On 23 July 1993, a group of 30 were invited to the South African embassy to hear Waldo Stumpf, CEO of the *Atomic Energy Corporation of South*

D. Hafemeister, *Nuclear Proliferation and Terrorism in the Post-9/11 World*,
DOI 10.1007/978-3-319-25367-1

Africa, give the historical perspective. He claimed South Africa's main national security concerns were the Cuban troops in Angola and the presence of the Soviet Union forces. South Africa hoped to draw the United States into deterring the Soviets if the Soviets considered attacking South Africa. It didn't seem that nuclear weapons would be useful in a war in Angola. And South Africa is a long distance from the Soviet Union to attack without ICBMs. South Africa's motives to obtain the bomb were weak. Did South African military leaders promote nuclear weapons, without a strong purpose. Stumpf quoted the IAEA's Hans Blix, saying "The IAEA team has found no evidence to suggest the original inventory is not complete."

1979 nuclear explosion? On 22 September 1979, a Vela satellite bhang-meter picked up a signal over the South Atlantic that appeared to be a 3-kiloton explosion. On 15 July 1980, the *Presidential Office of Science and Technology Policy* concluded the light signals recorded by Vela were *probably not from a nuclear explosion*. Others disagree with this conclusion and believe that Israel with South African help carried out a nuclear test.

Iraq and Iran. Saddam Hussein invaded Kuwait on 14 September 1990, which was followed by the US-led *Desert Storm* invasion of 16 January 1991. The nongovernmental community had warned that Saddam was working towards the bomb, but the intelligence community didn't pick up on it. The invasion uncovered a massive nuclear program, relying more on uranium enrichment than on plutonium reprocessing. Hussein spent billions of dollars, but made only slow progress. He was forced to dismantle the program, which he did. Unfortunately, bad intelligence and politics claimed Iraq had an active nuclear program, leading to the second invasion of Iraq on 19 March 2003.

Iran had ambitions for nuclear weapon dating back to the Shah's commissioned research, starting in the early 1970s. In 2015, Iran had some 10,000 operating centrifuges which produced 200 kg of 20 % enriched uranium. There is hope that the compromise solution can obtain the lifting of UN-sponsored economic sanctions in trade for severe limitations on LEU and HEU stockpiles, limits on the centrifuge program, redesigning the reactor, and robust monitoring. Israel, Saudi Arabia and 50 % of the US Congress are greatly concerned about allowing Iran to have any enrichment capability. Saudi Arabia claims it will have the same nuclear capabilities as Iran, stimulated by the Sunni–Shiite rivalry.

A.Q. Kahn's nuclear Walmart. In 1978 we were briefed that A.Q. Kahn took URENCO centrifuge plans from the Netherlands to Pakistan. The Intelligence Community knew this was happening, but failed to act in a timely manner. Using *worst-case analysis,* a first weapon was possible within two years. Pakistan exploded its first device in 1998, obtaining a weapon in one decade. Kahn later created the nuclear Wal-Mart, selling centrifuge designs and parts to Iran, North Korea, and Libya, after being rejected by Iraq. This had a major impact; Kahn proved uranium enrichment is a safer, faster path to weapons than plutonium. This shifts the main thrust of the Carter nonproliferation policy from reactors and plutonium to centrifuges and uranium. This shift makes controlling proliferation more difficult because reactors are easy to detect from orbit, while centrifuges use small amounts of power and can be hidden easily. Technology advances make nuclear proliferation control more difficult, our main hope in response is that of multilateral diplomacy.

B.2 Nuclear Proliferation Policy

A main conclusion of the 1977 Office of Technology Assessment study on nuclear proliferation was that unilateral US action would not be successful to halt proliferation, and only a multi-tasked, multi-nation approach of incentives and disincentives would work. *OTA concluded* with the following:

> What influence can the United States exert upon potential weapon states? *Finding.* In the long run two general rules apply: (a) Solutions to the proliferation problem will have to be found primarily, though not exclusively, through multilateral actions, and (b) the extent of U.S. influence will vary from country to country.

The main nonproliferation players in the Executive Branch were the State Department, the Arms Control and Disarmament Agency (ACDA) and the Department of Energy; this was the small group. A larger group consisted of a dozen agencies including the Intelligence Community, the Nuclear Regulatory Agency and the Departments of Defense and Commerce and others. On foreign-policy matters, Congress only has the heavy hammers of sanctions, financial constraints and setting criteria. If you have a problem with German foreign policy, you would visit the Foreign Ministry and not the Parliament. Chap. 8 lists the many proliferation policy options, here we examine a few examples.

Proliferation policy is more complex than *strategic arms policy* with the Soviet Union. A main difference in these two areas is that uranium and plutonium have economic consequences since large amounts of energy are needed to fuel modern economies. There are 190 parties to the NPT treaty, whereas the START and SALT treaties have but two. The seminal discussion of the Carter non-proliferation policy was written by its lead person, Joseph Nye in 1978 in *Foreign Affairs*. Nye describes the incentives and disincentives that help to dissuade nations from starting a nuclear weapons program. I think it stood the test of time, but at times it was under harsh attack from the nuclear industry, foreign governments, nations and states with plutonium projects, and the abolitionists. Perhaps the policy was mostly correct since it has mostly stood the test of time. We saved the US billions of dollars by abandoning the breeder, reprocessing and mixed oxide (MOX) fuels. Over time, most countries followed our path by abandoning the breeder and MOX fuels. The US raised the standard on exports on nuclear fuel and equipment.

Nuclear supplier's group. In the Ford and Carter administrations, Joe Nye and Lou Nosenzo devised ways to stiffen foreign-nation standards for nuclear exports, outside of the IAEA. This was done in the Nuclear Suppliers Group of 40 nations involved in nuclear commerce. The Suppliers Group required full-scope safeguards and constrained exports of enrichment and reprocessing technologies. The initial nuclear supplier guidelines were forwarded to the IAEA on 11 January 1978. Since then, exceptions were given to India that weakened the process.

Does NEPA apply abroad? Article IV of the NPT obligates the five nuclear weapon states to assist the Non-NWS with their peaceful nuclear programs. In the 1970s, when the US was the predominate supplier of nuclear fuel, this promise was taken seriously. At that time the Justice Department was concerned the Supreme Court might extend the 1969 National Environmental Policy Act to include nuclear exports. For this reason, secret domestic negotiations took place in the Carter Administration for two weeks during 1978. After two weeks, this process created the concept of *Environmental Reports* on significant nuclear exports. Environmental Reports had no previous status in law, they were to differ from Environmental Impact Statements. The Environmental Reports had no previous status in law. This was a compromise between rigorous Environmental Impact Statements and the present law that has no analytical requirements. President Carter could not implement this as a law, so it was implemented with an Executive Order. President Reagan removed this requirement with an Executive Order upon taking office.

Born secret. Howard Morland's article, *the H-Bomb Secret*, was published in the November-1979 *Progressive*. At that time most of us did not know that information on hydrogen weapons had been declassified. I thought Hans Bethe was correct in opposing publication of these proliferating concepts, but publication was legal because *UCRL-4725* and other sensitive documents were already in the public domain. Taylor was shocked by the declassification of *UCRL-4725*, stating that "the erroneous declassification of *UCRL-4275*…is the most serious break of security I am aware in this country's post-World War II nuclear weapons development programs."

Progressive editor, Erwin Knoll, justified publication as follows: "We hope the debate over [the H-Bomb Secret] will be a beginning of a process in which all of the nuclear policies pursued by our Government will be held up to public scrutiny and review." Morland's article did not advance the debate over nuclear weapons, excessive secrecy prevented good debates on issues, such as the Excalibur X-ray laser pumped with a nuclear weapon, the airborne laser, the number of nuclear weapons needed for deterrence, the capability of Soviet defenses, compliance data and so forth. Closed government bodies, such as the Defense Science Board and the DOE/NNSA, often fail the nation by failing to be critical when physics, strategy and economics of projects are untenable. Classification keeps bad science hidden from public view.

B.3 Nuclear Export Legislation

Joint committee on atomic energy. JCAE defied the intentions of the Constitution as it was populated equally by members of the House and Senate. It met in the Capitol, to avoid the normal hearing rooms in the Senate and House Office Buildings. When JCAE reported a bill, it was the same bill reported to both the House and the Senate. That is not the way the system is supposed to work. Normally, the Senate and House committees report separate bills to their legislative bodies, where they are amended and voted on. The two bills passed by the two bodies are then combined and compromised in the House-Senate Conference Committees, and then returned to the legislative bodies.

Because nuclear physics is mysterious and nuclear weapons are secret and powerful, the Congress deferred to the bad JCAE process. The Senate and House Foreign Relations Committees deferred to JACE on proliferation issues. SFRC worked only on strategic weapon treaties and not much on proliferation prior to 1978. This process, coupled with the fact that the Joint Committee was rather permissive, allowed a system that was soft on proliferation, allowing easier export of dangerous materials. The standing law of the time, the *Atomic Energy Act of 1954*, was the focus of our investigations and hearings. The Joint Committee was swept away after the election of 1976, putting the Senate Committees of Government Affairs and Foreign Relations in a shared leadership role on proliferation issues.

The Governmental Affairs Committee began its hearings in January 1976. The Indian explosion of 1974 was the energizing focus to pass new legislation. Our first witness was David Lilienthal, the first chair of the AEC from 1947 to 1950. Lilienthal had recently published an op-ed piece saying that mankind was being overtaken by nuclear proliferation that would destroy civilization. Lilienthal issued a newsworthy call to slow proliferation at the hearings:

> I therefore propose as a private citizen that this committee, with its great prestige, call upon the Congress and the President to order a complete stop of the export of all nuclear devices, and all nuclear material, that it be done now, and done unilaterally. Further, unilaterally, the United States should without delay proceed by lawful means to revoke existing American licenses and put an end to the future or pending licensing to foreign firms and governments of American know-how and the facilities paid for and created by American taxpayers' funds and American brains.

> This action taken alone will not solve the whole problem of proliferation. But it will put an end to our major part in it. For the fact is that we, the United States, our public agencies and our private manufacturers, have been and are the world's major proliferators. I consider the historical perspective, which I later summarize, to be distinctly relevant to the consideration of what we do now. It is the now, the present, and the immediate future, that interest me the most, and obviously of the most interest to the committee and to our fellow Americans in the cities, the towns and the farms of our country. Admittedly, any immediate embargo on the export of nuclear technology materials and plants would be drastic action, so drastic as to raise the question of America's moral position to take such a course. That moral position is a strong one as the history of American efforts demonstrates.

The hearings went on for months, producing a volume five inches thick. Hans Bethe initially agreed with Lilienthal, but then reversed himself. Another high point of the hearings had to do with the French export of a reprocessing plant to South Korea. The lead State Department proliferation administrator, Myron Kratzer, testified that South Korea had canceled the import of the French reprocessing plant. This miffed the French, who had championed the sanctity of contracts. Ad-Hoc Chair Glenn interrupted Kratzer and asked if it was true that South Korea had came to their senses and canceled this reprocessing plant sale? Glenn stated that "some sense of sanity" has returned to the nuclear field.

Secretary of State Henry Kissinger ducked the issue as to whether the US should have responded strongly when India exploded a nuclear device in May 1974. India obtained US heavy-water to produce plutonium in its Canadian reactor. There is a tendency for countries to want to get along with each other to obtain agreement on other near-term issues, so the US did nothing. The Senate bill ultimately strengthened the export criteria by requiring *safeguards on all nuclear facilities* to obtain nuclear export licenses. This criterion is referred to as *full-scope safeguards*. This is equivalent to signing the Nonproliferation Treaty, without signing the NPT, to show strong national sovereignty intentions, while complying with IAEA safeguards.

Nuclear non-proliferation act of 1978. India had nuclear weapons facilities and commercial power facilities outside safeguards. US contracts to sell nuclear fuel to India in 1978 did not consider full-scope safeguards. US exports were legal, but unwise. If the new law required all facilities to be under safeguards, US exports to India would not be legal.

Can you add new requirements on former contracts with other countries? The answer is no, but Congress can do whatever it wants and then the courts will have to sort it out. In our Congressional hearings we learned from Adrian "Butch" Fisher, former Deputy Director of ACDA, the phrase, "sic ribicus stantus." This means the following: If a change in circumstances took place that has a large impact on US national security, previous contracts can be nullified to protect the nation. The Executive-Branch lawyers did not like "sic ribicus stantus" and voted to allow exports to India to continue because it was guaranteed by previous contracts. It was clear that the new law would require full-scope safeguards on all nuclear facilities in non-nuclear weapon states, but what about the old contacts? The Congress ultimately decided to disallow nuclear fuel exports to India, phasing in a ban over 18 months. Later the US cushioned the blow to the India by helping to find a new supplier, France.

A second issue was to determine which nations would be allowed to reprocess US-origin spent fuel to obtain plutonium. Here again, the executive branch legal panel came up with softer criteria. Joe Nye gave several speeches, pointing out that only mature nations would be allowed to reprocess since they would have larger nuclear power needs to be fulfilled. Senator Glenn inserted the criteria *timely warning* into the legislation, requiring the Secretary of State to certify the following: For nations that were about to build nuclear weapons, the US must have sufficient time to respond militarily before the US could be meaningfully threatened. This was a tighter standard and useful. The NNPA, as amended by Glenn, states the timely warning requirement:

> Among all the factors in making this judgment, foremost consideration will be given to whether or not the reprocessing or retransfer will take place under conditions that will ensure timely warning to the United States of any diversion well in advance of the time at which the non-nuclear-weapon state could transform the diverted material into a nuclear explosive device;

President Carter released the findings of the proliferation interagency process on 7 April 7, 1977 and on *Nuclear Power Policy* on April 27. He discussed the proposed legislation, the "Nuclear Non-Proliferation Policy Act (NNPA)." Carter's statement discussed termination of the plutonium economy of breeders and reprocessing, but also for the US to continue as a reliable supplier of low enriched uranium exports, pursuant to the promises in the NPT. However future exports of US origin nuclear fuel would be diminished because of the higher US export criteria for nuclear exports. An example of this was a modest shift from light water reactors (LWR) that needed enriched fuel to the Canadian CANDU reactors that used natural uranium fuel with heavy water reactors (HWR). There were considerable complaints about NNPA-1978, but after 30 years it has remained without too much softening, except for the softened export criteria for India.NNPA required that Euratom and the US begin negotiations, or Europe would not get nuclear fuel. This did not go down well in Europe, which did not believe that the US Congress should or could unilaterally require negotiations to change conditions in existing contracts. The signing of NNPA by President Carter created an embargo on nuclear materials for four months in 1978. Carter and French President Valery d'Estaing agreed to a compromise (NY Times, 29 June 1978):

> the Common Markets Executive Commission can open preliminary talks with the Carter administration on bringing the existing supply agreement into line with the recently approved US NNPA…. Under the new agreement, the Common Market Commission will be barred from even discussing the real source of the quarrel between the United States and Europe over nuclear fuel, which is Europe's determination to develop a nuclear industry based on plutonium in defiance of President Carter's stated wishes.

The idea of turning off European lights by denying US contracts on reactor fuel to coerce tightened laws on global plutonium was a daydream. Only 3 % of France's electricity, 8 % of Germany's electricity and 9 % of Japan's electricity could be affected, and only in the near-term. And how would Japan and Germany take this US hardball play? Japan had only recently joined the NPT in 1976, and expected the US nuclear-umbrella to remain after they joined the NPT. But, the embargo forced Europe and Japan to focus on plutonium and ultimately tighten their export policies. Later the Europeans and others agreed that we should have higher standards for nuclear exports.

B.4 Sanctions Legislation

Over the years, Congress passed legislative sanctions to attempt to change nuclear behavior of other nations. This is a complex and dicey issue. Usually this assumes that the Non-Proliferation Treaty is the gold standard for the world. But the NPT allows the five nuclear-weapon states to have nuclear weapons while 185 other nations cannot have them. Is this a stable system? Probably not, but the alternative solutions don't have much traction. It is difficult to understand what other solution we can have in the near-term. Prior to 1976, the Congress mostly abstained from nuclear-sanction legislation because the State Department always wanted flexibility to cut a new deal. This is easily understood because once you use legislative sanctions and the other side doesn't comply, there's not much you can do without flexibility. However, when the Indian explosion of 1974 took place, the US government took no serious action. The combination of the 1974 Indian bomb and the 1974 resignation of Richard Nixon decreased the power of the presidency and increased the opportunity that Congress would pass significant sanction legislation. It sometimes happens that those who pass legislation then have to implement it in the Executive Branch. This is precisely what happened a year later when Congress passed further sanction amendments, and I was working in the State Department.

Enhanced non-proliferation sanctions. The 1976 Glenn-Symington amendment cut military and economic aid to nonnuclear weapon states that imported enrichment facilities that were not part of an international regime. Glenn added another sanction in 1977 that sanctioned non-nuclear nations that tested a nuclear device. Were these sanctions useful? They had merit in that they painted Congress as the bad cop, which helped State diplomacy, as the good cop. India and Pakistan did not test until 1998, 24 years after India's 1974 explosion. Sanctions were useful in holding back the bomb for that period of time. I agree there are limits to sanctions but they can be effective if you obtain all the major nations working together.

It took several years before the Nuclear Proliferation Prevention Act of 1994 became law. The sanction list was more inclusive than in Glenn-Symington sanctions, as it included a ban on government finance and loans, and in doing business in the US. The NPPA contained IAEA reforms, which are difficult to adopt in practice. In 1998, India and Pakistan each exploded about 5 devices, triggering Glenn-Pell-Helms sanctions, costing India billions of dollars. During the first year, most other nations joined the US sanctions, but this weakened with time. After a year, President Clinton started softening the sanctions, and President George W. Bush removed them entirely after entering office.

B.5 Nuclear Fuel Cycle Issues

Enrichment Capacity. It is useful to maintain guaranteed, stable supplies of uranium fuel for reactors, reducing the need for countries to enrich uranium or reprocess plutonium to obtain reactor fuel. Gaseous diffusion plants are inefficient and out-of-date. In 1977 there was a push for considerable nuclear power. Both domestic and international orders for nuclear reactors and fuel were in play. The legislation on nuclear exports reduced the need for US enrichment as our raised export standards lowered marketability of US fuel. The US was in a bind on this issue. NPT nuclear weapon states are supposed to help nuclear-power programs in non-nuclear weapon states. This is a controversial bargain because, on the one hand, the Eisenhower Atoms for Peace Program increased nuclear technology transfer and gave a political cover to hide clandestine weapon programs. On the other hand, it established the NPT Treaty and the International Atomic Energy Agency (IAEA) regime. Though flawed, the NPT is better than nothing. Since nuclear technology keeps advancing and getting easier to do, we are better off with the NPT/IAEA regime as compared to no relevant treaties and laws. It is true that technology transfer is enhanced with nuclear cooperation agreements, but ultimately political barriers exceed the technical barriers to the bomb as technology improves. From 1995 to 2015, we got 50 % of our uranium reactor fuel from Russia's dismantled nuclear weapons.

Nuclear economics. The three biggest economic factors for plutonium nuclear electricity are as follows:

1. Uranium resources available as a function of cost for thermal reactors,
2. Breeder capital cost, compared to capital cost of light water reactors, and
3. Growth rate of nuclear power.

Gregory Canavan originated a model with three intersecting graphs: comparative economics, geological resources, and nuclear growth. If uranium was cheap and plentiful and there was little growth in nuclear power, the breeder would not be competitive for hundreds of years. If uranium was in short supply and expensive and the breeder capital cost was cheap, the breeder would be competitive sooner. Lastly, the growth rate of nuclear power is relevant. A high growth rate of nuclear power lowers the commercialization time of the breeder, while a low nuclear growth rate lengthens the commercialization time. My estimates at the time were that breeders wouldn't be commercially competitive until about 2050 using the parameters of 1978. If these calculations were done today, the commercialization date would be much later than 2050 because other energy sources (efficiency, natural gas and alternate energy) have made nuclear power economics more difficult.

Spent fuel storage. Nuclear power made up 8 % of the 1978 US grid, it is now 20 %. Without reprocessing and the breeder, the Carter Administration concluded that the US should increase storage capacity of spent fuel ponds at reactor sites. Safety analysis raised no alarms. It was assumed that the coming underground geological storage would keep spent fuel ponds un-crowded, avoiding excessive heat if cooling water was lost. The Carter solution was to obtain funds for the underground repository by taxing nuclear electricity by 0.1 cents/kWh. If nuclear power at the busbar costs 5 cents/kWh, this is a 2 % tax. Since policy was driven initially by foreign affairs concerns on plutonium, it was necessary to discuss proliferation problems that were on the horizon. By paying this tax, the US was to take legal possession of the spent fuel. After 40 years, Yucca Mountain is not accepting spent fuel and the future is uncertain as where to place it. Compared to other dangers from energy production, the danger of placing older, spent fuel in Yucca Mountain is minimal. Russia indicated it would receive nuclear waste for burial beyond the Urals. The offer seems serious, but Russia wants to retain the right to reprocess spent fuel for plutonium without a US veto, an issue that is unacceptable to the US.

Forty years have passed with an over-accumulation of spent fuel. There is a potential problem if young spent fuel is placed in over-crowded ponds, giving more heat and trapping more heat. If a pond lost its cooling water when densely-packed, it could cause melting and perhaps a uranium fire, spreading radioactivity. This issue was discussed by Robert Alverez et al. (*Science and Global Security*, 2003). The National Academy of Sciences agreed with the Alvarez results, pushing the US to move older spent fuel rods into above ground storage that is air-cooled. The nuclear waste issue has become so politicized that the National Academy of Sciences has been hired to do the analysis, in place of the Department of Energy. Above-ground storage is an interim solution, which is generally accepted. Because spent-fuel rods contain 1 % plutonium, we need a better place to keep plutonium away from mischief for the much longer term. As long as we have 20 % nuclear power, in an era of climate change from burning carbon, it is likely the US will continue the present nuclear plants until they are decommissioned. In the meantime, explore the alternatives.

B.6 End of US Plutonium Economy

Grades of plutonium. All reactors produce plutonium. A typical 1-GW PWR power reactor produces 250 kg of plutonium per year. This plutonium is *reactor-grade plutonium* since it contains 20 % ^{240}Pu, with the remainder mostly ^{239}Pu. The ^{240}Pu in reactor-grade plutonium makes it much less useful for nuclear weapons, as it emits copious spontaneous neutrons that cause *pre-initiation*, similar to *pre-ignition* in cars. However, reprocessing facilities that separate reactor-grade plutonium can also separate *weapons-grade plutonium* with less than 6 % ^{240}Pu from fuel that remained a shorter time in the reactor. It is too expensive for commercial light water reactors to produce weapons-grade plutonium, but research reactors, heavy water reactors, and breeder reactors all naturally produce weapons-grade plutonium. Thus, under the guise of *peaceful nuclear power*, a program can be developed for *military nuclear power*. This is the approach that India and Israel have taken with their research reactors.

Plutonium economy. Even if nations follow rules of the road, it is possible for plutonium to be misplaced to end up in the wrong hands. After the oil embargo of 1973–74, many felt that nuclear power was going to alleviate the oil embargo situation. President Nixon made the liquid metal fast breeder reactor a major response to the oil embargo. This is ridiculous because oil is for cars and not electricity, which at that time, was not available for cars.

With this backdrop, the Carter administration addressed the plutonium economy. The February 11, 1977 issue of *Energy Daily* concluded the following.

> State Department strikes at the breeder. Schlesinger's independence and energy policy is threatened. A bruising fight has erupted inside the Carter administration on the fast breeder program. The contest pits the arms-control community, entrenched in the State Department and the Arms Control and Disarmament Agency against James Schlesinger and his energy policy team at the White House. It is widely believed that the end result will be further reductions in the amount the government will spend on the breeder reactor, cutting perhaps into money allotted for the Fast Flux Test Facility and the Clinch River demonstration plant.

The almost-complete Barnwell reprocessing facility in South Carolina was owned by Allied General Nuclear Services Corporation (AGNS). We flew in the AGNS corporate jet with State paying our airfare. The PUREX process is used to separate plutonium from uranium and its fission fragments. The AGNS design was different from that at military reprocessing plants, as it was less flexible and more contained. As a warm-up we visited the military reprocessing plant at Savanna River. The plant was a sight to behold, with huge caverns, controlled at great distances to protect the operators. The Barnwell facility had not yet been activated, so we could see it very closely. The Barnwell facility never operated because two presidents opposed it. On 28 October 1976, President Ford opposed reprocessing, and on 7 April 1977, President Carter made a stronger statement. The history of commercial reprocessing in the United States does not inspire confidence, as there were radioactive spills and facilities that never opened. Over the years only three countries continue to reprocess commercially: France, Britain and Japan, with India and Russia keeping the option open.

Senate acts on breeders and reprocessing. A letter from Secretary Vance to Senate Majority Leader Robert Byrd gave Carter's nonproliferation positions:

> The President's April 7 and April 27 statements on nonproliferation and nuclear power established for the first time a realistic basis for controlling the spread of nuclear weapons capability abroad. One of the crucial elements of this policy is a clear and unambiguous decision by the United States to cancel all construction and licensing of the Clincher River Breeder Reactor project. It is our recommendation that we pause before commercializing this next generation of nuclear technology, which in its current form would greatly increase the risk of nuclear weapons proliferation. We have launched a major diplomatic effort to persuade others to similarly avoid premature commercialization, and to join in an International Nuclear Fuel Cycle Evaluation program in which we would jointly explore ways to make future breeders more proliferation-resistant or find other alternatives that

> would satisfy our collective energy and national security concerns. I'm encouraged by the initial responses of the countries which have been invited to join the evaluation program.

The Carter Administration won on the Bumpers Amendment to block reprocessing at Barnwell. It was easier to block noxious plutonium reprocessing than to shift energy production from the breeder. The Kennedy amendment, four years after the oil embargo of 1973–4, failed to block the breeder reactor. Ultimately the breeder survived the House and Senate compromise, setting up Carter's first veto. A *New York Times* editorial commented:

> Capitulation to Congress now would open the entire policy to charges of hypocrisy.... [Halting the] Clinch River Breeder Reactor is a symbol of the American commitment to this cause. It demonstrates that Washington is not motivated by commercial considerations, as some Europeans have charged. The fact that a few nations are going ahead with small demonstration projects and that France is even proceeding with plans for the world's first commercial size breeder are not good reasons for the United States to continue work on a breeder reactor, already regarded as obsolete, unduly expensive and probably unnecessary.

After 40 years later, what has happened? It's clear we saved the US billions of dollars. The most successful breeder program is that of France with its *Phoenix* and *Super Phoenix* reactors, but both are shuttered. The French, Russian, German and Japanese breeder programs did not produce significant electricity. Reprocessing has occurred over time, primarily because countries don't know what to do with their spent fuel. If they reprocess it, the radioactivity is placed in vitrified glass, but still you must bury it. France and the UK reprocessed Japanese fuel rods, but at a very high cost. The Carter administration agreed in 1977 to the opening of the small Tokai Reprocessing plant, keeping Japan in the NPT regime. After decades of work, the Japanese are completing their large Rokkasho reprocessing plant.

B.7 Response to US Non-proliferation Policy

There were many high-level meetings with countries that wanted to reprocess or enrich nuclear fuels. Each negotiation was different: Japan, Germany, Romania, South Korea, Taiwan, India, Pakistan, Brazil, Argentina, South Africa, and others. Today's conflicts with Iran, North Korea, India, Pakistan, Israel and others show these issues continuing. European reaction was mostly unfavorable towards the Carter policies. Joe Nye wrote a letter to his counterpart in the British government, Patrick Moberly, requesting Britain to go slow on plutonium. The British House of Commons then voted resoundingly to approve the Windscale nuclear reprocessing plant. UK Foreign Secretary David Owen rejected the call for the United Kingdom to defer a decision on plutonium reprocessing until after the International Nuclear Fuel Cycle Evaluation (INFCE) was completed.

International nuclear fuel cycle evaluation. Study was needed to sort out the proliferation issues. This began with the introduction of INFCE at the US State Department in October 1977. President Carter and Secretary Vance welcomed the technical experts from 40 nations, which were separated into eight study groups: (1) natural resources, (2) enrichment, (3) fuel assurances, (4) reprocessing, (5) breeders, (6) spent fuel, (7) fuel conservation and (8) advanced reactor concepts.

As the State Department representative to INFCE-8, we examined fuel cycles such as the thorium cycle and research reactors. INFCE-8 was chaired by South Korea, Romania, and the US. This required the bureaucracy to produce technical papers from within and outside the government. We had formal meetings in Vienna and in Washington DC. The final results of INFCE-8 concluded that the thorium cycle had non-proliferation advantages, but considerable investment would be needed to make thorium competitive with the uranium cycle. That is the way things stand today, namely thorium is only used in few situations. The *Energy Daily* of July 13, 1978 described the debate:

> Joseph Nye, Deputy Undersecretary for Security Assistance, Science and Technology at the State Department, is an elegant and urbane man. Although he has been a diplomat only in the Carter Administration and is soon to return to a teaching position at Harvard, he is the archetypal diplomat. Ready smile, a quick wit and a pervasive charm serve him well as the chief spokesman for President Carter's anti-proliferation policies, policies, which have infuriated US allies around the world. Walter Marshall, Deputy Director of the UK Atomic Energy Authority, is a big bluff, untidy roly-poly man in appearance, and to some extent in style the antithesis of Nye. They also do not agree on very much when it comes to nuclear power in general, and plutonium in particular. At this week's Uranium Institute meeting in London, they squared off before a large gathering of nuclear representatives from around the world in a confrontation that promised high drama. Both men are too civilized for direct verbal fisticuffs, but the confrontation was the highlight of the two-day Uranium Institute meeting, and the difference in style as well as philosophy gave an air of theater to the exchanges on proliferation.

There were many estimates of how long US uranium reserves would supply the fleet of US light water reactors (LWR). The *Energy Daily*, March 9, 1978 quotesd our results:

> Assuming a price of $50 per pound, he calculates that probable reserves of 2.4 million tons will sustain 480 LWR's at currently projected nuclear capacities, leveling out after 2000, the uranium would last through 2011. If uranium reserves are larger, as he believes they are, 4.3 million tons of potential reserves will sustain 860 reactors to the year 2044. In addition, uranium reserves can be stretched by advanced thermal fuel cycles as well as by lower tails assays.

These estimates look too cautious today since the US peaked at 100 reactors, much less than the 480 or 860 reactors given above. And now considerable uranium is imported from decommissioned Russian warheads turned into nuclear fuel.

NPT Problems. Nervous nations want to attain nuclear weapons. The NPT regime has problems and is under attack. The US is part of the problem. If we were totally compliant to the spirit of the NPT process it would not solve all problems, but the situation could be better. It may be asking too much to have five nations with nuclear weapons, while 180 nations do not have any. Perhaps this concern causes former political leaders to believe the US should abstain and go to zero nuclear weapons. Why else would Paul Nitze, George Kennan, Henry Kissinger, George Shultz, Sam Nunn and Bill Perry promote a world without nuclear weapons. No one knows the path through the clouds to the top of the mountain without nuclear weapons. Nuclear weapons are small and can be hidden.

Much of the technical ideas are available, but it still takes nations a long time to develop nuclear weapons. Iran started its path toward the bomb in 1975 under the Shah of Iran, Mohammed Reza Pahlavi. Forty years later, Iran has gotten closer but they have not yet succeeded. Saddam Hussein spent $10 billion and failed, trying many uranium enrichment technologies and reactors to produce plutonium. However, South Africa was successful with a much smaller program. Nevertheless, the barriers to the bomb get lower every year as advancing technology continually makes the path easier.

What has changed? The basic solutions put forward by the Carter administration are mostly followed today, in spite of complaints. Breeder reactors and plutonium are not used today for economic and proliferation reasons. These decisions saved the nation billions of dollars. The Reagan administration initially said it was going to make changes (Science, 25 June 1982, pg. 1388). "President Reagan has approved a significant revision of US nuclear nonproliferation policy through a broadening of the terms under which US allies may reprocess nuclear fuel supplied by the United States." The only country that has had its US-origin spent fuel significantly reprocessed is Japan. This was done at a huge economic penalty to Japan as it paid France and UK large sums to do this. The breeder reactor programs in France, UK, Germany, Japan, and Russia are moribund. A big change in policy took place, when President George W. Bush agreed to give India the quasi-status of nuclear-weapon state, to obtain nuclear fuel without all of its facilities under safeguards. If this precedent is not repeated, this will not be overly dangerous, but will it be repeated? There are many other concerns. The proliferation problems of North Korea and Iran are not based on peaceful power programs. There has been progress as South Africa, Argentina, Brazil, Ukraine, Belarus and Kazakhstan pulled back from their nuclear weapons status. The programs in India and Pakistan were slowed by sanctions, but now each has 100 nuclear weapons. Israel's program has been in existence since 1960 and remains a major inconsistency for the United States. The world is more aware about technical aspects of proliferation through Carter's International Nuclear Fuel Cycle Evaluation and its successors. The IAEA is very slowly strengthening it standards for declarations and inspections on plutonium and enriched uranium. I am concerned that political differences will prevent the implementation of a consistent, cooperative nonproliferation policy between the global nations.

Appendix C
Websites

General

American Geophysical Union: http://sites.agu.org/
American Institute of Physics: aip.org/history/
American Physics Society: aps.org/policymakers/index.cfm
APS Forum on Physics and Society: aps.org/units/fps
Census Bureau: www.census.gov/population/
Congressional Budget Office: cbo.gov
Congressional Record: Gpoaccess.gov/crecord/indiex.html
DOE Information Bridge: osti.gov/bridge/
DOE Labs: XX.gov; XX = anl, bnl, lanl, lbl, llnl, ornl, pnl, sandia, y12
Executive Office of the President: whitehouse.gov/administration/eop
Federation Amer. Scientists: fas.org/publications
General Accountability Office: gao.gov
Government Printing Office: wgpo.gov/fdsys/
National Academy Press: nap.edu
Office of Technology Assessment Legacy: princeton.edu/~ota/
Science: sciencemag.org
White House: whitehouse.gov

Nuclear Arms

Argentina/Brazil: ABACC.org.br
Arms Control Association: armscontrol.org
Arms Control Wonk: armscontrolwonk.com
Atomic Archive: atomicarchive.com
Ballistic Missile Defense Agency: mda.mil
Bulletin of Atomic Scientists: thebulletin.org
B. Atom. Sci. Nuc. Notebook: thebulletin.org/collection/nuclear-notebook
Carnegie Endowment for Int. Peace: carnegieendowment.org
Center for Arms Control and Non-Proliferation: armscontrolcenter.org
Center for Defense Information: cdi.org

D. Hafemeister, *Nuclear Proliferation and Terrorism in the Post-9/11 World*,
DOI 10.1007/978-3-319-25367-1

Central Intelligence Agency: cia.gov/index.html
Chemical/Biological Defense: jpeocbd.osd.mil
Chuck Hanson: uscoldwar.com
Council for a Livable World: clw.org
CRS on CTBT by Medalia: fas.org/sgp/crs/nuke/RL33548.pdf
CTBT Organization: ctbto.org
Disarmament Diplomacy: acronym.org.uk
DoD Defense Threat Reduction Agency: dtra.mil
Federation of American Scientists: fas.org
FAS Security Security Blog: fas.org/blog/ssp
Garwin Archive: fas.org/rlg
Global Security (John Pike): globalsecurity.org
Harvard Kennedy School: belfercenter.ksg.harvard.edu
Institute for Science and International Security: isis-online.org
International Atomic Energy Agency: iaea.org
International Institute for Strategic Studies: iiss.org
International Panel on Fissile Materials: fissilematerials.org
IRIS Seismology: iris.edu
Los Alamos, Earth-Environmental Sciences: ees.lanl.gov
Los Alamos Study Group: lasg.org
MIIS Ctr. Nuclear Studies: cns.miis.edu
National Nuclear Security Agency: nnsa.doe.gov
Natural Resources Defense Council: www.nrdc.org
Nuclear Explosion Monitoring: nemre.nnsa.doe.gov
Non Proliferation Policy Education Center: npolicy.org
Nuclear Threat Initiative: www.nti.org
Pantex: pantex.com
Princeton: princeton.edu/sgs
Russia: armscontrol.ru
Russian Strategic Nuclear Forces: russianforces.org
Science and Global Security: princeton.edu/sgs/publications/sgs
SearchMil: searchmil.com, globemaster.de/govsearch/
Stanford CISAC: cisac.stanford.edu
State Department: www.state.gov/t/XX/, XX = isn, pm, vc, trty
Stimson Center: stimson.org
Stockholm International Peace Research Institute: sipri.se
Strategic Air Command: www.stratcom.mil
UK Atomic Weapons Establishment: awe.co.uk
Union of Concerned Scientists: www.ucsusa.org
United Nations: un.org/depts/
UN Institute for Disarmament Research: unidir.org
US Geological Survey: usgs.org
Verification Research, Training and Information Centre: vertic.org

Nuclear Energy and Environment

American Nuclear Society: ans.org
Center for Disease Control: cdc.gov
DOE Nuclear Waste: energy.gov/em/services/waste-management
Energy Information Agency: eia.doe.gov
Environmental Protection Agency: epa.gov
EPA Air and Radiation: epa.gov/air
EPA Water: water.epa.gov
Intergovernmental Panel on Climate Change: www.ipcc.ch
International Energy Agency: iea.org
National Cancer Institute: cancer.gov
NAS Nuclear and Radiation Board: dels.nas.edu/nrsb
Nuclear Waste Technology Review Board: nwtrb.gov
Michigan Radiation/Health: umich.edu/~radinfo
Nuclear Power: web.mit.edu/nuclearpower/
Nuclear Regulatory Commission: www.nrc.gov
Pacific Gas and Electric: www.pge.com
Princeton Plasma Physics Lab: www.ppnl.gov
World Bank: publications.worldbank.org

Glossary

Arms Control Issues

ablative shock	X-rays penetrate a short distance, vaporize surface volume, shocking the object.
ABM	Anti-Ballistic Missile Treaty constrained US/Soviet strategic defenses, 1972–2002.
adaptive optics	submicron mirror movements improves some monitoring and defense systems.
adaptive preferential targeting	SDI targeting maximizes defense effectiveness if all incoming trajectories are known.
anti-satellite weapon	easier to attack satellites than missiles, but threatens all satellites.
anti-submarine warfare	ASW consists of ocean sensors, trailing submarines, depth charges, and homing torpedoes.
Auxiliary Seismic Network	network of 120 IMS backup-stations assist 50 primary stations.
bare sphere	critical mass of spherical fissile weapon without reflectors, tampers and compression.
Becker nozzle	separates isotopes by passing a gas through a tiny tube around sharp corners.
bhangmeter	optical detector in orbit to detect nuclear explosions and determine yields.
bomblet	10 kg reentry vehicles, released at end of boost phase to carry biological weapons.
boost-phase	first minutes of a missile trajectory, before midcourse/ MIRV and reentry
boosted primary	neutrons from $^2H/^3H$ fusion increases fission efficiency, skips n generations.
breakout	potential massive violation of strategic arms treaty, threatening treaty partners.

D. Hafemeister, *Nuclear Proliferation and Terrorism in the Post-9/11 World*,
DOI 10.1007/978-3-319-25367-1

breeder reactor	a more compact reactor breeds more Pu than fuel it consumes.
burn hole in the atmosphere	heating a tube of atmosphere to reduce air density, allowing electron beam to pass.
chaff	counter measure to spoof the defense with small metallic wires that reflect radar.
countermeasures	protection of attacking missiles by blocking the defense's sensors.
coupled explosion	nuclear explosion joined to the earth by tamped soil/rock (see decoupled).
CTBT	Comprehensive Test Ban Treaty bans all nuclear explosions forever/ anywhere; defeated by US Senate in 1999.
critical mass	amount of fissile material that produces slightly more neutrons than are lost by absorption or leakage.
cookie-cutter	simplified zero to one probability, a step function, for silo destruction.
cyber war	ability of computers to obtain secrets and demolish systems and power grids.
decoupled explosion	small nuclear explosion in a large cavity; observed yield can be reduced up to a factor of 70, detection limit is better than 1–2 kton if detecting mature nuclear nation.
denatured uranium	adding natural uranium to weapons grade ^{235}U to remove its weapon capability; cannot be done with Pu.
diminishing returns	reduced destruction per warhead with increased numbers of warheads.
DoD and **DOE**	US Departments of Defense and Energy.
downblend U	mix 90 % enriched U with natural or 1.5 % to obtain 4.4 % reactor fuel.
download warheads	reduce number of MIRVed warheads, Minuteman's 3 to 1 reduction.
DPS	Defense Program Satellites at GEO with IR sensors to detect launches of missiles.
drawdown curve	number of surviving warheads as a function of size of attacking warheads.
drones	unmanned aerial vehicles (UAV) with sensors and kill capacity.
duty factor	defensive weapons in orbit have a duty factor of 5–10 % above the opposition's warheads.
earth-penetrating warhead	B61/11 bomb; increases coupling to earth, but the 5-kton version was rejected by congress. See RNEP.
effective verification	ability to respond to cheating before significant military loss, and to observe smaller cheating.
endoatmospheric	a location above the atmosphere, at an altitude above 10 km, approximately.

EMP electromagnetic pulse caused by asymmetrical electron currents in Earth's magnetic field.

EMP attack on power grid a 10 Mtons attack over US can generate 25,000 V/m and shut down the power grid.

exoatmospheric a location in the atmosphere.

exotic nuclear weapons see reliable replacement warhead, robust nuclear Earth-penetrating warhead, thorium to ^{233}U warheads, trans-actinide warheads, X-ray laser pumped with nuclear explosion.

first strike pre-emptive start to nuclear war.

fission splitting of ^{235}U or ^{239}Pu into two fission fragments, yielding 17 kton/kg.

fluence total number of particles or energy traversing a unit area.

flux flow of particles, fluid or energy per unit area/second.

fratricide destruction of one warhead by another warhead from the same side.

fuel cycles *once-through* fuel cycle uses nuclear fuel without reprocessing; *closed cycle* entails reprocessing to obtain Pu.

fusion the combining of ^{3}H and ^{2}H to give He and a neutron, yielding 85 kton/kg.

GEO geosynchronous orbit at 40,000 km radius, stationary with respect to Earth for DPS and other satellites.

GPS Global Positioning Satellites allows missile targeting to 10 cm or so, at 26,600 km radius.

gravitational bias Earth's 0.3 % oblateness considered for high accuracy trajectories.

hard target silo or command centers with a hardness of 1000–5000 psi.

Hiroshima Japanese city, targeted by Thin Man, uranium bomb, 13-kton kills some 75,000.

HEU highly enriched uranium, over 20 % ^{235}U, not weapons usable below some 60 %.

homeland security 9/11 attack launched DHS to protect U.S. from terrorists.

hydrogen bomb radiative compression from fission primary heats ^{2}H–^{3}H to fuse.

IHE insenstiive high explosives make warheads safer but take up extra space, compared to HE.

implosion quickly assembles critical mass by explosion to avoid pre-initiation from spontaneous neutrons.

interferometric synthetic aperture radar InSAR measures subsidence to 0.2 mm accuracy.

irradiance power per unit solid angle in watts/steradian.

isomer-EMP Bomb experiments with ^{178m}Hf failed.

Joint Actinide Shock Physics Experiment JASPER projects high velocity Pu on Pu targets to get Pu equation of state to improve implosion physics.

kill probability silo destruction depends on hardness, yield, accuracy, reliability, warhead numbers, bias, etc.

KKV kinetic kill vehicles collide and destroy missiles and reentry vehicles in midcourse phase.

Lanchester, Frederich developed conventional-arms battle models, favoring large forces by strength squared.

launch on warning an option in the SIOP to launch missiles, when electronic sensors perceive an attack is coming.

launch weight typically 20–30 times the throw-weight of a missile.

layered defense multiple layers in boost, midcourse and reentry phases to enhance kill probability.

legacy warheads seven warhead types to be maintained under CTBT, 2 for SLBMs, 2 for ICBMs, 2 for bombers, 1 for cruise missiles.

LEO low Earth orbit at altitudes of 100–1000 miles with a period of 1.5 h.

LEU low enriched uranium with less than 20 % ^{235}U.

lithium deuteride stable solid that contains deuterium and produces tritium in a neutron bath for quick-assembly fusion.

MAD Robert MacNamara's mutual assured destruction would destroy more than 25 % of population and 50 % of industry of the attacking nation.

minimum deterrence Robert Macnamara's assured destruction level for MAD.

Minuteman vulnerability worst case scenario was 80–90 % vulnerable, the untested attack would have to operate very well. MMan is only part of the triad.

MIRV multiple, independently targetable reentry vehicles with 10 RV's per SS-18/Peackeeper.

MOX for light water reactor fuel, combination of 5 % reprocessed Pu with 95 % natural uranium.

Nagasaki Japanese city targeted by Fat Man plutonium-implosion bomb with a 22-kton yield

neutron bomb 1–10 kton bomb with 50 % fusion uses fast neutrons to kill tank drivers quickly but buildings survive.

NWS/NNWS under NPT, big-five nuclear weapon states (NWS) may have nuclear weapons and the 180 non-NWSs may not.

NPT the 1970 Nuclear Nonproliferation Treaty is the controlling legal document for weapon possession and safeguards.

nuclear winter	massive burning from nuclear attack lofts carbon into stratosphere, cools Earth.
off-alert	minimize accidents, delaying technologies used to prevent rapid response.
Okla, Gabon	1.8 billion years ago, a natural reactor in Okla burned 100,000 years, lowering ^{235}U content to 0.4 %.
overpressure	2000 psi destroys most silos and 5 psi destroys most buildings.
popup missiles	fast rise missiles on subs based close to attacker, difficult in hostile regions.
primary	first stage of nuclear weapon, fissions ^{239}Pu or ^{235}U in gun or implosion weapons.
Pu pit lifetime	measurements by DOE and analysis by DOE/JASON gave over 100 years.
quick from the dead	Bernard Baruch's 1946 charge to the UN to control nuclear weapons.
railgun	based in space, uses electromagnetic forces to fire mass at attacking reentry vehicles.
reactor-grade Pu	plutonium that remains a longer time in reactor has over 20 % ^{240}Pu.
reentry vehicle	warhead packaged in heat resistant material for safe reentry to the atmosphere.
Richardson, Lewis	developed action–reaction spending model for nations fearing survival.
Robust Nuclear Earth Penetrating Warhead (RNEP)	terminated by Congress.
Reliable Replacement Warhead (RRW)	terminated by Congress.
rocket equation	Newton's second law combines exhaust gas thrust with variable mass; can include gravity and drag.
safeguard technologies	used by IAEA and nations for nuclear materials accounting.
SALT/START/SORT/ New-START	strategic arms control treaties in force between US and USSR/Russia between 1972–2013.
scaling laws	useful to understand basic physics and relative sizes of nuclear warheads.
science court	a process to improve the search for truth on S/T, tried with ABM in 1985.
second strike	response to a first strike, US exceeds the minimum deterrence level.
secondary	ignited by the fission primary, contains ^{6}LiD and uranium to give fusion and fission.
SDI	President Reagan's Strategic Defense Initiative, 1983–1990; elements still researched.

sine qua non	a necessary condition, space-based weapons are sine qua non for successful SDI.
SIOP	standard integrated operation plan, the US nuclear war plan.
SNM	special nuclear material encompasses weapons isotopes of ^{235}U, ^{233}U, and Pu.
soft target	buildings with hardness of 5–100 psi overpressure
specific impulse I_{sp}	rocket fuel exhaust velocity divided by *g*, in units of seconds.
spontaneous neutrons	spontaneous fission of ^{240}Pu gives neutrons that can reduce yield.
stockpile stewardship	DOE program to maintain reliability of nuclear warheads over time.
strategic weapons	nuclear weapons on ICBMs, SLBMs and bombers, range over 5500 km.
SWU	separated work unit, the work needed to raise enrichment levels of uranium.
technology transfer	many ways the secrets and materials move to aspiring states.
throw-weight	typically 3–5 % of an ICBM/SLBM's launch weight.
theater ballistic missile	capable of 2000-km range, less than intermediate range 5500 km (SS-20 and Pershing-II in INF).
time line	critical time pattern for decisions and operations for ABM systems.
total information awareness	big computers sift big data to estimate probability individuals might act badly.
Trajectory, flat Earth	simplification retains much of the physics of accurate trajectories.
Trajectory, spherical Earth	trajectories with minimum energy at 22° above horizon for 10,000-km range.
Trajectory, oblate Earth	requires Runge-Kutta solutions, or perturbation approximations.
triad	consists of ICBMs, SLBMs and heavy bombers, plus theater weapons.
tritium	^{3}H and ^{2}H fuse in boosted primaries, secondaries, or initiators, 12-year half life.
upload warheads	to increase number of warheads on an ICBM.
weapons-grade HEU/Pu	uranium enriched to over 80 % and Pu with less than 6 % ^{240}Pu (but all Pu can be used).
weapons yield	varies from 1 lb to 58 Mton (partial yield of 100–150 Mton Soviet weapon).

X-ray laser	laser, pumped with a nuclear weapon, pops-up from close-in subs, but difficult to accomplish and hard to deploy successfully.
yield excursion	beginning nuclear weapon state can't accurately predict yields.

Environmental Issues

BEIR	Biological Effectiveness of Ionizing Radiation of the National Research Council.
BWR	boiling water reactors use steam directly from the pressure vessel to drive turbines.
capacity factor	fraction of time that power plant is operating at rated capacity.
carbon tax	discourages carbon usage, but with uncertain impact.
closed fuel cycle	spent fuel is reprocessed, to obtain plutonium that is used in reactors.
duty/capacity factor	the fraction of time that a power plant is operating at full power.
ECCS	emergency core cooling system replaces lost coolant in a core after a pipe break.
EPA	US Environmental Protection Agency.
fault tree analysis	accident-probability of planes/reactors estimated with parallel/series logic.
ICRP	International Commission on Radiological Protection.
hormesis	controversial theory that small doses or radiation are beneficial.
HTGR	high temperature gas reactors use carbon as a moderator, which has more capacity to store heat but is flammable.
ICF	inertial confinement fusion contains hot plasma fuel only because it fuses quickly.
Lawson criterion	product of confinement time and particle density/cm^3 for energy break-even (with 1/3 recovery heat).
LOCA	loss-of-coolant accident; melting of the reactor core from un-cooled beta-decay heat from fission fragments.
LWR	light water reactors use 3–5 % enriched uranium and light water coolant/moderator in PWRs and BWRs.
HWR	heavy water reactors use natural uranium and heavy water coolant/moderator; makes weapons grade plutonium, which LWRs do not.

mean life the average life of radioactivity, which is 44 % longer than its half-life.

National Ignition Facility 200 beam laser, 1.8 MJ pulses for stockpile stewardship at LLNL.

NRC US Nuclear Regulatory Commission.

once-through fuel cycle reactor spent fuel is not reprocessed, but stored in a repository.

passively safe reactors remain safe with loss of coolant/power, requires smaller nuclear plants.

person-sievert units of radiation received by populace (1 person Sv = 100 person-rem).

plumes SO_2, ^{137}Cs, and Pu plumes depend on wind velocity/turbulence, source strength, deposition velocity and response.

potassium iodide KI pills greatly slow absorption of radioactive iodine from reactor accidents.

probabilistic risk analysis uses fault tree analysis with serial/parallel logic to quantify risk.

PWR pressurized-water reactor uses a heat exchanger to transfer heat from the reactor to make steam to drive turbines.

radiation dose energy absorbed (1 gray = 100 rad) and effective dose (1 sivert = 100 rem).

radon daughters ^{218}Po, ^{214}Po, ^{214}Pb and ^{214}Bi from decay of ^{222}Rn with a 3.8 day half-life.

rational actors assumption that nations and citizens decide issues rationally.

relative biological effectiveness factor between 1–20, takes into account the type of radiation.

relative risk disease rate of those exposed divided by rate of those not exposed.

rem Roentgen equivalent man; radiation dose of 0.0087 J/kg times a RBE factor for the type of radiation.

spent-fuel ponds reactor spent-fuel cools 5-10 years, before surface/underground storage.

thorium cycle alternative nuclear fuel cycle that uses thorium to produce ^{233}U nuclear fuel.

thermal rise time exponential rise time or the time it takes for a system to exceed safety.

thermal efficiency ratio of useful output to energy input.

warhead accident non-nuclear accident, could spread Pu, linked to nuclear testing issue.

Yucca Mountain Nevada storage repository for commercial spent fuel, now closed.

Proliferation Issues

ABACC	Brazilian Argentine Agency for Accounting and Control for Nuclear Materials, a true success and a model for others.
Article VI of NPT	Suggests nuclear weapon states move towards nuclear disarmament.
Atoms for Peace	Eisenhower's 1953 UN speech which began nuclear power growth, but under IAEA rules, a tradeoff of technology transfer versus government controls.
Counter-Proliferation Attacks	Israeli destroyed Iraq's Osirak reactor, shutting down Iraq's plutonium economy, not as likely to work on Iran's centrifuges.
Defacto 4	Who, after Israel, India, Pakistan and North Korea?
EMIS	Electromagnetic isotope separation improved for Iraq in the 1990s.
Enrichment of Uranium	Separation by gravity, diffusion, centrifuges, lasers, nozzles, chemicals and cyclotron resonance.
Export Criteria	1978 Non-Nuclear Proliferation Act prevents nuclear exports to nations that don't have all nuclear facilities under IAEA Safeguards, and they need US agreement if the spent fuel is to be reprocessed.
Fissile Material Control Treaty	The P5 no longer produced substantial amounts of HEU and Pu, the FMCT hopes to expand this behavior to the next generation of nuclear weapon states.
Fuel Bank	The IAEA and some nations guarantee an assured supply of low-enriched nuclear fuel if they play by the rules.
Horizontal versus Vertical Proliferation	A nuclear weapon state getting more nuclear weapons (vertical proliferation) versus a non-nuclear weapon state getting its first nuclear weapon (horizontal proliferation).
IAEA burden	IAEA both promotes peaceful nuclear power and regulates it.
incentives and disincentives	NPT policy is incremental, trying with deep reason.
JFK's 20 NWSs	Kennedy estimates in 1960 that 20 nations would have nuclear weapons.
megatons to megawatts	500 tonnes of Soviet HEU down-blended to reactor fuel for US cash.

Missile Technology Control Regime	A non-treaty regime that constrains commerce in missiles capable of carrying nuclear weapons and BW/CW.
nth Country Experiment	Three young physicists figured it out from the open literature in the 1960s.
Non-Proliferation Treaty	is the legal basis for global nonproliferation policy. NPT does not constrain enrichment and reprocessing, unless a nation cheats. It is the best we can do now.
North Korea nuclear tests	Small tests in 2006, 2009 and 2013. Will China reign in DPRK?
NPT Review Conference	The disagreements are strong as NPT lacks a consensus.
NPT and CTBT, a tale of two treaties	A way to strengthen the regime.
Nuclear Suppliers Group	A group of about 40 nuclear supplier nations promulgate rules needed to obtain nuclear imports.
Nuclear Wal-Mart of AQ Kahn	Iran, North Korea, Libya and Iraq approached with designs and centrifuges for a price.
P5	The nuclear weapon states that also have UN veto power; US, Russia, UK, France, China.
peaceful versus military fuel cycle	Is the barrier between cycles robust?
phishing	a way to trick you to respond to a computer trap.
plutonium economy	During 1976–78, Presidents Ford and Carter moved nuclear power away from an unencumbered commercial use of plutonium.
Plutonium, weapons-grade versus reactor grade	Weapons prefer 6 % ^{240}Pu, while reactor grade Pu remains in reactors longer, producing more spontaneous neutrons.
Safeguards	Nuclear detection methods that can be used to make quantitative measurements of nuclear supplies and presented, operated by the IAEA.
sanctions	Starting with Glenn-Symington in 1976, Pakistan and others lost economic and military aid if they moved towards the bomb. To be effective, sanctions must be adopted by most nations.
S. Africa/Ukraine/Khazakhstan/Belarus	turn the weapons in success story.
Zero Nuclear Weapons	The non-nuclear weapon states obtain encouragement if the P5 move towards far fewer weapons.

Terrorism

attribution	can we determine who sent those 0's and 1's?
authority to kill	if there is an imminent threat.
BWC	Biological Weapons Convention, without verification, but mostly followed.
CWC	Chemical Weapons Convention, with verification, Egypt/Israel not signed up.
denial of service attack	A multiplying attack to block your servers.
Department of Homeland Security	mega-department controlling the judicial levers of power.
dirty bombs	use ^{60}Co, ^{90}Sr and ^{137}Cs to pollute the neighborhood.
drones	can be used to gain information and to attack with a variety of methods.
enhanced interrogation technologies	such as water boarding, to obtain the truth or not.
FISA Court	The Foreign Intelligence Surveillance Act (1978) and the Patriot Act (2001) gives power to the FISA Court on terrorism issues.
FitBit	For $90 it will keep track of your energy, distance, steps, elevation and sleep.
Goggle Glass	glasses that film and record all your conversations, useful in court.
going dark	when encryption and new methods hide the spies, going dark.
infrastructure	Computer weaknesses in the banking, transport, electric grid, health care and energy structures.
insider threat	the design basis threat becomes more serious if there are two insiders.
Nuclear Emergency Support Team	NEST examines the device and takes it apart, on leave from the weapon's laboratories.
nuclear forensics	determines isotope production methods by counting their gamma rays.
perfect storm vulnerabilities	A combination of vulnerable infrastructures, available technologies and massive urban populations.
privacy	is not as private as you once thought, making a difference they say.
risk	is the probability of an attack, times the severity of the attack.
Snowden, Edward	Living abroad since 6 June 2013, formerly worked on NSA secrets.

Stuxnet	Invited in 2007 to attack Iran's centrifuges. Will it spread?
Sunni versus Shiite	A strong civil war being fought between fundamentals on both sides in ISIS versus Iraq/Iran.
Total Information Awareness	advantages of Moore's law of greatly reduced costs of computers and memory to obtain the correlations in our lives, to learn our truths.
Transportation Security Administration	Your friends in the airport, making you safe.